Current Progress in Stem Cells Research

Current Progress in Stem Cells Research

Edited by Mark Walters

hayle medical

New York

Hayle Medical,
750 Third Avenue, 9th Floor,
New York, NY 10017, USA

Visit us on the World Wide Web at:
www.haylemedical.com

ISBN: 978-1-63241-489-2

The publisher's policy is to use permanent paper from mills that operate a sustainable forestry policy. Furthermore, the publisher ensures that the text paper and cover boards used have met acceptable environmental accreditation standards.

Printed in China.

Cataloging-in-Publication Data

Current progress in stem cells research / edited by Mark Walters.
p. cm.
Includes bibliographical references and index.
ISBN 978-1-63241-489-2
1. Stem cells. 2. Stem cells--Research. I. Walters, Mark.
QH588.S83 C87 2018
616.027 74--dc23

Table of Contents

Preface ... IX

Chapter 1 **Long-Term MRI Cell Tracking after Intraventricular Delivery in a Patient with Global Cerebral Ischemia and Prospects for Magnetic Navigation of Stem Cells within the CSF** ... 1
Miroslaw Janowski, Piotr Walczak, Tomasz Kropiwnicki, Elzbieta Jurkiewicz, Krystyna Domanska-Janik, Jeff W. M. Bulte, Barbara Lukomska, Marcin Roszkowski

Chapter 2 **Label-Free Morphology-Based Prediction of Multiple Differentiation Potentials of Human Mesenchymal Stem Cells for Early Evaluation of Intact Cells** ... 8
Hiroto Sasaki, Ichiro Takeuchi, Mai Okada, Rumi Sawada, Kei Kanie, Yasujiro Kiyota, Hiroyuki Honda, Ryuji Kato

Chapter 3 **Tracking Transplanted Bone Marrow Stem Cells and their Effects in the Rat MCAO Stroke Model** ... 22
Gregory V. Goldmacher, Rena Nasser, Daniel Y. Lee, Sargon Yigit, Robert Rosenwasser, Lorraine Iacovitti

Chapter 4 **Generation and Characterization of Functional Cardiomyocytes Derived from Human T Cell-Derived Induced Pluripotent Stem Cells** ... 31
Tomohisa Seki, Shinsuke Yuasa, Dai Kusumoto, Akira Kunitomi, Yuki Saito, Shugo Tohyama, Kojiro Yae, Yoshikazu Kishino, Marina Okada, Hisayuki Hashimoto, Makoto Takei, Toru Egashira, Masaki Kodaira, Yusuke Kuroda, Atsushi Tanaka, Shinichiro Okata, Tomoyuki Suzuki, Mitsushige Murata, Jun Fujita, Keiichi Fukuda

Chapter 5 **Behavioral and Histopathological Assessment of Adult Ischemic Rat Brains after Intracerebral Transplantation of NSI-566RSC Cell Lines** ... 38
Naoki Tajiri, David M. Quach, Yuji Kaneko, Stephanie Wu, David Lee, Tina Lam, Ken L. Hayama, Thomas G. Hazel, Karl Johe, Michael C. Wu, Cesar V. Borlongan

Chapter 6 **Ontological Differences in First Compared to Third Trimester Human Fetal Placental Chorionic Stem Cells** ... 47
Gemma N. Jones, Dafni Moschidou, Tamara-Isabel Puga-Iglesias, Katarzyna Kuleszewicz, Maximilien Vanleene, Sandra J. Shefelbine, George Bou-Gharios, Nicholas M. Fisk, Anna L. David, Paolo De Coppi, Pascale V. Guillot

Chapter 7 **A Novel Berbamine Derivative Inhibits Cell Viability and Induces Apoptosis in Cancer Stem-Like Cells of Human Glioblastoma, via Up-Regulation of miRNA-4284 and JNK/AP-1 Signaling** ... 62
Fan Yang, Sangkil Nam, Christine E. Brown, Robin Zhao, Renate Starr, David A. Horne, Linda H. Malkas, Richard Jove, Robert J. Hickey

Chapter 8 **Cell-Based Regenerative Strategies for Treatment of Diabetic Skin Wounds, a Comparative Study between Human Umbilical Cord Blood-Mononuclear Cells and Calves' Blood Haemodialysate** **71**
Hala O. El-Mesallamy, Mohamed R. Diab, Nadia M. Hamdy, Sarah M. Dardir

Chapter 9 **Influence of Delivery Method on Neuroprotection by Bone Marrow Mononuclear Cell Therapy following Ventral Root Reimplantation with Fibrin Sealant** **81**
Roberta Barbizan, Mateus V. Castro, Benedito Barraviera, Rui S. Ferreira Jr., Alexandre L. R. Oliveira

Chapter 10 **Baculovirus as an Ideal Radionuclide Reporter Gene Vector: A New Strategy for Monitoring the Fate of Human Stem Cells *In Vivo*** **94**
Yu Pan, Shuai Liu, Haifei Wu, Jing Lv, Xiaoqian Xu, Yifan Zhang

Chapter 11 **Oxygen Transport and Stem Cell Aggregation in Stirred-Suspension Bioreactor Cultures** **104**
Jincheng Wu, Mahboubeh Rahmati Rostami, Diana P. Cadavid Olaya, Emmanuel S. Tzanakakis

Chapter 12 **Loss of Sirt3 Limits Bone Marrow Cell-Mediated Angiogenesis and Cardiac Repair in Post-Myocardial Infarction** **116**
Heng Zeng, Lanfang Li, Jian-Xiong Chen

Chapter 13 ***In Vivo* Tracking and Comparison of the Therapeutic Effects of MSCs and HSCs for Liver Injury** **126**
Qiang Li, Xinmin Zhou, Yongquan Shi., Jinge Li, Linhua Zheng, Lina Cui, Jun Zhang, Lu Wang, Zheyi Han, Ying Han, Daiming Fan

Chapter 14 **n-Butylidenephthalide (BP) Maintains Stem Cell Pluripotency by Activating Jak2/Stat3 Pathway and Increases the Efficiency of iPS Cells Generation** **137**
Shih-Ping Liu, Horng-Jyh Harn, Ying-Jiun Chien, Cheng-Hsuan Chang, Chien-Yu Hsu, Ru-Huei Fu, Yu-Chuen Huang, Shih-Yin Chen, Woei-Cherng Shyu, Shinn-Zong Lin

Chapter 15 **Reversal of Ischemic Cardiomyopathy with Sca-1^{+} Stem Cells Modified with Multiple Growth Factors** **149**
Ning Li, Zeeshan Pasha, Muhammad Ashraf

Chapter 16 **Intramyocardial Transplantation and Tracking of Human Mesenchymal Stem Cells in a Novel Intra-Uterine Pre-Immune Fetal Sheep Myocardial Infarction Model: A Proof of Concept Study** **160**
Maximilian Y. Emmert, Benedikt Weber, Petra Wolint, Thomas Frauenfelder, Steffen M. Zeisberger, Luc Behr, Sebastien Sammut, Jacques Scherman, Chad E. Brokopp, Ruth Schwartländer, Viola Vogel, Peter Vogt, Jürg Grünenfelder, Hatem Alkadhi, Volkmar Falk, Andreas Boss, Simon P. Hoerstrup

Chapter 17 **Testing Stem Cell Therapy in a Rat Model of Inflammatory Bowel Disease: Role of Bone Marrow Stem Cells and Stem Cell Factor in Mucosal Regeneration** **174**
Bo Qu, Guo-Rong Xin, Li-Xia Zhao, Hui Xing, Li-Ying Lian, Hai-Yan Jiang, Jia-Zhao Tong, Bei-Bei Wang, Shi-Zhu Jin

Chapter 18 **Interactions of Anaerobic Bacteria with Dental Stem Cells: An *In Vitro* Study**........................184
Anne Biedermann, Katja Kriebel, Bernd Kreikemeyer, Hermann Lang

Chapter 19 **Identification of Rabbit Annulus Fibrosus-Derived Stem Cells**........................191
Chen Liu, Qianping Guo, Jun Li, Shenghao Wang, Yibin Wang, Bin Li, Huilin Yang

Chapter 20 **Paracrine-Mediated Neuroprotection and Neuritogenesis of Axotomised Retinal Ganglion Cells by Human Dental Pulp Stem Cells: Comparison with Human Bone Marrow and Adipose-Derived Mesenchymal Stem Cells**........................199
Ben Mead, Ann Logan, Martin Berry, Wendy Leadbeater, Ben A. Scheven

Chapter 21 **A Novel *In Vitro* Method for Detecting Undifferentiated Human Pluripotent Stem Cells as Impurities in Cell Therapy Products using a Highly Efficient Culture System**........................210
Keiko Tano, Satoshi Yasuda, Takuya Kuroda, Hirohisa Saito, Akihiro Umezawa, Yoji Sato

Permissions

List of Contributors

Index

Preface

Over the recent decade, advancements and applications have progressed exponentially. This has led to the increased interest in this field and projects are being conducted to enhance knowledge. The main objective of this book is to present some of the critical challenges and provide insights into possible solutions. This book will answer the varied questions that arise in the field and also provide an increased scope for furthering studies.

Stem cell therapy has tremendous scope of treating a range of diseases. The most commonly used stem cell therapy is the transplantation of the bone marrow. It is used in treating patients with diseases like leukemia and lymphoma. Recent research focuses on studying stem cells in order to use them to treat diseases. The objective of this book is to give a general view of the different areas of stem cells research and its applications. Scientists and students actively engaged in this field will find this book full of crucial and unexplored concepts.

I hope that this book, with its visionary approach, will be a valuable addition and will promote interest among readers. Each of the authors has provided their extraordinary competence in their specific fields by providing different perspectives as they come from diverse nations and regions. I thank them for their contributions.

Editor

Long-Term MRI Cell Tracking after Intraventricular Delivery in a Patient with Global Cerebral Ischemia and Prospects for Magnetic Navigation of Stem Cells within the CSF

Miroslaw Janowski[1,2,3,4], Piotr Walczak[3,4,5], Tomasz Kropiwnicki[6], Elzbieta Jurkiewicz[7], Krystyna Domanska-Janik[1], Jeff W. M. Bulte[3,4,8,9], Barbara Lukomska[1], Marcin Roszkowski[6]*

1 NeuroRepair Department, Mossakowski Medical Research Centre, Polish Academy of Sciences, Warsaw, Poland, **2** Department of Neurosurgery, Mossakowski Medical Research Centre, Polish Academy of Sciences, Warsaw, Poland, **3** Russell H. Morgan Department of Radiology and Radiological Science, Division of Magnetic Resonance Research, The Johns Hopkins University School of Medicine, Baltimore, Maryland, United States of America, **4** Cellular Imaging Section and Vascular Biology Program, Institute for Cell Engineering, The Johns Hopkins University School of Medicine, Baltimore, Maryland, United States of America, **5** Department of Radiology, Faculty of Medical Sciences, University of Warmia and Mazury, Olsztyn, Poland, **6** Department of Neurosurgery, The Children's Memorial Health Institute, Warsaw, Poland, **7** Department of Radiology, Magnetic Resonance Unit, The Children's Memorial Health Institute, Warsaw, Poland, **8** Department of Chemical and Biomolecular Engineering, The Johns Hopkins University School of Medicine, Baltimore, Maryland, United States of America, **9** Department of Biomedical Engineering, The Johns Hopkins University School of Medicine, Baltimore, Maryland, United States of America

Abstract

Background: The purpose of the study was to evaluate the long-term clinical tracking of magnetically labeled stem cells after intracerebroventricular transplantation as well as to investigate *in vitro* feasibility for magnetic guidance of cell therapy within large fluid compartments.

Method: After approval by our Institutional Review Board, an 18-month-old patient, diagnosed as being in a vegetative state due to global cerebral ischemia, underwent cell transplantation to the frontal horn of the lateral ventricle, with umbilical cord blood-derived stem cells labeled with superparamagnetic iron oxide (SPIO) contrast agent. The patient was followed over 33 months with clinical examinations and MRI. To evaluate the forces governing the distribution of cells within the fluid compartment of the ventricular system *in vivo*, a gravity-driven sedimentation assay and a magnetic field-driven cell attraction assay were developed *in vitro*.

Results: Twenty-four hours post-transplantation, MR imaging (MRI) was able to detect hypointense cells in the occipital horn of the lateral ventricle. The signal gradually decreased over 4 months and became undetectable at 33 months. *In vitro*, no significant difference in cell sedimentation between SPIO-labeled and unlabeled cells was observed (p = NS). An external magnet was effective in attracting cells over distances comparable to the size of human lateral ventricles.

Conclusions: MR imaging of SPIO-labeled cells allows monitoring of cells within lateral ventricles. While the initial biodistribution is governed by gravity-driven sedimentation, an external magnetic field may possibly be applied to further direct the distribution of labeled cells within large fluid compartments such as the ventricular system.

Editor: Cesar V. Borlongan, University of South Florida, United States of America

Funding: R21NS081544, RO1 NS076573, MSCRFII0193, MSCRFII190, MSCRFE0178, MJ was supported by a Mobility Plus Fellowship from the Polish Ministry of Science and Education. The funders had no role in study design, data collection and analysis, decision to publish, or preparation of the manuscript.

Competing Interests: The authors have declared that no competing interests exist.

* E-mail: marcroszkowski@gmail.com

Introduction

Stem and progenitor cell-based therapy is considered a new avenue for the treatment of various diseases for which there is no effective cure [1,2]. Neurological diseases pose a special challenge due to the complexity of the central nervous system (CNS) [3,4]. There have been a few reports on successful, open-label cell therapy trials for Parkinson's disease, [5,6]. However, double-blind trials failed to reveal a statistically significant improvement, which was in part due to the high variability of the obtained outcomes [7–9]. Nevertheless, cell transplantation experiments are being performed preclinically and clinically in dozens of otherwise untreatable neurological disorders [10]. Intraparenchymal stereotaxic injection has initially been the method of choice for targeting cells toward well-defined anatomical locations. Systemic (i.v.) injections have also been used in several clinical trials [11,12]. A major obstacle in the evaluation of these clinical trials is the uncertainty if cells are delivered correctly at the desired location and/or reach their target successfully. For intracebroventricular (ICV) injections, non-invasive visualization of cells is of particular importance as the cell dispersion is dictated by cerebro-spinal fluid (CSF)-driven flow mechanisms where the distribution of injected cells can be highly variable.

MRI cell tracking has recently gained attention as a clinically applicable tool to track cells non-invasively in real-time [13]. These initial clinical studies, performed in patients with cancer [14], brain trauma [15], multiple sclerosis [16], and diabetes [17] have demonstrated proof of feasibility of clinical detection. The very rigorous study performed on healthy volunteers has just confirmed safety of cell labeling by super-paramagnetic iron oxide SPIO [18]. For these studies, the longest time frame for follow upis 6 months [16]. The early outcome in a severely, globally ischemic patient who was transplanted ICV with autologous cord-blood-derived, SPIO-labeled neural progenitors, was reported previously [19]. In this study, we present a long-term imaging evaluation where the patient was followed for 33 months. Since only 20 percent of transplanted cells were labeled in this clinical experiment, additional *in vitro* fluid-phase studies modeling the movements of SPIO-labeled and unlabeled cells were conducted to gain a better insight about the fate of transplanted cells *in vivo*. CSF circulation directs the flow from the lateral ventricle to the subarachnoid space within the posterior fossa; this flow can potentially affect the distribution of injected cells. SPIO-labeling and subsequent MRI visualization may be used to evaluate such a possible CSF-driven biodistribution. Since sedimentation resulting from the gravity can also be a leading factor determining cell distribution within a fluid compartment [20], we performed an *in vitro* assay to compare the speed of sedimentation of SPIO-labeled vs. non-labeled cells. We also demonstrate here the potential for guiding the ICV distribution of SPIO-labeled cells with the use of an external magnetic field.

Materials and Methods

2.1 Patient History

A nine-month-old patient was in a vegetative state as a result of global cerebral ischemia. An extensive rehabilitation program over three months did not result in any recovery, and a permanent vegetative state was diagnosed [21]. MR imaging revealed a mild global atrophy without focal lesions. Experimental cell therapy was considered due to extremely poor prognosis. The patient's own cord blood was deposited at birth in a private blood bank; the parents of the patient decided to store his cord blood and covered all expenses related to it. The access to patient's own source of stem cells facilitated the decision on cell transplantation. The parents provided written informed consent to include the patient in the study and have potentially personally identifying information published. The clinical study was conducted in Warsaw after approval by the Institutional Review Board (Bioethics Committee) at the Children's Memorial Health Institute, Warsaw, Poland.

Briefly, autologous cord blood nucleated cells obtained during full-time delivery (2.4×10^7 cells/ml stored in 10% DMSO) were thawed and cultured for 10 days in previously described neurogenic conditions [22] in a GMP facility. A total of 3.6×10^7 cells were delivered in three equal doses, with the injections performed at one-month intervals. For the first dose of 1.2×10^7 cells, 20% of cells were labeled with Feridex/PLL as previously described [19]. The two other injections were given with unlabeled cells only. The transplantation procedure was performed under general anesthesia and 0.5 ml of cells in saline was delivered to the anterior horn of the right lateral ventricle using a standard shunt drain. The patient was operated in supine position, typical for shunt drain placement. MR images were obtained before and at 1 day, 1 week, 1 month, 2 months, and 4 months after the first injection using a 1.5 T scanner equipped with an 8-channel phased-array head coil (Magnetom Sonata Maestro Class, Siemens, Germany). For the detection of SPIO signal, a SWI imaging sequence was used: slabs = 1, phase R-L, slice over-sampling = 11, slices per slab = 72, FOV 230 mm, slice thickness 1.6 mm, dist factor 20%, averages = 1, TR/TE = 49/40, flip angle = 15, matrix 168×256, voxel size 1.2×0.9×1.6 mm. A neurological examination was performed six months after the first transplantation. The patient was then admitted to the hospital for clinical and radiological evaluation up to 33 months after the first cell transplantation. Osirix (Pixmeo) and Amira (Visage Imaging) software were used for image processing. The entire procedure was performed primarily for therapeutic purposes and the use of labeled cells was only a side-aspect of the whole treatment.

2.2 Cell Sedimentation Assay

Non-transformed line of neural stem cells derived from human umbilical cord blood (HUCB-NSC) was used for *in vitro* study [23]. HUCB-NSC were cultured in DMEM/F12 supplemented with 2% fetal bovine serum, 1% insulin-transferrin-selenium, and 1% antibiotic/antimycotic solution (Invitrogen). Fifty percent of cells were labeled for two days with Molday ION Rhodamine B-conjugated iron oxide nanoparticles (25 μg/ml, MIRB, BioPAL) and the transfection agent poly-L-Lysine (Sigma P-1524, 375 ng/ml) [24]. The unlabeled fraction of cells was counterstained with CellTracker cGreen (Molecular Probes). For sedimentation experiments, both red (SPIO-labeled) and green (unlabeled) cells were mixed to reach a final working solution of 2,000 cells/μl.

Sedimentation chambers were designed by removing the convex tip of 10 mm NMR tubes (New Era Enterprises, INC). The cylinders were then placed vertically on round glass-bottom, 1.5-inch cell culture dishes (RD-1.5-UNC, LiveAssay), and were thermally glued to form a waterproof seal. The resulting sedimentation chambers had a height of 15 cm and a diameter of 1 cm. They were filled with 10 mM PBS and seeded with a small amount of Hoechst-labeled cells in order to determine the bottom level of the sedimentation chamber for focusing of the microscope. A Zeiss inverted microscope (Axiovert 200 M) was programmed to acquire both red and green channel images automatically at one-minute intervals. Ten μl of 50/50 labeled/unlabeled cell suspension (20,000 cells) was placed on top of the sedimentation chamber and allowed to sediment over 30 minutes. Images were acquired at 20x magnification. Independent experiments were repeated six times. Both green and red cells were counted manually and normalized to the total number of cells that sedimented during the 30 min period. Both the percentage of cells that sedimented at five-minute intervals and the time required for half the cells to sediment (sedimentation half-time) were determined.

2.3 External Magnetic Field Assay

The direct visualization of cell distribution within the large fluid compartments is challenging, and BLI signal dramatically facilitates cell quantitation in these conditions. Thus for visualization and quantification purposes, HUCB-NSCs were transduced with a lentivector encoding the firefly luciferase under a constitutive CMV promoter as previously described [25]. With a transduction efficiency of nearly 100%, further selection was not needed. Cells were labeled with rhodamine-SPIO as described above, and unlabeled cells were used as control. For these experiments, cells were suspended in culture medium at a density of 10^4 cells/μl. NMR tubes (New Era Enterprises, INC) were filled with 10 mM PBS and placed vertically on a stand. A permanent neodymium magnet with a diameter of 2 inches (Magnet4less, INC) was mounted on a separate stand at the mid-height level of the NMR tube. The distance from the NMR tube to the magnet was adjusted during the experiment, ranging from 0 to 14 cm at 2 cm increments. The external field strength (FS) of the permanent magnet was measured at different distances from the magnet surface using a gaussmeter (GM2, AlphaLab Inc.). With both the NMR tube and the magnet fixed at the desired distance, luciferin (300 ng/ml in 10 mM PBS) was added to the NMR tube, and then 15 μl of 10 mM PBS containing 1.5×10^5 cells was gently layered on the surface of the 10 mM PBS and allowed to sediment over 15 minutes. After that time, the NMR tube was gently removed from the stand, and placed horizontally in a Xenogen IVIS Spectrum optical imager (Caliper Life Sciences, Billerica, MA) for cell location and quantification measurements. For quantification, the lower half of the NMR tubes was divided into three sections (upper = section A, middle = section B, and lower = section C) and the respective BLI signal was measured for each section depending on the distance from the magnet, and then subjected to regression analysis.

2.4 Statistical Analysis

All statistical calculations were performed using PROC MIXED (SAS 9.2). The lowest mean square (LMS) difference test was employed for comparison between means, and the results of the regression analysis were reported as Type III Tests of Fixed Effects, and presented as regression lines.

Results

3.1 Clinical MRI Cell Tracking after ICV Injection

Cell transplantation into the right frontal horn of the lateral ventricle resulted in a focal deposition of cells within the ipsilateral occipital horn, as detected on 24 h post-operative MR scans (Fig. 1 A,B,C, E). Over a period of four months, the cell location within the occipital horn did not change, although a gradual decrease of signal was apparent (Fig. 1 D–H). Very few hypointense regions were observed in other parts of the ventricular system, including the contralateral lateral ventricle or the fourth ventricle. At the long-term follow-up visit (33 months post transplantation), the SPIO-related hypointense signals were no longer detectable (Fig. 1I). MRI at that time did not detect any abnormality that could be attributed to tumor or overgrowth of transplanted cells. There was no major change in the patient's neurological status over a period of 33 months after transplantation. Despite the poor prognosis, the patient survived without deterioration for an additional three years.

3.2 Effect of SPIO-labeling on HUCB-NSC Sedimentation

The number of sedimented cells measured at five-minute intervals did not differ between SPIO-labeled and unlabeled HUCB-NSC at any given time point (p = NS, Fig. 2A). The sedimentation half-time amounted to 7.5±3.3 minutes for unlabeled HUCB-NSC and 9.8±5.9 minutes for cells labeled with SPIO (p = NS). Minute-by-minute evaluation revealed that the majority of cells sedimented to the bottom of the chamber within the first seven minutes. Regression analysis did not also reveal a significant difference in the speed of cell sedimentation between SPIO-labeled HUCB-NSC and cells not labeled with SPIO (Num DF = 1, Den DF = 357, F = 2.33, p = 0.13, Fig. 2B), while, as expected has shown decrease of number of sedimented cells over time, with no further sedimentation observed after 30 minutes (Num DF = 1, Den DF = 357, F = 946.1, $p<0.0001$).

3.3 The Distribution of SPIO-labeled HUCB-NSC can be Guided by Applying an External Magnet

We observed that unlabeled HUCB-NSC are not affected by an external magnetic field in terms of sedimentation rate or direction (Fig. 3A). When the magnet was placed directly at the sedimentation tube (FS = 5400 Gs), it enabled us to position the cells exactly at the height corresponding to the center of the magnet (Fig. 3B). When the magnet was placed 2 cm away from the tube containing SPIO-labeled HUCB-NSC (FS = 1400 Gs) it was apparent that most of the cells localized at the magnet level, but were more dispersed (Fig. 3C). Increasing the distance between the tube and the magnet to 4 (FS = 700 Gs) and 6 (FS = 300 Gs) cm still showed a strong pulling effect on labeled cells (Fig. 3D,E). At a distance of 8 (FS = 160 Gs) and 10 (FS = 100 Gs) cm, the effect was reduced, with the cells located between the lower limit of the magnet and the bottom of the tube (Fig. 3F,G), but the force was still sufficient for the cells to prevent sedimentation. At a distance of 12 (FS = 60 Gs) and 14 (FS = 40 Gs) cm the cells sedimented normally (Fig. 3H,I). The strength of magnetic field at each distance was also shown for reference (Fig. 3J). For comparison to the strength of MR scanners the highest used *in vitro* field of 5400 Gs equals 0.54T. Regression analysis confirmed that the distance between the magnet and the chamber had a statistically significant impact on cell location within sections A (upper section) (Num DF = 1, Den DF = 6, F = 17.45, p = 0.0058) and C (lower section) (Num DF = 1, Den DF = 6, F = 6.20, p = 0.0471) (Fig. 4).

Discussion

Stem cell therapy for neurological disorders is still in its infancy [26]. Nevertheless, it is being offered to patients in several centers located in less-developed countries worldwide [27–32]. Such commercial, profit-driven initiatives do not meet the criteria for evidence-based medicine, and this was part of the reason that stem cell transplant services provided by a private clinic in Germany were banned (http://www.xcell-center.com). The International Society for Stem Cell Research recently stepped up efforts to provide guidelines for both patients and clinicians. It is now commonly agreed that the implementation of non-invasive cell tracking techniques is highly desirable, as it could provide a means by which to correlate the presence of transplanted cells with functional improvement. Not including monitoring of cell delivery and subsequent cell distribution in clinical trials has been criticized [33].

This study is one of the first clinical MRI cell tracking trials that aims to demonstrate the feasibility of this approach. This study was particularly focused on the special circumstances associated with an ICV delivery route and the specific benefits of applying MRI cell tracking in this clinical scenario. Due to the absence of focal brain injury, intraparenchymal cell injection was not an option in this patient and alternative routes were considered. The long period from ischemic insult (six months) did not favor an intravascular route, as a potential restoration of the disrupted of the blood-brain barrier (BBB) was expected to occur at this time point. A CSF-mediated route of cell transplantation was considered superior, as it can deliver cells beyond the BBB while still ensuring cell access to a large area of the CNS. CSF is produced mainly in the lateral ventricles by the choroid plexus and flows toward all areas of the ventricular system, emptying within the posterior cranial fossa into the subarachnoid space, and is then finally absorbed into vascular system. In normal physiological conditions, the CSF never returns from the subarachnoid space to the ventricular system. An important implication for cell

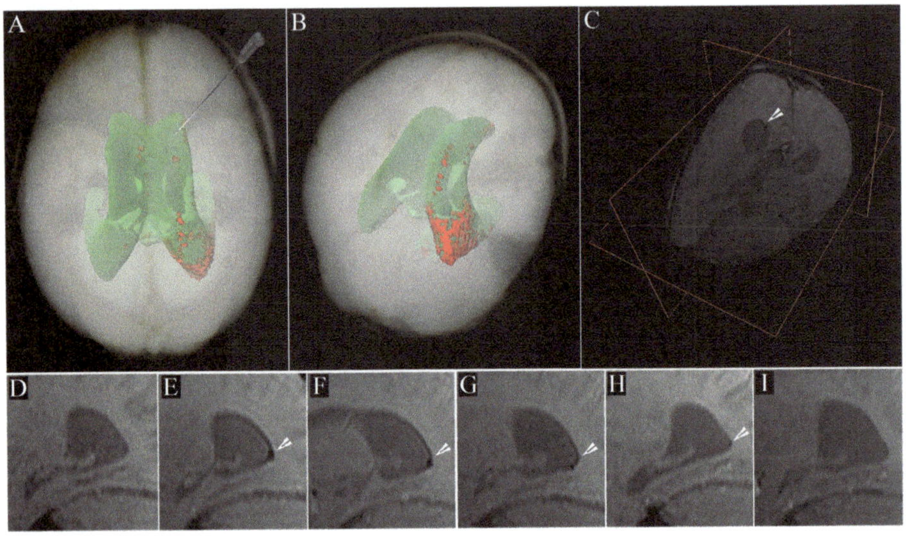

Figure 1. Imaging of SPIO-labeled autologous cord blood derived cells in a patient with global cerebral ischemia. A) Volume rendering of MRI data of the patient's head obtained 24 hours post transplantation. Semi-automatic segmentation is based on pixel intensity, showing the projection of the ventricular system (*green*) and the distribution of the SPIO signal from the transplanted cells within the occipital horn of the right ventricle (*red*). Note the supine configuration of the head, corresponding to positioning during surgery. The route and trajectory of cell transplantation via the frontal horn is represented by the needle. **B**) Postero-superior view of the patient's head, emphasizing the location of the hypointense SPIO signal from autologous cord blood derived cells transplanted within the occipital horn. **C**) T2*-weighted image with an orthogonal view centered on the cellular SPIO signal in the occipital horn (white arrowhead). **D–I**) Sagittal T2*-weighted MRI scans showing a longitudinal dispersion of SPIO signal within the occipital horn (white arrowheads); **D**) pre-transplantation, **E**) 24 hours post transplantation (PT), **F**) 7 days PT, **G**) 2 months PT, **H**) 4 months PT, and **I**) 33 months PT.

transplantation is that the ventricular system is lined with an ependyma, with this interface being highly permeable to proteins and passable by cells [34], making the ventricular system an excellent gateway to the brain parenchyma. In contrast, CSF in the subarachnoid space is separated from the brain by the pia mater, a membrane composed of fibrous, impermeable tissue [35]. Thus, lumbar and sub-occipital punctures, although being considered safer procedures, were abandoned in favor of ICV cell delivery. Although access to the ventricles is not entirely non-invasive, it offers an avenue to deliver cells throughout the entire neuroaxis to large areas of the brain.

It has never been reported in humans whether therapeutic cells deposited within the ventricular system are carried along with the CSF flow and distribute throughout the entire neuroaxis. We selected the frontal horns for the site of cell injection, as this is the most commonly used site for ventricular puncture, with a minimal risk for neurosurgical complications. After cell delivery, we could easily identify the hypointense SPIO signal of transplanted cells within the occipital horn, which was the lowest point with the patient in the supine position. Unexpectedly, cells did not distribute throughout the entire brain, but were found to sediment locally in the CSF. Longitudinal imaging over 33 months demonstrated a gradual decrease of signal, with a complete disappearance of the signal at 33 months follow-up. The biological significance of rather short-term retained SPIO signal has yet to be explained. It was shown in rodent studies that neural progenitors transplanted to the lateral ventricles can lose SPIO label before entering the brain parenchyma [34]. In addition, cell death is

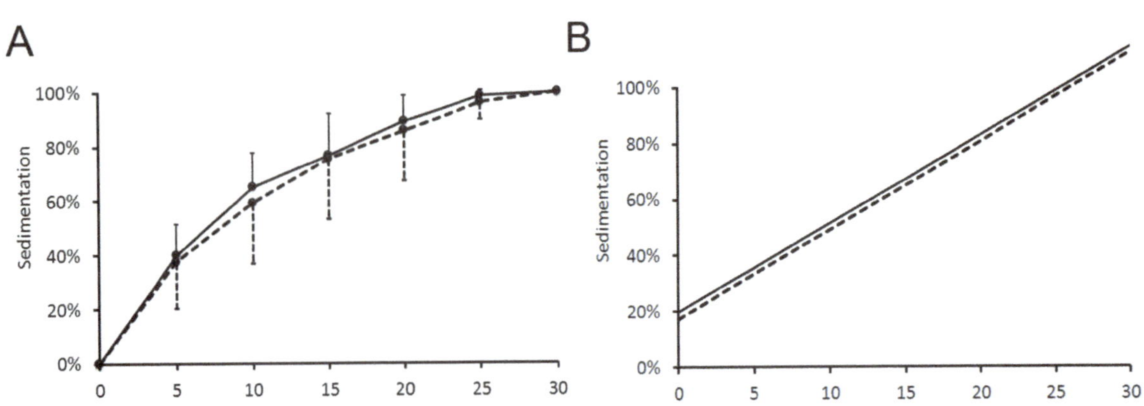

Figure 2. Time course of sedimentation for SPIO-labeled (dashed line) and unlabeled (solid line) HUCB-NSC (n = 6) presented as means±SD (A) and regression lines (B). No statistically significant difference was observed between SPIO-labeled and unlabeled cells.

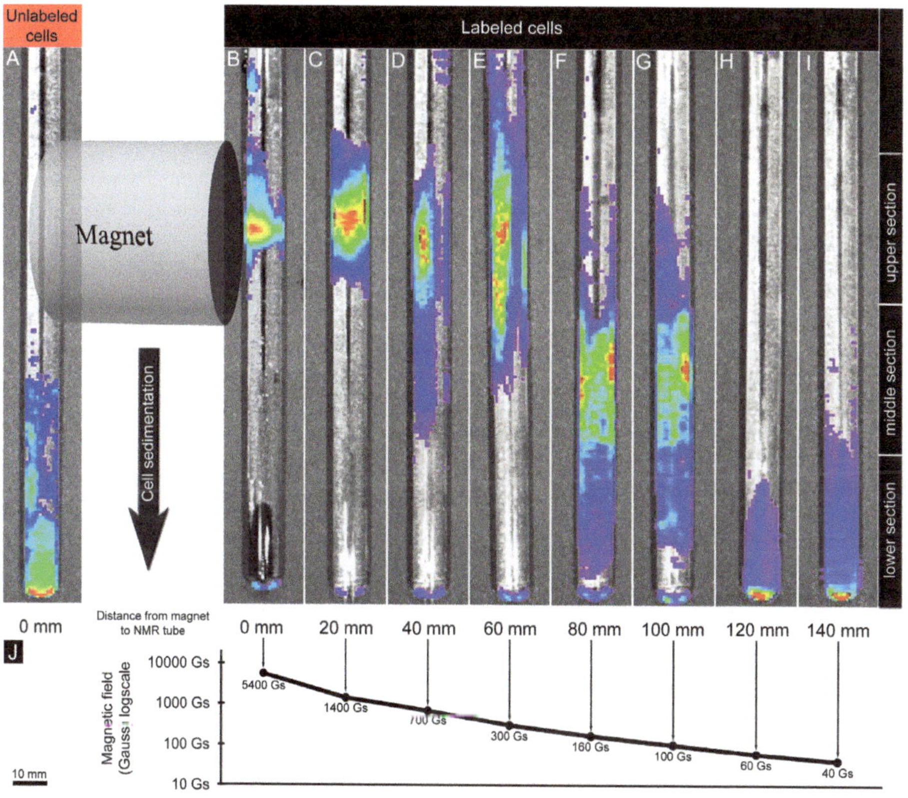

Figure 3. The effect of applying an external magnetic field on HUCB-NSC sedimentation process. Scale bar for the magnet and the tubes is indicated on bottom left.

associated with a longer persistence of SPIO signal as compared to surviving, proliferating cells. There, the persistent hypointense signal was derived from amorphous extracellular iron complexes, presumably released from dying cells [36]. Thus, a gradual decrease of signal after transplantation may indicate cell survival and proliferation or even migration into the brain parenchyma.

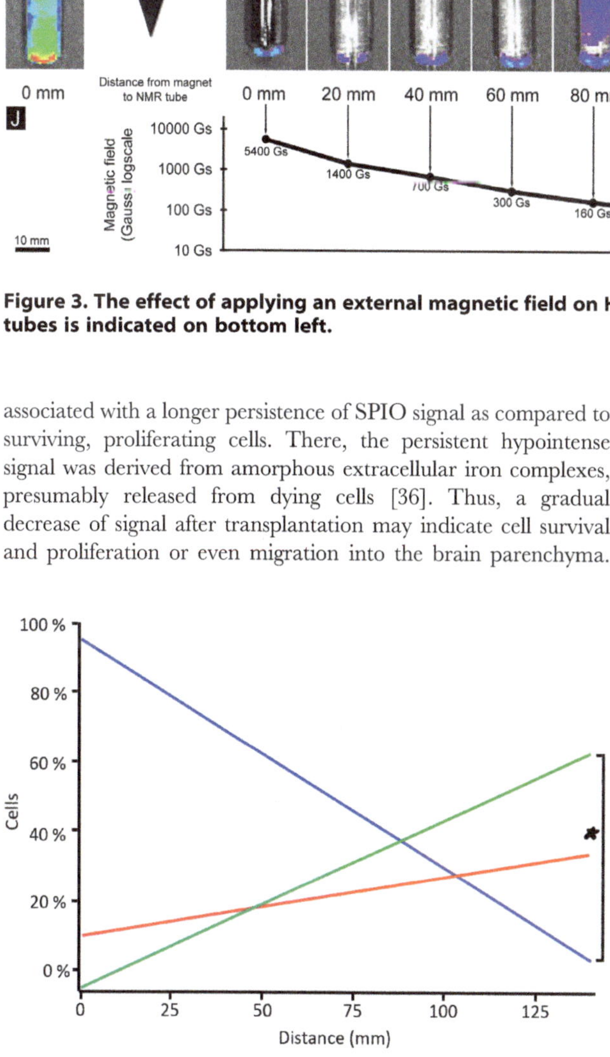

Figure 4. Regression lines show the presence of cells within particular sections in a function of distance between the magnet and the chamber. Statistically significant difference was observed between the cell presence in the upper and lower sections (marked by asterisk). Upper section = green line, middle section = red line, lower section = green line.

The detection of widely distributed transplanted cells within the brain parenchyma might be a barrier and since iron oxide provides a very strong signal, probably the limit of MRI has been reached. Due to a very high sensitivity nuclear medicine might be used in future, but no satisfactory PET or SPECT markers for cell detection beyond the blood-brain barrier (BBB) have been found as yet.

The surprising localization of transplanted cells by MRI spurred the *in vitro* experiments geared toward the explanation of observed phenomenon. Although SPIO-labeling has in general no adverse effect on cell function, we minimized any possible side effects by labeling only 20% of transplanted cells. From the MR images, it was apparent that cells rapidly sedimented into the occipital horn. We hypothesized that perhaps SPIO-labeled cells were heavier than unlabeled cells and would sediment with gravity more easily, even against the forces of CSF circulation. Thus, we compared the *in vitro* sedimentation behavior of both SPIO-labeled and unlabeled cells using human umbilical cord blood derived neural stem cells (HUCB-NSC) within a fluid compartment, and found no difference between the two cell populations, both reaching the bottom of a 15 cm-high chamber within one minute. We therefore conclude that, for future studies, only a small subset of transplanted cells can be labeled in order to provide adequate information about the distribution of the entire cell population. The first clinical MRI cell tracking using cancer vaccines also used partial SPIO-labeling, i.e. 50% of all cells [37].

As we have shown here with an *in vitro* assay, directing the position of SPIO-labeled HUCB-NSC against the force of gravity is feasible with the use of an external magnetic field. Even a readily available, inexpensive neodymium magnet is sufficient to remotely direct the cells over a distance compatible with the size of the lateral ventricles in the human brain. It may be possible that applying an external magnetic field may be used to direct cells within the ventricular fluid compartment toward the desired part of the ventricular wall located closest to the focal lesion. A stable pattern of the exact location of hypointensities in our patient over time suggests that, once cells adhere to the ependyma, they do not redistribute. As a consequence, any manipulation of cell distribution should be performed at the time of cell injection, with a small window of opportunity of about 10 minutes following transplantation. This very narrow time frame results from observation that cell sedimentation is a rapid process, which is completed in a few minutes.

The observation that SPIO-labeled HUCB-NSC are strongly attracted by an external magnetic field gives rise to the question if the high magnetic field generated by the MR scanner may also have contributed to the observed local cellular distribution. Nonetheless, in case of the patient, the cellular location did not change over time between any of the multiple MR imaging sessions. In addition, with cord blood derived cells deposited within the occipital horn, the magnetic field of a 1.5 T MR scanner would pull the cells rostrally toward the body of the lateral ventricle, which was not observed, as the cells remained within the occipital horn. Thus, it appears highly unlikely that the MR imaging procedure itself had any effect on the distribution of SPIO-labeled cord blood derived stem cells.

In conclusion, we have shown that MRI cell tracking is a powerful clinical tool to assess the distribution of transplanted cells in vivo. Following transplantation to the frontal horn, autologous cord blood derived cells unexpectedly remained localized within a well-defined region of the ependyma in the occipital horn for at least 4 months post transplantation. An external magnet may possibly be used to direct SPIO-labeled cells within a fluid compartment such as the ventricular system. For future implementation, MRI cell tracking will be mandatory to evaluate the effectiveness and outcome of this approach.

Acknowledgments

The authors thank Kshitiz Gupta for his assistance with using the Axiovert 200 microscope and Mary McAlister for editorial assistance.

Author Contributions

Conceived and designed the experiments: MJ PW KD-J JWMB BL MR. Performed the experiments: MJ PW TK EJ MR. Analyzed the data: MJ PW TK EJ JWMB BL MR. Wrote the paper: MJ PW JWMB BL MR.

References

1. Lin TC, Lee OK (2008) Stem cells: a primer. Chin J Physiol 51: 197–207.
2. Placzek MR, Chung IM, Macedo HM, Ismail S, Mortera Blanco T, et al. (2009) Stem cell bioprocessing: fundamentals and principles. J R Soc Interface 6: 209–232.
3. Farin A, Liu CY, Elder JB, Langmoen IA, Apuzzo ML (2009) The biological restoration of central nervous system architecture and function: part 1-foundations and historical landmarks in contemporary stem cell biology. Neurosurgery 64: 15–39; discussion 34.
4. Hess DC, Borlongan CV (2008) Stem cells and neurological diseases. Cell Prolif 41 Suppl 1: 94–114.
5. Lindvall O, Rehncrona S, Gustavii B, Brundin P, Astedt B, et al. (1988) Fetal dopamine-rich mesencephalic grafts in Parkinson's disease. Lancet 2: 1483–1484.
6. Widner H, Tetrud J, Rehncrona S, Snow B, Brundin P, et al. (1992) Bilateral fetal mesencephalic grafting in two patients with parkinsonism induced by 1-methyl-4-phenyl-1,2,3,6-tetrahydropyridine (MPTP). N Engl J Med 327: 1556–1563.
7. Mackay-Sim A, Feron F, Cochrane J, Bassingthwaighte L, Bayliss C, et al. (2008) Autologous olfactory ensheathing cell transplantation in human paraplegia: a 3-year clinical trial. Brain 131: 2376–2386.
8. Olanow CW, Goetz CG, Kordower JH, Stoessl AJ, Sossi V, et al. (2003) A double-blind controlled trial of bilateral fetal nigral transplantation in Parkinson's disease. Ann Neurol 54: 403–414.
9. Freed CR, Greene PE, Breeze RE, Tsai WY, DuMouchel W, et al. (2001) Transplantation of embryonic dopamine neurons for severe Parkinson's disease. N Engl J Med 344: 710–719.
10. Janowski M, Walczak P, Date I (2010) Intravenous route of cell delivery for treatment of neurological disorders: a meta-analysis of preclinical results. Stem Cells Dev 19: 5–16.
11. Battistella V, de Freitas GR, da Fonseca LM, Mercante D, Gutfilen B, et al. (2011) Safety of autologous bone marrow mononuclear cell transplantation in patients with nonacute ischemic stroke. Regen Med 6: 45–52.
12. Savitz SI, Misra V, Kasam M, Juneja H, Cox CS Jr, et al. (2011) Intravenous autologous bone marrow mononuclear cells for ischemic stroke. Ann Neurol 70: 59–69.
13. Bulte JW (2009) In vivo MRI cell tracking: clinical studies. AJR Am J Roentgenol 193: 314–325.
14. de Vries IJ, Lesterhuis WJ, Barentsz JO, Verdijk P, van Krieken JH, et al. (2005) Magnetic resonance tracking of dendritic cells in melanoma patients for monitoring of cellular therapy. Nature biotechnology 23: 1407–1413.
15. Zhu J, Zhou L, XingWu F (2006) Tracking neural stem cells in patients with brain trauma. N Engl J Med 355: 2376–2378.
16. Karussis D, Karageorgiou C, Vaknin-Dembinsky A, Gowda-Kurkalli B, Gomori JM, et al. (2010) Safety and immunological effects of mesenchymal stem cell transplantation in patients with multiple sclerosis and amyotrophic lateral sclerosis. Arch Neurol 67: 1187–1194.
17. Toso C, Vallee JP, Morel P, Ris F, Demuylder-Mischler S, et al. (2008) Clinical magnetic resonance imaging of pancreatic islet grafts after iron nanoparticle labeling. Am J Transplant 8: 701–706.
18. Richards JM, Shaw CA, Lang NN, Williams MC, Semple SI, et al. (2012) In vivo mononuclear cell tracking using superparamagnetic particles of iron oxide: feasibility and safety in humans. Circ Cardiovasc Imaging 5: 509–517.
19. Jozwiak S, Habich A, Kotulska K, Sarnowska A, Kropiwnicki T, et al. (2010) Intracerebrobentricular Transplantation of Cord Blood-Derived Neural Progenitors in a Child with Severe Global Brain Ischemic Injury. Cell Medicine 1: 71–80.
20. Janowski M, Kuzma-Kozakiewicz M, Binder D, Habisch HJ, Habich A, et al. (2008) Neurotransplantation in mice: the concorde-like position ensures minimal cell leakage and widespread distribution of cells transplanted into the cisterna magna. Neurosci Lett 430: 169–174.
21. Bates D (2005) The vegetative state and the Royal College of Physicians guidance. Neuropsychol Rehabil 15: 175–183.
22. Habich A, Jurga M, Markiewicz I, Lukomska B, Bany-Laszewicz U, et al. (2006) Early appearance of stem/progenitor cells with neural-like characteristics in human cord blood mononuclear fraction cultured in vitro. Exp Hematol 34: 914–925.
23. Buzanska L, Jurga M, Stachowiak EK, Stachowiak MK, Domanska-Janik K (2006) Neural stem-like cell line derived from a nonhematopoietic population of human umbilical cord blood. Stem Cells Dev 15: 391–406.
24. Bulte JW, Arbab AS, Douglas T, Frank JA (2004) Preparation of magnetically labeled cells for cell tracking by magnetic resonance imaging. Methods Enzymol 386: 275–299.
25. Gorelik M, Janowski M, Galpoththawela C, Rifkin R, Levy M, et al. (2012) Non-Invasive Monitoring of Immunosuppressive Drug Efficacy to Prevent Rejection of Intracerebral Glial Precursor Allografts. Cell Transplant in press.
26. Janowski M, Date I (2009) Systemic neurotransplantation–a problem-oriented systematic review. Rev Neurosci 20: 39–60.
27. Geffner LF, Santacruz P, Izurieta M, Flor L, Maldonado B, et al. (2008) Administration of autologous bone marrow stem cells into spinal cord injury patients via multiple routes is safe and improves their quality of life: comprehensive case studies. Cell Transplant 17: 1277–1293.
28. Huang H, Chen L, Xi H, Wang H, Zhang J, et al. (2008) Fetal olfactory ensheathing cells transplantation in amyotrophic lateral sclerosis patients: a controlled pilot study. Clin Transplant 22: 710–718.
29. Huang H, Chen L, Xi H, Wang Q, Zhang J, et al. (2009) [Olfactory ensheathing cells transplantation for central nervous system diseases in 1,255 patients]. Zhongguo Xiu Fu Chong Jian Wai Ke Za Zhi 23: 14–20.
30. Deda H, Inci MC, Kurekci AE, Sav A, Kayihan K, et al. (2009) Treatment of amyotrophic lateral sclerosis patients by autologous bone marrow-derived hematopoietic stem cell transplantation: a 1-year follow-up. Cytotherapy 11: 18–25.
31. Saberi H, Moshayedi P, Aghayan HR, Arjmand B, Hosseini SK, et al. (2008) Treatment of chronic thoracic spinal cord injury patients with autologous Schwann cell transplantation: an interim report on safety considerations and possible outcomes. Neurosci Lett 443: 46–50.
32. Tupitsyn NN, Yaryghin VN, Bryukhovetskiy AS, Grivtsova LY, Mentkevich GL, et al. (2006) Immunophenotypic peculiarities of mobilized stem (CD34+) cells in blood from patients with severe spinal cord injury. J Biol Regul Homeost Agents 20: 36–40.

33. Ashwal S, Obenaus A, Snyder EY (2009) Neuroimaging as a basis for rational stem cell therapy. Pediatr Neurol 40: 227–236.
34. Walczak P, Kedziorek DA, Gilad AA, Barnett BP, Bulte JW (2007) Applicability and limitations of MR tracking of neural stem cells with asymmetric cell division and rapid turnover: the case of the shiverer dysmyelinated mouse brain. Magn Reson Med 58: 261–269.
35. Weller RO (2005) Microscopic morphology and histology of the human meninges. Morphologie 89: 22–34.
36. Berman SC, Galpoththawela C, Gilad AA, Bulte JW, Walczak P (2011) Long-term MR cell tracking of neural stem cells grafted in immunocompetent versus immunodeficient mice reveals distinct differences in contrast between live and dead cells. Magn Reson Med 65: 564–574.
37. de Vries IJ, Lesterhuis WJ, Barentsz JO, Verdijk P, van Krieken JH, et al. (2005) Magnetic resonance tracking of dendritic cells in melanoma patients for monitoring of cellular therapy. Nat Biotechnol 23: 1407–1413.

2

Label-Free Morphology-Based Prediction of Multiple Differentiation Potentials of Human Mesenchymal Stem Cells for Early Evaluation of Intact Cells

Hiroto Sasaki[1], Ichiro Takeuchi[2], Mai Okada[3], Rumi Sawada[4], Kei Kanie[1,3], Yasujiro Kiyota[5], Hiroyuki Honda[1], Ryuji Kato[1,3*]

1 Department of Biotechnology, Graduate School of Engineering, Nagoya University, Nagoya, Aichi, Japan, **2** Department of Computer Science/Scientific and Engineering Simulation, Graduate School of Engineering, Nagoya Institute of Technology, Nagoya, Aichi, Japan, **3** Department of Basic Medicinal Sciences, Graduate School of Pharmaceutical Sciences, Nagoya University, Nagoya, Aichi, Japan, **4** Division of Medical Devices, National Institute of Health Sciences, Setagaya-ku, Tokyo, Japan, **5** Nikon Corporation, Chiyoda-ku, Tokyo, Japan

Abstract

Precise quantification of cellular potential of stem cells, such as human bone marrow–derived mesenchymal stem cells (hBMSCs), is important for achieving stable and effective outcomes in clinical stem cell therapy. Here, we report a method for image-based prediction of the multiple differentiation potentials of hBMSCs. This method has four major advantages: (1) the cells used for potential prediction are fully intact, and therefore directly usable for clinical applications; (2) predictions of potentials are generated before differentiation cultures are initiated; (3) prediction of multiple potentials can be provided simultaneously for each sample; and (4) predictions of potentials yield quantitative values that correlate strongly with the experimental data. Our results show that the collapse of hBMSC differentiation potentials, triggered by *in vitro* expansion, can be quantitatively predicted far in advance by predicting multiple potentials, multi-lineage differentiation potentials (osteogenic, adipogenic, and chondrogenic) and population doubling potential using morphological features apparent during the first 4 days of expansion culture. In order to understand how such morphological features can be effective for advance predictions, we measured gene-expression profiles of the same early undifferentiated cells. Both senescence-related genes (p16 and p21) and cytoskeleton-related genes (*PTK2*, CD146, and CD49) already correlated to the decrease of potentials at this stage. To objectively compare the performance of morphology and gene expression for such early prediction, we tested a range of models using various combinations of features. Such comparison of predictive performances revealed that morphological features performed better overall than gene-expression profiles, balancing the predictive accuracy with the effort required for model construction. This benchmark list of various prediction models not only identifies the best morphological feature conversion method for objective potential prediction, but should also allow clinicians to choose the most practical morphology-based prediction method for their own purposes.

Editor: Christina Chan, Michigan State University, United States of America

Funding: The authors are grateful for financial support from the New Energy and Industrial Technology Development Organization (NEDO), which provided the Grant for Industrial Technology Research (Financial Support to Young Researchers, 09C46036a); the Japan Society for the Promotion of Science (JSPS) no. 25•8678; the Hori Sciences and Arts Foundation; and the Program for Leading Graduate Schools "Integrative Graduate Education and Research in Green Natural Sciences" from the Ministry of Education, Culture, Sports, Science and Technology (MEXT) of Japan. The funders had no role in study design, data collection and analysis, decision to publish, or preparation of the manuscript.

Competing Interests: The authors declare there are no competing interests except for the patents and employment stated below. Author RK received a research grant from Nikon Corporation for collaborative research titled "Development of quality assurance technology using cellular images" (18000 US dollars/year). Also, RK is one of the main investigators of 12 patents (pending) related to basic technologies for "algorithms of cellular image processing". Author YK is the manager of the Business Planning Department, Instrument Company, Nikon Corporation. He is a member of a collaborative project with Nagoya University and is also one of the inventors of 12 related patents. Author HS is one of the main investigators of 2 patents in 12 related patents. Related patents pending: WO 2010/098105, PCT/JP2010/005119, 2010-100162, 2010-100163, 2010-100164, 2010-100165, 2010-100166, 2010-100167, 2011-042210, 2011-096258, 2011-501512, US-2012-0092478-A1.

* E-mail: kato-r@ps.nagoya-u.ac.jp

Introduction

The application of recent advances in cell technologies in regenerative medicine holds great promise for revolutionizing conventional medical therapies [1]. However, the lack of assessment technology for quantitatively evaluating cell quality, in particular for revealing both the current properties and the future potentials of intact cells, is a technical obstacle to the development of quality-assured cellular products for medical use [2,3]. Conventional methods for cellular assessment using standard techniques of molecular biology are incompatible with satisfying clinical requirements, because these methods damage cultured cells. As a result, manual microscopic monitoring, the basic and the most traditional scheme for maintaining cells, is still the most practical quality-control method for facilities that distribute regenerative cell therapies [4–6].

Technological advances in optical systems and image-processing technologies have changed the status of image-based data from an art, available only to experts, to a technique that can be used to generate unbiased data. Many high-content image-analysis methodologies based on imaging and image-processing technologies, especially those focused on fluorescently labeled images, have contributed to advances in drug discoveries [7–10]. In the field of applied cell therapy, several reports have indicated that cellular morphological information, combined with sophisticated computational modeling approaches, can serve as a descriptive indicator in evaluations of stem cells [11–14]. However, to fulfill the clinical requirements for producing intact cells for therapies, wider use of cell-morphology analysis methodologies that focus on label-free images should be encouraged.

In an effort to overcome these limitations of conventional methods using fluorescently labeled images, we previously performed a model case study of the label-free morphology-based prediction of the osteogenic differentiation potential of human bone marrow–derived stem cells (hBMSCs) [15,16], using a technique that combines an automatic cell monitoring system with effective computational modeling [17]. Statistically extracted features of cellular morphologies clearly indicated that their information content can satisfactorily train computational models, not only to quantitatively evaluate current cellular status, but also to quantitatively forecast their future status, i.e., their potentials. The greatest advantage of our proposed morphology-based cell quality assessment is its non-invasiveness. As a result of this feature, our method has benefits that cannot be achieved by conventional techniques for producing cells for clinical regenerative medicine: (1) elimination of risk factors, e.g., contamination and mishandling by the operator; (2) synchronic and flexible scheduling of culture and clinical operations, for the best timing of cellular activity; and (3) repeated assessment of the same sample, by multiple criteria and at multiple times, yielding data that better reflects the complex and dynamic features of the samples. Such intelligent control of culture processes is also a key technology for process automation [18].

In this work, we expanded our previous efforts to predict single-lineage differentiation potentials [17] by pursuing five important aims: (I) Confirmation of the robustness of our method for adapting to the practical cellular variation. In our earlier work, it was not clear whether our original methodology was applicable to wider ranges of cellular variations. To investigate this issue, our data were expanded to cover eight continuous passages, ranging from very recently derived cells to those that had completely lost their doubling potential. Since a computational modeling solution for adapting to cellular variations resulting from patient diversity was already proposed in our previous work [17], our experimental design in this work was focused on cellular variations affected by culture processes, because these are the most difficult aspect of stem cells to evaluate daily. (II) Investigation of the possibility of shifting the prediction timing to the very early stage. Our previous prediction required 2 weeks of image acquisition after the differentiation process began [17]. In this study, however, we investigated whether much earlier and shorter periods were possible. In this work, only four images, obtained from the same sample repeatedly with a 24-hour interval during the first 4 days of expansion before differentiation culture, were used in the predictions. (III) Multiplication of the variations of *in silico* predictions. Compared to the previous prediction scheme [17], which could predict osteogenic differentiation potential from the same image, in this study we attempted to predict four types of potentials (osteogenic/adipocyte/chondrocyte differentiation, and population doubling time (PDT)) from the same image. Such simultaneous prediction of multiple potentials for the same cells can be achieved by processing the same image data, although the predictions are performed by four types of differently trained prediction models running in parallel. Thus, this is a trial of "overlapping" computational evaluation that can compensate for multiple immunohistochemical staining. (IV) Establishment of new conversion schemes of morphological feature usage that can achieve high predictive performance. Morphological features are the essential information generated from imaging data, and use of this information is critical in imaging-based applications. To date, however, there have been few comprehensive studies that compare the effects of different conversions of morphological features, especially in the context of label-free time-course imaging data. To reveal differences resulting from the use of various morphological features, we proposed six types of novel morphological feature conversion methods, and then compared their prediction performances in detail. To interpret the patterns of morphological features engaged in high-performance models in each differentiation lineage, we selected LASSO regression as a modeling method. (V) Quantitative comparison of morphology and gene expression in prior prediction of differentiation potential. Although morphological information has long been used as an indicator for cellular evaluation, it has remained unclear how descriptive such information really is. To quantitatively compare the performance of morphological and biological information, we directly compared the performances of predictive models using morphological features, gene expression, or both in predicting differentiation potentials from the undifferentiated state. This comparison provides a performance benchmark for our proposed morphology-based cellular potential prediction methodology, enabling complete, non-invasive, daily cellular evaluations that could support or complement evaluations that rely on conventional biomarkers.

Results

Construction of a dataset that relates hBMSC morphological information with differentiation potential, for the purpose of developing a model for early prediction using undifferentiated status images

To construct the morphology-based cell-quality prediction model, we first designed to prepare the dataset of hBMSCs images and their experimentally determined differentiation potential data. To assemble this dataset, three lots of hBMSCs were continuously cultured (8 passages) until their growth terminated. The range of cells was intended to mimic the wide variations in cell qualities of clinical hBMSCs. At each passage, each sample was divided into three groups: passage sample (SEED group), pre-differentiation sample (PRE group), and differentiation sample (DIFF group) (Fig. 1). Because the diversity of our cell samples was intended to mimic the clinical situation, in which a minimum cell yield is often required to meet the production criteria, the passage timing was controlled by confluency. Specifically, passage was performed when confluency exceeded 80%. Continuous passage was maintained using the SEED group. Meanwhile, the PRE group was subjected to phase-contrast microscopic image acquisition (4 days, 24-h intervals), and the DIFF samples were differentiated into three mesenchymal lineages (osteogenic, adipogenic, and chondrogenic). After long-term differentiation into the three lineages, cells were evaluated for their differentiation rate and PDT; these data were taken to represent the biological differentiation potentials. In the dataset, these potentials were linked to the morphological features measured from images in the PRE groups by machine learning using the LASSO model. Because we sought to investigate the possibility of extremely early prediction of stem cell differentiation potentials for clinical applications, we acquired our image data, which we expected to contain predictive information, before the differentiation process began. Ultimately, the full hBMSC dataset contained 24 samples of cell variants (3 lots×8 passages [P2–P9]); 80 images (5 fields of view×4 wells×4 time points) from each PRE group; and 296 experimentally determined differentiation values (=2×[18 images×8 passages]+[1 image×8 passages]) (Table S1).

By monitoring morphological changes in response to continuous passage, we observed a clear morphological transition from a spindle shape into a flat and polygonal shape, a typical indication of decay of differentiation potential (Fig. 2A). The morphological information was quantitatively extracted as morphological features, as described in the Materials and Methods section.

Values related to differentiation into the three mesenchymal lineages revealed that continuous passage severely reduced the differentiation potential of hBMSCs (Fig. 2B–D, Fig. 3A–C). However, the transition patterns of differentiation potentials for the three lineages varied in a complex manner. Potentials to differentiate into adipogenic (Fig. 3B) and chondrogenic (Fig. 3C) lineages dropped rapidly, but these potentials were not correlated with the osteogenic differentiation potential (Fig. 3A). There were also variations in the changes in differentiation potentials that could be attributed primarily to differences among patients. Lot A retained its chondrogenic differentiation potential for a relatively long period, but suddenly lost it after P7 (Fig. 3C). In Lots B and C, osteogenic differentiation potential changed dramatically during continuous passage (Fig. 3A). In Lot C, adipogenic differentiation potential was sustained in any passages (Fig. 3B). Therefore, as a summary of Fig. 3B, it was realized that the tendency of differentiation potentials between patient cells can be drastically disturbed by the effect of culture process. This result indicates that simple categorization of cells by "patient information" is not effective in the clinical cell production processes, and their daily evaluation is essential.

PDT changed relatively slowly between P2 and P8 in Lots B and C (Fig. 3D). By contrast, in the case of Lot A, a rapid increase in PDT (i.e., reduction in growth rate) was observed starting at P6, indicating that this lot was sensitive to passage-related stresses (known as culture process–derived stress) triggered by both manipulations and *in vitro* culture conditions [19]. The irregular PDT increase in Lot A could be an indication of loss of

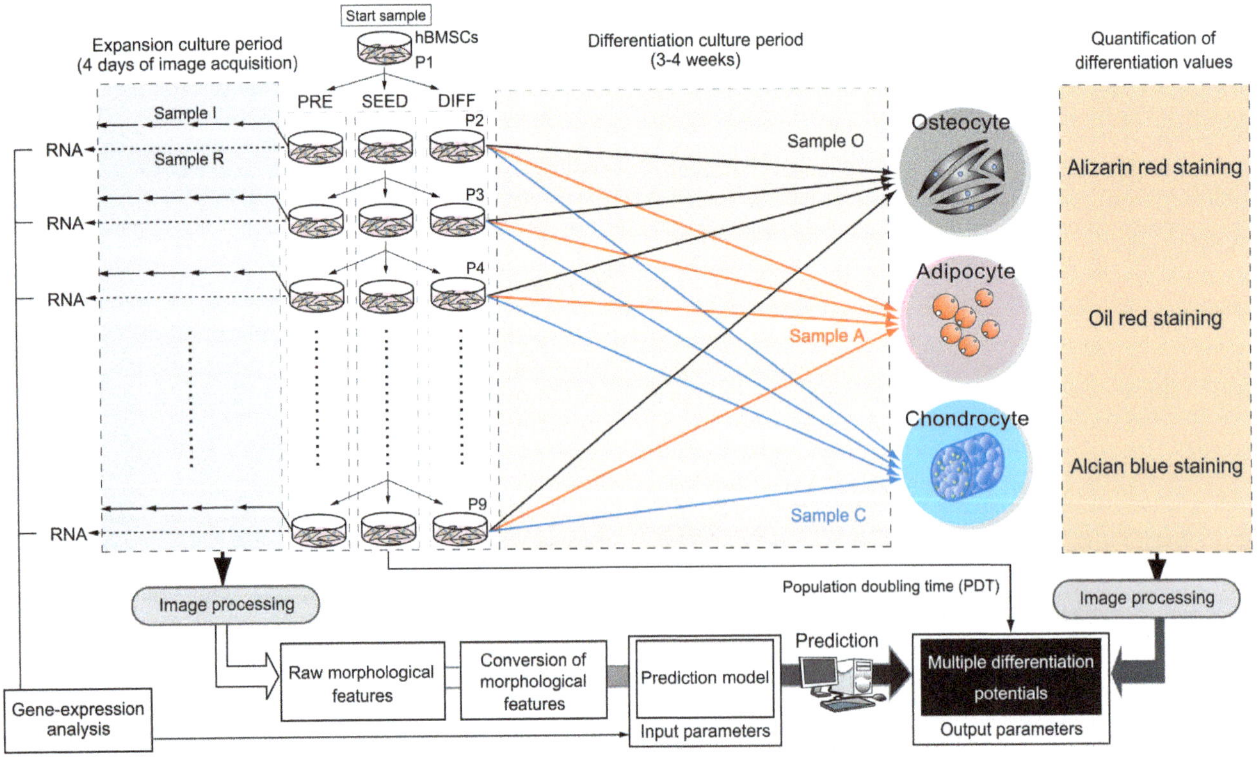

Figure 1. Schematic illustration of experimental setup for dataset construction for morphology-based prediction model construction. Figure S1 shows the illustration of usage of the objective morphology-based prediction model, and its major technological achievements using this dataset. The initial sample (P1) was divided into three separate culture samples (SEED, PRE, and DIFF) at each passage. SEED samples were mainly used for the continuous-passage culture until termination of growth (P9). From the cell yield at each passage of the SEED samples, population doubling time (PDT) was calculated, and taken as the experimentally determined potential. DIFF samples derived from each passage were divided into three differentiation cultures (samples O, A, and C for osteogenic, adipogenic, and chondrogenic differentiation, respectively) and grown under the indicated conditions for 3–4 weeks. The differentiation values of samples O, A, and C were experimentally quantified by individual staining protocols. The staining results were then converted by image-processing analysis to obtain the experimentally determined differentiation potentials. The three types of differentiation potentials together with the population doubling potential (population doubling time: PDT) were designated as "multiple differentiation potentials" of the hBMSCs. PRE samples consisted of sample I (for imaging) and sample R (for RNA extraction). From sample I in each passages, phase-contrast image were acquired at 24 h intervals over 4 days. Acquired images were then converted by image processing to obtain morphological features from every cell in all images (see also Fig. S2 and S3 for the details of image processing). Morphological features were statistically processed to yield transformed morphological features through data cleansing and statistical calculations, and the results were used as the input features. Sample R were subjected to total RNA extraction for gene-expression analysis. Either or both morphological features or/and gene-expression data were combined (input parameters), and arranged with the experimentally determined potentials of the hBMSCs (output parameters) to constitute training data for construction of prediction models.

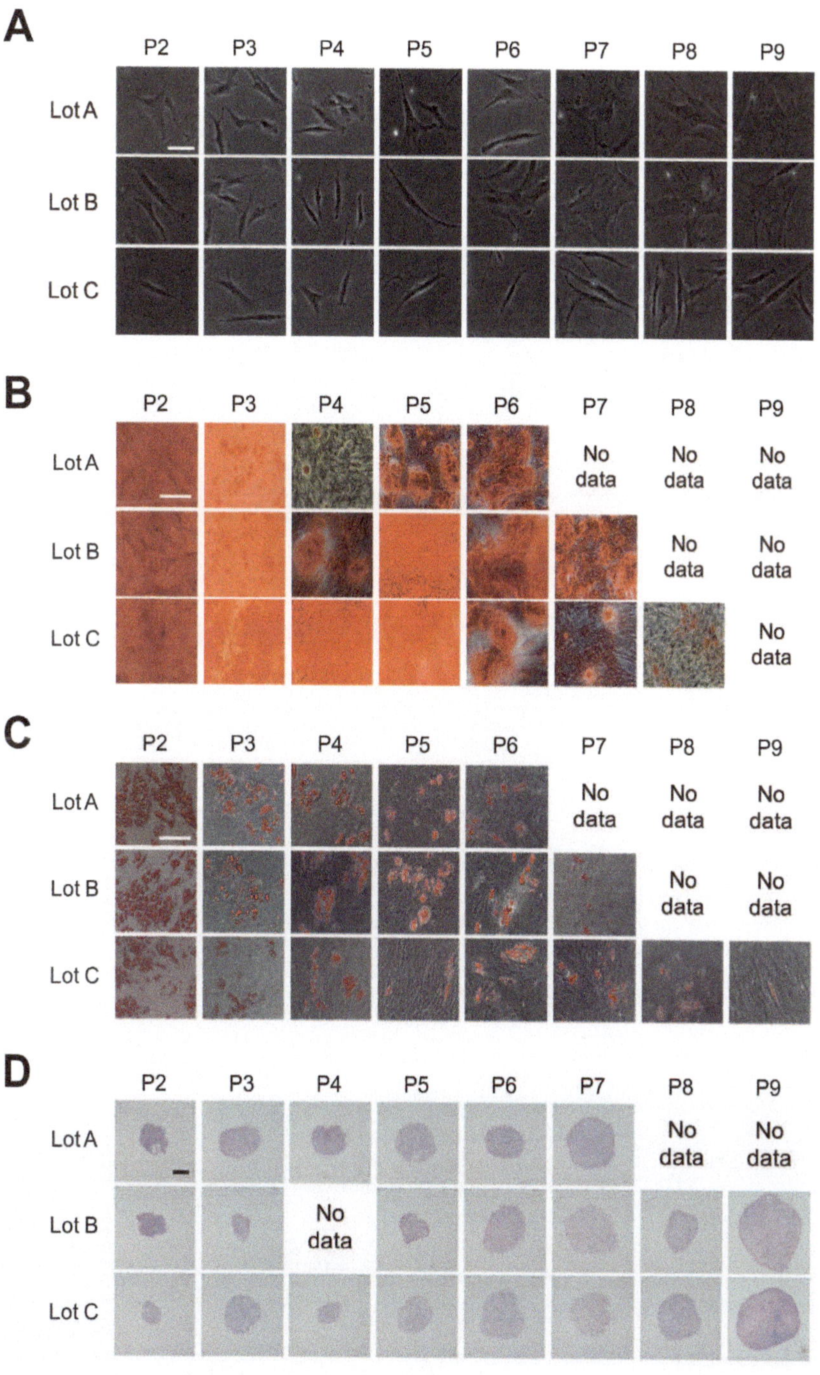

Figure 2. Representative morphological images of continuously passaged hBMSCs. Columns indicate passage numbers, indicated as P-number. Rows indicate hBMSC lot names. (A) Phase-contrast microscopic images (10×) prior to differentiation culture (sample I). Scale bar, 50 μm. (low-resolution cellular images shown in Figure S4) (B) Alizarin red staining after 2 weeks of osteogenic differentiation culture (sample O). Scale bar, 200 μm. (C) Oil red staining after 3 weeks of adipogenic differentiation culture (sample A). Scale bar, 200 μm. (D) Alcian blue staining after 4 weeks of chondrogenic differentiation culture (sample C). Scale bar, 200 μm. From P7–P9, near the termination of growth, differentiation samples could not be prepared for (B) and (C) because of the lack of cell numbers. In (D), when the pellet sizes were smaller than 200 μm, we declined to produce specimens from the sample on the grounds that the differentiation culture had not been successful.

differentiation potential; however, such an indication does not explain the early change in adipogenic and chondrogenic differentiation potentials in Lots B and C. These results reveal that there are no simple correlations between passage number and transition patterns. If passage numbers or PDT do not reflect changes in differentiation potential, then this information would never be sufficient to avoid a sudden quality collapse or insufficiency of cellular potential.

From gene-expression profiles of the earliest stage of expansion culture prior to differentiation, we found that most of the conventional differentiation markers did not exhibit clear synchronization with passage number (Fig. 4). Most of the

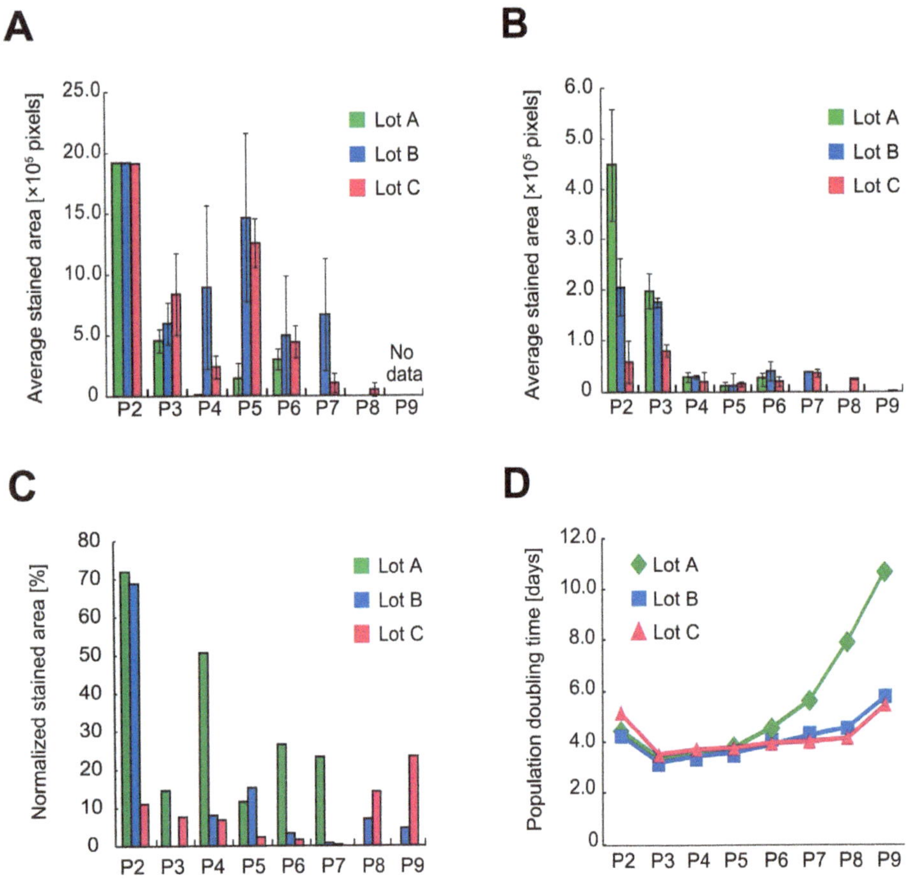

Figure 3. Quantified experimentally determined differentiation values and population doubling times of hBMSCs. Green bar, Lot A; blue bar, Lot B; pink bar, Lot C. Passage numbers are indicated as P2–P9. (A) Bar plots of average stained areas of Alizarin red–stained samples ($n = 6$). (B) Bar plots of average stained areas of Oil red–stained samples ($n = 6$). (C) Bar plots of stained areas in Alcian blue–stained samples ($n = 1$), normalized by their pellet size. (D) Line plots of PDT. Green diamonds, Lot A; blue squares, Lot B; pink triangles, Lot C. Error bars indicate standard deviation (s.d.).

clustered genes indicated the expression transition characteristic to certain cell lot. Therefore, there were several clusters, which partially showed synchronization to passage number in certain lot was considered to be more genes that reflect patient specific response to passages. However, in clustered genes which indicated clear synchronization with passage number (correlation coefficient>0.673) among all cell lots, either cellular senescence–related genes (*CDKN1A* [p21], *CDKN2A* [p16]) or cytoskeleton-related genes (*PTK2*, CD146 [*MCAM*], and CD49 [*ITGA1*]) were included. Since passage number significantly correlated with the decrease of differentiation potentials (Fig. 3), such passage number synchronizing gene expressions commonly observed in all cell lots were considered to be the "genetic signature of potential collapse".

Comparison of performances of prediction models to achieve the most balanced performance

In our previous study, we found that time courses of morphological features of cultured hBMSCs were informative in the construction of computational models aimed at forecasting future osteogenic differentiation [17]. To evaluate the multiple potentials of hBMSCs in practice, our concept of prediction had to be expanded from single-lineage to multi-lineage differentiation while retaining the ability to adapt to wider cellular variations. However, we hypothesized that in order to predict multi-lineage differentiation potential, informative morphological features and combinations thereof should be optimized for each type of differentiation. In addition, in order to increase the clinical applicability of this approach, our conceptual prediction models had to balance the accuracy with the effort (time, cost, and computational memory size) required for model preparation. Therefore, we have set our goal to define the effective construction scheme yielding the optimized prediction performance for each four different types of hBMSC potentials: potential I, osteogenic differentiation rate after 2 weeks of differentiation; potential II, adipogenic differentiation rate after 3 weeks of differentiation; potential III, chondrogenic differentiation rate after 4 weeks of differentiation; and potential IV, PDT of cells after the passages. Our objective prediction model with the newly developed techniques in this work is illustrated in Fig. S1.

To achieve the best prediction models for these objectives, we examined nine patterns (Models 1–9) of input data usages, which critically change the users' efforts for data preparation. Model 1 was designed to be the negative control, and Models 2 and 3 were designed to compare gene expression–based predictions compared to morphology-based models. Models 4–9, consisting of five model patterns (M-patterns), were designed to compare morphological feature conversion methods by investigating the various conversion concepts and time-course data usage in morphological features (Fig. 5). The comparison of morphological features was

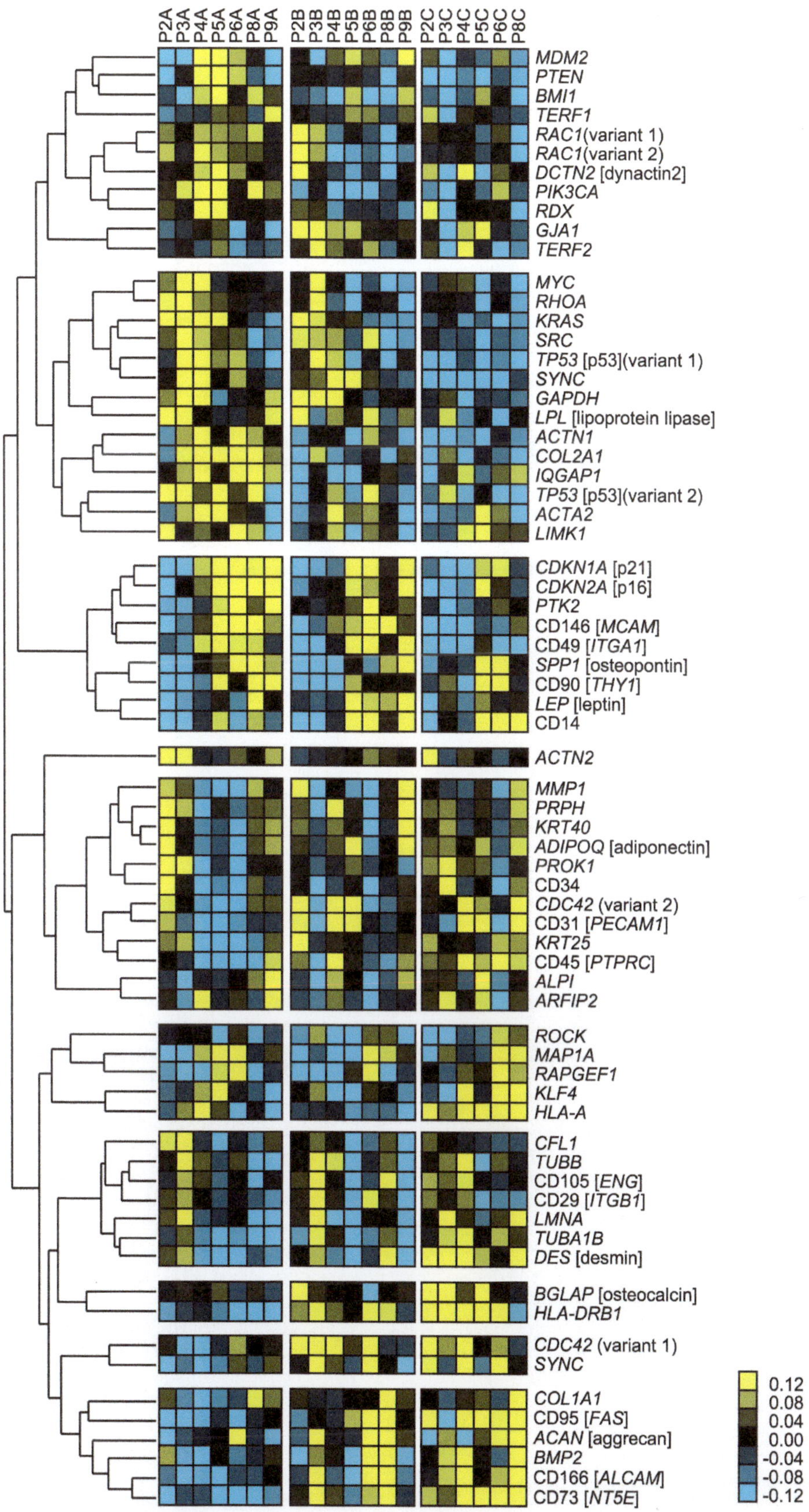

Figure 4. Heat map of gene-expression transitions and passage numbers. Genes were clustered by hierarchical clustering for indicating clusters that correlate to the passage number increases. The red boxed cluster is the cluster of genes that correlate to passage number within all cell lots, indicating non–patient-specific passage-related genes. The relationship between colors and normalized values of gene expression is illustrated in the explanatory heat map at lower right.

deepened by balancing accuracy vs. feasibility of usage in the clinic. Therefore, M-patterns were numbered in the order of higher cost performance, considering the time, cost, and computational memory size involved in the model-construction process. All model performances were compared by two criteria, "scaled error rate" and "correlation coefficient". The scaled error rate indicates the median value of prediction errors among all the samples, normalized by the actual experimental values. Therefore, low scaled error rate indicates that prediction values are relevant to the experimental values. However, usage of the scaled error rate alone is vulnerable to accidental prediction noises. Therefore, we introduced the second criterion, correlation coefficient, which evaluates the combined correlations of experimentally determined values and predicted values among all samples. These criteria are complementary: scaled error reflects differences between plots, however discards information about overall plot accuracy, whereas correlation coefficient reflects the overall similarity of measurements and predictions, however is sensitive to outliers. The combination of low scaled error rate and high correlation coefficient indicates stable performance of a given model.

Figure 6 depicts all prediction results. The data indicate that by using only the prior morphologies before the differentiation process, future collapses in all of differentiation potentials (potential I–IV) under continuous passage stresses can be predicted in advance. Comparisons of the transition patterns of the experimentally determined and predicted values (blue line plots and red line plots, respectively, depicted in Fig. 6) revealed that all cellular properties were predicted with reasonable accuracy. Furthermore, in contrast to our previous study that used all morphological data from 14 days of differentiation culture period [17], the predictive performance was enhanced in this study using morphological data collected only from the first 4 days before the differentiation.

For prediction of potential I (osteogenic differentiation rate), the best prediction accuracy was achieved by Model 3, which utilizes both morphological features (M-pattern 1) and gene-expression profiles. Compared to the scaled error rate of the NULL model (Model 1), the performance of Model 3 can be expressed as 2.6-fold more accurate. In the sense of cost-efficiency of model construction, Model 8, which utilizes only the morphological features from 24 h, had a reasonably high predictive performance (scaled error rate = 0.338).

For predicting potential II (adipogenic differentiation rate), morphology-based models such as Models 4, 6, 7, and 9 yielded extremely high predictive performance. Model 4 achieved the best accuracy, and Model 7 was the best model at the lowest cost (scaled error rate = 0.954).

For predicting potential III (chondrogenic differentiation rate), most of the models could not significantly outperform the NULL model. However, Model 9 had fairly accurate predictive performance.

For predicting potential IV (PDT of cells after repeated passages), most of the models had very high predictive accuracies (scaled error rate$<$0.09). The best performance was achieved by Model 4, which utilizes only morphological features.

Although morphology-based prediction models (Models 4–9; M-Patterns 1–5) had consistently high overall performances in predicting various potentials, the use of direct biological information (i.e., gene-expression information including conventional differentiation markers) did not dramatically improve the predictive performance (Model 2 in Fig. 6). From the interpretation of parameter usages in LASSO models (Table S4–S7), we found that cytoskeleton-related genes were more frequently involved than differentiation markers in the prediction models (Models 2 and 3). This result is a biological confirmation that morphological genes are more informative than our selected subset of differentiation marker genes for prediction of differentiation potential, and explains the high performances of models that use only morphological data (Models 4–9).

Discussion

To replace human estimations of cell quality in the production of cells for cell-based therapies, we examined the performances of machine-learning models in predicting the quantitative rates of multi-lineage differentiation after long-term differentiation, using data from undifferentiated label-free images of hBMSCs. The novel advancing technological points achieved in this work are illustrated in Fig. S1. From images collected during the first 4 days of expansion culture before differentiation, the morphological features of each cell in the images were individually measured and converted into various morphological metrics that represented the statistical morphological profiles of the group of cells. These features were then used to train computational models that forecast the experimental results collected 2–4 weeks after the differentiation. Advancing from our previous success in predicting the single-lineage differentiation potentials of hBMSCs [17], here we showed that the best predictive results for all differentiation potentials (i.e., the differentiation rates into three lineages and their PDTs) can be obtained at the same time, even in the early stage before differentiation, using selected morphological features. In these comparisons, we addressed three technical questions, with the aim of identifying the most practical scheme for obtaining such cell-quality prediction models in clinical facilities. First, can morphology-based prediction methods be expanded to the prediction of multiple differentiation potentials? Second, is morphological information (i.e., indirect phenotypic signals) of greater use than gene-expression information (i.e., direct biological signals) in predicting the qualities of hBMSCs? Third, how far can we optimize model performance by selecting the appropriate conversion and combination of information from the time-course morphological features?

To our great surprise, considering the current lack of comparable evaluation methods, most of the examined prediction models using only morphological features showed practically useful performance in multiple predictions (Fig. 5 and 6). Even with the Model 9 (M-pattern 6, using morphological features obtained only from the first day of expansion culture), the multi-lineage potential prediction was available. Practically, potential II (adipogenic differentiation rate after 3 weeks) can be predicted with high accuracy using only morphological data from the first 4 days of culture. Both potentials I and III (osteogenic and chondrogenic differentiation rates) could also be predicted with reasonable accuracy from the early morphological data. In addition to differentiation rates, future PDT following repeated passages can also be predicted with high accuracy using only morphological features. These results strongly indicate that it will be possible to develop practical methods for cell assessment that are multiple, rapid, cheap, non-invasive, and significantly more effective than conventional staining-based assessment techniques. Our models' performance indicate that such novel predictive methods will enjoy several advantages: (1) non-invasiveness, i.e., avoiding damage to patients' cells; (2) synchronism, repeated quality evaluation throughout the culture period for all patients; and (3) multivalent consideration of the same sample, i.e., multiple quality assessments can be performed with the same sample, which is not possible when using data obtained by destructive methods such as fluorescently labeled imaging analysis.

M-pattern	Parameter conversion profile									Parameter usage
	Statistical values from cellular objects	Linkage of parameters through time points	Time points [h]				Number of parameters			
			24	48	72	96	Morphological parameters	Counts	Total	
1	Average	Non-linked	•	•	•	•	36	4	40	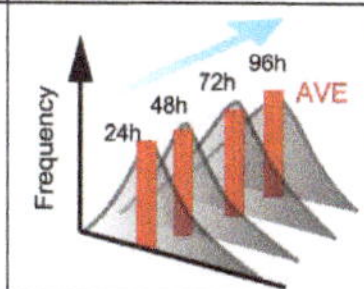
2	Quantile points	Non-linked	•	•	•	•	180	4	184	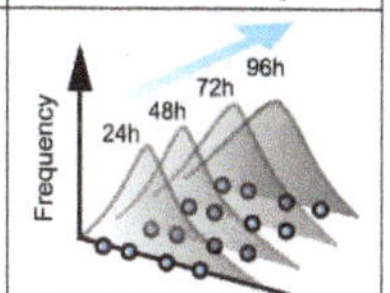
3	Quantile points	Linked	▶	▶	▶	▶	135	3	138	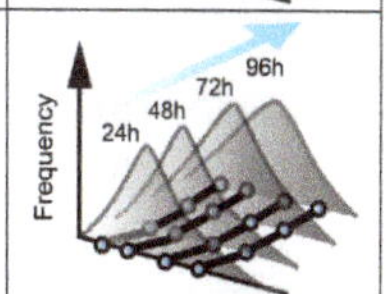
4	Distribution representative (median, interqurtile range, robust skewness, robust kurtosis)	Non-linked	•	•	•	•	144	4	148	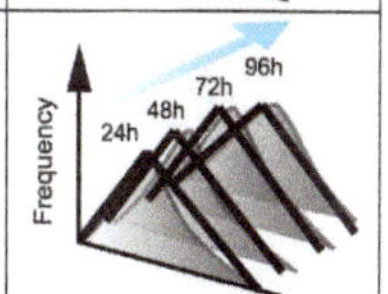
5	Quantile points	Linked	━	━	━	▶	45	1	46	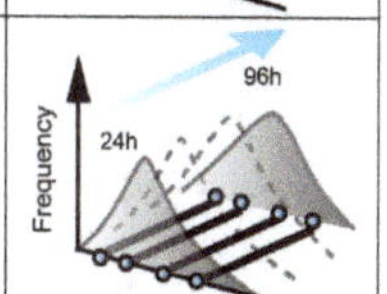
6	Quantile points	Non-linked	•				45	1	46	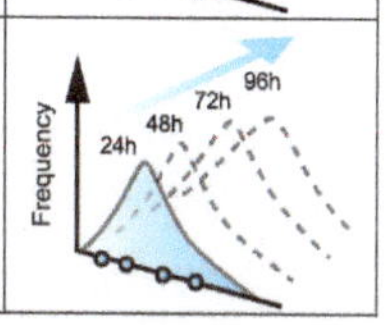

Figure 5. Schematic illustration matrix of prediction feature data profile and usage concepts of prediction models. Six types of morphological feature conversion methods are proposed as M-patterns. Briefly, M-patterns are numbered in order of the amount of efforts required to prepare for model construction. M-patterns 1–4 require four images at 24-hour intervals; M-pattern 5 requires two images each on days 1 and 4; and M-pattern 6 requires only one image on the first day. For parameters described as "linked", each morphological feature is not only used as the data for each time point, but this information is also converted into the changing ratio between time points. For "non-linked" parameters, morphological features are used as they are. Averages, quintile points, and groups of distribution representatives were compared to find the best statistical parameter to represent the morphological features measured in all individual cells in an image. Therefore, M-patterns 1–4 were designed to increase the amount of information about cellular distribution for incrementing the heterogeneity of cells.

The quantitative predictions made possible by these methods will permit prior evaluation of cellular fate, which will in turn facilitate scheduling of cell-therapy operations in the clinic. As shown in Fig. 3, most of the transition events in hBMSC potentials were abrupt, and would be nearly impossible to estimate the future linearly from the present result plots. Therefore, conventional cell-assessment techniques could never outperform quantitative prediction methods for hBMSC quality assessment. Our results thus provide a successful example of the use of machine-learning models to model biological information and generate output that can overcome a major practical problem in clinical cell therapy.

Taken together with the non-linear correlation of conventional marker gene-expression levels with passage numbers (Fig. 4) and the predictive performance of models (Models 2 and 3 in Fig. 6), we concluded that morphological data from the early stage of culture are more useful than measurements of conventional markers in forecasting future quality disruptions. In some cases, gene-expression measurement enhanced morphological predictions, when an early gene marker such as *SPP1* [osteopontin] occasionally function as extreme early osteogenesis predictor (Model 2, Potential I prediction in Fig. 6). However, differentiation gene markers are not always promising to function as extreme early predictor in the undifferentiation stage. By introducing LASSO modeling into this work, the combinational effects of parameters can be interpreted in our models (Fig. 6). In particular, by interpreting the parameter usages chosen through automatic

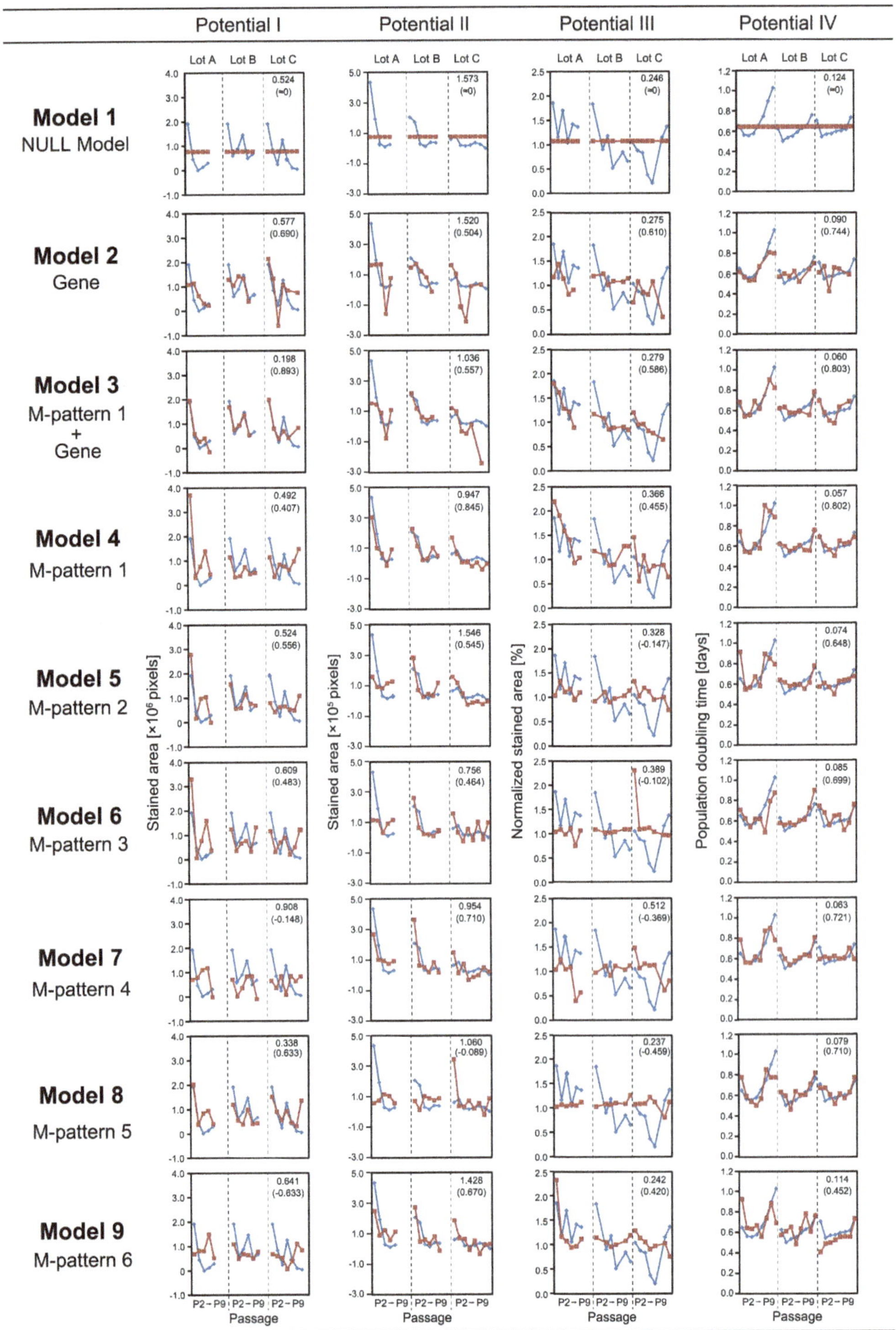

Figure 6. Comparisons of prediction models. For the detailed definition of "M-pattern", see Material and Methods. Conceptual illustration of M-pattern appears in Figure 4. Potential I, osteogenic differentiation rate; potential II, adipogenic differentiation rate; potential III, chondrogenic differentiation rate; potential IV, PDT. Each matrix involving line plots consists of three columns, separated by dotted lines, representing differences among lots (Lots A, B, and C). In each column, horizontal axis represents passage numbers, from P2 on the left to P9 on the right. Upper number at the shoulder of each matrix indicates scaled error rate, i.e., the median value of prediction errors among all the samples, normalized by the experimental values. The lower number at the shoulder of each matrix indicates the correlation coefficient. Blue line plot represents the value of experimentally determined values. Red line plot represents the prediction values from the prediction models. Greater overlap between blue and red line plots and minimum differences across passages and lot differences corresponds to higher predictive performance, represented by lower scaled error rate and higher correlation coefficient.

exploration of the best LASSO model formula, it was possible to clearly detect the correlation of cytoskeleton-related and senescence-related genes with the decrease of potentials (Table S4–S7). Such correlation was supported by previous studies showing that the TGF-β signaling cascade links cellular quality collapse with morphological changes [20–22].

Compared to the successful predictions of adipogenic differentiation potential and PDT, the predictive performance for osteogenic and chondrogenic differentiation could not be increased by altering the modeling techniques. We believe that the main reason for this limitation on performance did not reflect a shortcoming of our method. Machine-learning performance relies heavily on the quality of training data. In this work, we are uncertain of the quality of our 'teacher signal' data, i.e., the converted data from conventional staining assays for evaluations of differentiation potential. In practice, the staining technique is usually used as only one aspect of differentiation confirmation, but is not commonly used for strict quantitative analysis. A growing body of evidence describes the quantitative use of immunohistochemical staining results in high-content analysis, analogous to the way in which we converted the staining results into numerical values. However, our machine-learning model results show that only Oil red [23], but not other staining values, results in excellent performances. In addition we have identified a few critical sources of experimental noise that might partially explain the comparatively poorer performance of osteogenic and chondrogenic differentiation. For example, in osteogenic staining with Alizarin red [24], small parts of stained cells tend to be ripped from the plate during the washing process, resulting in larger deviations within some wells. Similarly, in chondrogenic staining, the pellet size and its slice position greatly affect the staining level, resulting in larger deviations within some samples. Hence, the large difference between prediction models suggests that reproducibility and signal-to-noise ratio of staining results must be carefully examined in order for modeling to be effective. In other words, if one can introduce more stable staining, the prediction models should perform better.

Our expression data regarding other types of genes lead us to expect that effective gene combinations could be defined as new quality assessment markers (Fig. 4B, 4C). The interpretation of weights of LASSO regression models can provide insights regarding essential parameters that contributed to successful predictions (Table S4–S7). For prediction of potential I (osteogenic differentiation rate), a combination of morphological features from the whole pre-differentiation period (days 1–4), together with expression of the cytoskeleton-related genes (*RAC1* [25] and *RHOA* [26]) and the early osteogenic marker *SPP1* [osteopontin] [27], were weighted. The decision to weight these genes reflected previous reports of interactions between osteogenic marker genes and cytoskeleton genes [27]. Together, the selected morphological features support the commonly observed flat and expanded cellular morphology of hBMSCs, known as an indication of bone differentiation. For prediction of potential II (adipogenic differentiation rate), relative hole area (the morphological feature that describes the "roughness" of the cell surface) and inner radius (a reflection of polygonal and tentacle-like features in the cellular periphery) from the whole pre-differentiation period (days 1–4) were weighted. This result can be interpreted to mean that the continuous evolution of hBMSCs during the expansion period toward a jagged morphology is the signature of adipogenic potential. For prediction of potential III (chondrogenic differentiation rate), a shape factor (specifically, the roundness of cells) on day 1, but not throughout the whole period of expansion culture, was weighted. This can be interpreted to mean that very early roundness of hBMSCs can indicate the potential for chondrogenesis. For prediction of differentiation potential before induction, morphogenic markers are sufficient because these data contain both time-course and multi-parametric information. However, in microarray experiments, we identified four genes related to osteogenic differentiation (*ALPI*, *BMP2*, *BGLAP* [osteocalcin], *SSP1* [osteopontin]), three genes related to adipogenic differentiation (*ADIPOQ* [adiponectin], *LEP* [leptin], and *LPL* [lipoprotein lipase]), and one gene related to chondrogenic differentiation (*ACAN* [aggrecan]). From this background, these gene profiles were not sufficient to explain the difference in differentiation potential. As it happens, in prediction of potential I, *PIK3CA* and *SPP1* [osteopontin] were useful markers even before differentiation induction; however, for the other two cell types, there were no critical marker genes. For prediction of potential IV (PDT of cells after repeated passages), both gene-expression profiles and morphological features were weighted. Cell cycle–related genes such as *TP53* [p53] [28] and *CDKN1A* [p21] [29,30], actin-related genes such as *ACTA2* [31] and *IQGAP1* [32], and the cellular senescence–related gene *CDKN2* [p16] [33–35] were heavily weighted in the best prediction models. Such gene selection reflects reports that indicate a correlation between senescence and cytoskeleton gene cascades in hBMSCs [19,36]. The weighted morphological features (total area, inner radius, and fiber length) reflect the common culture sense of regular size and slenderness during the expansion process are markers of active hBMSCs [19,37].

To confirm the utility of our proposed method in clinics, its adaptive performance in the context of the cellular diversity is an important criterion. Cellular diversity derived from patient diversity should be the first concern; we have previously investigated a new modeling scenario, designated as the "ongoing patient scenario", as one strategy for adapting to such diversity [17]. In this scenario, we proved that inclusion of new patients' own morphological features in prediction models can greatly enhance the prediction accuracy for new patients. Thus, this proposed method avoids attempting to adapt to all cellular variations arising due to patient diversity, which is ideal but impossible, and instead seeks to 're-train' the prediction model upon the arrival of each new patient, thereby allowing the system to adapt to new morphologies. Such re-training is feasible with our proposed regression models, and it is also feasible at various times in the clinic, because there are various opportunities to obtain images of primary cells before making critical potential predictions. Therefore, in this study, we attempted to optimize the adaptive performance of our model in the context of cellular diversity arising due to culture processes. In contrast to patient diversity, such culture process–derived diversity may expand during the cell production process, and may therefore require daily monitoring with non-invasive methods. Our results showed that even an extraordinarily diverse group of samples could be modeled feasibly using only morphological features. In addition, because our model performance is a summary of detailed cross-validation results combined with our ongoing patient scenario, we believe that our results provide a reliable performance benchmark reflecting robustness.

Throughout this work, we have successfully demonstrated models for predicting multiple future qualities of hBMSCs. Such models address an urgent need on the part of facilities that provide cell-based therapies. With similar objectives, Moghe's group and Gantenbein-Ritter's group have reported encouraging studies that reveal the effectiveness of multidimensional morphological parameter modeling in the evaluation of stem cell differentiation potentials [11–13]. However, in spite of the effectiveness of these strategies, such quantitative and bioinformatic technologies are still not fully utilized in clinical cell therapies. To advance image-based cellular evaluation technologies as a reliable supportive option in regenerative medicine, it will be necessary to perform more studies that connect computational technology with stem cell biology. We believe our morphology-based modeling approach will contribute to new technological developments in regenerative medicine. It has been known that cell shape takes a part in the regulation of biological processes, such as proliferation and differentiation [27,38–40]. Although our main focus was to utilize such morphological information as "a signature of biological reflection" instead of investigating its meaning, our gene expression analysis and LASSO model interpretation have revealed the involvement of previously known cell shape-regulatory proteins, such as RHO-related proteins [27]. Therefore, we believe that our morphology-based modeling approach will contribute not only to the new technological developments in regenerative medicine, but also for deeper understanding of morphology effect in stem cell biology.

Materials and Methods

Cells and cell culture

Three lots of hBMSCs were purchased from Lonza (Walkersville, MD, USA): Lot A (lot number 8F3211, Black, Male, 18 year-old), Lot B (lot number 8F3434, Caucasian, Male, 22-year-old), and Lot C (lot number 8F3560, Hispanic, Female, 24-year-old). All cultures were maintained in MSCGM (Lonza) supplemented with BulletKit (Lonza), and sub-culture was performed with 0.025% trypsin solution (Life Technologies, Inc., Carlsbad, CA, USA). To expand the cell populations, each lot was continuously sub-cultured at 7–21-day intervals until the termination of cell growth. Passage timing was controlled by confluency. Specifically, passage was performed when confluency exceeded 80%, and we obeyed the supplier-recommended cell seeding density for Poietics^{TM} human mesenchymal stem cells (Lonza). From all cell variants, P2–P9 samples were used. At each passage, cells were divided into three groups: passage sample (SEED group), pre-differentiation sample (PRE group), and differentiation sample (DIFF group). Sample I was subjected to an image-acquisition step. Sample R was subjected to RNA extraction for gene-expression profiling. The PRE groups represent the intact differentiation potentials of each hBMSC sample before the subsequent long-term differentiation process. The DIFF group was subdivided into three differentiation samples: Sample O for osteogenic differentiation, sample A for adipogenic differentiation, and sample C for chondrogenic differentiation. All differentiation cultures followed the protocol for Poietics Human Mesenchymal Stem Cells (Lonza). Briefly, samples were differentiated for 2 weeks (sample O), 3 weeks (sample A), or 4 weeks (sample C). See also Fig. 1 for the experimental scheme.

Image acquisition

For sample I, 10× magnification images at the exactly same positions (five fields at the center and four neighboring images separated by vertical displacements of 2.2 mm) were obtained every 24 hours for 4 days by phase-contrast microscopy on an X71 instrument regulated by an electric x-y stage (Olympus, Tokyo, Japan) (Fig. 1). The image data were gray-scale, 8-bit, 1360×1024 TIFF format. Focusing was semi-automatically defined by the original regulation journal run by MetaMorph (Molecular Devices, Sunnyvale, CA, USA). In total, 1,920 images (=5 view fields×4 replicates of wells×3 lots×4 time points×8 passages) were stored in the image database.

Image processing

All phase-contrast microscopic images were processed using MetaMorph (Molecular device) with an original combination of image-processing filter sets (Figure S2 and Table S3) to measure morphological features of cells. In the binarization process, a single universal threshold was applied to all images. The universal threshold was defined as the threshold that provided the minimum error between manually determined cell number and the recognized total object counts (obtained using that threshold) among 30 images picked randomly from all lots and time points. After binarization, all individual objects in each image, consisting of cells and noise (non-cell objects), were measured using the integrated morphometry analysis function to measure nine morphological features: (1) breadth, (2) elliptical form factor, (3) fiber breadth, (4) fiber length, (5) hole area, (6) inner radius, (7) relative hole area, (8) shape factor, and (9) total area (detailed in Figure S3). In addition to morphological features, cellular object counts were added as the tenth feature. The morphological features were carefully selected in the MetaMorph measurement settings by clustering analysis, in order to eliminate problems arising from multi-colinearity. From this data, which consisted of object IDs and the corresponding morphological features, the noise (non-cell objects) was then removed by a noise-reduction algorithm developed in-house before subsequent analysis. The object morphological data obtained from each PRE sample summarized nearly 20 images consisting of 2,000–5,000 cellular objects. The morphological feature information representing these groups of objects was characterized by various statistical values, including average (AVE), quantile points (10, 25, 50, 75, and 90%), median (MED), interqurtile range (INT), robust skewness (SKEW), and robust kurtosis (KURT) (see detail in Fig. 5). Such statistics related to morphological values were obtained from four time points (24, 48, 72, and 96 hours). Six patterns of morphological feature sets (M-patterns 1–6) were designed to examine their effects on predictive performance (Fig. 5). M-pattern 1 (non-continuous average information), 40 features: AVE of nine morphology features at all four time points. M-pattern 2 (non-continuous quantile information), 184 features: five quantile points of nine morphology features at all four time points. M-pattern 3 (short continuous quantile information), 138 features: ratio of five quantile points of nine morphological features within three intervals between four time points. M-pattern 4 (non-continuous distribution pattern), 148 features: MED, INT, SKEW and KURT of nine morphological features at all four time points. M-pattern 5 (long continuous quantile information), 46 features: ratio of five quantile points of nine morphological features within one interval between 0 h and 96 h. M-pattern 6 (short non-continuous quantile information), 46 features: five quantile points of nine morphological features at 24 h only.

Quantitation of osteogenic differentiation rate

Calcium accumulation of differentiated samples was determined by Alizarin red staining [24] with some modifications. Cells were fixed with 70% ethanol for 1 hour, washed, and stained for 10 min with 40 mM alizarin red S solution (pH: 4.2). The quantitation of staining results was summarized by image-based measurements of the red pixels from 18 images (3 view fields×6 replicate wells, 1600×1200, color, RGB, jpg file) collected using a DP21 CCD camera (Olympus). The color image measurements were processed using MetaMorph (Molecular Devices) to extract the red pixels. The count of red pixels was taken as the quantitated value of the osteogenic differentiation rate.

Quantitation of adipogenic differentiation rate

Samples subjected to adipogenic differentiation were stained with Oil red S [23]. Briefly, cells were fixed with 4% paraformaldehyde (PFA) for 1 hour, washed, and stained for 30 min with Oil red S solution. The quantitation of staining results was summarized by image-based measurements of the area of stained droplets from 18 images (3 view fields×6 replicate wells, 1600×1200, color, RGB, jpg file) using a DP21 CCD camera. The stained lipid droplets were identified using MetaMorph (Molecular Devices), and their total pixel area was taken as the quantitated value of the adipogenic differentiation rate.

Quantitation of chondrogenic differentiation rate

Pellet samples differentiated into chondrocytes cultured under the chondrogenic differentiation condition were stained with Alcian blue [41]. Briefly, cells were fixed with 4% PFA for 1 hour, embedded in paraffin, and sliced into 20-μm sections. All sliced tissue samples were stained on the same day according to a conventional Alcian blue protocol. The quantitation of chondrogenic differentiation rate was obtained from image-based measurements of 1 image (1600×1200, color, jpg file) collected using a DP21 CCD camera. The area of Alcian blue staining was identified using MetaMorph (Molecular Device), and the total number of pixels was measured. To control for pellet size, the number of pixels staining positive for Alcian blue were normalized against the total pellet size (in pixels), and the resultant value was taken as the quantitated value of the chondrogenic differentiation rate.

Quantitation of population doubling time (PDT)

At each passage, the total cell number was counted to obtain the PDT [42].

Gene expression measurement

Total RNA was extracted using the RNAprotect Cell Reagent (Qiagen, Hilden, Germany) from sample R at confluence. RNA samples from P2–P9 were applied to the custom designed gene chip GenoPerl (Mitsubishi Rayon, Kyoto, Japan) (gene list in Table S2). The gene chip assay and analysis were carried out according to the manufacturer's protocols, and the data was used for prediction modeling. For global gene-expression analysis, expression levels of all genes were scaled by standard normalization between arrays and genes. The clustering heat map was created by Cluster 3.0 (http://bonsai.hgc.jp/~mdehoon/software/cluster/software.htm#ctv) and Java Tree View (http://jtreeview.sourceforge.net) with some modifications.

Prediction model construction

The LASSO regression model was selected for modeling the relationships between morphological features and experimentally determined differentiation potentials. The detailed modeling process was previously described [17]. Briefly, the LASSO regression model is a penalized regression model that is widely used in the statistics and machine-learning literatures [43]. By using LASSO, one can find the linear combination of input features that best predict the teaching signal. LASSO tends to induce a sparse linear model, i.e., it can also select a set of input features that are useful for predictive purposes. The entire LASSO model-building process is automatic, including the model-selection process using leave-one-out cross-validation, and the relationships between morphological features and experimentally determined values can be represented as a simple and robust linear model. For model training, a total of 24 samples (P2–P9 in all three lots) were used as the dataset. For input features, six patterns of morphological features (M-patterns 1–6) were assigned as non-invasively obtained information for cell-quality prediction. For teaching signal information, four types of experimentally determined differentiation potentials were used to train models: Potential I, osteogenic differentiation potential (Osteo), osteogenic differentiation rate determined by Alizarin red; potential II, adipogenic differentiation potential (Adipo), adipogenic differentiation rate determined by Oil red staining; potential III, chondrogenic differentiation potential (Chondro), chondrogenic differentiation rate determined by Alcian blue staining; and potential IV, population doubling time (PDT)[42] from the growth rate over continuous passages. For the relative comparison of prediction performance, a NULL prediction model (NULL model) was constructed. The NULL model is defined as the prediction model that sets the average value of all experimentally determined values as the threshold. This imitates the case of a poor prediction, and is easily established in practical experiments. Simply put, the NULL model is a negative-control model, similar to a random-guessing model restricted to one threshold value. Furthermore, gene-expression data were also assigned in the prediction model, as rivals of morphology-based cell-quality prediction models. From the custom microarray measurements, 69 gene-expression profiles were used as additional input features in the prediction modeling. Two types of prediction models were constructed: expression data for 69 genes in combination with the M-pattern 1 feature set, and the other was the expression data for 69 genes without morphological features. In total, 36 prediction models (9 types of input feature sets×4 types of differentiation potential predictions [3 lineages and population doubling time]) were constructed (Fig. 5).

Supporting Information

Figure S1 Conceptual illustration of usage and technological achievements of label-free morphology-based prediction of multiple differentiation potentials. A user can obtain three advantageous profits from our investigated method; (1) An early prediction, even from the images from the undifferentiation period to predict the final result after differentiation. Such prediction timing is designed to be fastened in this work, by the examination of the effect of early and sparse cellular images for future prediction. (2) An effective morphological feature conversion method, which can maximize the objective prediction of certain potential. Such morphological feature conversion method is comprehensively examined in this work, to reflect the meaning of heterogeneous nature of cells and their time-course changes by various ideas of morphological feature calculations. (3) A multiple simultaneous prediction for same image. In our method, four types of potential prediction model are constructed, and provide results at the same time for one image. Such paralleled prediction concept enables "overlapping" multiple evaluations of cells with non-invasive manner.

Figure S2 Schematic procedure of image processing and data processing. (A) The procedure listed as Filter 1–4 in Table S3 is illustrated. (B) The procedure listed as Filer 5–6 in Table S3 is illustrated. Especially, the illustration describes the detail of cell measurement and their data processing scheme followed by Filter 6 processing. As shown in the figure, all cells in the images are measured as data consist of "group of cells", and their distribution is used to calculate statistic values to describe such "group of cells". Through the process, ells are measured individually by morphological indices; however our final morphological features reflect the information of "group of cells". In other words, our morphological features contain the information of "heterogeneity" of "group of cells", which strengthen our prediction model.

Figure S3 Detailed definition of the basic nine morphological features used for cell measurements. Formulas and schematic illustration of morphological features are presented in detail. The nine features are carefully selected to represent independent information, with the aspect of low-correlating parameters, for stabilizing prediction models which utilize morphological features from label-free phase contrast images.

Figure S4 Low-magnification images of cells in Figure 2A. The low-magnification images provides overall image of cellular morphological profile and its distribution.

Table S1 Dataset profile used for prediction model construction.

Table S2 Gene list in custom designed gene chip microarray.

Table S3 Image processing scheme.

Table S4 Parameters and weights explored to contribute in prediction models for Potential I.

Table S5 Parameters and weights explored to contribute in prediction models for Potential II.

Table S6 Parameters and weights explored to contribute in prediction models for Potential III.

Table S7 Parameters and weights explored to contribute in prediction models for Potential IV.

Acknowledgments

We greatly thank Asuka Miwa, Wakana Yamamoto, Yoshihide Nagura, Kazuhiro Mukaiyama for establishing the basal analysis protocol and algorithms for the image-analysis procedure.

Author Contributions

Conceived and designed the experiments: HS RS KK RK. Performed the experiments: HS MO. Analyzed the data: HS IT RK. Contributed reagents/materials/analysis tools: YK HH RK. Wrote the paper: HS IT RK.

References

1. Arthur A, Zannettino A, Gronthos S (2009) The therapeutic applications of multipotential mesenchymal/stromal stem cells in skeletal tissue repair. J Cell Physiol 218: 237–245.
2. Braude P, Minger S, Warwick R (2005) Stem cell therapy: hope or hype? Safety and quality must be assured before this treatment can really benefit patients. Br Med J 330: 1159–1160.
3. Wang Y, Han ZB, Song YP, Han ZC (2012) Safety of mesenchymal stem cells for clinical application. Stem Cells Int 2012: 652034.
4. Werbowetski-Ogilvie T, Bosse M, Stewart M, Schnerch A, Ramos-Mejia V, et al. (2009) Characterization of human embryonic stem cells with features of neoplastic progression. Nat Biotechnol 27: 91–97.
5. Settleman J (2004) Tension precedes commitment - even for a stem cell. Mol Cell 14: 148–150.
6. Docheva D, Padula D, Popov C, Mutschler W, Clausen-Schaumann H, et al. (2008) Researching into the cellular shape, volume and elasticity of mesenchymal stem cells, osteoblasts and osteosarcoma cells by atomic force microscopy. J Cell Mol Med 12: 537–552.
7. Zhang L, Yu J, Pan H, Hu P, Hao Y, et al. (2007) Small molecule regulators of autophagy identified by an image-based high-throughput screen. Proc Natl Acad Sci USA 104: 19023–19028.
8. Carpenter A, Jones T, Lamprecht M, Clarke C, Kang I, et al. (2006) CellProfiler: image analysis software for identifying and quantifying cell phenotypes. Genome Biol 7: R100.
9. Futamura Y, Kawatani M, Kazami S, Tanaka K, Muroi M, et al. (2012) Morphobase, an encyclopedic cell morphology database, and its use for drug target identification. Chem Biol 19: 1620–1630.
10. Misselwitz B, Strittmatter G, Periaswamy B, Schlumberger MC, Rout S, et al. (2010) Enhanced CellClassifier: a multi-class classification tool for microscopy images. BMC Bioinformatics 11.
11. Treiser MD, Yang EH, Gordonov S, Cohen DM, Androulakis IP, et al. (2010) Cytoskeleton-based forecasting of stem cell lineage fates. Proc Natl Acad Sci U S A 107: 610–615.
12. Kim JJ, Vega SL, Moghe PV (2013) A high content imaging-based approach for classifying cellular phenotypes. Methods in molecular biology 1052: 41–48.
13. Vega SL, Liu E, Patel PJ, Kulesa AB, Carlson AL, et al. (2012) High-content imaging-based screening of microenvironment-induced changes to stem cells. J Biomol Screen 17: 1151–1162.
14. Seiler C, Gazdhar A, Reyes M, Benneker LM, Geiser T, et al. (2012) Time-lapse microscopy and classification of 2D human mesenchymal stem cells based on cell shape picks up myogenic from osteogenic and adipogenic differentiation. J Tissue Eng Regen Med.
15. Friedenstein A, Gorskaja U, Kulagina N (1976) Fibroblast precursors in normal and irradiated mouse hematopoietic organs. Exp Hematol 4: 267–274.
16. Caplan AI, Bruder SP (2001) Mesenchymal stem cells: building blocks for molecular medicine in the 21st century. Trends in molecular medicine 7: 259–264.
17. Matsuoka F, Takeuchi I, Agata H, Kagami H, Shiono H, et al. (2013) Morphology-based prediction of osteogenic differentiation potential of human mesenchymal stem cells. PloS ONE 8: e55082.
18. Konstantinov B, Zhou W, Golini F, Hu W-S (1994) Expert-system in the control of animal-cell culture processes - potentials, functions, and perspectives. Cytotechnology 14: 233–246.
19. Røsland GV, Svendsen A, Torsvik A, Sobala E, McCormack E, et al. (2009) Long-term cultures of bone marrow derived human mesenchymal stem cells frequeantly undergo spontaneous malignant transformation. Cancer Res 69: 5331–5339.
20. Ito T, Sawada R, Fujiwara Y, Tsuchiya T (2008) FGF-2 increases osteogenic and chondrogenic differentiation potentials of human mesenchymal stem cells by inactivation of TGF-beta signaling. Cytotechnology 56: 1–7.
21. Wang DJ, Park JS, Chu JSF, Krakowski A, Luo KX, et al. (2004) Proteomic profiling of bone marrow mesenchymal stem cells upon transforming growth factor beta 1 stimulation. J Biol Chem 279: 43725–43734.
22. Massague J, Xi Q (2012) TGF-beta control of stem cell differentiation genes. FEBS letters 586: 1953–1958.
23. Kinkel A, Fernyhough M, Helterline D, Vierck J, Oberg K, et al. (2004) Oil red-O stains non-adipogenic cells: a precautionary note. Cytotechnology 46: 49–56.
24. Gregory C, Gunn W, Peister A, Prockop D (2004) An Alizarin red-based assay of mineralization by adherent cells in culture: comparison with cetylpyridinium chloride extraction. Anal Biochem 329: 77–84.
25. Jordan P, Brazao R, Boavida MG, Gespach C, Chastre E (1999) Cloning of a novel human Rac1b splice variant with increased expression in colorectal tumors. Oncogene 18: 6835–6839.
26. Jaffe AB, Hall A (2005) Rho GTPases: biochemistry and biology. Annu Rev Cell Dev Biol 21: 247–269.
27. McBeath R, Pirone D, Nelson C, Bhadriraju K, Chen C (2004) Cell shape, cytoskeletal tension, and RhoA regulate stem cell lineage commitment. Dev Cell 6: 483–495.
28. Zuckerman V, Wolyniec K, Sionov R, Haupt S, Haupt Y (2009) Tumour suppression by p53: the importance of apoptosis and cellular senescence. J Pathol 219: 3–15.
29. Motaln H, Schichor C, Lah TT (2010) Human mesenchymal stem cells and their use in cell-based therapies. Cancer 116: 2519–2530.
30. Ong CC, Jubb AM, Haverty PM, Zhou W, Tran V, et al. (2011) Targeting p21-activated kinase 1 (PAK1) to induce apoptosis of tumor cells. Proc Natl Acad Sci USA 108: 7177–7182.
31. Comer K, Dennis P, Armstrong L, Catino J, Kastan M, et al. (1998) Human smooth muscle alpha-actin gene is a transcriptional target of the p53 tumor suppressor protein. Oncogene 16: 1299–1308.

32. Weissbach L, Settleman J, Kalady M, Snijders A, Murthy A, et al. (1994) Identification of a humana RASGAP-related protein containing calmodulin-binding motifs. J Biol Chem 269: 20517–20521.
33. Jacobs J, Kieboom K, Marino S, DePinho R, van Lohuizen M (1999) The oncogene and polycomb-group gene bmi-1 regulates cell proliferation and senescence through the ink4a locus. Nature 397: 164–168.
34. Janzen V, Forkert R, Fleming HE, Saito Y, Waring MT, et al. (2006) Stem-cell ageing modified by the cyclin-dependent kinase inhibitor p16(INK4a). Nature 443: 421–426.
35. Shibata KR, Aoyama T, Shima Y, Fukiage K, Otsuka S, et al. (2007) Expression of the p16INK4A gene is associated closely with senescence of human mesenchymal stem cells and is potentially silenced by DNA methylation during in vitro expansion. Stem Cells 25: 2371–2382.
36. Bin Noh H, Ahn H, Lee W, Kwack K, Do Kwon Y (2010) The molecular signature of in vitro senescence in human mesenchymal stem cells. Genes Genomics 32: 87–93.
37. Ito T, Sawada R, Fujiwara Y, Seyama Y, Tsuchiya T (2007) FGF-2 suppresses cellular senescence of human mesenchymal stem cells by down-regulation of TGF-beta 2. Biochem Biophys Res Commun 359: 108–114.
38. Chen CS, Mrksich M, Huang S, Whitesides GM, Ingber DE (1997) Geometric control of cell life and death. Science 276: 1425–1428.
39. Roskelley CD, Desprez PY, Bissell MJ (1994) Extracellular Matrix-Dependent Tissue-Specific Gene-Expression in Mammary Epithelial-Cells Requires Both Physical and Biochemical Signal-Transduction. Proc Natl Acad Sci USA 91: 12378–12382.
40. Watt FM, Jordan PW, Oneill CH (1988) Cell-Shape Controls Terminal Differentiation of Human Epidermal-Keratinocytes. Proc Natl Acad Sci USA 85: 5576–5580.
41. Steedman H (1950) Alcian blue 8 GS - A new stain for mucin. Q J Microsc Sci 91: 477–479.
42. Ben Azouna N, Jenhani F, Regaya Z, Berraeis L, Ben Othman T, et al. (2012) Phenotypical and functional characteristics of mesenchymal stem cells from bone marrow: comparison of culture using different media supplemented with human platelet lysate or fetal bovine serum. Stem Cell Res Ther 3: 6–10.
43. Tibshirani R (1996) Regression shrinkage and selection via the Lasso. J Roy Stat Soc Series B Stat Methodol 58: 267–288.

3

Tracking Transplanted Bone Marrow Stem Cells and Their Effects in the Rat MCAO Stroke Model

Gregory V. Goldmacher[1], Rena Nasser[1,2], Daniel Y. Lee[1], Sargon Yigit[1,2], Robert Rosenwasser[3], Lorraine Iacovitti[1,2,3]*

1 Farber Institute for Neurosciences, Thomas Jefferson University, Philadelphia, Pennsylvania, United States of America, **2** Department of Neuroscience, Thomas Jefferson University, Philadelphia, Pennsylvania, United States of America, **3** Department of Neurological Surgery, Thomas Jefferson University, Philadelphia, Pennsylvania, United States of America

Abstract

In this study, rat bone marrow stromal stem cells (BMSCs) were tracked after IV administration to rats with experimental stroke caused by middle cerebral artery occlusion (MCAO). In addition, the effects of BMSC treatment on blood cell composition, brain glia and sensorimotor behavior was studied and compared to that which occurred spontaneously during the normal recovery process after stroke. We found that the vast majority of radiolabeled or fluorescently labeled BMSCs traveled to and remained in peripheral organs (lungs, spleen, liver) 3 days after IV injection in the MCAO rat. Once in the circulation, BMSCs also produced rapid alterations in host blood cell composition, increasing both neutrophil and total white blood cell count by 6 hours post-injection. In contrast, few injected BMSCs traveled to the brain and almost none endured there long term. Nonetheless, BMSC treatment produced dramatic changes in the number and activation of brain astroglia and microglia, particularly in the region of the infarct. These cellular changes were correlated with a marked improvement in performance on tests of sensory and motor function as compared to the partial recovery of function seen in PBS-injected control rats. We conclude that the notable recovery in function observed after systemic administration of BMSCs to MCAO rats is likely due to the cellular changes in blood and/or brain cell number, activation state and their cytokine/growth factor products.

Editor: Cesar V. Borlongan, University of South Florida, United States of America

Funding: The authors gratefully acknowledge The Joe and Marie Field Foundation for their generous support of this work. The funders had no role in study design, data collection and analysis, decision to publish, or preparation of the manuscript.

Competing Interests: The authors have declared that no competing interests exist.

* E-mail: lorraine.iacovittti@jefferson.edu

Introduction

Currently, stroke is the leading cause of disability and the second leading cause of death worldwide. With the advent of anti-thrombosis treatment in the last decade [1,2], the mortality rate associated with stroke has declined somewhat (12%). However, since only a small fraction of ischemic patients benefit from anti-thrombotic therapy, the overall incidence of stroke continues to climb, as does its enormous burden on society (for review, [3]. Therefore, developing new ways to treat stroke remains of utmost importance.

Stem cells, with their ability to both stimulate endogenous repair mechanisms and replace dying neurons offer great promise as a potential stroke therapy [4,5]. Indeed, studies over the last decade have shown a marked improvement in brain structure and function following stem cell transplant, regardless of the route of cell delivery (intraparenchymal, intraventricular, intravenous; IV) [6,7] or the type of stem cell administered (bone marrow, umbilical cord blood, ES cells) (for review, [8,9]). Despite this progress, there remain fundamental gaps in our understanding regarding the fate of stem cells once transplanted and the mechanism of action underlying their beneficial effects.

In this study, we have employed imaging techniques to track the journey followed by radiolabeled or fluorescently labeled rat bone marrow stromal stem cells (BMSCs) after IV injection into rats with experimental stroke caused by middle cerebral artery occlusion (MCAO). In addition, the effects of BMSC treatment on blood cell composition, brain glia and sensorimotor behavior have been chronicled during an extended period following infarction and compared to those that occur during normal spontaneous recovery from stroke. We will show that while brain structure and function can partially recover from experimental stroke, IV administration of BMSCs greatly increases the extent of that recovery. The latter occurs despite the fact that the vast majority of BMSCs remain in the periphery after systemic injection into the MCAO rat where they produce changes in blood cell composition. Likewise, in the brain where few injected BMSCs endure long term, there are dramatic changes in the number and activation of astroglia and microglia in the region of the infarct that is correlated with an improvement in behavioral measures. We conclude that the sustained recovery in sensorimotor function observed in BMSC-treated MCAO rats is likely due to the changes in blood and/or brain cells and/or their cytokine/growth factor products.

Materials and Methods

Animals

Ethics Statement. All procedures in this study were carried out in accordance with the recommendations in the Guide for the Care and Use of Laboratory Animals of the National Institutes of Health. The protocol was approved by the IACUC Committee of the Thomas Jefferson University (Protocol Number: 457I). All surgery was performed under anesthesia, and all efforts were made to minimize suffering.

MCAO lesion. Adult, male, Sprague-Dawley rats weighing 275–350 grams were anesthetized using 60 mg/kg SQ ketamine hydrochloride, 10 mg/kg xylazine and 5 mg/kg acepromazine maleate. When necessary, depending on a reflex withdrawal response and breathing rate, maintenance doses of 10 mg/kg SQ ketamine hydrochloride, 1.5 mg/kg xylazine and 0.75 mg/kg acepromazine maleate were administered to maintain anesthesia. Atropine (0.054 mg/kg, SQ) was given during the procedure to decrease bronchial secretions and keep up the heart rate. Core body temperature was monitored with a rectal temperature probe, and maintained with a heating pad and/or a small fan to within 0.5°C.

Rats were shaved and artificial tears ointment was applied to the animal's eyes for protection and lubrication. The surgical site was scrubbed with Betadine, Chlorhexidine, and 70% ethyl alcohol. The animal was positioned in a prone position. A midline incision was made and the skull is thinned over the right parietal cortex (5 mm lateral and 1 mm posterior to Bregma) without injury to the dura mater with the burr hole. The laser Doppler was attached to the skull with dental cement. Cerebral blood flow was measured starting before, during and after ischemia (Blood Flow Meter, AD Instruments, Colorado Springs, CO). Then the animal was placed supine, and the MCAO performed. Briefly, the right common carotid (CCA) and external and internal carotid arteries (ECA, ICA) were exposed through a 2 cm midline incision. Using an operating microscope, the right CCA was ligated at the proximal end. Then the right ECA was ligated just distal to the branch of occipital artery. A microvascular clamp was used to temporarily occlude the pterygoplatine artery. The occipital artery was separated and cauterized. Then a 3-0 silk suture was used to place a slack knot around the right CCA between the first ligature and the bifurcation. Right ICA blood flow was then interrupted by clamping with a microvascular clip at its origin. A tiny opening on the right CCA was made between the first ligature and the slack knot. A silicone rubber-coated monofilament nylon suture (diameter with coating is 0.39 mm, Doccol) was then inserted into the lumen of the CCA through this opening. The slack silk suture was then constricted around the CCA over the nylon suture to avoid bleeding. The clamps were removed and the nylon suture was carefully advanced into the ICA until it obstructed the MCA (18 mm from the bifurcation of the CCA). Occlusion produced on average a 40–70% decline in blood flow as detected by laser Doppler. Two hours later, the constricted suture over the CCA was loosened and the filament removed for reperfusion of the brain. Sham operated animals were anesthetized and underwent surgery and brain microdissection as described above but without tying off of the arteries.

BMSC harvest and cell culture. Rat BMSCs were harvested from SD rats (Harlan) or from a transgenic rat cell line in which the human placental alkaline phosphatase (Alk phos) gene was placed under the control of the human ROSA26 regulatory element (Invitrogen). In brief, bone marrow cells were flushed from the shaft of the femur bone with PBS. Cells were then plated into medium containing 15% FCS, 1% nonessential amino acids, 1% glutamine, 1% pen-strep in DMEM. Cells were passaged 6–8 times before use in transplantation studies.

Cell labeling and transplantation procedures. For isotope labeling, cells were harvested in 200 uL of PBS. 1.1 mCi of In-111 oxine in 1 ml of PBS was added. Cells were incubated with tracer for 30 minutes at RT, followed by repeated rinses to remove unbound activity. BMSCs were isolated from the Alk phos transgenic rat line, allowing for the histochemical visualization of individually tagged BMSCs. Alk-phos-labeled BMSCs were harvested from culture using standard methods and tracked in vivo via histochemistry using the ELF® 97 Endogenous Phosphatase Detection Kit (Invitrogen) and visualized using an Olympus IX81 Image System. Labeled or unlabeled rat BMSCs were harvested from culture, centrifuged (1000 RPM) and the pellet resuspended in sterilized PSB and injected 5×10^5 cells in 0.5 mL of sterile saline into the rat tail vein after MCAO surgery.

Imaging Methods. To assess the functional deficits associated with infarction, we injected ^{18F}-Fluorodeoxyglucose (FDG) intravenously into rats one day post-MCAO. Animals were fasted for 6 hours prior to each scan to eliminate any elevations in blood glucose at the time of the scan. In brief, the rat was anesthetized with Ketamine-Xylazine-Acepromizine mixture i.p. before the injection of tracer to avoid unnecessary uptake of tracer into the muscles due to the movement of the rat. We injected approximately 350 μCi of ^{18}F-FDG intravenously through tail vein. The rat continued to be under the effect of anesthesia for another 90 minutes, during which time imaging procedures were performed. An Inveon pre-clinical PET scanner (Siemens, Knoxville, TN) was used for PET imaging. A 10 min PET acquisition was performed at 45 min from the time of injection of tracer in addition to the 10 min transmission imaging. A CT scan was performed in MicroCAT II CT scanner (Siemens, Knoxville, TN) after the PET imaging. After the scan, the animals were placed in quarantine in a biohazard hood. Rats were not returned to the animal colony until the radiation had completely decayed to background (24 hours for ^{18}F).

For real-time tracking of injected BMSCs, rats were transported to the TJU small animal imaging facility. Usually 1–2 but no more than 4 animals were studied in any one session. To trace the fate of injected In-111 labeled BMSCs, cells were resuspended and slowly infused into the tail vein of a rat 24 hrs after MCAO surgery, followed by 0.5 ml of PBS "chaser" (final injected activity was approximately 570 uCi). CT-imaging was performed using a gamma camera. (each scan lasting up to 30 minutes) at 1 hour, 2 hours, 4 hours, 8 hours, 12 hours, 18 hours, 24 hours, 36 hours, and 48 hours after injection. Whole body images were obtained using a flat collimator and a pinhole collimator was used to capture brain images. The animals were anesthetized using 2% isoflurane for imaging and were allowed to recover between imaging time points any time the imaging interval was greater than 1 hour. The rats were monitored in the animal imaging facility until fully awake and recovered. Imaging was repeated for 3 sham-lesioned and 3 MCAO lesioned rats.

Blood Cell Analysis. To track changes in circulating blood cell populations following IV injection of BMSC's, host blood was collected from the tail vein before and 24 hr after MCAO and 6 and 24 hr after BMSC treatment. Blood was collected in 2 ml tubes spray-dried with 1.2–2.0 mg EDTA/1 mL of blood (Fischer Scientific Cat No. 22-040-101), samples were kept at 4°C and sent them to Bioreliance (Rockville, MD) for blood cell analysis. Briefly, cells were separated using an automated system that employs the ADVIA 120 Hematology analyzer to perform differential CBC assay using cytochemical light scatter and light absorption measurements and LIMS software analysis [10].

Behavioral Tests. To evaluate neurological function before and after stem cell transplant, rats were subjected to a battery of tests before and after MCAO and transplantation. Motor deficits were evaluated using a modified Neurological Severity Score (mNSS) [11] which includes fore and hind limb flexion; walking, righting, placing, balance, body resistance, paralysis). Using this scale, one point was given for the inability to perform a test. Consequently, the higher the score, the more severe is the motor deficit (maximum score = 18). In addition, motor function was assessed using a forelimb use asymmetry test [12–14], wherein rats were placed in a plexiglass cylinder and the number of attempts to climb the sidewalls with impaired (I), unimpaired (U) and both (B) limbs was quantified during a 10 min trial. The asymmetry index was calculated as [U/(U+I+B)] – [I/(U+I+B)] [12–14]. The higher the index, the greater is the asymmetry in use. We also used a corner test wherein rats were placed in a 30 degree corner and the percentage of turns to the right (unimpaired side) was calculated for 15 trials [12,15]. Since normal rats turn 50% of the time towards the right and 50% toward the left side, the higher the score, the greater the MCAO-induced deficit. Behavior was assessed at regular intervals by an observer blinded to treatment status.

Immunocytochemistry. Transplanted rat brains were perfused with 4% cold (4°C) periodate-lysine-paraformaldehyde, sectioned at 30 μm on a freezing microtome and processed for immunocytochemistry using the immunofluorescence staining method described previously [16]. Cultures or sections through brain, spleen, liver and heart were stained with one or more of the following antibodies: mouse anti-human nuclear antigen (HNA) from Millipore at 1:40; rabbit antibodies to GFAP, 1:1000; monoclonal mouse antibodies to CD11, 1:25. All secondary antibodies were Alexa Fluor antibodies from Invitrogen used at 1:200–1:300. Cultures or sections were analyzed using a Nikon-Scanalytics Image System or an Olympus IX81 Image System. After immunostaining, five regions of interest surrounding the infarct on the side of MCAO, and comparable regions in the contralateral and PBS control striatum, were quantified in 10 sections through each forebrain. Image J software (http://rsb.info.nih.gov/ij/) was used to quantify pixel intensity values and percent relative values determined. The data were analyzed using GraphPad Prism software with a non-paired Student's *t*-test.

TTC Staining method. Following sacrifice, fresh (unfixed) brain was removed from the skull, washed in iced PBS and placed in a brain mold. Coronal sections 1 mm in thickness were cut through the cerebrum and placed in 1% 2,3, 5 triphenyltetrazolium chloride (TTC; Sigma) for 30 min in a 3°C incubator. TTC solution was drained and sections were immersed in 4% formalin in 0.1 m PB at 4°C and stained as described previously [16].

Statistical Analysis. Data were statistically analyzed by one-way analysis of variance. When $P<0.05$, then the *F*-test was followed by the two-tailed Students t-test to compare the statistical significance between control and experimental groups (ANOVA with subsequent post-hoc analysis using Bonferroni correction). Statistical differences were considered significant only when the *P* value was<0.05.

Results

Assessing stroke and its effects

To assess functional change in the brain and body, fluorodeoxyglucose (^{18}F) (FDG) utilization was imaged in rats for up to a month following experimental stroke and compared to sham-operated controls (n = 3). Reconstruction was performed with 3D-OSEM algorithm to get high resolution images of brain metabolism. A CT scan was performed after the PET imaging and image registration was used to align CT images and PET images. PET-CT fusion images were used to identify anatomical landmarks. Regions of interest (ROI) were drawn on the PET images to measure the amount of glucose metabolism in different parts of the brain. From these, we calculated the size of the infarct and the improvement in uptake at different time points (1, 4 weeks). We found that using FDG, one could image the region and functional loss after MCAO and even assess the improvement which occurs spontaneously over time (example provided in **Fig. 1**). Note that over a 4 week recovery period, slice 3, near the greatest area of infarct, spontaneously recovered only 4% while slices 2 and 4–6 showed greater recovery (8–22%) and slices 1 and 7 where there is collateral circulation from the anterior and posterior cerebral were less affected (10%) by occlusion of the MCA.

Importantly, the improvement reported above was accompanied by a substantial degree of spontaneous recovery in some but not all sensorimotor behaviors. In studies done to date on MCAO rats, the modified Neurological Severity Score (mNSS) has been the most widely used scale of behaviors to assess changes in neurological function [11]. As anticipated, after a major infarct in the brain, there was a significant decline in sensorimotor function on mNSS score (Mean score of 14/18) at 1 day post-MCAO (set to 100%) (N = 5–8). However, 1 week later, deficits were markedly reduced (27% improvement in the mNSS score; **Fig. 2A**) in infarcted rats receiving only palliative care, due particularly to improvements in gait, body resistance and spontaneous movement. By 2–4 weeks after injury, there was further recovery that plateaued when the mNSS score reached approximately 50% of the original 1 day post-MCAO score. In contrast, other behaviors, such as, use of the impaired forelimb in climbing (**Fig. 2B**), and ability to turn toward both left and right in the corner test (**Fig. 2C**), which relies on both sensory function of the vibrissae and motor function on the impaired side [12,15], did not show substantial spontaneous recovery in function over the course of a month.

In vivo tracking of radioisotope-labeled BMSCs in MCAO rats. As little is yet known about the path that BMSCs travel after their injection into the bloodstream, we tracked the journey of radioisotope-labeled BMSCs in rats with moderate to severe stroke. To do so, BMSCs (8.1×10^6 cells) were harvested and grown as described by us previously [16]. For isotope labeling, BMSCs were harvested in 200 uL of PBS. 1.1 mCi of In-111 oxine in 1 ml of PBS was added. Cells were incubated with tracer for 30 minutes at RT, followed by repeated rinses to remove unbound activity. Cells were re-suspended and slowly infused into the tail vein of a rat 24 hrs after MCAO surgery, followed by 0.5 ml of PBS "chaser" (final injected activity was approximately 570 uCi). Imaging was performed using a gamma camera. Whole body images were obtained using a flat collimator, at 4, 20, 44 and 70 hrs after injection, with 350,000 counts collected for each image. We found that the vast majority of IV-injected cells traveled to the lungs where they remained 3 days later. In addition, activity in the liver and spleen increased over time (**Fig. 3A**). Images of the brain were obtained using a pinhole collimator at 20, 44 and 70 hours, with 100,000 counts collected for each image. By 44 and 70 hours, higher brain activity was observed on the side ipsilateral to the infarct, suggesting that MSCs had in fact homed to the site of brain injury (**Fig. 3B**).

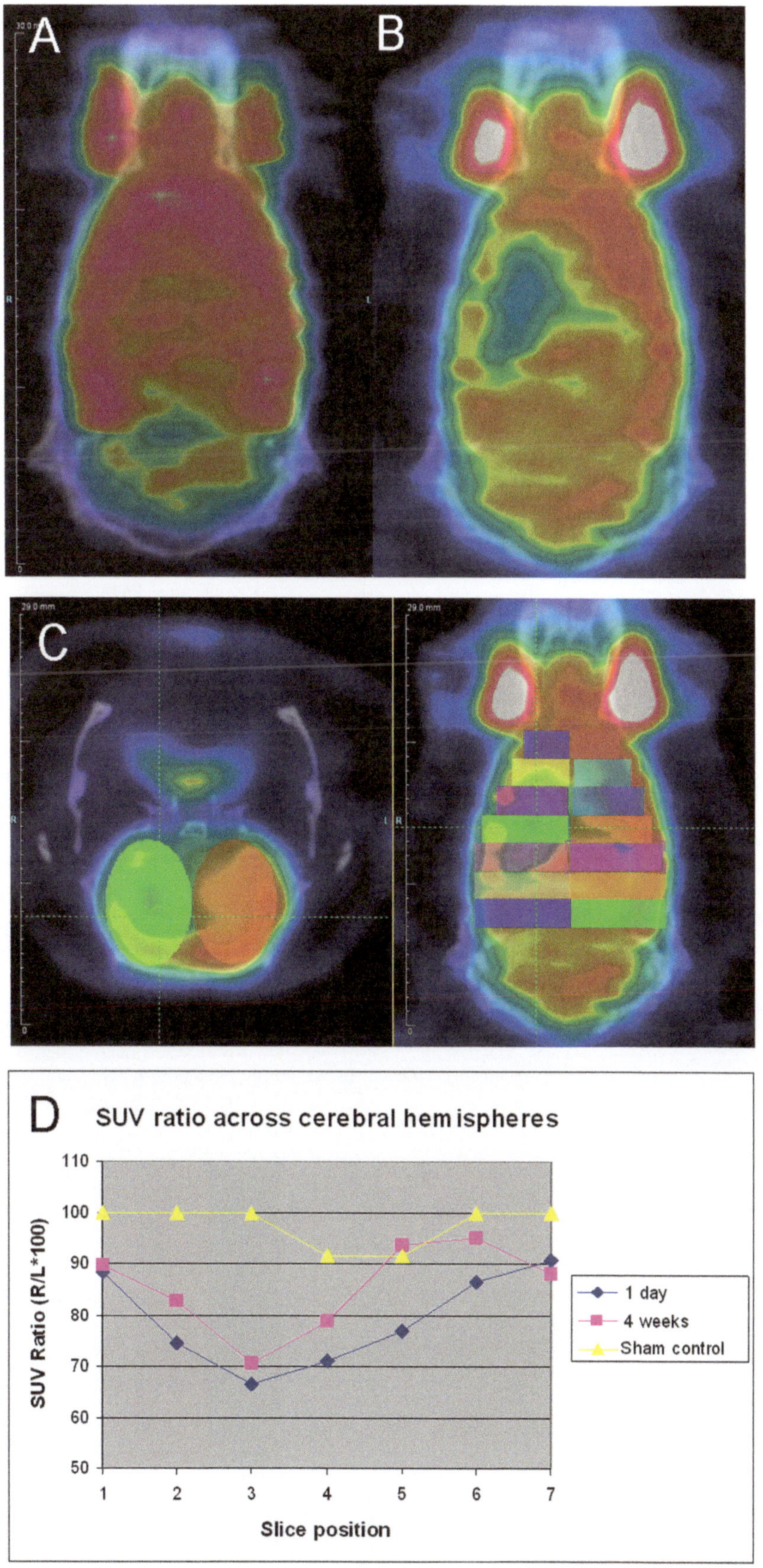

Figure 1. Analysis of infarct severity and extent on PET/CT imaging in representative sham and MCAO brains. A. Coronal view of a sham-operated animal showing normal symmetrical glucose uptake. B. One day post-MCAO, there is decreased uptake on the side of the infarct. C.

Axial and coronal images showing placement of 2 mm thick ovoid regions of interest (ROIs) [symmetrical across the midline, arbitrary colors] across the cerebral hemispheres. D. Mean standardized uptake value (SUV) ratios (infarcted to contralateral side ROI) are plotted against slice position comparing infarcted and sham-operated control rat shown in A–C. The severity of the infarct is reflected both by the number of slices affected, and the degree to which activity is suppressed (relative to normal) in each slice. Over 4 weeks, there was a modest spontaneous improvement in FDG uptake, which correlated with the improvement seen in mNSS scores from 6 to 4 (not shown). Imaging was repeated for 3 sham-lesioned and 3 MCAO lesioned rats.

Tracking transplanted alkaline phosphatase labeled BMSCs in MCAO rats

Although the radioisotope imaging technique is useful as a first level of investigation, resolution at the cellular level requires additional analysis. Therefore, we used a labeling technique which allows for histochemical visualization of individual Alk phos tagged BMSCs (Invitrogen). BMSCs (derived as described in Methods) robustly expressed Alk phos during cell expansion and differentiation in culture (**Fig. 4A inset**). After IV injection into the MCAO rat tail vein, Alk phos-tagged cells were readily detected in the lung (**Fig. 4A**) and in the brain at days **3** post-transplant (**Fig. 4B**). However, most labeled BMSCs in the brain had vanished by 8 days post-transplant although some remained at the edge of the infarcted region (**Fig. 4C**).

Changes in host cells after BMSC treatment in MCAO rats: Analysis of Host Blood Cell Composition

Although many labs have examined the effects of BMSCs in the brain, no one has tracked their effects on host blood cells. To assess changes in blood cell populations following IV injection of BMSC's, the blood was collected from the tail vein before and 24 hr after MCAO and 6 and 24 hr after BMSC injection. We found that MCAO procedure alone produced an increase in neutrophils, a decrease in lymphocytes and no change in other cell types or the total white blood cell count (**Fig. 5**). However, by 6 hours after BMSC treatment, there was a further amplification in the number of neutrophils, resulting in an increase in the total while blood cell count despite the fact that lymphocytes remained unchanged from MCAO-induced levels (**Fig. 5**). Importantly, by 24 h hours after stem cell treatment, all cell counts reverted to pretreatment or post-MCAO levels.

Analysis of BMSC Effects on the Brain

As shown previously [16–18], BMSC treatment produces significant changes in local brain cells beyond that which occurs due to MCAO. Thus, by 1 week (data not shown) and continuing at 4 weeks post-injection (**Fig. 6A**), rats administered BMSCs 24 hrs after MCAO exhibited a marked increase in the number of reactive GFAP$^+$ astrocytes (412%±29% relative fluorescence intensity) and CD11$^+$ microglia (487%±17% relative fluorescence intensity) in the area surrounding the infarct compared to the contralateral side (100%±9% relative fluorescence intensity) or PBS-treated MCAO rats (98%±12%). In addition, glia and microglial cells in the stem cell-treated rats exhibited an activated morphology, shown in **Fig. 6B** insets, at 4 weeks, long after their counterparts in untreated MCAO rats had returned to the quiescent state (**Fig. S1**), as noted previously [16].

Recovery of function after BMSCs treatment in MCAO rats

Importantly, changes seen in the brain after BMSC treatment were correlated with functional recovery of sensorimotor behaviors when compared to control rats (PBS-treated). Thus, stem cell treatment resulted in a significant improvement in neurologic function (ie. mNSS score) (**Fig. 7A**) over the partial recovery seen spontaneously after MCAO (**Fig. 2A**). In addition, BMSC treatment significantly improved other long-lasting measures of sensorimotor function, such as cylinder climbing capacity and left corner turning ability, both of which did not show signs of spontaneous recovery (**Figs. 7B,C**). In all of these cases, BMSC-induced recovery in behavioral deficits did not reach significance until 4 weeks after treatment was initiated.

Discussion

Numerous studies have chronicled the remarkable ability of systemic BMSCs to produce recovery of brain structure and function after experimental stroke in rats [9,16,19–23]. However, few if any studies have adequately assessed the degree to which these indices improve spontaneously over time without interventional treatment. In addition, relatively few studies have actually

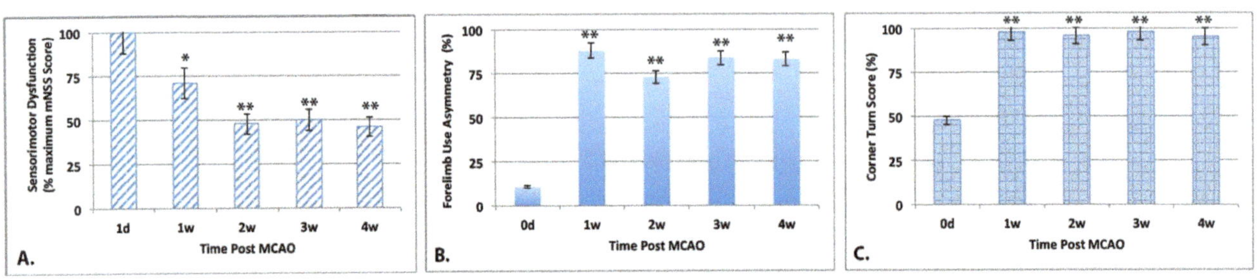

Figure 2. Spontaneous recovery of behavioral deficits after MCAO. Panel A represents sensorimotor loss as measured by the mNSS score. mNSS scores are normalized to a maximum of 18 on the mNSS scale (100% deficit). Note that following MCAO there was substantial spontaneous recovery in the mNSS score. N = 4–6, *p<0.01 and **p<0.001 compared to scores at 1 d post-MCAO. Panel B represents the rats' ability to climb the sidewall of a cylinder using the affected limb after MCAO. All rats were tested prior to MCAO and at 1, 2, 3 and 4 weeks after MCAO and the asymmetry index calculated (See Methods). Note that the decline in forelimb climbing remained even at 4 weeks post-MCAO. N = 4–6, ** p<0.001. Panel C represents the ability of the test animal to turn toward the unimpaired right side when placed in a corner. All rats are tested prior to MCAO, turning 50% of the time to the right and 50% of the time to the left. Values represent % of trials in which the rat turned toward the unimpaired side after MCAO. Note that following MCAO, this ability declined dramatically. ** p<0.001. In all 3 panels, the higher the value, the greater the deficit. (N = 4–8 ± SEM).

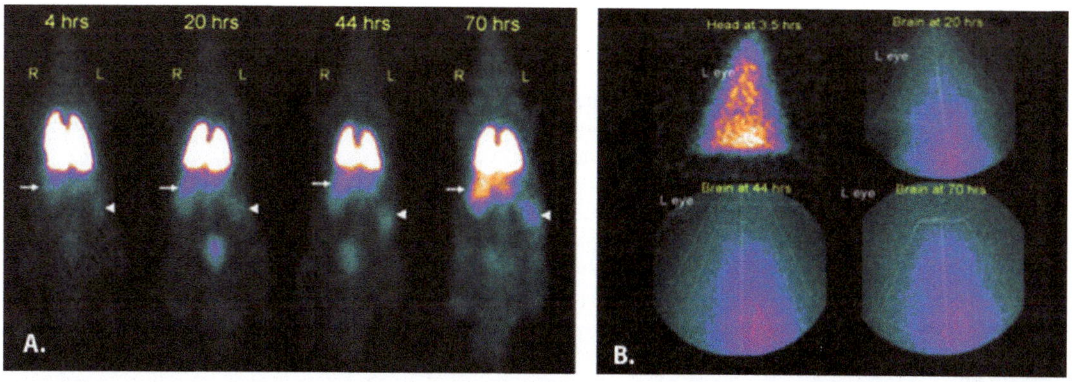

Figure 3. Whole body and brain images following IV injection of In-111 oxine–labeled BMSCs. (A) Images show persistent activity in the lungs, with relative activity in the liver (arrows) and spleen (arrowheads) increasing over time. Bladder activity is also seen. (B) Brain images at various times after injection. Larger magnification through a pinhole collimator with increasing proximity to the forebrain in later images (20 hrs, 44 hrs, 70 hrs) showed higher activity on the side of the infarct. Approximate position of left eye is marked in each image. (N = 4).

tracked the journey taken by stem cells after their systemic administration and documented their effects in blood and brain during the recovery period [24–26]. The current study demonstrates that the brain is indeed capable of considerable spontaneous recovery in structure and function after experimental stroke but that systemic BMSC treatment, which affects both blood and brain cell composition, provides added benefit, improving all indices examined, particularly at later post-recovery times.

Until now, the role of spontaneous recovery of function after MCAO has not been well-investigated. The present study however found that even after a large MCAO-induced infarction, there was a partial but significant restoration in brain glucose function. In fact, the high demand for energy in the brain is almost exclusively provided by the oxidative metabolism of glucose. Importantly, brain regions at greater distances from the infarct showed considerable (22%) spontaneous recovery of glucose utilization, while contrastingly, areas of greatest infarct showed much less recovery of glucose function (4%). These findings suggest that most rescue in glucose function occurs in the penumbra region surrounding the infarct, as has been noted previously [27–29]. Moreover, regions with greater collateral blood circulation from the anterior and posterior cerebral arteries are able to sustain a drop in blood flow without significant decline in glucose utilization.

In examining how functional changes in glucose uptake correlated with changes in sensorimotor function, a somewhat different profile emerged, depending on the behavioral assessment. We found that during the first 1–2 weeks after MCAO, there was a substantial degree of recovery in the mNSS score, which assesses a wide range of motor and sensory skills. The recovery stemmed mainly from improvements in gait, body resistance and spontaneous activity in MCAO rats. However, by, 2 weeks post-MCAO, this recovery plateaued, with no further improvement at later times. In contrast to the mNSS score, other sensorimotor tests affected by MCAO remained unchanged one month later. Thus, there was no spontaneous recovery in the ability to use the affected limb to climb the sidewalls of a cylinder and virtually no improvement in the rats' ability to turn towards the impaired side in the corner test one month after MCAO. While both of the latter tests assess motor function, the corner test also depends upon sensory stimulation of the vibrissae to induce turning behavior [12,15]. Thus, in judging the success of new therapies for stroke, caution must be exercised as some behaviors partially improve spontaneously over time after MCAO, while others do not.

Additionally, our study examined whether therapeutic treatment with systemic BMSCs could ameliorate the deficits in brain structure and function caused by MCAO, enhancing recovery beyond that which occurs spontaneously. Despite the wide array of studies describing stem cells and their effects after experimental stroke in the rat [9,16,19,20,22,23,30], little is known about the route that BMSCs take in the body and brain when administered directly into the peripheral circulation. In our imaging experi-

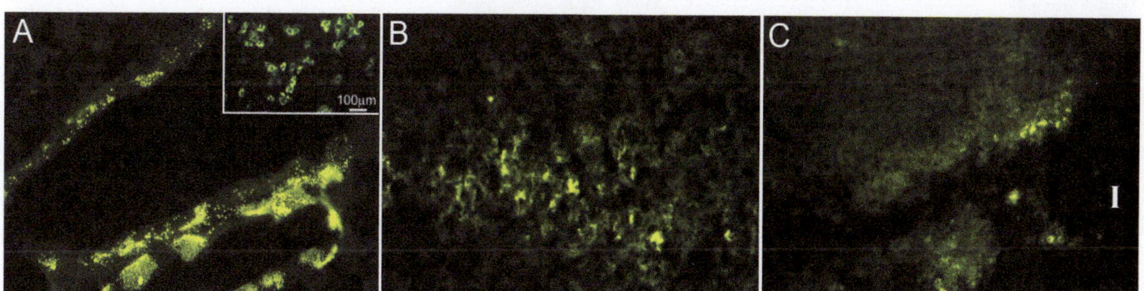

Figure 4. Alkaline phosphatase labeled rBMSCs in vitro and after IV administration into MCAO rats. Panel A.: Lung sections 3 days after transplantation; BMSCs show intense activity in the small airways. Note distinct green punctate Alk Phos histochemical staining. Inset: Human alk phos in BMSCs in culture. B. Brain sections from the same animal showing activity in the peri-infarcted tissue. C. Brain at 8 days after transplantation shows some activity at the edge of the infarct zone ("I" = infarct, with no intact tissue). (N = 3) Scale bars in A,C = 100 um; B = 50 um.

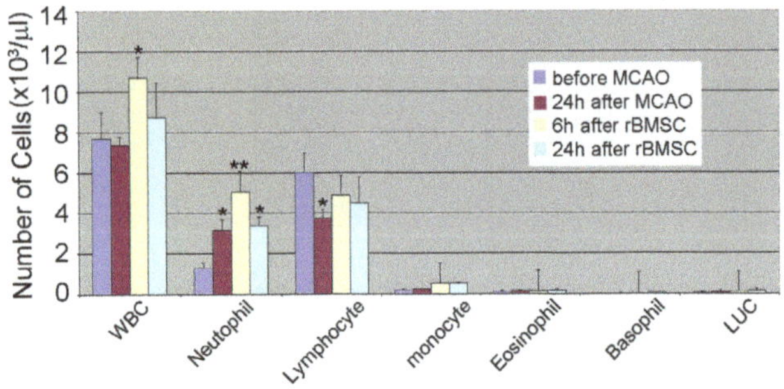

Figure 5. Analysis of blood composition before and after MCAO and rat BMSC treatment. Blood was collected from the tail vein before and 24 hr after MCAO and 6 and 24 hours after IV injection of BMSCs and quantitatively analyzed for counts of white blood cells (WBC), neutrophils, lymphocytes, monocytes, eosinophils, basophils and large unstained cells (LUC) by Bioreliance (Rockville, MD). (N = 3) $^{*}p<0.05$ and $^{**}p<0.01$.

ments, we found that most IV-injected BMSCs remained in the periphery, lodging in the lung, the first capillary bed reached after tail vein injection and in the liver and spleen. More importantly, when cells were labeled with Indium-111 and scanned 3 days later or when Alk phos-labeled cells were imaged by fluorescence 8 days after transplantation, relatively few BMSCs were found in the brain, suggesting that the vast majority of cells either did not successfully cross the infarct-damaged blood-brain-barrier [31] or perished soon after their arrival there. Similarly, in a recent study tracking radiolabeled cells after intra-arterial infusion into the external carotid artery, the majority of cells that made it to the brain disappeared within 24 hours [24]. Although the brain is thought to be immune-privileged, it is likely that BMSCs were eliminated via immune rejection mechanisms [32]. Interestingly, recent studies using GFP-labeled BMSCs injected directly into brain parenchyma contrastingly found labeled cells in the brain many weeks after transplantation [33,34], suggesting that the absence of BMSCs in the brain in our study may be initiated by rejection-related mechanisms in the periphery.

Thus, in this regard, it was particularly interesting to find that the IV injection of BMSCs into the circulation resulted in a rapid (within 6 hours of treatment) alteration in the composition of blood. In particular, we noted a significant increase in host neutrophils, the white blood cells which are the usual first responders to inflammation [35]. This correlates well with changes in blood cytokines (ie. VEGF, MIP, MMP2) noted previously after BMSC treatment by us and others [16,35–37]. Although we did not look for the enhanced presence of neutrophils in the brain, the fact that these cells readily migrate across blood vessels suggests that the increase in circulating neutrophils observed after systemic BMSCs could readily result in greater numbers of neutrophils or their products in the brain [31,35]. Also, in regions where there has been a significant breakdown in the BBB integrity due to the infarct, white blood cells could gain easy access to the brain, causing cytokines and chemokines to be released into the parenchyma [31,32].

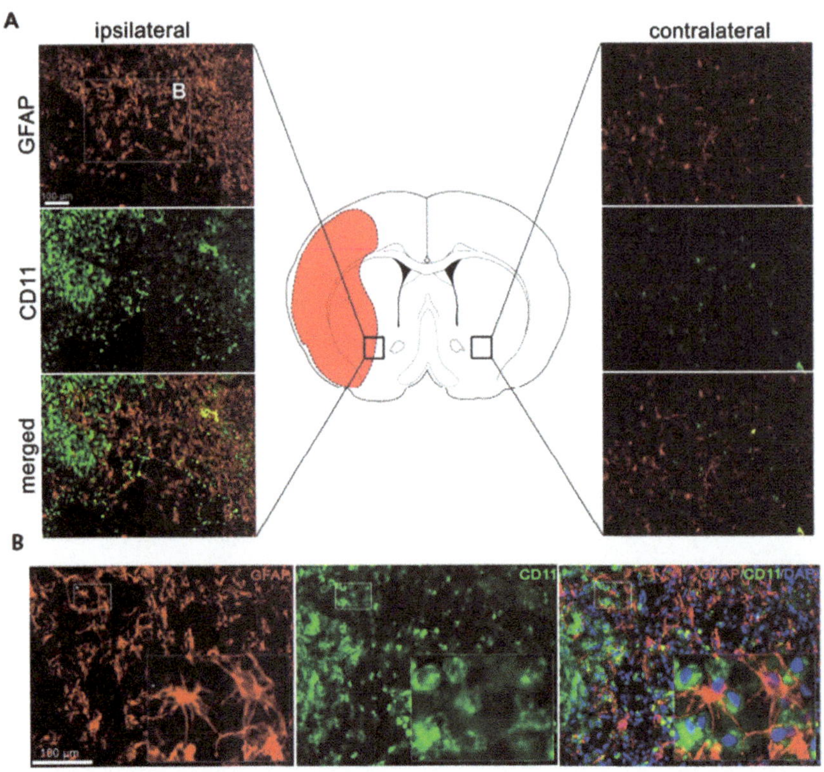

Figure 6. Immunocytochemical analysis of reactive glia and activated microglia in the region of the infarct 4 weeks after BMSC treatment in MCAO rats. Panel A. Representative section through the brain of MCAO rat (center) showing area of infarction (red). Area in rectangle on ipsilateral (left) or contralateral (right) sides of MCAO shown enlarged after staining with GFAP (reactive astrocytes) and CD11 (activated microglia) or merged images in side panels. Panel B. Area labeled "B" in rectangle of Panel A is shown at higher power after staining for GFAP or CD11; Insets: magnified small rectangles of B. Bars in A and B = 100 μm. (N = 4).

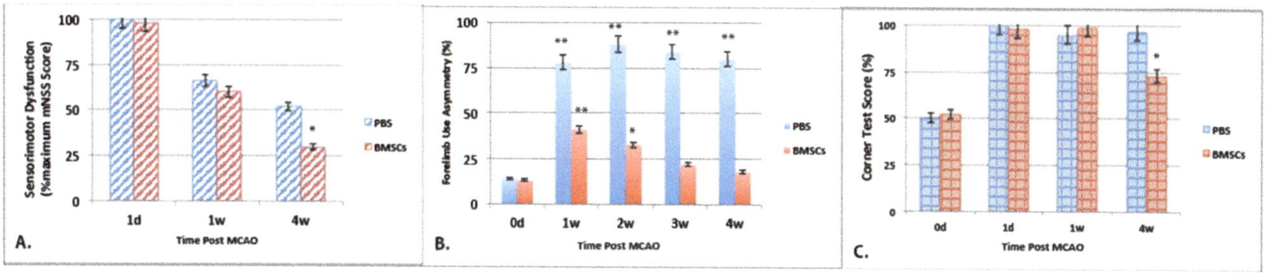

Figure 7. Correlation of BMSC treatment with a significant improvement in behavior over time. All behavioral assessments as described in Fig. legend 2. Panel A. Sensorimotor behavior as assessed on the mNSS scale 1 or 4 weeks after PBS or BMSC treatment (compared to mNSS score at 1 day post-MCAO set to 100%). Panel B. Climbing ability as assessed on the forelimb asymmetry test at 1, 2, 3 and 4 weeks after PBS or BMSC treatment. When compared to pre-MCAO performance on day 0, PBS groups wer significantly different at all treatment times while those treated with BMSCs differed at 1 and 2 weeks post-treatment but normalized towards control at later time points. Importantly, forelimb asymmetry was significantly different ($p<0.001$) between BMSC and PBS treatment groups at all 4 treatment times. Panel C. Corner test evaluated at 1 day, 1 and 4 weeks after PBS or BMSC treatment (compared to pre-MCAO performance on day 0). N = 6, *$p<0.01$ and **$p<0.001$.

How these stem cell-induced changes modify the well-documented inflammatory response which occurs in the host brain after experimental stroke [18,35,38,39] is only now being explored. We found that BMSC treatment in MCAO rats indeed results in a marked up-regulation in the number and in the activation state of local CD11$^+$ microglia and GFAP$^+$ astrocytes, particularly in the penumbra surrounding the infarct. These changes persist even at one month after MCAO, that is, several weeks after the MCAO-induced glial response has returned to normal [16,40] and the infarcted region has been walled off by astrocytes in the ischemic brain [16]. Thus, BMSC-induced cellular alterations in the host glia, combined with the reported increase in vasculogenesis [41–44], suggests that stem cells and/or their products play a pivotal role in partitioning off damage, safe-guarding tissue integrity and/or possibly promoting regeneration in the brain after experimental stroke.

Of significance, these BMSC-induced cellular changes in brain were associated with a substantial improvement in all behavioral indices examined, including those that had partially recovered spontaneously after MCAO. Thus, at 4 weeks post-MCAO, performance on the mNSS, climbing and corner turning tests were significantly improved in the BMSC treatment group compared to PBS controls. Since 4 weeks was the latest time point examined, in the future, it will be important to extend the time course of these studies to determine whether even greater recovery of function is possible at later time points or after multiple treatments with BMSCs. A similar lag in functional recovery in electrophysiology, FMRI and behavioral parameters has indeed been reported previously [45]. Whether these delayed changes in brain structure and function owe to direct effects from the increase in blood neutrophils or from the few BMSCs found in the brain or instead from secondary processes initiated by their cell products (cytokines/chemokines/growth factors) remains to be elucidated.

Supporting Information

Figure S1 Immunocytochemical analysis of reactive GFAP+ glia and activated CD11+ microglia in the striatum on the side of the infarct at 1 (A, B) and 4 (C,D) weeks after MCAO in PBS-infused rats. Note that in control (IV infusion of PBS) rats, there is considerable microglial activation and some astroglial reactivity in the areas adjacent to the infarct as a result of MCAO. However, by 4 weeks, much of this activation/reactivity due to MCAO had receded. Bar = 100 µm. (N = 4).

Author Contributions

Conceived and designed the experiments: LI RHR. Performed the experiments: GG DYL RN SY. Analyzed the data: LI GG RN DYL SY RHR. Wrote the paper: LI.

References

1. Vendrame M, Gemma C, De Mesquita D, Collier L, Bickford PC, et al. (2005) Anti-inflammatory effects of human cord blood cells in a rat model of stroke. Stem cells and development 14: 595–604. Available: http://www.ncbi.nlm.nih.gov/pubmed/16305344. Accessed 9 December 2012.
2. Lalancette-Hebert M, Phaneuf D, Soucy G, Weng YC, Kriz J (2009) Live imaging of Toll-like receptor 2 response in cerebral ischaemia reveals a role of olfactory bulb microglia as modulators of inflammation. Brain Available: 10.1093/brain/awn345.
3. Eissa A, Krass I, Bajorek B V (2012) Optimizing the management of acute ischaemic stroke: a review of the utilization of intravenous recombinant tissue plasminogen activator (tPA). Journal of clinical pharmacy and therapeutics 37: 620–629. Available: http://www.ncbi.nlm.nih.gov/pubmed/22708668. Accessed 9 December 2012.
4. Ramos-Cabrer P, Justicia C, Wiedermann D, Hoehn M (2010) Stem cell mediation of functional recovery after stroke in the rat. PloS one 5: e12779. Available: http://www.pubmedcentral.nih.gov/articlerender.fcgi?artid = 2943902&tool = pmcentrez&rendertype = abstract. Accessed 9 December 2012.
5. De Feo D, Merlini A, Laterza C, Martino G (2012) Neural stem cell transplantation in central nervous system disorders: from cell replacement to neuroprotection. Current opinion in neurology 25: 322–333. Available: http://www.ncbi.nlm.nih.gov/pubmed/22547103. Accessed 9 December 2012.
6. Li L, Jiang Q, Ding G, Zhang L, Zhang ZG, et al. (2010) Effects of administration route on migration and distribution of neural progenitor cells transplanted into rats with focal cerebral ischemia, an MRI study. Journal of cerebral blood flow and metabolism: official journal of the International Society of Cerebral Blood Flow and Metabolism 30: 653–662. Available: http://www.pubmedcentral.nih.gov/articlerender.fcgi?artid = 2844252&tool = pmcentrez&rendertype = abstract. Accessed 9 December 2012.
7. Zhang L, Li Y, Romanko M, Kramer BC, Gosiewska A, et al. (2012) Different routes of administration of human umbilical tissue-derived cells improve functional recovery in the rat after focal cerebral ischemia. Brain research 1489: 104–112. Available: http://www.ncbi.nlm.nih.gov/pubmed/23063717. Accessed 11 December 2012.
8. Zhang ZG, Chopp M (2009) Neurorestorative therapies for stroke: underlying mechanisms and translation to the clinic. Lancet neurology 8: 491–500. Available: http://www.pubmedcentral.nih.gov/articlerender.fcgi?artid = 2727708&tool = pmcentrez&rendertype = abstract. Accessed 18 November 2012.
9. Sanberg PR, Eve DJ, Metcalf C, Borlongan CV (2012) Advantages and challenges of alternative sources of adult-derived stem cells for brain repair in stroke. Progress in brain research 201: 99–117. Available: http://www.ncbi.nlm.nih.gov/pubmed/23186712. Accessed 9 December 2012.
10. Aulesa C, Mainar I, Prieto M, Cobos N, Galimany R (2003) Use of the Advia 120 hematology analyzer in the differential cytologic analysis of biological fluids (cerebrospinal, peritoneal, pleural, pericardial, synovial, and others). Laboratory hematology: official publication of the International Society for Laboratory

Hematology 9: 214–224. Available: http://www.ncbi.nlm.nih.gov/pubmed/14649464. Accessed 24 January 2013.
11. Sughrue ME, Mocco J, Komotar RJ, Mehra A, D'Ambrosio AL, et al. (2006) An improved test of neurological dysfunction following transient focal cerebral ischemia in rats. Journal of neuroscience methods 151: 83–89. Available: http://www.ncbi.nlm.nih.gov/pubmed/16476486. Accessed 10 December 2012.
12. Hua Y, Schallert T, Keep RF, Wu J, Hoff JT, et al. (2002) Behavioral tests after intracerebral hemorrhage in the rat. Stroke; a journal of cerebral circulation 33: 2478–2484. Available: http://www.ncbi.nlm.nih.gov/pubmed/12364741. Accessed 9 December 2012.
13. Schallert T, Fleming SM, Leasure JL, Tillerson JL, Bland ST (2000) CNS plasticity and assessment of forelimb sensorimotor outcome in unilateral rat models of stroke, cortical ablation, parkinsonism and spinal cord injury. Neuropharmacology 39: 777–787. Available: http://www.ncbi.nlm.nih.gov/pubmed/10699444. Accessed 26 April 2011.
14. Schallert T (2006) Behavioral tests for preclinical intervention assessment. NeuroRx: the journal of the American Society for Experimental NeuroTherapeutics 3: 497–504. Available: http://www.ncbi.nlm.nih.gov/pubmed/17012064. Accessed 9 December 2012.
15. Zhang L, Schallert T, Zhang ZG, Jiang Q, Arniego P, et al. (2002) A test for detecting long-term sensorimotor dysfunction in the mouse after focal cerebral ischemia. Journal of neuroscience methods 117: 207–214. Available: http://www.ncbi.nlm.nih.gov/pubmed/12100987. Accessed 1 November 2012.
16. Yang M, Wei X, Li J, Heine LA, Rosenwasser R, et al. (2010) Changes in host blood factors and brain glia accompanying the functional recovery after systemic administration of bone marrow stem cells in ischemic stroke rats. Cell transplantation 19: 1073–1084. Available: http://www.ncbi.nlm.nih.gov/pubmed/20412636. Accessed 10 December 2012.
17. Lalancette-Hebert M, Gowing G, simard A, Weng YC, Kriz J (2007) Selective ablation of proliferating microglial cells exacerbates ischemic injury in the brain. J Neurosci 27: 2596–2605.
18. Kriz J (2006) Inflammation in ischemic brain injury: timing is important. Crit Rev Neurobiol 18: 145–157.
19. Caplan AI (2007) Adult mesenchymal stem cells for tissue engineering versus regenerative medicine. Journal of cellular physiology 213: 341–347. Available: http://www.ncbi.nlm.nih.gov/pubmed/17620285. Accessed 3 November 2012.
20. Hess DC, Borlongan C V (2008) Cell-based therapy in ischemic stroke. Expert review of neurotherapeutics 8: 1193–1201. Available: http://www.pubmedcentral.nih.gov/articlerender.fcgi?artid=2753686&tool=pmcentrez&rendertype=abstract. Accessed 9 December 2012.
21. Li WY, Choi YJ, Lee PH, Huh K, Kang YM, et al. (2008) Mesenchymal stem cells for ischemic stroke: changes in effects after ex vivo culturing. Cell transplantation 17: 1045–1059. Available: http://www.ncbi.nlm.nih.gov/pubmed/19177841. Accessed 9 December 2012.
22. Li Y, Chopp M (2009) Marrow stromal cell transplantation in stroke and traumatic brain injury. Neuroscience letters 456: 120–123. Available: http://www.pubmedcentral.nih.gov/articlerender.fcgi?artid=3359793&tool=pmcentrez&rendertype=abstract. Accessed 9 December 2012.
23. Gutiérrez-Fernández M, Rodríguez-Frutos B, Alvarez-Grech J, Vallejo-Cremades MT, Expósito-Alcaide M, et al. (2011) Functional recovery after hematic administration of allogenic mesenchymal stem cells in acute ischemic stroke in rats. Neuroscience 175: 394–405. Available: http://www.ncbi.nlm.nih.gov/pubmed/21144885. Accessed 9 December 2012.
24. Mitkari B, Kerkelä E, Nystedt J, Korhonen M, Mikkonen V, et al. (2012) Intra-arterial infusion of human bone marrow-derived mesenchymal stem cells results in transient localization in the brain after cerebral ischemia in rats. Experimental neurology 239C: 158–162. Available: http://www.ncbi.nlm.nih.gov/pubmed/23059455. Accessed 9 December 2012.
25. Detante O, Moisan A, Dimastromatteo J, Richard M-J, Riou L, et al. (2009) Intravenous administration of 99mTc-HMPAO-labeled human mesenchymal stem cells after stroke: in vivo imaging and biodistribution. Cell transplantation 18: 1369–1379. Available: http://www.ncbi.nlm.nih.gov/pubmed/19849895. Accessed 22 January 2013.
26. Steiner B, Roch M, Holtkamp N, Kurtz A (2012) Systemically administered human bone marrow-derived mesenchymal stem home into peripheral organs but do not induce neuroprotective effects in the MCAo-mouse model for cerebral ischemia. Neuroscience letters 513: 25–30. Available: http://www.ncbi.nlm.nih.gov/pubmed/22342911. Accessed 22 January 2013.
27. Astrup J, Siesjö BK, Symon L (1981) Thresholds in cerebral ischemia - the ischemic penumbra. Stroke; a journal of cerebral circulation 12: 723–725. Available: http://www.ncbi.nlm.nih.gov/pubmed/6272455. Accessed 9 December 2012.
28. Heiss WD (1992) Experimental evidence of ischemic thresholds and functional recovery. Stroke; a journal of cerebral circulation 23: 1668–1672. Available: http://www.ncbi.nlm.nih.gov/pubmed/1440719. Accessed 9 December 2012.
29. Symon L (2007) The Ischemic Penumbra. 1st Editio. Donnan A. G, Baron J-C, Davis SM, Sharp FRS, editors New York: Informa Healthcare.
30. Li WY, Choi YJ, Lee PH, Huh K, Kang YM, et al. (2008) Mesenchymal stem cells for ischemic stroke: changes in effects after ex vivo culturing. Cell transplantation 17: 1045–1059. Available: http://www.ncbi.nlm.nih.gov/pubmed/19177841. Accessed 9 December 2012.
31. Takeshita Y, Ransohoff RM (2012) Inflammatory cell trafficking across the blood-brain barrier: chemokine regulation and in vitro models. Immunological reviews 248: 228–239. Available: http://www.ncbi.nlm.nih.gov/pubmed/22725965. Accessed 9 December 2012.
32. Miller DW (1999) Immunobiology of the blood-brain barrier. Journal of neurovirology 5: 570–578. Available: http://www.ncbi.nlm.nih.gov/pubmed/10602398. Accessed 9 December 2012.
33. Shichinohe H, Kuroda S, Lee J-B, Nishimura G, Yano S, et al. (2004) In vivo tracking of bone marrow stromal cells transplanted into mice cerebral infarct by fluorescence optical imaging. Brain research Brain research protocols 13: 166–175. Available: http://www.ncbi.nlm.nih.gov/pubmed/15296854. Accessed 9 December 2012.
34. Miyamoto M, Kuroda S, Zhao S, Magota K, Shichinohe H, et al. (2012) Bone Marrow Stromal Cell Transplantation Enhances Recovery of Local Glucose Metabolism After Cerebral Infarction in Rats: A Serial 18F-FDG PET Study. Journal of nuclear medicine: official publication, Society of Nuclear Medicine. Available: http://www.ncbi.nlm.nih.gov/pubmed/23204494. Accessed 9 December 2012.
35. Jin R, Yang G, Li G (2010) Inflammatory mechanisms in ischemic stroke: role of inflammatory cells. Journal of leukocyte biology 87: 779–789. Available: http://www.pubmedcentral.nih.gov/articlerender.fcgi?artid=2858674&tool=pmcentrez&rendertype=abstract. Accessed 13 November 2012.
36. Deng YB, Ye WB, Hu ZZ, Yan Y, Wang Y, et al. (2010) Intravenously administered BMSCs reduce neuronal apoptosis and promote neuronal proliferation through the release of VEGF after stroke in rats. Neurological research 32: 148–156. Available: http://www.ncbi.nlm.nih.gov/pubmed/19473555. Accessed 9 December 2012.
37. Zhao Y, Li Z, Wang R, Wei J, Li G, et al. (2010) Angiopoietin 1 counteracts vascular endothelial growth factor-induced blood-brain barrier permeability and alleviates ischemic injury in the early stages of transient focal cerebral ischemia in rats. Neurological research 32: 748–755. Available: http://www.ncbi.nlm.nih.gov/pubmed/19660197. Accessed 9 December 2012.
38. Lalancette-Hébert M, Gowing G, Simard A, Weng YC, Kriz J (2007) Selective ablation of proliferating microglial cells exacerbates ischemic injury in the brain. The Journal of neuroscience: the official journal of the Society for Neuroscience 27: 2596–2605. Available: http://www.ncbi.nlm.nih.gov/pubmed/17344397. Accessed 9 December 2012.
39. Kriz J, Lalancette-Hébert M (2009) Inflammation, plasticity and real-time imaging after cerebral ischemia. Acta neuropathologica 117: 497–509. Available: http://www.ncbi.nlm.nih.gov/pubmed/19225790. Accessed 9 December 2012.
40. Pekney M, Wilhemsson U, Bogestal YR, Pekna M (2007) The Role of Astrocytes and Complement System in Neural Pasticity. Int Rev Neurobiol 82: 95–111.
41. Chen J, Zhang ZG, Li Y, Wang L, Xu YX, et al. (2003) Intravenous administration of human bone marrow stromal cells induces angiogenesis in the ischemic boundary zone after stroke in rats. Circulation research 92: 692–699. Available: http://www.ncbi.nlm.nih.gov/pubmed/12609969. Accessed 9 December 2012.
42. Taguchi A, Soma T, Tanaka H, Kanda T, Nishimura H, et al. (2004) Administration of CD34+ cells after stroke enhances neurogenesis via angiogenesis in a mouse model. J Clin Invest 114: 330–338.
43. Jiang Q, Zhang ZG, Ding GL, Zhang L, Ewing JR, et al. (2005) Investigation of neural progenitor cell induced angiogenesis after embolic stroke in rat using MRI. NeuroImage 28: 698–707. Available: http://www.ncbi.nlm.nih.gov/pubmed/16112879. Accessed 9 December 2012.
44. Shyu WC, Lin SZ, Chiang MF, Su CY, Li H (2006) Intracerebral peripheral blood stem cell (CD34+) implantation induces neuroplasticity by enhancing beta1 integrin-mediated angiogenesis in chronic stroke rats. J Neurosci 26: 3444–3453.
45. Ramos-Cabrer P, Justicia C, Wiedermann D, Hoehn M (2010) Stem cell mediation of functional recovery after stroke in the rat. PloS one 5: e12779. Available: http://www.pubmedcentral.nih.gov/articlerender.fcgi?artid=2943902&tool=pmcentrez&rendertype=abstract. Accessed 9 December 2012.

4

Generation and Characterization of Functional Cardiomyocytes Derived from Human T Cell-Derived Induced Pluripotent Stem Cells

Tomohisa Seki[1,2], Shinsuke Yuasa[1], Dai Kusumoto[1], Akira Kunitomi[1], Yuki Saito[1], Shugo Tohyama[1,2], Kojiro Yae[1], Yoshikazu Kishino[1], Marina Okada[1], Hisayuki Hashimoto[1], Makoto Takei[1], Toru Egashira[1], Masaki Kodaira[1], Yusuke Kuroda[1], Atsushi Tanaka[1], Shinichiro Okata[1], Tomoyuki Suzuki[1], Mitsushige Murata[1], Jun Fujita[1], Keiichi Fukuda[1]*

1 Department of Cardiology, Keio University School of Medicine, Shinjuku-ku, Tokyo, Japan, **2** Japan Society for the Promotion of Science, Chiyoda-ku, Tokyo, Japan

Abstract

Induced pluripotent stem cells (iPSCs) have been proposed as novel cell sources for genetic disease models and revolutionary clinical therapies. Accordingly, human iPSC-derived cardiomyocytes are potential cell sources for cardiomyocyte transplantation therapy. We previously developed a novel generation method for human peripheral T cell-derived iPSCs (TiPSCs) that uses a minimally invasive approach to obtain patient cells. However, it remained unknown whether TiPSCs with genomic rearrangements in the T cell receptor (TCR) gene could differentiate into functional cardiomyocyte in vitro. To address this issue, we investigated the morphology, gene expression pattern, and electrophysiological properties of TiPSC-derived cardiomyocytes differentiated by floating culture. RT-PCR analysis and immunohistochemistry showed that the TiPSC-derived cardiomyocytes properly express cardiomyocyte markers and ion channels, and show the typical cardiomyocyte morphology. Multiple electrode arrays with application of ion channel inhibitors also revealed normal electrophysiological responses in the TiPSC-derived cardiomyocytes in terms of beating rate and the field potential waveform. In this report, we showed that TiPSCs successfully differentiated into cardiomyocytes with morphology, gene expression patterns, and electrophysiological features typical of native cardiomyocytes. TiPSCs-derived cardiomyocytes obtained from patients by a minimally invasive technique could therefore become disease models for understanding the mechanisms of cardiac disease and cell sources for revolutionary cardiomyocyte therapies.

Editor: Rajasingh Johnson, University of Kansas Medical Center, United States of America

Funding: This work was partly supported by the Support Program to break the bottlenecks at R&D Systems for accelerating the practical use of Health Research Outcome, the Highway Program for the Realization of Regenerative Medicine, and a Grant-in-Aid for JSPS Fellows. The funders had no role in study design, data collection and analysis, decision to publish, or preparation of the manuscript.

Competing Interests: The authors have declared that no competing interests exist.

* E-mail: kfukuda@a2.keio.jp

Introduction

Severe heart failure is a progressive and treatment-resistant disease for which the only complete cure is heart transplantation [1]. In addition, many of the etiological problems leading to heart failure remain unsolved. Direct reprogramming of somatic cells to produce induced pluripotent stem cells (iPSCs) through the forced expression of several embryonic stem (ESC)-specific transcription factors is a prominent recent advance in stem cell biology [2], and human iPSCs hold great promise as new tools for investigating mechanism of disease and as cell sources for transplantation therapy [3]. Human iPSC-derived cardiomyocytes (iPSC-CMs) are thus potential future cell sources for novel cardiomyocyte transplantation therapies for severe heart failure [4,5].

Successful differentiation of human iPSCs into cardiomyocytes was first reported in 2009 [6]. Subsequently, human iPSC-CMs have been developed as in vitro models for cardiac electrophysiological studies and drug screening [7]. Disease modeling of cardiac disease using patient-derived iPSC-CMs was also reported in the field of inherited arrhythmia diseases [8–12] and inherited cardiomyopathies [13–15]. Furthermore, iPSC-CMs have been engrafted successfully into the hearts of experimental animals [16] and used to improve cardiac function after ischemic cardiomyopathy in a porcine model [17]. Thus, research using iPSC-CM will be important in clarifying disease mechanisms and establishing novel therapies [18–20].

A wide range of donor cell types, gene-introducing vehicles, and combinations of reprogramming factors have been used for generating iPSCs [21]; however, minimally invasive methods have the significant practical advantage for generating iPSCs from patients [22]. We previously reported the generation of T cell-derived iPSCs (TiPSCs), which involves a non-invasive method and requires only small volumes of peripheral blood, which means an increased number of eligible patients [23]. Further to this, the present study sought to clarify whether TiPSCs with genomic T cell receptor (TCR) gene rearrangements could differentiate into functional cardiomyocyte in vitro, by examining the morphology, gene expression pattern, and electrophysiological properties of TiPSC-CMs.

Materials and Methods

Cell culturing

The human TiPSC lines, T05 and T07 were generated from human peripheral T cells by introducing *OCT3/4*, *SOX2*, *KLF4*, and *c-MYC* with Sendai virus vectors [23]. The human iPSC line, F16 was generated from human fibroblasts by introducing *OCT3/4*, *SOX2*, *KLF4*, and *c-MYC* with retrovirus vectors [24]. The human ESC line (KhES-2) [25] was obtained from the Department of Development and Differentiation, Institute for Frontier Medical Sciences, Kyoto University and used in conformity with the Guidelines for Derivation and Utilization of Human Embryonic Stem Cells of the Ministry of Education, Culture, Sports, Science, and Technology, Japan. These cell lines were cultured in human iPSC medium consisting of DMEM/F12 medium (Invitrogen) supplemented with 20% Knock-out Serum Replacement (KSR; Invitrogen), 1 mM L-glutamine, 1 mM non-essential amino acids, 0.1 mM β-mercaptoethanol, 50 U penicillin and 50 mg/ml streptomycin (Invitrogen), and 4 ng/ml basic fibroblast growth factor (bFGF; WAKO). The human iPSC medium was changed every other day until the colonies were picked. The iPSCs and ESCs were maintained on irradiated mouse embryonic fibroblast (MEF) feeder cells [26] from wild-type ICR mice in human iPSC medium which was changed every 2–3 days, and the cells were passaged using 1 mg/ml collagenase IV (Invitrogen) every 5–6 days.

In vitro differentiation of iPSCs and ESCs

The iPSCs and ESCs were harvested using 1 mg/ml collagenase IV, and transferred to ultra-low attachment plates (Corning) in differentiation medium. The differentiation medium consisting of Minimum Essential Medium Alpha Medium (Gibco) supplemented with 2 mM L-glutamine (Invitrogen), 0.1 mM non-essential amino acids (Sigma), 0.1 mM 2-mercaptoethanol, 50 U/ml penicillin and 50 mg/ml streptomycin (Invitrogen), and 20% fetal bovine serum (Gibco). The medium was replaced every second or third day. To promote cardiac differentiation, recombinant human Wnt3a was added to the differentiation medium for the first 4 days of culture at 100 ng/ml [27]. The time window of differentiation for analyzing the beating embryoid bodies (EBs) and cardiomyocytes was 30–60 days from starting the differentiating conditions.

Reverse transcription-polymerase chain reaction (RT-PCR)

Total RNA was isolated from cells using TRIZOL reagent (Invitrogen) and RNasefree DNase I (Qiagen), according to the manufacturer's instructions. The concentration and purity of the RNA were determined using an ND-1000 spectrophotometer (Nanodrop). The cDNA was synthesized using the Superscript First-Strand Synthesis System (Invitrogen). PCR was performed with rTaq (TAKARA). The primers used for PCR are listed in Table S1.

Quantitative RT-PCR analysis

Total RNA was isolated and used for reverse transcription with random primers as described by the manufacturer. Real-time quantitative RT-PCR was performed using MX3000P (Stratagene), with SYBR Premix ExTaq (TAKARA). The amount of mRNA was normalized to GAPDH mRNA. Primer sequences and cycling conditions are listed in Table S1.

Immunofluorescence

The differentiated cardiomyocytes plated on fibronectin-coated dishes were washed once with PBS, and fixed with 4% Paraformaldehyde Fixative (MUTO Pure Chemicals) at 4°C for 30 min. After fixation, cells were treated with 0.4% Triton X-100 in PBS for 15 min at room temperature. After blocked with ImmunoBlock (DS Pharma Biomedical) for 5 min three times, cells were incubated at 4°C overnight with the primary antibodies, followed by washing with the blocking medium and incubation at room temperature for 60 min with the corresponding secondary antibodies. The immunostaining was performed using the following primary antibodies and reagents: anti-α-Actinin (A7811, Sigma-Aldrich), anti-ANP (sc-20158, Santa Cruz), anti- Troponin-I (ab52862, Abcam), anti-GATA4 (sc-1237, Santa Cruz), anti-Nkx2.5 (sc-8697, Santa Cruz), anti-Titin (HPA007042, Sigma-Aldrich), anti-SERCA2 (MAB2636, Millipore), anti-Connexin43(c6219, Sigma-Aldrich), anti-Na/Ca exchanger(MA3-926) and 4′,6-Diamidino-2-Phenylindole (DAPI, Molecular Probes). The secondary antibodies used were anti-rabbit IgG and anti-mouse IgG or IgM conjugated with Alexa Fluor 488 or Alexa Fluor 568 (Molecular Probes). Signal was detected using a conventional fluorescence laser microscope equipped with a colour charge-coupled device camera (BZ-9000, KEYENCE).

Electron microscopy

Contracting EBs were fixed with 2.5% glutaraldehyde in 60 mM HEPES, pH 7.4 for 120 min at room temperature, and washed three times with 0.2 M phosphate buffer. Cells were post-fixed in 1% OsO_4 in 60 mM HEPES, pH 7.4 for 120 min at room temperature, and stained en bloc in 2% uranyl acetate for 30 min at room temperature. Tissues were dehydrated with gradually increasing concentrations of ethanol for resin embedding, before thin sectioning and staining.

Field potential recordings using the on-chip multi-electrode array system

Multi-electrode array (MEA) chips from Multi Channel Systems were coated with fibronectin (Sigma-Aldrich). Beating EBs were plated and incubated at 37°C, and MEA measurements were performed at 37°C, as previously described [7]. The signals were initially processed, and the obtained data were subsequently analyzed with MC Rack (Multi Channel Systems). Data for analysis were extracted from 2–5 min of the obtained data. The recorded extracellular electrograms were used to determine local field potential duration (FPD), defined as the time interval between the initial deflection of the field potential(FP) and the maximum local T wave. FPD measurements were normalized (corrected FPD: cFPD) to the activation rate using Bazett's correction formulae: $cFPD = FPD/(RR\ interval)^{1/2}$, where RR represents the time interval (in seconds) between two consecutive beats. Isoproterenol hydrochloride (Sigma-Aldrich), E4031 (Sigma-Aldrich), and verapamil hydrochloride (Sigma-Aldrich) were prepared as 1 or 10 mM stock solutions. The FPs were recorded for 5 min, and then drug was added to the medium. After 2–3 min of incubation, the FPs were again measured for 5 min.

Ca^{2+} Imaging

For calcium imaging, beating EBs were manually dissected and digested with collagenase 4 (Gibco) at 0.1 mg/ml in ADS buffer (116.4 mM NaCl, 5.4 mM KCl, 5.6 mM dextrose, 10.9. mM NaH_2PO_4, 405.7 μM $MgSO_4$, and 20 mM Hepes, pH 7.3) and plated onto gelatin-coated dishes in differentiation medium. After 7 days in culture the beating cardiomyocyte clusters were loaded with 6 μg/ml Fluo-4 AM (Invitrogen) for 10 min at 37°C. Ca^{2+} transients were recorded in iPSC-CMs using a LSM-710 laser

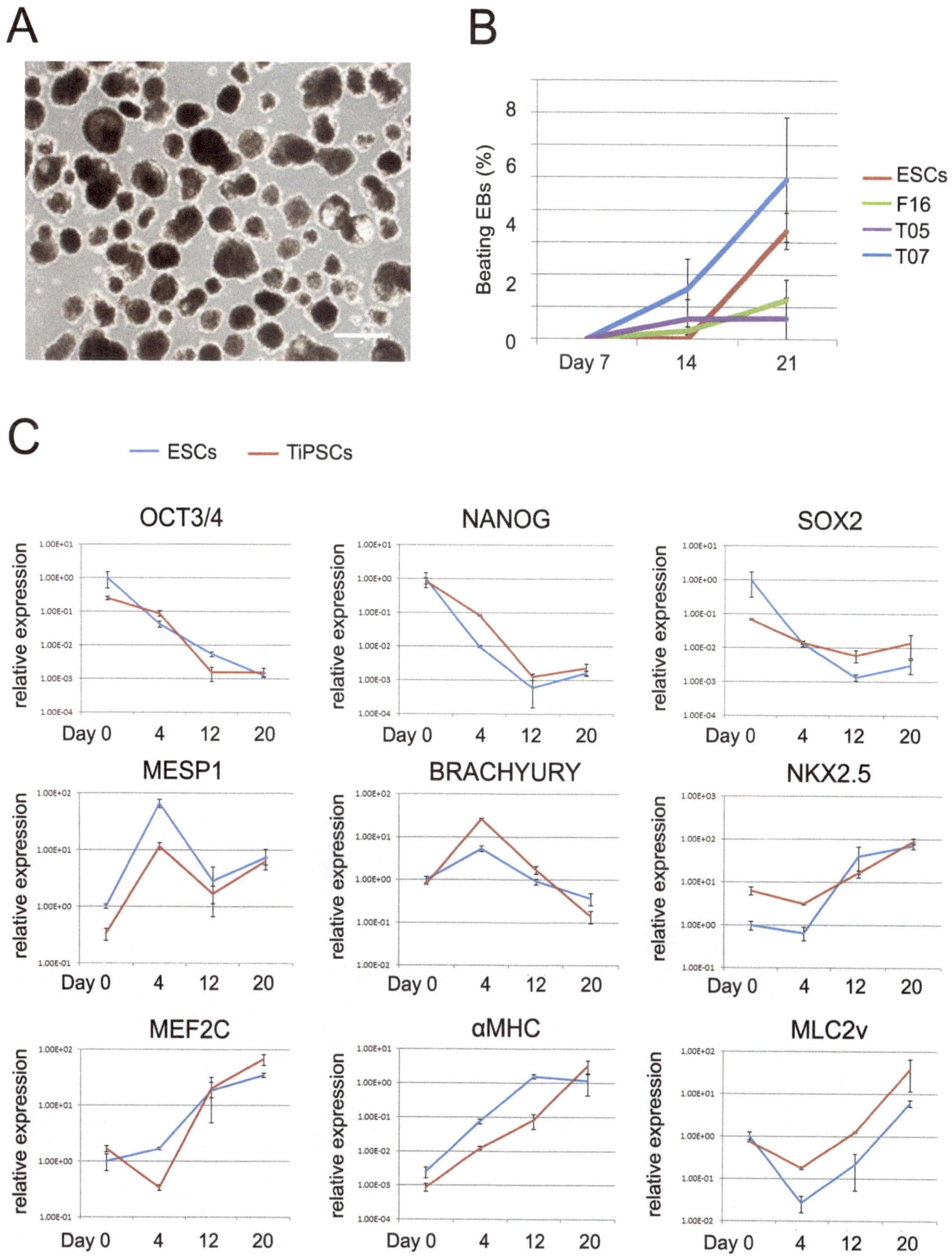

Figure 1. Characterization of in vitro cardiac-differentiation ability of TiPSCs. (A) Morphologies of EBs derived from T07. Scale bar shows 500 μm. (B) Rate of beating EBs derived from ESCs, fibroblast-derived iPSCs line F16, and TiPSC lines T05 and T07 on day 7, day 14, and day 21 of suspension culture. (C) Gene-expression patterns during cardiomyocyte differentiation of TiPSC line T07 and ESCs. Results of QT-PCR studies showed

that cardiomyocyte differentiation was characterized by a continuous decrease in the expression of pluripotent markers (*OCT4*, *NANOG*, and *SOX2*) coupled with an initial increase in mesoderm and cardiomesoderm markers (*MESP1* and *Brachyury*). This was followed by the expression of cardiac-related transcription factors (*NKX2.5* and *MEF2C*), and finally by cardiac-specific structural genes (*αMHC* and *MLC2V*). The graph shows an average of three independent examinations. Error bars show means ± s.d.

scanning confocal microscope (Carl Zeiss), and images were acquired in the line-scan mode.

Results

Characterization of in vitro cardiac differentiation ability of TiPSCs

First, we examined the in vitro cardiac differentiation ability of TiPSCs. We used suspension culture to induce the TiPSCs to form beating EBs, as shown previously for ESCs and iPSCs [28]. The TiPSCs made beating EBs in the same way as ESCs and fibroblast-derived iPSCs (Figure 1A, 1B, Movie S1), confirming that TiPSCs could differentiate into beating cardiomyocytes in suspension culture in vitro.

We also examined the temporal gene expression pattern of derivatives of TiPSCs and ESCs during the in vitro differentiation process. While the expression of undifferentiated pluripotent markers *OCT3/4*, *NANOG*, and *SOX2* decreased similarly, the cardiac-associated transcription factors, *NKX2.5* and *MEF2C*, and the cardiac-specific structural genes, *αMHC* and *MLC2V*, showed high expression in the late stages of floating culture (Figure 1C). On the other hand, primitive streak, mesoderm, and cardiomesoderm markers, *MESP1* and *BRACHYURY*, showed transient expression at the early stage of differentiation (Figure 1C). These results indicated that the expression patterns of undifferentiated and cardiac markers in the derivatives of TiPSCs and ESCs were similar to that observed during embryonic development.

Molecular and cellular characterization of TiPSC-CMs

We next examined expression of the cardiac-specific markers of TiPSC-CMs by further analyzing the derived beating EBs. RT-PCR revealed that our TiPSC-CMs properly expressed cardiomyocyte markers and showed decreased pluripotent marker expression (Figure 2A). Immunostaining confirmed the proper expression of cardiomyocyte markers, α-Actinin, Troponin-I, Titin, ANP, GATA4, and Nkx2.5 in the TiPSC-CMs (Figure 2B).

We then checked the expression of proteins associated with the functionality of cardiomyocytes other than sarcomeric components in TiPSC-CMs. Connexin43 is the most abundant connexin in the ventricular myocardium; it constitutes gap junctions, which are localized to the cardiomyocyte plasma membrane and electrically couple cardiomyocytes to orchestrate action potential propagation [29]. Immunostaining revealed that Connexin43 to the cardiomyocyte plasma membrane and to the intercellular junctional regions in TiPSC-CMs, suggesting the proper development of gap junctions and intact electric coupling (Figure 3A). The activity of the cardiac sarcoplasmic reticulum Ca^{2+} transport ATPase (SERCA2a) and Na^{+}-Ca^{2+} exchanger control cardiomyocyte relaxation and contraction by regulating intracellular Ca^{2+} [30]. In our TiPSC-CMs, SERCA2a was localized to the cytoplasm and Na^{+}-Ca^{2+} exchanger was localized to the plasma membrane (Figure 3A). Electron microscopy of the TiPSC-CMs revealed subcellular morphology identical to that observed in native cardiomyocytes such as the sarcomeric organization and gap junctions (Figure 3B). Ca^{2+} oscillations are part of the fully

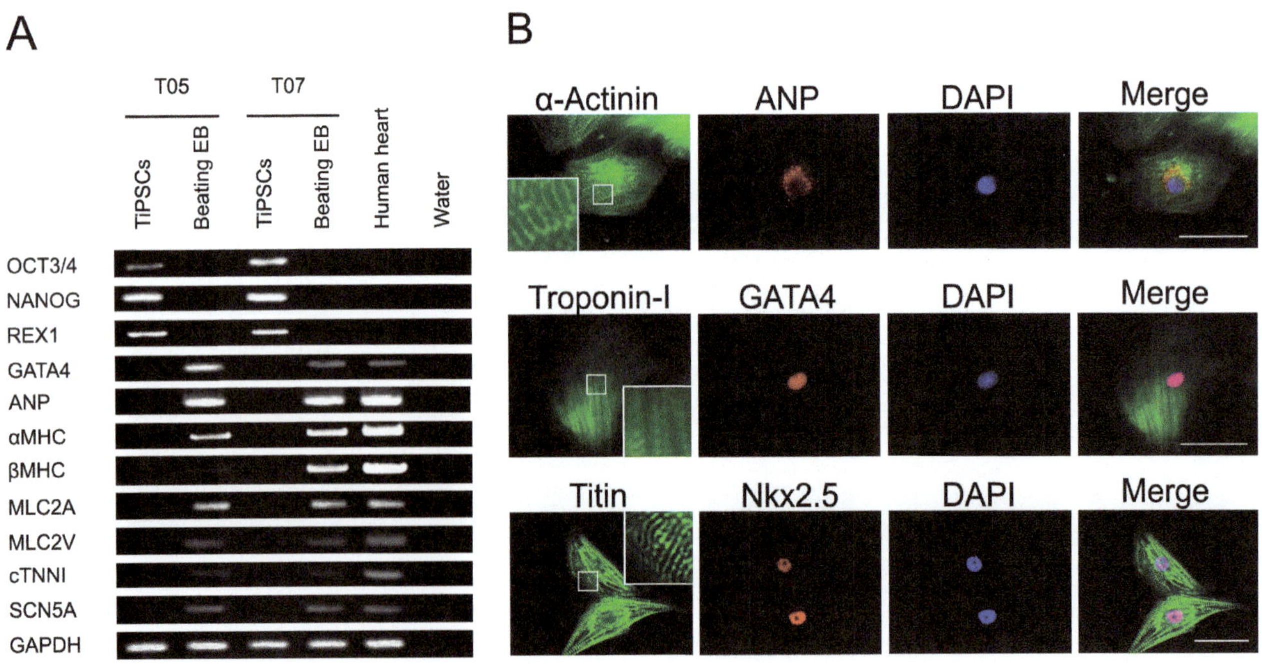

Figure 2. Cardiac-specific markers in TiPSC-CMs. (A) RT-PCR analyses of beating EBs derived from T05 and T07 for the pluripotent marker genes *OCT3/4*, *NANOG*, and *REX1*, the cardiac-related transcription factors *GATA4* and *MEF2C*, and the cardiac-specific marker genes *ANP*, *αMHC*, *βMHC*, *MLC2A*, *MLC2V*, *cTNNI*, and *SCN5A*. (B) Immunofluorescence staining for cardiac-specific markers in cardiomyocytes derived from T07. Scale bar shows 50 μm.

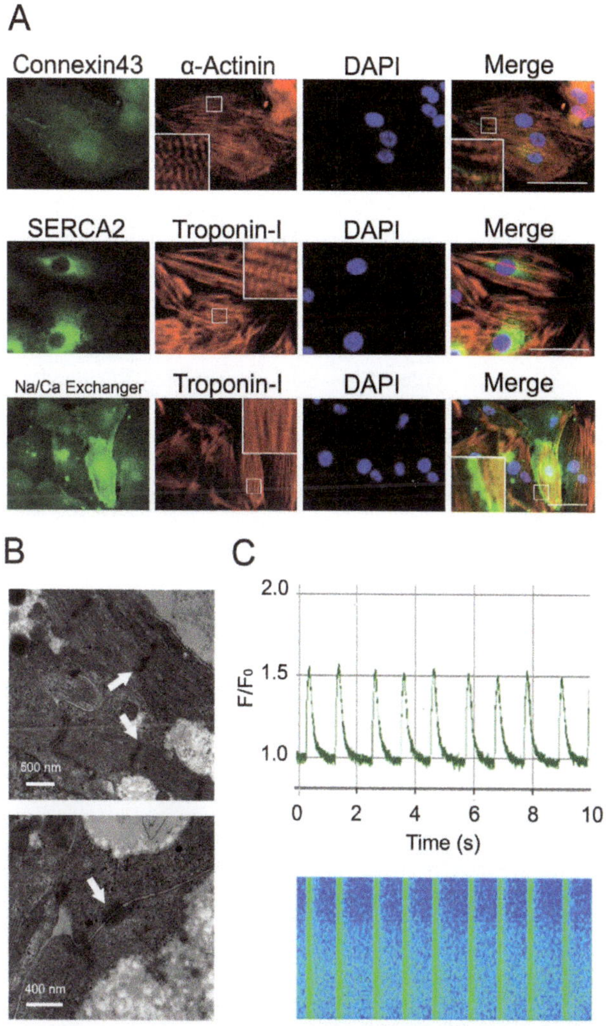

Figure 3. Molecular and cellular characterization of TiPSC-CMs. (A) Immunostaining of proteins involved in cardiac functionality in cardiomyocytes derived from T07. Scale bar shows 50 μm. (B) Electron microscopy analysis of cardiomyocytes derived from T07. Upper micrograph shows sarcomere structures of cardiomyocytes derived from T07. White arrows indicate the Z lines of sarcomere structures. Scale bar shows 500 nm. Lower micrograph shows gap junctions in cardiomyocytes derived from T07 (white arrow). Scale bar shows 400 nm. (C) Line-scan image of Ca^{2+} transient(lower figure) and its average fluorescence intensity in cardiomyocytes derived from T07 (upper graph) were shown. F/F_0 means fluorescence (F) normalized to baseline fluorescence (F_0).

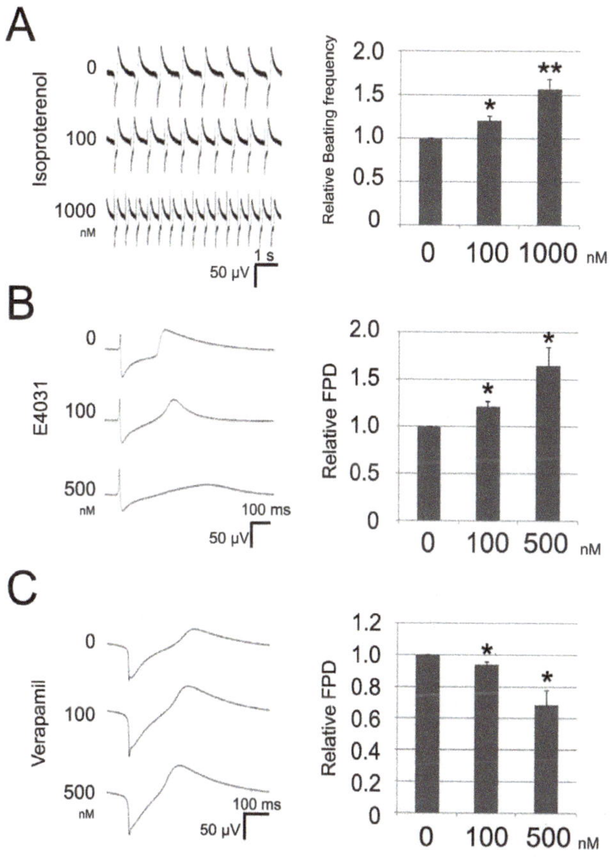

Figure 4. Pharmacological characterization of TiPSC-CMs. (A) Effect of isoproterenol on beating frequency of EBs derived from T07 was shown. The left traces show representative FPs for the various concentrations of isoproterenol. The right graph summarizes the changes in beating frequency by isoproterenol. Error bars show means ± s.d. (B) Effect of E4031 on FP wave form of beating EBs derived from T07 was shown. The left traces show representative FPs for the various concentrations of E4031. The right graph summarizes the changes in FPD by E4031. Error bars show means ± s.d. (C) Effect of verapamil on FP wave form of beating EBs derived from T07 was shown. The left traces show representative FPs for the various concentrations of verapamil. The right graph summarizes the changes in FPD by verapamil. Error bars show means ± s.d. Statistical analyses were performed using the unpaired two-tailed Student's t test (*$P<0.05$; **$P<0.01$).

differentiated cardiac phenotype and key to controlled cardiomyocyte relaxation and contraction. To evaluate intracellular Ca^{2+} behavior, Ca^{2+} transients were measured from spontaneously contracting TiPSC-CMs using Fluo-4AM. Line-scan images of Ca^{2+} transient and its average fluorescence intensity in TiPSC-CMs showed Ca^{2+} oscillations (Figure 3C) that indicate electrophysiological function typical of native cardiomyocytes. These findings indicated that the generated TiPSC-CMs were functional in regulating intracellular Ca^{2+}.

Pharmacolobical characterization of TiPSC-CMs

To further evaluate the functionality of TiPSC-CMs, we examined the pharmacological effects of various cardioactive drugs by measuring FPs of the spontaneously beating EBs (Movie S2). To check the response to adrenergic stimulation, we added isoproterenol to the beating EBs derived from TiPSCs, and found a dose-dependent increase in beating frequency, indicating an intact β-adrenergic signaling cascade (Figure 4A). Next, we investigated the effects of ion channel-specific inhibitors on the FP waveform, to determine whether TiPSC-CMs show a native cardiomyocyte-like response to ion channel-specific inhibitors. The IKr blocker E4031 dose-dependently prolonged FPDs in the TiPSCs-CMs (Figure 4B), and the L-type calcium channel blocker, verapamil dose-dependently shortened the length of FPD (Figure 4C). These results indicated that the TiPSC-CMs have identical electrophysiological responses to native cardiomyocytes.

Discussion

The cardiac differentiation ability of iPSCs and functionality of human iPSCs-derived cardiomyocytes have been characterized in previous reports [6,7,31,32]. However, the only evaluation of functionality in derivatives of human pluripotent stem cells that contain genomic rearrangements in the TCR gene locus was in hepatic cells derived from human TiPSCs [33]. TiPSC differentiation into the three germ-cell layers was shown for in vivo teratoma formation [23,34–37], and mouse mature T cells that contained genomic rearrangements in the TCR gene locus successfully generated monoclonal mice by the nuclear transfer method [38]. However, the functionality of cardiomyocytes differentiated from human TiPSCs in vitro has not been evaluated. In the present study, we demonstrated several characteristics in TiPSC-CMs that are seen in native cardiomyocytes, and also were previously described as cardiac-specific and essential characters in ESC-CMs and iPSC-CMs without genomic rearrangement [6,39]. Our results thus confirm the potential application of TiPSCs to clinical regenerative therapy for cardiovascular disease and to cardiac disease modeling.

For cardiac cell-based drug screens and cardiac disease research, the minimally invasive methods for tissue sampling involved in generating TiPSCs would be useful for increasing the number of patients who could be subjected to iPSC generation. Our results confirmed the potential of TiPSC-CMs as a cell source of cardiomyocytes with electrophysiological features resembling those of native cardiomyocytes, Therefore, TiPSC-CMs are possible and appealing cell sources for cardiac cell-based drug screens and cardiac disease research.

On another front, TiPSC-CMs have two advantages for transplantation therapy. First, the descendents of TiPSCs that transform into malignant tumors are easily identifiable by their rearrangement patterns. When iPSC-CMs obtained from dermal fibroblasts or other somatic cells are transplanted into diseased patients, there is no good procedure for following the progenies. In animal models, several marker genes can be used to chart the progression and consequences of iPSC-CM transplantation, e.g., GFP and luciferase genes are easily detected and frequently used in experiments with animal models. However, it is not desirable to insert exogenous marker genes into the genomes of human iPSCs for clinical use. In contrast, TiPSCs already have a signature in their specific TCR gene rearrangement, and this could be tracked in TiPSCs descendents [23].

Second, the minimally invasive methods for tissue sampling involved in generating TiPSCs would be especially appealing to allogenic iPSC derivative transplantation because increasing the number of patients subjected to iPSC generation is important for providing HLA-matched cells. Indeed, an HLA-haplotype bank of pluripotent stem cell lines would also be beneficial for reducing the high cost and time spent in generating iPSCs from individual patients [40]. Because TiPSCs can be generated using less invasive tissue sampling compared to dermal fibroblasts by skin biopsy sampling, they are well suited for providing such HLA-matched cell lines for the target population.

In conclusion, TiPSC-CMs have the potential to provide a unique and valuable cell source for clinical regenerative therapy in cardiovascular disease and for cardiac disease modeling.

Supporting Information

Movie S1 The beating EB derived from TiPSCs in suspension culture.

Movie S2 The beating EB derived from TiPSCs on MEA chips.

Table S1 Oligonucleotide primers used for PCR.

Acknowledgments

We thank Toshihiro Nagai in the Electron Microscope Laboratory, Keio University School of Medicine for technical assistance with electron microscopy.

Author Contributions

Conceived and designed the experiments: T. Seki SY KF. Performed the experiments: T. Seki. Analyzed the data: T. Seki. Contributed reagents/materials/analysis tools: DK AK YS ST KY Y. Kishino MO HH MT TE MK Y. Kuroda AT SO T. Suzuki MM KF. Wrote the paper: T. Seki SY JF.

References

1. Roger VL, Go AS, Lloyd-Jones DM, Adams RJ, Berry JD, et al. (2011) Heart disease and stroke statistics–2011 update: a report from the American Heart Association. Circulation 123: e18–e209.
2. Takahashi K, Yamanaka S (2006) Induction of pluripotent stem cells from mouse embryonic and adult fibroblast cultures by defined factors. Cell 126: 663–676.
3. Bellin M, Marchetto MC, Gage FH, Mummery CL (2012) Induced pluripotent stem cells: the new patient? Nat Rev Mol Cell Biol 13: 713–726.
4. Yuasa S, Fukuda K (2008) Recent advances in cardiovascular regenerative medicine: the induced pluripotent stem cell era. Expert Rev Cardiovasc Ther 6: 803–810.
5. Choi SH, Jung SY, Kwon SM, Baek SH (2012) Perspectives on stem cell therapy for cardiac regeneration. Advances and challenges. Circ J 76: 1307–1312.
6. Zhang J, Wilson GF, Soerens AG, Koonce CH, Yu J, et al. (2009) Functional cardiomyocytes derived from human induced pluripotent stem cells. Circ Res 104: e30–41.
7. Tanaka T, Tohyama S, Murata M, Nomura F, Kaneko T, et al. (2009) In vitro pharmacologic testing using human induced pluripotent stem cell-derived cardiomyocytes. Biochem Biophys Res Commun 385: 497–502.
8. Malan D, Friedrichs S, Fleischmann BK, Sasse P (2011) Cardiomyocytes obtained from induced pluripotent stem cells with long-QT syndrome 3 recapitulate typical disease-specific features in vitro. Circ Res 109: 841–847.
9. Novak A, Barad L, Zeevi-Levin N, Shick R, Shtrichman R, et al. (2012) Cardiomyocytes generated from CPVTD307H patients are arrhythmogenic in response to beta-adrenergic stimulation. J Cell Mol Med 16: 468–482.
10. Moretti A, Bellin M, Welling A, Jung CB, Lam JT, et al. (2010) Patient-specific induced pluripotent stem-cell models for long-QT syndrome. N Engl J Med 363: 1397–1409.
11. Ma D, Wei H, Lu J, Ho S, Zhang G, et al. (2013) Generation of patient-specific induced pluripotent stem cell-derived cardiomyocytes as a cellular model of arrhythmogenic right ventricular cardiomyopathy. Eur Heart J 34: 1122–1133.
12. Yazawa M, Hsueh B, Jia X, Pasca AM, Bernstein JA, et al. (2011) Using induced pluripotent stem cells to investigate cardiac phenotypes in Timothy syndrome. Nature 471: 230–234.
13. Sun N, Yazawa M, Liu J, Han L, Sanchez-Freire V, et al. (2012) Patient-specific induced pluripotent stem cells as a model for familial dilated cardiomyopathy. Sci Transl Med 4: 130ra147.
14. Lan F, Lee AS, Liang P, Sanchez-Freire V, Nguyen PK, et al. (2013) Abnormal calcium handling properties underlie familial hypertrophic cardiomyopathy pathology in patient-specific induced pluripotent stem cells. Cell Stem Cell 12: 101–113.
15. Carvajal-Vergara X, Sevilla A, D'Souza SL, Ang YS, Schaniel C, et al. (2010) Patient-specific induced pluripotent stem-cell-derived models of LEOPARD syndrome. Nature 465: 808–812.
16. Zwi-Dantsis L, Huber I, Habib M, Winterstern A, Gepstein A, et al. (2013) Derivation and cardiomyocyte differentiation of induced pluripotent stem cells from heart failure patients. Eur Heart J 34: 1575–1586.
17. Kawamura M, Miyagawa S, Miki K, Saito A, Fukushima S, et al. (2012) Feasibility, safety, and therapeutic efficacy of human induced pluripotent stem cell-derived cardiomyocyte sheets in a porcine ischemic cardiomyopathy model. Circulation 126: S29–37.

18. Sinnecker D, Goedel A, Laugwitz KL, Moretti A (2013) Induced pluripotent stem cell-derived cardiomyocytes: a versatile tool for arrhythmia research. Circ Res 112: 961–968.
19. Knollmann BC (2013) Induced pluripotent stem cell-derived cardiomyocytes: boutique science or valuable arrhythmia model? Circ Res 112: 969–976; discussion 976.
20. Egashira T, Yuasa S, Fukuda K (2013) Novel insights into disease modeling using induced pluripotent stem cells. Biol Pharm Bull 36: 182–188.
21. Gonzalez F, Boue S, Izpisua Belmonte JC (2011) Methods for making induced pluripotent stem cells: reprogramming a la carte. Nat Rev Genet 12: 231–242.
22. Yamanaka S (2010) Patient-specific pluripotent stem cells become even more accessible. Cell Stem Cell 7: 1–2.
23. Seki T, Yuasa S, Oda M, Egashira T, Yae K, et al. (2010) Generation of induced pluripotent stem cells from human terminally differentiated circulating T cells. Cell Stem Cell 7: 11–14.
24. Egashira T, Yuasa S, Suzuki T, Aizawa Y, Yamakawa H, et al. (2012) Disease characterization using LQTS-specific induced pluripotent stem cells. Cardiovasc Res 95: 419–429.
25. Suemori H, Yasuchika K, Hasegawa K, Fujioka T, Tsuneyoshi N, et al. (2006) Efficient establishment of human embryonic stem cell lines and long-term maintenance with stable karyotype by enzymatic bulk passage. Biochem Biophys Res Commun 345: 926–932.
26. Conner DA (2001) Mouse embryo fibroblast (MEF) feeder cell preparation. Curr Protoc Mol Biol Chapter 23: Unit 23 22.
27. Tran TH, Wang X, Browne C, Zhang Y, Schinke M, et al. (2009) Wnt3a-induced mesoderm formation and cardiomyogenesis in human embryonic stem cells. Stem Cells 27: 1869–1878.
28. Kehat I, Kenyagin-Karsenti D, Snir M, Segev H, Amit M, et al. (2001) Human embryonic stem cells can differentiate into myocytes with structural and functional properties of cardiomyocytes. J Clin Invest 108: 407–414.
29. Smyth JW, Shaw RM (2012) The gap junction life cycle. Heart Rhythm 9: 151–153.
30. Schaub MC, Hefti MA, Zaugg M (2006) Integration of calcium with the signaling network in cardiac myocytes. J Mol Cell Cardiol 41: 183–214.
31. Zwi L, Caspi O, Arbel G, Huber I, Gepstein A, et al. (2009) Cardiomyocyte differentiation of human induced pluripotent stem cells. Circulation 120: 1513–1523.
32. Germanguz I, Sedan O, Zeevi-Levin N, Shtrichman R, Barak E, et al. (2011) Molecular characterization and functional properties of cardiomyocytes derived from human inducible pluripotent stem cells. J Cell Mol Med 15: 38–51.
33. Kajiwara M, Aoi T, Okita K, Takahashi R, Inoue H, et al. (2012) Donor-dependent variations in hepatic differentiation from human-induced pluripotent stem cells. Proc Natl Acad Sci U S A 109: 12538–12543.
34. Staerk J, Dawlaty MM, Gao Q, Maetzel D, Hanna J, et al. (2010) Reprogramming of human peripheral blood cells to induced pluripotent stem cells. Cell Stem Cell 7: 20–24.
35. Loh YH, Hartung O, Li H, Guo C, Sahalie JM, et al. (2010) Reprogramming of T cells from human peripheral blood. Cell Stem Cell 7: 15–19.
36. Kunisato A, Wakatsuki M, Shinba H, Ota T, Ishida I, et al. (2011) Direct generation of induced pluripotent stem cells from human nonmobilized blood. Stem Cells Dev 20: 159–168.
37. Brown ME, Rondon E, Rajesh D, Mack A, Lewis R, et al. (2010) Derivation of induced pluripotent stem cells from human peripheral blood T lymphocytes. PLoS One 5: e11373.
38. Hochedlinger K, Jaenisch R (2002) Monoclonal mice generated by nuclear transfer from mature B and T donor cells. Nature 415: 1035–1038.
39. Mummery C, Ward D, van den Brink CE, Bird SD, Doevendans PA, et al. (2002) Cardiomyocyte differentiation of mouse and human embryonic stem cells. J Anat 200: 233–242.
40. Nakatsuji N, Nakajima F, Tokunaga K (2008) HLA-haplotype banking and iPS cells. Nat Biotechnol 26: 739–740.

5

Behavioral and Histopathological Assessment of Adult Ischemic Rat Brains after Intracerebral Transplantation of NSI-566RSC Cell Lines

Naoki Tajiri[1], David M. Quach[2], Yuji Kaneko[1], Stephanie Wu[2], David Lee[2], Tina Lam[2], Ken L. Hayama[2], Thomas G. Hazel[2], Karl Johe[2], Michael C. Wu[3], Cesar V. Borlongan[1]*

1 Center of Excellence for Aging & Brain Repair, Department of Neurosurgery and Brain Repair, University of South Florida College of Medicine, Tampa, Florida, United States of America, **2** Neuralstem, Inc., Rockville, Maryland, United States of America, **3** Neurodigitech, LLC., San Diego, California, United States of America

Abstract

Stroke is a major cause of death and disability, with very limited treatment option. Cell-based therapies have emerged as potential treatments for stroke. Indeed, studies have shown that transplantation of neural stem cells (NSCs) exerts functional benefits in stroke models. However, graft survival and integration with the host remain pressing concerns with cell-based treatments. The current study set out to investigate those very issues using a human NSC line, NSI-566RSC, in a rat model of ischemic stroke induced by transient occlusion of the middle cerebral artery. Seven days after stroke surgery, those animals that showed significant motor and neurological impairments were randomly assigned to receive NSI-566RSC intracerebral transplants at two sites within the striatum at three different doses: group A (0 cells/μl), group B (5,000 cells/μl), group C (10,000 cells/μl), and group D (20,000 cells/μl). Weekly behavioral tests, starting at seven days and continued up to 8 weeks after transplantation, revealed dose-dependent recovery from both motor and neurological deficits in transplanted stroke animals. Eight weeks after cell transplantation, immunohistochemical investigations via hematoxylin and eosin staining revealed infarct size was similar across all groups. To identify the cell graft, and estimate volume, immunohistochemistry was performed using two human-specific antibodies: one to detect all human nuclei (HuNu), and another to detect human neuron-specific enolase (hNSE). Surviving cell grafts were confirmed in 10/10 animals of group B, 9/10 group C, and 9/10 in group D. hNSE and HuNu staining revealed similar graft volume estimates in transplanted stroke animals. hNSE-immunoreactive fibers were also present within the corpus callosum, coursing in parallel with host tracts, suggesting a propensity to follow established neuroanatomical features. Despite absence of reduction in infarct volume, NSI-566RSC transplantation produced behavioral improvements possibly via robust engraftment and neuronal differentiation, supporting the use of this NSC line for stroke therapy.

Editor: Eva Mezey, National Institutes of Health, United States of America

Funding: CVB received research support from Neuralstem Inc for this study. DMQ, SW, DL, TL, KLH, TGH and KJ are employed by the funder of this study, Neuralstem Inc. CVB is supported by National Institutes of Health, National Institute of Neurological Disorders and Stroke 1R01NS071956-01 and James and Esther King Foundation for Biomedical Research Program. The funder provided support in the form of salaries for authors DMQ, SW, DL, TL, KLH, TGH and KJ, but did not have any additional role in the study design, data collection and analysis, decision to publish, or preparation of the manuscript. The specific roles of these authors are articulated in the 'author contributions' section.

Competing Interests: I have read the journal's policy and have the following conflicts: CVB received research support from Neuralstem Inc for this study. DMQ, SW, DL, TL, KLH, TGH and KJ are employed by the funder of this study, Neuralstem Inc. MCW is employed by Neurodigitech, LLC. CVB is supported by National Institutes of Health, National Institute of Neurological Disorders and Stroke 1R01NS071956-01 and James and Esther King Foundation for Biomedical Research Program.

* E-mail: cborlong@health.usf.edu

Introduction

Stroke is a major unmet clinical need with only one current FDA-approved drug, the tissue plasminogen activator (tPA) [1–5]. The efficacy of tPA is limited to 4.5 hours after stroke onset and benefits only about 3% of ischemic stroke patients [6–8]. The advent of stem cell therapy opens the possibility of regenerating the injured brain and may prove effective in stroke beyond the acute phase of the disease [9–13]. With the increasing diversity of stem cell sources emerging for donor cells in transplantation therapy, many laboratory-to-clinic translational factors must first be considered, dynamics such as the source of the cells, ease of extraction, immunogenicity, capacity for proliferation, and cell yield [14–16]. These concerns may serve as potential limitations respective to the donor cell origin being considered, necessitating the need for a particular stem cell source to be more suitable for a specific disease.

Because stroke is a major cause of death and disability, any treatment that would help stroke patients recover some of the lost motor or cognitive function, would substantially improve their quality of life. Cell-based therapies have emerged as potential methods to treat several neuropathological diseases and injuries, including stroke [1–5,9–13]. Laboratory studies and limited clinical trials have shown that transplantation of neural stem cells (NSCs) in stroke is safe and effective [17–20]. The mechanism of action of stem cell therapy for stroke remains not fully understood,

Table 1. Experimental Treatment Conditions.

Group	Group Size	Test Article	Dose	Survival Time
A	8	Vehicle	0 cells/μL×3 μl/site×3 sites	56+/−4 days
B	8	NSI-566RSC	5,000 cells/μL×3 μl/site×3 sites	56+/−4 days
C	8	NSI-566RSC	10,000 cells/μL×3 μl/site×3 sites	56+/−4 days
D	8	NSI-566RSC	20,000 cells/μL×3 μl/site×3 sites	56+/−4 days

Treatment Summary Table. Table 1 summarizes the 4 treatment groups of randomly assigned stroke animals.

but the two major postulated reparative pathways involve cell replacement and secretion of growth factors [1–5,9–13,21,22]. To date, graft survival and integration with the host remain pressing concerns with cell-based treatment options. The current study set out to investigate those very issues using a human NSC line, NSI-566RSC, in a rat model of ischemic stroke.

Preclinical evidence has demonstrated the safety and efficacy of NSI-566RSC in animal models of the motor neuron disease amyotrophic lateral sclerosis (ALS) [23–26], spinal cord injury [27], and ischemic paraplegia [28]. Larger animal models have also been used to assess safety of NSI-566RSC for CNS transplantation [29,30]. Functional recovery observed in these animal models has been ascribed to neuronal differentiation capacity of NSI-566RSC [25,31], which parallels extensive *in vitro* characterization of these cells similarly demonstrating the cells' ability to display neuronal phenotypic features (i.e., functional motoneurons) [32,33]. The need for immunosuppression in order to augment graft survival and functional effects has been indicated in relevant ALS animal models [30,34]. This translational research portfolio forms the basis for a clinical trial of transplanting NSI-566RSC in ALS patients [35].

Our long-standing interest in stem cell therapy for stroke prompted us to examine the efficacy of NSI-566RSC in an animal model of cerebral ischemia. We report here that intracerebral transplantation of NSI-566RSC at subacute phase of stroke produced behavioral improvements accompanied by graft survival and neuronal differentiation, extending the application of NSI-566RSC for transplantation therapy in ischemic stroke.

Materials and Methods

Subjects

All experiments were conducted in accordance with the National Institute of Health Guide and Use of Laboratory Animals, and were approved by the Institutional Animal Care and Use Committee of the University of South Florida, Morsani College of Medicine. Rats were housed two per cage in a temperature- and humidity-controlled room that was maintained on 12/12 hour-light/dark cycles. They had free access to food and water. All necessary steps were performed to minimize animal pain and stress throughout the study.

Stroke and Transplant Surgery

A total of 32 male Sprague-Dawley rats received middle cerebral artery occlusion (MCAo), a well-established stroke model, as described previously [36–40]. Animals were anesthetized using isoflurane (1.5%–2.5% with oxygen). The scalp skin was shaved and scrubbed with alcohol and chlorhexidine surgical scrub. The animal was fixed in stereotaxic apparatus, then starting slightly behind the eyes, a midline sagittal incision about 2.5 cm long is

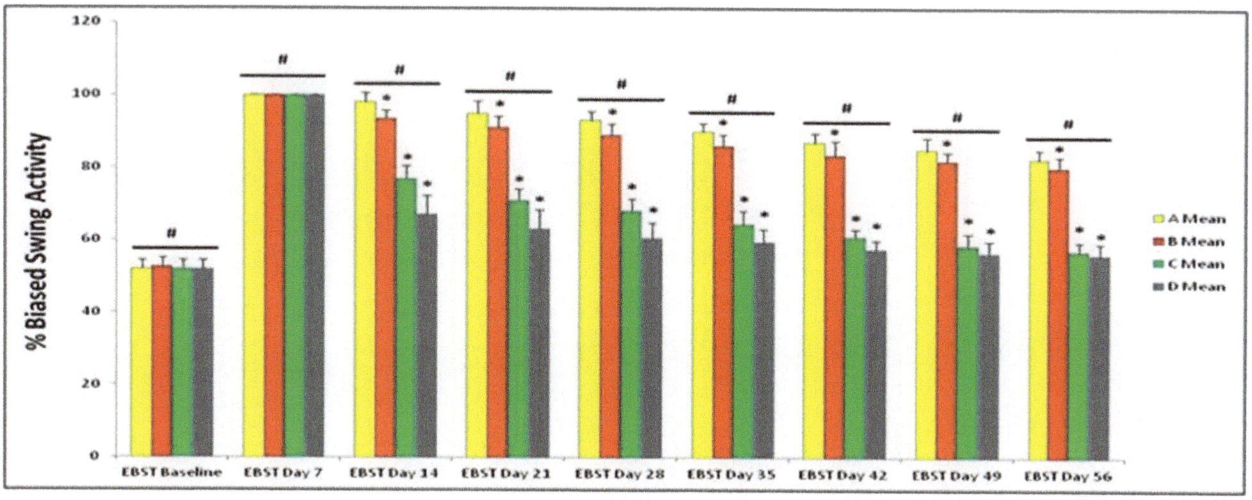

Figure 1. NSI-566RSC cell grafts ameliorate stroke-induced motor deficits. Motor performance was measured by EBST (Figure 1). All animals displayed normal motor behavior at Baseline (i.e., prior to stroke). At day 7 post-stroke, all animals exhibited 100% biased swing behaviors, indicating that all animals received successful stroke. At 14 days post-stroke onwards, dose-dependent and timing-dependent effects of the treatment were recognized, in that the improvement in behavioral performance was in the order of high dose to zero dose as follows: 20,000 cells/μl (D) >10,000 cells/μl (C) >5,000 cells/μl (B)>vehicle infusion only (A). In addition, over time there was a trend of better improvement, with the most significant improvement seen at 56 days post-stroke. *significant <0.05 vs. other treatment groups within time point; #significant <0.05 vs. other time points.

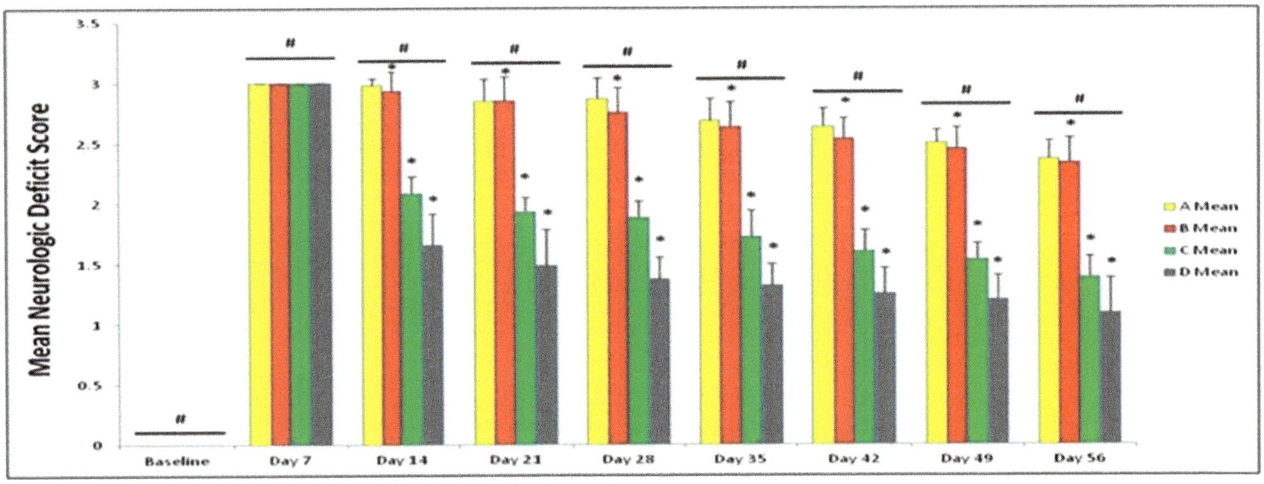

Figure 2. NSI-566RSC cell grafts attenuate stroke-induced neurologic impairments. Neurological function was assessed using a battery of neurological tests (Figure 2). All animals displayed normal neurological function at Baseline (i.e., prior to stroke). At day 7 post-stroke, all animals exhibited significant impairment in neurological function, indicating that all animals received successful stroke. At 14 days post-stroke onwards, dose-dependent and timing-dependent effects of the treatment were recognized, in that the improvement in neurological performance was in the order of high dose to zero dose as follows: 20,000 cells/µl (D) >10,000 cells/µl (C) >5,000 cells/µl (B)>vehicle infusion only (A). In addition, over time there was a trend of better improvement, with the most significant improvement seen at 56 days post-stroke. *significant <0.05 vs. other treatment groups within time point; #significant <0.05 vs. other time points.

made. With the rounded end of a spatula, the skull area was exposed. Using the bregma as reference point (i.e., prior to stroke surgery) laser Doppler recordings were obtained from the following coordinates (AP: +2.0, ML: ±2.0) at pre, during, and post surgery to reveal successful MCAo. The skin on the ventral neck was shaved from the jaw to the manubrium and scrubbed with alcohol and chlorhexidine surgical scrub. A skin incision was made over the right common carotid artery. The external carotid was isolated and ligated as far distally as possible. An incision using a pair of microscissor was made in the stump of the external carotid and a 4-0 nylon filament with a pre-fabricated end was inserted and passed up into the internal carotid artery until resistance was felt (approximately 15–17 mm), which effectively blocked the middle cerebral artery (MCA). The isoflurane was discontinued and the animal placed in a recovery cage over a warming blanket. After 60 minutes, the animal was anesthetized again with isoflurane and the incision opened. The filament causing the occlusion was removed and the stump of the external carotid ligated close to the carotid bifurcation. The skin incision was closed with staples. Finally, the animal was placed in a cage over a warming blanket until full recovery from anesthesia. Forty stroke animals were randomly divided into four Groups A–D (Table 1). They were tested for behavioral and neurological deficits on Day 6 post stroke (baseline), and then treated on Day 7 post stroke. Those animals that showed significant motor and neurological impairments were randomly assigned to receive NSI-566RSC intracerebral transplants at two sites within the striatum at three different doses: group A (0 cells/µl), group B (5,000 cells/µl), group C (10,000 cells/µl), and group D (20,000 cells/µl). The animal was fixed to a stereotaxic apparatus (Kopf Instruments). A 26-gauge Hamilton syringe was then lowered into a small burred skull opening (transplant coordinates were adjusted to 0.5 mm

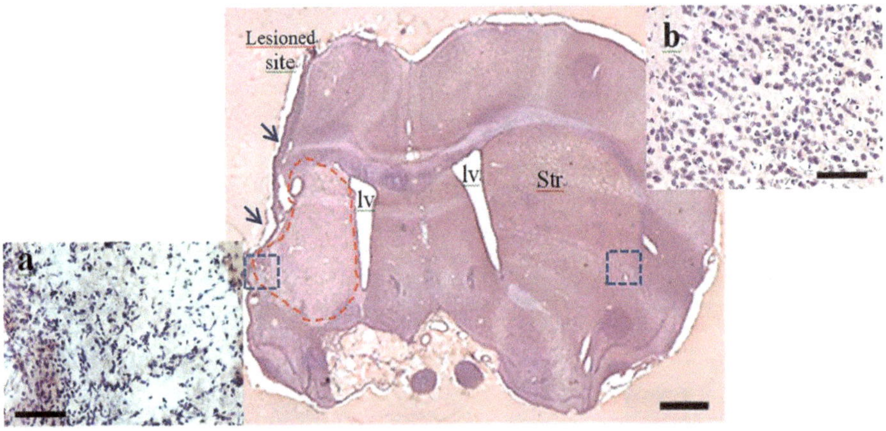

Figure 3. A typical infarct lesion in the striatum and cortex. The tissue damage covers the cortex and infarct striatum (dash area) and areas lateral to infarct core lesion (not shown here). Note the shrinkage and dissolution of the cells in the penumbra (box) compared to the corresponding contralateral side of the brain (Insets a and b). Scale bar = 1.0 mm [lv; lateral ventricles, Str: striatum].

Table 2. Summary of histological assessment of infarct brains transplanted with NSI-566RSC (n = 10 per group)*.

	Stains	Group A-Vehicle	Group B-Low	Group C-Medium	Group D-High	Mean-Total
Freq. of the infarct lesion**	*H&E*	9 (90%)	9 (90%)	9 (90%)	10 (100%)	93%
Freq. of surviving grafts in the striatum	*HuNu*	Not present	10 (100%)	9 (90%)	9 (90%)	93%
	hNSE	Not present	10 (100%)	9 (90%)	8 (80%)	90%
Est. graft size (mm^3)	*HuNu*	Not present	0.85±0.37	0.19±0.05	0.63±0.22	0.56±0.15
	hNSE	Not present	0.99±0.50	0.16±0.04	0.54±0.33	0.58±0.21

Table 2 summarizes the histological assessment of infarct brains that received three different doses of NSI-566RSC cell lines (Group B, C, and D) in comparison with the vehicle (Group A).

*Quantitative analysis did not find statistical difference in graft size across the three dose groups, either by HuNu or by hNSE stain ($p = 0.2002$).

**Freq. of infarct lesion was based on evidence of tissue damage and cellular abnormality such as cell loss, apoptotic bodies in penumbra regions.

anterior and 2.8 mm lateral to bregma and 5.0 mm below the dural surface [41]. Within this single needle pass, 3 deposits of the test articles including 2 into the striatum (5.0 mm and 4.0 mm below the dural surface) and 1 into the cortex (3.0 mm below the dural surface) were made. The target area was the medial striatum and medial cortex which corresponds to the ischemic peri-infarct (or penumbra) area, based on previously established target sites for similar stereotaxic implants [42]. Each deposit consisted of 3 μl volume infused over a period of 3 minutes. All animals received daily immunosuppression (tacrolimus) 1 mg/kg/day IP. All animals were monitored weekly post-grafting for behavioral and neurological outcomes. Animals were euthanized at Day 56 days post-grafting and transcardially perfused with saline and 4% paraformaldehyde.

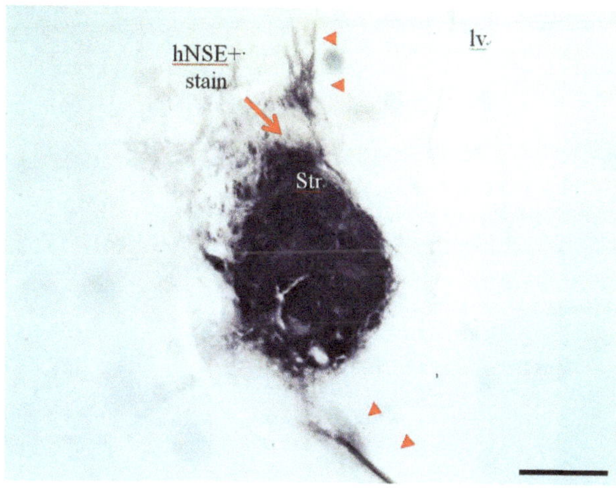

Figure 5. A representative image of hNSE staining in the striatum. Human graft cells were densely confined to the striatum (arrow) with small amounts of hNSE+fibers that extended dorsally and ventrally from the striatum (arrowheads). Scale bar = 0.5 mm [lv; lateral ventricle, Str: striatum].

Behavioral Tests

All investigators testing the animals were blinded to the treatment condition. Animals were subjected to elevated body swing test (EBST) and neurological exam. EBST involved handling the animal by its tail and recording the direction of the swings [36–40]. The test apparatus consisted of a clear Plexiglas box (40×40×35.5 cm). The animal was gently picked up at the base of the tail, and elevated by the tail until the animal's nose was at a height of 2 inches (5 cm) above the surface. The direction of the swing, either left or right, was counted once the animals head moved sideways approximately 10 degrees from the midline position of the body. After a single swing, the animal was placed back in the Plexiglas box and allowed to move freely for 30 seconds prior to retesting. These steps were repeated 20 times for each animal. Intact rats displayed a 50% swing bias, that was, the same number of swings to the left and to the right. A 75% swing bias was used as criterion of successful MCAo. About one hour after the EBST, neurological exam was conducted as previously described [36–40] with minor modifications. Neurologic score for each rat was obtained using 3 tests which included (1) forelimb retraction, which measures the ability of the animal to replace the forelimb after it was displaced laterally by 2 to 3 cm, graded from 0 (immediate replacement) to 3 (replacement after several seconds or no replacement); (2) beam walking ability, graded 0 for a rat that readily traversed a 2.4-cm-wide, 80-cm-long beam to 3 for a rat unable to stay on the beam for 10 seconds; and (3) bilateral forepaw grasp, which measured the ability to hold onto a 2-mm-diameter steel rod, graded 0 for a rat with normal forepaw grasping behavior to 3 for a rat unable to grasp with the forepaws. The scores from all 3 tests were averaged to give a mean neurologic deficit score (maximum possible score, 9 points divided

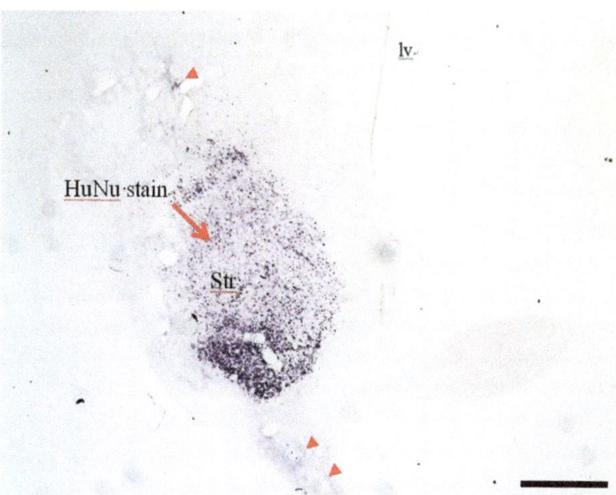

Figure 4. A representative image of HuNu staining in the striatum. HuNu+cells were confined to the striatum of rats receiving the lowest dose of graft cells (Group B). Note that small amounts of HuNu staining extended both dorsally and ventrally from the striatum. Scale bar = 0.5 mm [lv; lateral ventricle, Str: striatum].

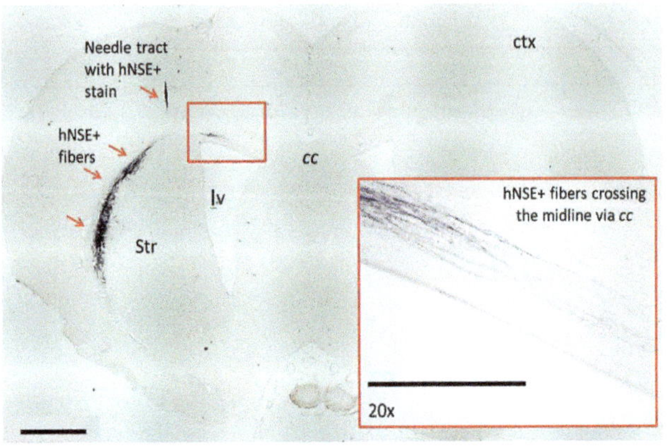

Figure 6. Migration of hNSE+fibers toward the contralateral side of non-infarct region. Needle tract with hNSE+stain was found in the cortex. Additionally, small amounts of hNSE+fibers crossed the *cc contralaterally (inset).* Scale bar = 1.0 mm [lv; lateral ventricle, Str: striatum, ctx: cortex, *cc*: corpus callosum].

by 3 tests = 3). Animals were subjected to both tests at baseline (prior to stroke), then at 7 days after stroke (prior to transplantation) and at weekly intervals over an eight-week period.

Histology

All histological analyses were performed at Neuralstem Inc. (San Diego, CA). Brains were cryoprotected in PBS containing 30% sucrose and later sectioned at 40 µm thickness using a freezing sliding microtome. Every sixth section through the brain was selected for staining.

Immunohistochemistry/Histology

Sections were treated with 3% hydrogen peroxide for 15 minutes, washed 3 times with PBS, and further treated with 0.3% Triton X-100 for 30 minutes followed by wash. Tissue was then blocked in normal horse serum for 90 minutes, and incubated with either mouse anti human nuclei (HuNu; MAB1281; Chemicon International, Temecula, CA; 1:2000) or chicken anti human-specific neuron-specific enolase (hNSE; AB-9698; Chemicon International, Temecula, CA 1:2000) at room temperature overnight. Sections were washed and again blocked with horse serum for 90 minutes, then treated with biotinylated donkey anti-mouse or biotinylated donkey anti-chicken secondary antibody (Jackson ImmunoResearch, West Grove, PA; 1:2000) for 90 minutes. Following wash, sections were incubated with streptavidin conjugated to horseradish peroxidase (Jackson ImmunoResearch, West Grove, PA; 1:5000) for 90 minutes, and then developed with diaminobenzidine containing nickel chloride for 1 minute. The sections were then mounted on gelatin-coated slides, dried, labeled, and cover-slipped. In addition, sections adjacent to sections (1-in-6 series) stained for human stem cells were processed with hematoxylin and eosin (H&E). All slides were examined and scanned using Nikon AZ100 Multizoom microscope (Nikon Instruments, Inc, Japan) for further qualitative and semi-quantitative analyses.

Measurements of Infarct Size

At least 4 coronal sections per brain were processed for H&E. The indirect lesion area, in which the intact area of the ipsilateral hemisphere was subtracted from the area of the contralateral hemisphere, was calculated to reveal cerebral infarction. Scores for infarct lesion were rated accordingly: LOW <30%, MEDIUM 30–60%, HIGH >60%, while scores for volume were rated as follows: Low <1 mm^3, Medium 1–2 mm^3, High >2 mm^3. For calculation of area of infarction in mm2, a separate cohort of animals (n = 8 per group) subjected to the same experimental paradigm were processed for Triphenyltetrazolium chloride (TTC) staining. Anesthetized animals were perfused intracardially with saline. The brain tissue was then removed, immersed in cold saline for 5 minutes, and sliced into 2.0-mm thick sections. The brain slices were incubated in a 2% triphenyltetrazolium chloride (Sigma) dissolved in normal saline for 30 minutes at 37°C and then transferred into a 5% formaldehyde solution for fixation. The area of infarction in each slice was measured using a digital scanner and the Image Tools program (University of Texas Health Sciences Center in San Antonio). The area of infarction in each animal was obtained from largest infarcted brain slice.

Cell Engraftment and Neuronal Phenotype Assessment

The slides were examined at 10X magnification of Nikon AZ100 Multizoom microscope and digitized using a PC-based Image Tools computer program (Nikon, NIS Elements software). Brain sections were blind-coded in assessing cell engraftment index for NSI-566RSC graft survival based on HuNu stained and hNSE stained slides, respectively.

Statistical Analysis

The data were evaluated statistically using either Two-way analysis of variance (ANOVA), or repeated measure of ANOVA and subsequent post hoc compromised t-tests for behavior. These statistical analyses employed a StatView 5.0 software (USA). Statistical significance was preset at $p<0.05$.

Results

NSI-566RSC Cell Grafts Promote Behavioral Recovery in Stroke Animals

Behavioral analyses revealed that all animals included in this study displayed normal behaviors at baseline prior to stroke surgery. At 7 days post-stroke, all animals exhibited the typical stroke-induced behavioral deficits characterized by significantly biased swing activity and impaired neurological performance. However, at 14 days post-stroke (i.e., 7 days post-transplantation onwards), significant recovery of motor (Figure 1) and neurological (Figure 2) functions in stroke animals that received intracerebral transplantation of high doses of 10,000 cells/µl (C) and 20,000 cells/µl (D) of NSI-566RSC compared to vehicle-infused stroke animals (A) or those stroke animals that received the low dose of 5,000 cells/µl (B). ANOVA revealed statistically significant treatment effects (p's<0.01). Post hoc compromised t-tests also revealed dose-dependent treatment effects and post-transplantation timing effects (p's<0.05). There was stable improvement in those stroke transplanted with high doses of cells (Groups C and D) which showed a trend of better improvement over time up to the 56-day study period. Although there was slight improvement in the behavioral performance of vehicle-infused stroke animals and stroke animals that received the low dose of 5,000 cells/µl, the recovery of motor and neurological tests demonstrated by high dose transplanted stroke animals was significantly better throughout the 56-day period compared to vehicle-infused stroke animals or those stroke animals that received the low dose of 5,000 cells/µl (p's<0.05). There were significant dose-dependent differences in the behavioral improvement across treatment groups at post-transplantation periods with the highest dose of 20,000 cells/µl showing the most significant improvement in both motor and

Table 3. Summary of infarct size and graft size.

ID	H&E-infarct	HuNu-Volume	hNSE-Volume
Group A Vehicle	STR+CTX	STR+CTX	STR+CTX
A02	LOW	None	None
A04	LOW	None	None
A13	LOW	None	None
A14	LOW	None	None
A15	HIGH	None	None
A18	LOW	None	None
A19	LOW	None	None
A20	LOW	None	None
A25	LOW	None	None
A26	None	None	None
Group B (5 K cells/μl)			
B01	None	LOW	LOW
B03	LOW	LOW	LOW
B05	HIGH	HIGH	HIGH
B06	MEDIUM	HIGH	LOW
B07	LOW	LOW	LOW
B08	LOW	LOW	LOW
B09	LOW	LOW	LOW
B10	LOW	HIGH	HIGH
B11	LOW	LOW	LOW
B12	LOW	LOW	MEDIUM
Group C (10 K cells/μl)			
C21	LOW	LOW	LOW
C22	LOW	LOW	LOW
C23	LOW	LOW	LOW
C24	MEDIUM	LOW	LOW
C01	LOW	LOW	LOW
C02	LOW	LOW	LOW
C06	MEDIUM	LOW	LOW
C27	LOW	LOW	LOW
C28	LOW	LOW	N/A
C29	None	LOW	LOW
Group D (20 K cells/μl)			
D05	LOW	LOW	LOW
D08	LOW	None	None
D11	HIGH	MEDIUM	HIGH
D16	HIGH	HIGH	LOW
D19	MEDIUM	LOW	LOW
D21	LOW	LOW	LOW
D22	MEDIUM	LOW	LOW
D23	LOW	LOW	LOW
D24	MEDIUM	LOW	LOW
D30	LOW	None	None

Table 3 summarizes the semi-quantitative rating of the infarct size and the size of HuNu (nuclei staining) or hNSE (neuropil staining) positive grafts.
1. Scores for Infarct lesion: LOW <30%, MEDIUM 30–60%, HIGH >60%.
2. Scores for Volume: Low <1 mm^3, Medium 1–2 mm^3, High >2 mm^3.
3. Abbreviations: STR: striatum, CTX: cerebral cortex.

neurological tests (p's<0.05). Similarly, there were significant differences in the behavioral performance among treatment groups at post-transplantation periods with the most significant improvement in both motor and neurological tests seen at day 56 post-transplantation (p's<0.05).

NSI-566RSC Cell Grafts do not Reduce Infarct Size

To confirm infarcts in stroke animals, H&E was used. Infarcts were confirmed in 93% of the animals; infarct size was similar across all groups ($p = 0.2676$) (Figure 3). A summary of infarct frequency for each treatment group is presented for clarity (Table 2). Additionally, we rated the size of the infarct in all animals to get a better approximation of any infarct variability that might have been affected by the grafts, but did not detect any as significant number of outliers across the treatment groups (i.e., only 1 animal each from Groups A–C). Based on these data, infarct size was not reduced by NSI-566RSC cell grafts.

NSI-566RSC Cell Grafts Survive and Display Neuronal Phenotype

To identify the cell graft, and estimate volume, immunohistochemistry was performed using two human-specific antibodies: one to detect all HuNu (Figures 4 & 5), and another to detect hNSE (Figure 6). Tabular summaries of the frequency of surviving grafts and estimated graft size (Table 2 and Table 3) are presented for clarity. Surviving cell grafts were confirmed in 10/10 animals of group A, 9/10 group B, and 9/10 in group C. hNSE and HuNu staining revealed similar graft volume estimates (HuNu volume: $0.56 \pm 0.15\ mm^3$ and hNSE volume: $0.58 \pm 0.21\ mm^3$). hNSE-immunoreactive fibers were also present within the corpus callosum, coursing in parallel with host tracts, suggesting a propensity to follow established neuroanatomical features. Supplementary files are also provided to show individual animal's histology capturing representative images of HuNu and hNSE (Figures S1, S2, S3, S4, S5, S6, S7, S8, and S9).

Discussion

Here, we demonstrated that intracerebral transplantation of NSI-566RSC, a spinal cord-derived NSC line, reduced behavioral deficits associated with ischemic stroke. Significant improvements in both motor and neurological tests were detected in the NSI-566RSC-treated stroke animals. In addition, the results revealed significant dose-dependent differences in the behavioral improvement across treatment groups at post-transplantation periods with the highest NSI-566RSC dose showing the most significant improvement in both motor and neurological tests. Similarly, there were significant differences in the timing of behavioral performance among treatment groups at post-transplantation periods with the most significant improvement in both motor and neurological tests seen at day 56 post-transplantation, indicating that the transplanted stroke animals showed improved recovery over graft maturation period. We observed though that a few transplanted animals did not contain surviving grafts. The reason why some transplanted animals do not have detectable transplanted cells is unknown but it is likely because of graft failure due to immunosuppressant being ineffective to those particular animals.

We used a potent immunosuppressive agent (tacrolimus) to avoid rejection of the human cells, but are aware of the action of this drug in the brain response to ischemia itself or a possible synergic effect with the cell-based therapy proposed in our study. However, in our previous studies, using cyclosporine and tacrolimus [43–45], we came into the conclusion that these immunosuppressant drugs, albeit with neuoroprotective effects, display limited therapeutic window of opportunity, in that the drugs must be administered prior to, during, and/or immediately after the injury. Beyond this very acute time point, such as the present two weeks post-stroke period, we and several others have found that these drugs are not effective. In addition, a vis-à-vis comparison between non-immunosuppressed transplanted stroke animals versus immunosuppressed transplanted stroke animals revealed that immunosuppressed transplanted stroke animals significantly produced functional recovery compared to immunosuppression treatment alone [46]. Moreover, the choice to deliver the cocktail of immunosuppressants in the present study was based on a series of preclinical studies that supported the ongoing clinical trials of the same cells in ALS patients. This cocktail was proven effective in facilitating the graft effects in animal models of ALS, and approved by the FDA for this clinical trial. Accordingly, as we translate these NSCs for stroke clinical application, in keeping with the FDA's requirement and also based on preclinical studies from our group and several others, we maintain the transplant protocol with the immunosuppresion regimen. However, if the transplant procedure was done acutely in stroke, we would be compelled to conduct another vis-à-vis study comparing NSCs with vehicle versus NSCs with immunosuppressants. For the present delayed transplantation regimen, and with the FDA requiring that the same protocol done in ALS be followed in the stroke arena in order to gain entry into the clinic, both scientific and clinical bases support our decision to maintain the NSCs with immunosuppressant regimen.

This study was designed to evaluate the potential therapeutic value of intracerebral dosing of NSI-566RSC in an animal model of adult ischemic stroke. The recovery of motor and neurological tests demonstrated by high dose transplanted stroke animals was significantly better throughout the 56 day study period compared to vehicle-infused stroke animals, or low-dosed transplanted animals. Moreover, there was stable improvement in the high dose animals, and they showed a trend of better improvement over time.

These results demonstrated safety and efficacy of NSI-566RSC in a subacute model of ischemic stroke in rats, indicating a strong proof-of-principle data that NSI-566RSC are potent cell donors for transplantation therapy to treat paralysis in stroke patients. A Phase I trial testing the safety of NSI-566RSC has recently been completed in the treatment of ALS [35]. Our current observations of safety and efficacy of NSI-566RSC in stroke animals mimic similar therapeutic readouts following transplantation of NSI-566RSC in animal models of ALS [23–26], spinal cord injury [27], and ischemic paraplegia [28]. Testing the efficacy of NSI-566RSC in larger stroke animals may be needed, although a safety portfolio already exists in mini-pigs for NSI-566RSC transplantation in CNS [29,30]. That the present transplanted stroke animals exhibited improvement in motor and neurological functions accompanied by NSI-566RSC graft survival and neuronal differentiation also resembled the histopathological outcomes seen in animal models of ALS, spinal cord injury, ischemic paraplegia demonstrating engraftment and neuronal phenotypic persistence in transplanted NSI-566RSC [25,31]. While graft survival and neuronal differentiation have been confirmed, on-going work is investigating the integration of graft into host tissue, determining the neuronal phenotypes of the grafted neurons, and confirming the low proliferation rates of the grafts as suggested from these preliminary histological studies.

Stroke remains a significant unmet clinical need [1–5] with tPA efficacy limited to 4.5 hours after stroke onset and benefits only about 3% of ischemic stroke patients [6–8]. The tPA's narrow therapetuc window has prompted the need for novel treatments, such as stem cell therapy, designed to abrogate stroke beyond the acute phase of the disease [9–13]. Current clinical trials of cell therapy for stroke have pursued adult stem cells (.e.g., bone marrow and placenta) and NSCs, based on ample laboratory evidence demonstrating safety and efficacy of these cells [4,5,42,47–49]. Optimization of clinical outcomes has been a key translational research effort for cell therapy for stroke [10,13,50–52]. With myriad stem cell sources in transplantation therapy, many laboratory-to-clinic translational factors have been recognized including well-defined donor stem cells, cell isolation and propagation, immunogenicity, capacity for proliferation, and cell yield [14–16]. Although NSC transplantation has been shown effective in stroke models [19,53,54], the mechanism of action remains elusive, in particular graft survival and integration with the host have yet to be fully understood in the stroke brain. Here, we provide evidence of a regenerative pathway involving robust NSI-566RSC engraftment and neuronal differentiation that accompanied behavioral improvements despite absence of cerebral infarct reduction. Axonal growth with long fibers from NSC grafts extended to the corpus callosum and traversed with host tracts, indicating graft-host integration recapitulating endogenous neuroanatomical landmarks. In addition, while on-going clinical trials of cell therapy have targeted the acute and chronic phase of stroke, the present study suggests the sub-acute phase as an equally effective therapeutic window for NSI-566RSC transplantation in ischemic stroke.

Supporting Information

Figure S1 HuNu staining in vehicle-infused group. No detectable HuNu staining is observed.

Figure S2 HuNu staining in low dose-transplanted group. HuNu staining is detected in all transplanted stroke brain.

Figure S3 HuNu staining in medium dose-transplanted group. HuNu staining is detected in all transplanted stroke brain except for one brain (C29).

Figure S4 HuNu staining in high dose-transplanted group. HuNu staining is detected in all transplanted stroke brain except for one brain (D30).

Figure S5 hNSE staining in vehicle-infused group. No detectable hNSE staining is observed.

Figure S6 hNSE staining in low dose-transplanted group. hNSE staining is detected in all transplanted stroke brain.

Figure S7 hNSE staining in medium dose-transplanted group. hNSE staining is detected in all transplanted stroke brain except for one brain (C29).

Figure S8 hNSE staining in high dose-transplanted group. hNSE staining is detected in all transplanted stroke brain except for two brains (D08 and D30).

Figure S9 Area of infarction. TTC staining reveals area of infarction (mm^2) in the laragest infarcted slice. *$p<0.05$ versus A or B.

Author Contributions

Conceived and designed the experiments: KJ CVB. Performed the experiments: NT DMQ YK SW DL TL KLH TGH KJ MCW CVB. Analyzed the data: NT KJ MCW CVB. Contributed reagents/materials/analysis tools: KJ MCW CVB. Wrote the paper: NT KJ CVB.

References

1. Tajiri N, Dailey T, Metcalf C, Mosley YI, Lau T, et al. (2013) In vivo animal stroke models: a rationale for rodent and non-human primate models. Transl Stroke Res 4: 308–321.
2. Kaneko Y, Pabon MM, Dailey T, Weinbren NL, Rizzi J, et al. (2013) The battle of the sexes for stroke therapy: female- versus male-derived stem cells. CNS Neurol Disord Drug Targets 12: 405–412.
3. Kaneko Y, Tajiri N, Shinozuka K, Glover LE, Weinbren NL, et al. (2012) Cell therapy for stroke: emphasis on optimizing safety and efficacy profile of endothelial progenitor cells. Curr Pharm Des 18: 3731–3734.
4. Borlongan CV, Glover LE, Tajiri N, Kaneko Y, Freeman TB (2011) The great migration of bone marrow-derived stem cells toward the ischemic brain: therapeutic implications for stroke and other neurological disorders. Prog Neurobiol 95: 213–228.
5. Borlongan CV (2011) Bone marrow stem cell mobilization in stroke: a 'bonehead' may be good after all! Leukemia 25: 1674–1686.
6. Tan Tanny SP, Busija L, Liew D, Teo S, Davis SM, et al. (2013) Cost-effectiveness of thrombolysis within 4.5 hours of acute ischemic stroke: experience from Australian stroke center. Stroke 44: 2269–2274.
7. Saver JL, Fonarow GC, Smith EE, Reeves MJ, Grau-Sepulveda MV, et al. (2013) Time to treatment with intravenous tissue plasminogen activator and outcome from acute ischemic stroke. JAMA 309: 2480–2488.
8. Fisher M, Albers GW (2013) Advanced imaging to extend the therapeutic time window of acute ischemic stroke. Ann Neurol 73: 4–9.
9. Chopp M, Steinberg GK, Kondziolka D, Lu M, Bliss TM, et al. (2009) Who's in favor of translational cell therapy for stroke: STEPS forward please? Cell Transplant 18: 691–693.
10. Stem Cell Therapies as an Emerging Paradigm in Stroke P (2009) Stem Cell Therapies as an Emerging Paradigm in Stroke (STEPS): bridging basic and clinical science for cellular and neurogenic factor therapy in treating stroke. Stroke 40: 510–515.
11. Hess DC, Borlongan CV (2008) Cell-based therapy in ischemic stroke. Expert Rev Neurother 8: 1193–1201.
12. Hess DC, Borlongan CV (2008) Stem cells and neurological diseases. Cell Prolif 41 Suppl 1: 94–114.
13. Borlongan CV, Chopp M, Steinberg GK, Bliss TM, Li Y, et al. (2008) Potential of stem/progenitor cells in treating stroke: the missing steps in translating cell therapy from laboratory to clinic. Regen Med 3: 249–250.
14. Sanberg PR, Eve DJ, Metcalf C, Borlongan CV (2012) Advantages and challenges of alternative sources of adult-derived stem cells for brain repair in stroke. Prog Brain Res 201: 99–117.
15. Sanberg PR, Eve DJ, Cruz LE, Borlongan CV (2012) Neurological disorders and the potential role for stem cells as a therapy. Br Med Bull 101: 163–181.
16. Rodrigues MC, Voltarelli J, Sanberg PR, Allickson JG, Kuzmin-Nichols N, et al. (2012) Recent progress in cell therapy for basal ganglia disorders with emphasis on menstrual blood transplantation in stroke. Neurosci Biobehav Rev 36: 177–190.
17. Borlongan CV, Tajima Y, Trojanowski JQ, Lee VM, Sanberg PR (1998) Transplantation of cryopreserved human embryonal carcinoma-derived neurons (NT2N cells) promotes functional recovery in ischemic rats. Exp Neurol 149: 310–321.
18. Nelson PT, Kondziolka D, Wechsler L, Goldstein S, Gebel J, et al. (2002) Clonal human (hNT) neuron grafts for stroke therapy: neuropathology in a patient 27 months after implantation. Am J Pathol 160: 1201–1206.
19. Stroemer P, Patel S, Hope A, Oliveira C, Pollock K, et al. (2009) The neural stem cell line CTX0E03 promotes behavioral recovery and endogenous neurogenesis after experimental stroke in a dose-dependent fashion. Neurorehabil Neural Repair 23: 895–909.
20. Sinden JD, Vishnubhatla I, Muir KW (2012) Prospects for stem cell-derived therapy in stroke. Prog Brain Res 201: 119–167.
21. Teng YD, Benn SC, Kalkanis SN, Shefner JM, Onario RC, et al. (2012) Multimodal actions of neural stem cells in a mouse model of ALS: a meta-analysis. Sci Transl Med 4: 165ra164.
22. Teng YD, Yu D, Ropper AE, Li J, Kabatas S, et al. (2011) Functional multipotency of stem cells: a conceptual review of neurotrophic factor-based

evidence and its role in translational research. Curr Neuropharmacol 9: 574–585.

23. Xu L, Shen P, Hazel T, Johe K, Koliatsos VE (2011) Dual transplantation of human neural stem cells into cervical and lumbar cord ameliorates motor neuron disease in SOD1 transgenic rats. Neurosci Lett 494: 222–226.
24. Xu L, Yan J, Chen D, Welsh AM, Hazel T, et al. (2006) Human neural stem cell grafts ameliorate motor neuron disease in SOD-1 transgenic rats. Transplantation 82: 865–875.
25. Xu L, Ryugo DK, Pongstaporn T, Johe K, Koliatsos VE (2009) Human neural stem cell grafts in the spinal cord of SOD1 transgenic rats: differentiation and structural integration into the segmental motor circuitry. J Comp Neurol 514: 297–309.
26. Hefferan MP, Galik J, Kakinohana O, Sekerkova G, Santucci C, et al. (2012) Human neural stem cell replacement therapy for amyotrophic lateral sclerosis by spinal transplantation. PLoS One 7: e42614.
27. van Gorp S, Leerink M, Kakinohana O, Platoshyn O, Santucci C, et al. (2013) Amelioration of motor/sensory dysfunction and spasticity in a rat model of acute lumbar spinal cord injury by human neural stem cell transplantation. Stem Cell Res Ther 4: 57.
28. Cizkova D, Kakinohana O, Kucharova K, Marsala S, Johe K, et al. (2007) Functional recovery in rats with ischemic paraplegia after spinal grafting of human spinal stem cells. Neuroscience 147: 546–560.
29. Raore B, Federici T, Taub J, Wu MC, Riley J, et al. (2011) Cervical multilevel intraspinal stem cell therapy: assessment of surgical risks in Gottingen minipigs. Spine (Phila Pa 1976) 36: E164–171.
30. Usvald D, Vodicka P, Hlucilova J, Prochazka R, Motlik J, et al. (2010) Analysis of dosing regimen and reproducibility of intraspinal grafting of human spinal stem cells in immunosuppressed minipigs. Cell Transplant 19: 1103–1122.
31. Yan J, Xu L, Welsh AM, Hatfield G, Hazel T, et al. (2007) Extensive neuronal differentiation of human neural stem cell grafts in adult rat spinal cord. PLoS Med 4: e39.
32. Lunn JS, Pacut C, Backus C, Hong Y, Johe K, et al. (2010) The pleotrophic effects of insulin-like growth factor-I on human spinal cord neural progenitor cells. Stem Cells Dev 19: 1983–1993.
33. Guo X, Johe K, Molnar P, Davis H, Hickman J (2010) Characterization of a human fetal spinal cord stem cell line, NSI-566RSC, and its induction to functional motoneurons. J Tissue Eng Regen Med 4: 181–193.
34. Yan J, Xu L, Welsh AM, Chen D, Hazel T, et al. (2006) Combined immunosuppressive agents or CD4 antibodies prolong survival of human neural stem cell grafts and improve disease outcomes in amyotrophic lateral sclerosis transgenic mice. Stem Cells 24: 1976–1985.
35. Glass JD, Boulis NM, Johe K, Rutkove SB, Federici T, et al. (2012) Lumbar intraspinal injection of neural stem cells in patients with amyotrophic lateral sclerosis: results of a phase I trial in 12 patients. Stem Cells 30: 1144–1151.
36. Garbuzova-Davis S, Rodrigues MC, Hernandez-Ontiveros DG, Tajiri N, Frisina-Deyo A, et al. (2013) Blood-brain barrier alterations provide evidence of subacute diaschisis in an ischemic stroke rat model. PLoS One 8: e63553.
37. Tajiri N, Acosta S, Glover LE, Bickford PC, Jacotte Simancas A, et al. (2012) Intravenous grafts of amniotic fluid-derived stem cells induce endogenous cell proliferation and attenuate behavioral deficits in ischemic stroke rats. PLoS One 7: e43779.
38. Borlongan CV, Kaneko Y, Maki M, Yu SJ, Ali M, et al. (2010) Menstrual blood cells display stem cell-like phenotypic markers and exert neuroprotection following transplantation in experimental stroke. Stem Cells Dev 19: 439–452.
39. Matsukawa N, Yasuhara T, Hara K, Xu L, Maki M, et al. (2009) Therapeutic targets and limits of minocycline neuroprotection in experimental ischemic stroke. BMC Neurosci 10: 126.
40. Borlongan CV, Hayashi T, Oeltgen PR, Su TP, Wang Y (2009) Hibernation-like state induced by an opioid peptide protects against experimental stroke. BMC Biol 7: 31.
41. Paxinos G, Watson C (2007) The Rat Brain in Stereotaxic Coordinates: Hard Cover Edition: Academic press.
42. Yasuhara T, Matsukawa N, Hara K, Maki M, Ali MM, et al. (2009) Notch-induced rat and human bone marrow stromal cell grafts reduce ischemic cell loss and ameliorate behavioral deficits in chronic stroke animals. Stem Cells Dev 18: 1501–1514.
43. Dailey T, Eve DJ, Tajiri N, Lau T, Mosley YI, van Loveren H, Sanberg PR, Kaneko Y, Borlongan CV. (2013) Different sources of stem cells for transplantation therapy in stroke. In: *Cell-Based Therapies in Stroke*. Jolkkonen J, Walczak P (eds). Springer. 29–46.
44. Borlongan CV, Yu G, Matsukawa N, Xu L, Hess DC, et al. (2005) Acute functional effects of cyclosporine-A and methylprednisolone treatment in adult rats exposed to transient ischemic stroke. Life Sci 76: 1503–1512.
45. Osman MM, Lulic D, Glover L, Stahl CE, Lau T, et al. (2011) Cyclosporine-A as a neuroprotective agent against stroke: its translation from laboratory research to clinical application. Neuropeptides 45: 359–368.
46. Borlongan CV, Tajima Y, Trojanowski JQ, Lee VM, Sanberg PR (1998) Transplantation of cryopreserved human embryonal carcinoma-derived neurons (NT2N cells) promotes functional recovery in ischemic rats. Exp Neurol 149: 310–321.
47. Mays RW, Borlongan CV, Yasuhara T, Hara K, Maki M, et al. (2010) Development of an allogeneic adherent stem cell therapy for treatment of ischemic stroke. J Exp Stroke Transl Med 3: 34–46.
48. Chen J, Shehadah A, Pal A, Zacharek A, Cui X, et al. (2013) Neuroprotective effect of human placenta-derived cell treatment of stroke in rats. Cell Transplant 22: 871–879.
49. Yang B, Xi X, Aronowski J, Savitz SI (2012) Ischemic stroke may activate bone marrow mononuclear cells to enhance recovery after stroke. Stem Cells Dev 21: 3332–3340.
50. Borlongan CV (2009) Cell therapy for stroke: remaining issues to address before embarking on clinical trials. Stroke 40: S146–148.
51. Savitz SI (2013) Cell therapies: careful translation from animals to patients. Stroke 44: S107–109.
52. Sinden JD, Muir KW (2012) Stem cells in stroke treatment: the promise and the challenges. Int J Stroke 7: 426–434.
53. Pollock K, Stroemer P, Patel S, Stevanato L, Hope A, et al. (2006) A conditionally immortal clonal stem cell line from human cortical neuroepithelium for the treatment of ischemic stroke. Exp Neurol 199: 143–155.
54. Guzman R, De Los Angeles A, Cheshier S, Choi R, Hoang S, et al. (2008) Intracarotid injection of fluorescence activated cell-sorted CD49d-positive neural stem cells improves targeted cell delivery and behavior after stroke in a mouse stroke model. Stroke 39: 1300–1306.

6

Ontological Differences in First Compared to Third Trimester Human Fetal Placental Chorionic Stem Cells

Gemma N. Jones[1], Dafni Moschidou[1], Tamara-Isabel Puga-Iglesias[1], Katarzyna Kuleszewicz[1], Maximilien Vanleene[2], Sandra J. Shefelbine[2], George Bou-Gharios[3], Nicholas M. Fisk[4], Anna L. David[5], Paolo De Coppi[6], Pascale V. Guillot[1]*

1 Institute of Reproductive and Developmental Biology, Imperial College London, London, United Kingdom, **2** Department of Bioengineering, Imperial College London, London, United Kingdom, **3** Kennedy Institute of Rheumatology, University of Oxford, London, United Kingdom, **4** UQ Centre for Clinical Research, University of Queensland, Brisbane, Queensland, Australia, **5** Prenatal Cell and Gene Therapy Group, Institute for Women's Health, University College London, London, United Kingdom, **6** Surgery Unit, UCL Institute of Child Health, London, United Kingdom

Abstract

Human mesenchymal stromal/stem cells (MSC) isolated from fetal tissues hold promise for use in tissue engineering applications and cell-based therapies, but their collection is restricted ethically and technically. In contrast, the placenta is a potential source of readily-obtainable stem cells throughout pregnancy. In fetal tissues, early gestational stem cells are known to have advantageous characteristics over neonatal and adult stem cells. Accordingly, we investigated whether early fetal placental chorionic stem cells (e-CSC) were physiologically superior to their late gestation fetal chorionic counterparts (l-CSC). We showed that e-CSC shared a common phenotype with l-CSC, differentiating down the osteogenic, adipogenic and neurogenic pathways, and containing a subset of cells endogenously expressing *NANOG, SOX2, c-MYC*, and *KLF4*, as well as an array of genes expressed in pluripotent stem cells and primordial germ cells, including *CD24, NANOG, SSEA4, SSEA3, TRA-1-60, TRA-1-81, STELLA, FRAGILIS, NANOS3, DAZL* and *SSEA1*. However, we showed that e-CSC have characteristics of an earlier state of stemness compared to l-CSC, such as smaller size, faster kinetics, uniquely expressing *OCT4A* variant 1 and showing higher levels of expression of *NANOG, SOX2, c-MYC* and *KLF4* than l-CSC. Furthermore e-CSC, but not l-CSC, formed embryoid bodies containing cells from the three germ layer lineages. Finally, we showed that e-CSC demonstrate higher tissue repair *in vivo*; when transplanted in the *osteogenesis imperfecta* mice, e-CSC, but not l-CSC increased bone quality and plasticity; and when applied to a skin wound, e-CSC, but not l-CSC, accelerated healing compared to controls. Our results provide insight into the ontogeny of the stemness phenotype during fetal development and suggest that the more primitive characteristics of early compared to late gestation fetal chorionic stem cells may be translationally advantageous.

Editor: Xing-Ming Shi, Georgia Health Sciences University, United States of America

Funding: This research was funded by Genesis Research Trust, Henry Smith Charity and Action Medical Research. GNJ was supported by the Medical Research Council. DM was supported by Kidney Research United Kingdom. KK was supported by Biotechnology and Biological Science Research Council. PDC was supported by Great Ormond Street Hospital Children's Charity. ALD is supported by the National Institute for Health Research (NIHR) through a Senior Lectureship and the Department of Health's NIHR Biomedical Research Centers funding scheme. The funders had no role in study design, data collection and analysis, decision to publish, or preparation of the manuscript.

Competing Interests: The authors have declared that no competing interests exist.

* E-mail: Pascale.Guillot@imperial.ac.uk

Introduction

Mesenchymal stromal/stem cells (MSC), isolated from a range of adult and fetal tissues, have generated substantial interest for use in cell therapy and tissue engineering due to their ability to migrate to sites of injury and regenerate and repair damaged tissues [1,2,3]. We [4] and others [5,6,7] have shown that human first trimester fetal MSC present advantages for regenerative medicine over adult and perinatal MSC, such as faster kinetics, greater expansion potential, smaller size, unique adhesion molecule profile, greater telomerase activity and broader differentiation potential.

First trimester human fetal MSC have now been applied with significant effect in both hereditary and acquired disease paradigms. Human fetal blood MSC transplanted in a mouse model of osteogenesis imperfecta substantially reduced long bone fracture rates, with donor cells engrafting at sites of bone formation and differentiating into functional osteoblasts, which modified bone matrix and reduced bone brittleness [8,9]. Similarly transplantation of fetal blood MSC in a mouse model of collagen type 1 deficiency led to improvement of glomerulopathy [10]. Finally tissue-engineered bone grafts seeded with first trimester bone marrow MSC resulted in closure of critical-sized femoral defects in rats by promoting woven and compact bone formation [11].

Collection of fetal blood and somatic tissues in the first trimester, however is technically challenging and usually requires pregnancy termination, an obstacle to autologous applications [12]. For allogeneic applications, the placenta in contrast is larger and easier to separate than other fetal tissues following pregnancy termination and hence a more realistic candidate for cell banking. In terms of supply, first trimester suction termination of pregnancy is the commonest operation in the world, but remains ethically contentious as a cell source. However fetal MSC can also be isolated in ongoing pregnancies from surplus tissues obtained during routine prenatal diagnostic procedures such as chorionic villous sampling [13,14,15] and amniocentesis [16,17,18]. Although amniotic fluid stem cells have therapeutic potential in the clinic, they are restricted to harvest in the mid-trimester, whereas the earlier gestational age of chorionic stem cells harvested at 10–12 weeks should be preferable for autologous prenatal cell therapy within the window of fetal immunological naivety. First trimester chorionic stem cells may also be developmentally-advantageous for allogenic and postnatal autologous use and have already been expanded large-scale in human serum with substantially more success than second trimester amniotic fluid stem cells [7].

Although there are substantial reports characterizing amniotic fluid stem cells, the literature on chorion-derived stem cells is insufficient with limited characterisation of immuno-phenotype and a focus primarily on isolation techniques and identification of their vascular niche within the placenta [13,19,20,21,22]. In addition many studies of term chorionic cells report cells of both fetal and maternal origin [22,23], with some groups isolating pure maternal origin cells only [21,24,25,26], and others failing to exclude maternal cell origin [19,20,27,28,29,30]. In contrast work on first trimester chorionic stem cells has shown fetal origin by karyotyping [13], although not carried out in all studies [14,15,31]. Notwithstanding this problem of contaminating decidual maternal stem cells, comparisons of term placenta/chorionic stem cells with adult bone marrow MSC [22,25] have indicated that these cells may be more primitive than adult MSC with evidence of greater self renewal [26], expression of some pluripotency markers [20,27,29] and potential to differentiate beyond mesodermal lineages to hepatogenic [28] and neurogenic lineages [15,31].

We compared the phenotype of first trimester and term fetal placental chorionic stem cells (e-CSC and l-CSC respectively) to investigate their suitability for cell-based therapies and tissue engineering applications. Here, we show that compared to l-CSC, e-CSC are smaller cells with faster growth kinetics, and higher levels of pluripotency marker expression. We also found that e-CSC uniquely expressed *OCT4A variant 1* and had potential to differentiate into lineages of the three germ layers *in vitro*. In addition e-CSC and l-CSC express markers associated with primordial germ cells (PGC) and thus may share a developmental origin with these cells. Finally, we showed that e-CSC demonstrate higher tissue repair *in vivo*. Together our results suggest that e-CSC are more primitive than l-CSC and give insight into the developmental ontogeny of fetal chorionic stem cells.

Results

e-CSC and l-CSC are of fetal origin

Both e-CSC and l-CSC were isolated based on plastic adherence and cultured in DMEM supplemented with 10% FBS. Both populations were c-KIT$^+$ (Fig. 1a), grew in monolayers and showed a spindle shaped cytoplasm morphology, (Fig. 1b). The fetal origin of the cells was confirmed by the presence of the *SRY* gene in male samples (Fig. 1c). Fluorescence *in situ* hybridisation (FISH) analysis showed that all nuclei contained the X and Y chromosome in both e-CSC and l-CSC populations, identifying cells as being of fetal origin, with no contaminating maternal cells (Fig. 1d).

e-CSC are smaller in size with faster growth kinetics than l-CSC

On average e-CSC were smaller in size than l-CSC (16.5 μm±0.1 s.e.m *vs.* 24.0 μm±0.1, P<0.01; Fig. 2a and 2b). Kinetic analysis revealed that e-CSC grew faster than l-CSC (population doubling time 48.6 h±3.7 s.e.m *vs.* 120.7 h±27.2, P<0.05; Fig. 2c) over a longer period of time when maintained at sub-confluence (17 *vs.* 11 passages without reduced growth; Fig. 2d) and generated greater cell numbers when grown beyond confluence (5.0 PD±0.6 s.e.m *vs.* 2.3 PD±1.3 over 288 hours, P<0.05; Fig. 2e).

e-CSC and l-CSC have osteogenic and adipogenic differentiation capacity

The e-CSC population showed higher osteogenic potential than l-CSC, as shown qualitatively by increased intensity of alizarin red and von kossa staining after culture in osteogenic permissive conditions for 14 days (Fig. 3a). We used quantitative real time PCR (primers in Table S2) to semi-quantify the upregulation of late osteogenic genes and bone matrix related genes in e-CSC and l-CSC after 2 weeks *in vitro* differentiation. Relative expression of each gene was normalized by its expression in non-differentiating growth medium (basal level) after normalization to *GAPDH* ($2^{-\Delta\Delta C_t}$). Results showed only differentiated e-CSC expressed osteocalcin (*OC*; 5.2%±4.6 s.e.m) and differentiated e-CSC compared to l-CSC had higher expression of osteopontin (*OP*; 17.5%±5.6 vs. 0.3%±0.3, P<0.05), bone sialo-protein-II (*BSP*; 67.8%±5.0 vs. 9.7%±11.1, P<0.05) and bone morphogenic protein-2 (*BMP2*; 7.5%±1.2 vs. 0.5%±0.3, P<0.01) (Fig. 3b). At the protein level, collagen type I (COL1) was absent in non-differentiated cells, but present in differentiated e-CSC and l-CSC samples, with e-CSC showing 3.2 times more protein (Fig. 3c).

Culture of the cells in adipogenic permissive medium for three weeks demonstrated that e-CSC produced considerably more Oil Red O-positive lipid vacuoles than l-CSC (Fig. 3d). Quantification of adipocyte differentiation regulator peroxisome proliferator-activated receptor gamma (*PPARγ*) using real time PCR analysis revealed greater expression in differentiated e-CSC than l-CSC (P<0.05) (Fig. 3e). Together, these results illustrate the higher adipogenic potential of e-CSC over l-CSC.

e-CSC and l-CSC differentiate along the neurogenic but not the hepatogenic lineages

We next investigated the capacity of e-CSC and l-CSC to differentiate into non-mesodermal lineages. When co-cultured with C17.2 mouse neural progenitor cells in chemically defined medium supplemented with Baicalin for 5 days, e-CSC and l-CSC changed morphology and expressed the neuronal markers β-tubulin, the microtubule-associated protein 2 (MAP2) and the NMDA receptor NR1 (Fig. 3f, positive control in Fig. S2a), indicating that both cell populations have the potential to differentiate into neural-like cells.

However, culture of e-CSC and l-CSC in hepatic-permissive medium supplemented with hepatocyte growth factor over 21 days failed to induce expression of the hepatic markers albumin and alpha-fetoprotein (*AFP*) as visualized by confocal immunostaining (Fig. S1a), with the cells not producing urea in the culture medium (Fig. S1b).

e-CSC and l-CSC share a common immunophenotype

Both e-CSC and l-CSC share a largely common adhesion molecule profile, expressing *CD11a*, *CD106*, *CD51/61*, *CXCR4*, integrin β3, *CD49f*, *CD49d* and αV integrin. However, only e-CSC expressed *CD62P* and integrin β7 and, when compared to l-CSC, e-CSC showed higher levels of *CD49b* (50.1%±3.9 s.e.m vs. 33.0%±4.5, P<0.05) and *CD49e* (81.1%±1.1 vs. 67.4%±4.2, P<0.05) (Fig. 4a). Adhesion assays showed that both e-CSC and l-

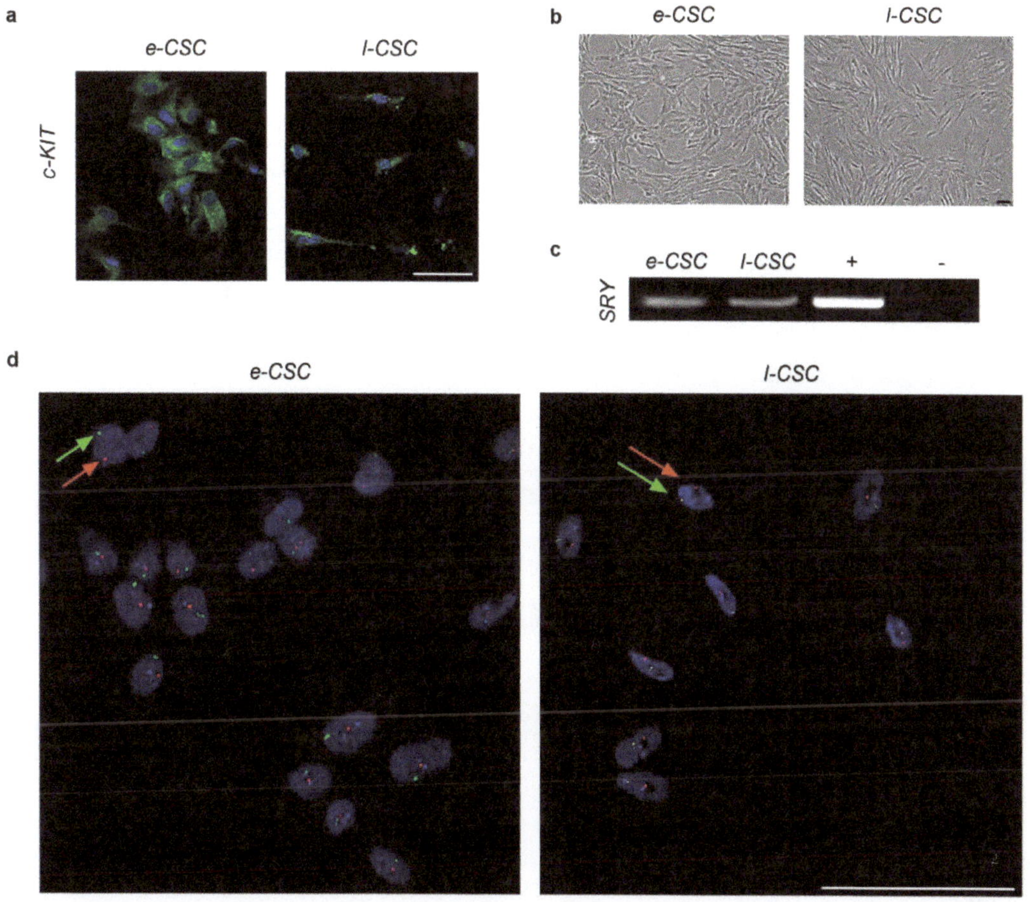

Figure 1. e-CSC and l-CSC are of fetal origin. (**a**) Confocal immuno-fluorescence for c-KIT (**b**) Cell morphology of early and late gestation chorionic stem cells (e-CSC and l-CSC respectively) passage 4–5. (**c**) PCR for Y chromosome specific SRY. Positive and negative controls shown. (**d**) FISH analysis for X (FITC) and Y (Texas Red) chromosomes indicated with green and red arrows respectively. 100 cells were counted. All scale bars 100 μm.

CSC exhibited equal adherence to human fibronectin α-chymotryptic 40 kDa (7.5%±0.7 s.e.m vs. 9.6%±0.4) and 120 kDa fragments (30.3%±9.3 vs. 32.6%±6.6), human plasma fibronectin (34.2%±1.0 vs. 48.2%±12.2), collagen type I (38.8%±8.1 vs. 45.9%±5.4), collagen type IV (42.6%±9.7 vs. 36.1%±5.6), human placenta laminin (39.8%±6.7 vs. 48.1%±15.8) and human VCAM-1 (42.9%±9.4 vs. 18.2%±5.8) (Fig. 4b).

Both e-CSC and l-CSC share characteristics with MSC [32] with absence of endothelial (*CD14*) and hematopoietic markers (*CD34* and *CD45*), and expression of MSC-associated markers *CD105*, *CD73*, *CD44* and the matrix protein vimentin as well as *CD29* and *CD90*, although *CD29* and *CD90* are also found in human embryonic stem cells (hESC) [33,34] (Fig. 4c). Positive controls are shown in Fig. S2b.

e-CSC uniquely express *OCT4A* variant 1 and show higher levels of pluripotency-associated markers

To avoid false positive detection of *OCT4* pseudogenes not involved in maintaining pluripotency [35,36,37,38], RT-PCR was performed with *OCT4A* variant 1 specific primers [35] and results showed expression in e-CSC, but not l-CSC (Fig. 5a). However, expression of the other variants of *OCT4A* was shown in both e-CSC and l-CSC when primers that also detected *OCT4A* variants 2 and 3 were used (Fig. 5b). Transcript levels of the other reprogramming factors *SOX2*, *c-MYC*, *KLF4* as well as *NANOG* were subsequently quantified and showed expression levels were considerably lower relative to hESC for *SOX2* (8.5%±1.0 s.e.m and 4.6%±0.5; Fig. 5b) and *NANOG* (2.5%±0.1 and 0.6%±0.1; Fig. 5c). However, expression levels were markedly higher in both e-CSC and l-CSC respectively relative to those found in hESC for *c-MYC* (2,472%±785.9 vs. 556.4%±106.7; Fig. 5d) and *KLF4* (2,906%±277.3 vs. 1,940%±305.8; Fig. 5e). When we compared e-CSC to l-CSC, all five pluripotency markers were expressed at greater relative levels in e-CSC; *OCT4A* ($P<0.05$), *SOX2* ($P<0.05$), *c-MYC* ($P<0.05$), *KLF4* ($P<0.05$) and *NANOG* ($P<0.001$).

We next used flow cytometry to determine whether lower expression in l-CSC was associated with fewer positive cells in the population than e-CSC. However, both populations contained a similar sized subset of CD24 positive cells (34.2%±3.5 s.e.m vs. 36.0%±10.0), a marker of undifferentiated hESC. In addition CD24 is not found expressed in MSC and hence we have not referred to the chorionic stem cells as MSC [33,39]. As well as CD24, a subpopulation of cells positive for OCT4A (63.1%±5.1 s.e.m vs. 70.2%±6.6) were identified in e-CSC and l-CSC respectively (Fig. 5b).

Both e-CSC and l-CSC also contained a similar sized subpopulation of cells expressing other markers typical of undifferentiated hESC [40,41]; SOX2 (62.2%±4.2 s.e.m vs. 66.1%±3.3), c-MYC (52.3%±11.2 vs. 61.6%±11.4), SSEA4

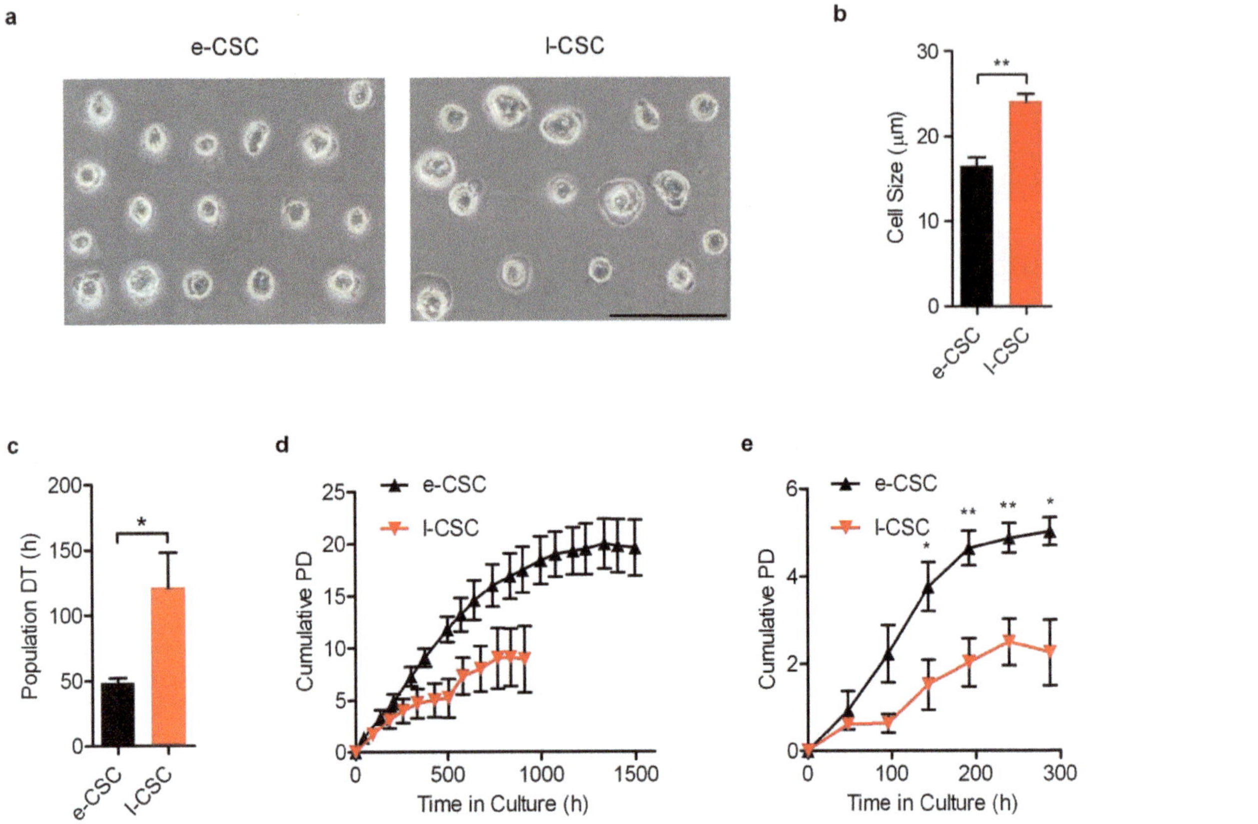

Figure 2. e-CSC are smaller in size and grow faster than l-CSC. (**a**) Image of cells in suspension used for cell size analysis. (**b**) Average cell size in µm of e-CSC (black) and l-CSC (red) when in suspension. (**c**) Average growth rate in hours taken for cell population to double during exponential growth phase; i.e. population doubling time (DT). e-CSC (black), l-CSC (red). (**d**) Cell expansion capacity over 1500 hours measured by average cumulative population doublings (Cumulative PD) of e-CSC (▲) and l-CSC (▼) when passaged at sub-confluence. (**e**) Cell kinetics measured by average cumulative population doublings of e-CSC (▲) and l-CSC (▼) when seeded at low density and grown beyond confluence for 288 hours. * $P<0.05$; ** $P<0.01$, Student's *t* test, $n=3$ different samples per cell group. Mean ± s.e.m. All scale bars 100 µm.

(83.1%±8.2 vs. 69.6%±22.3), SSEA3 (63.7%±6.6 vs. 66.1%±3.3), TRA-1-60 (59.4%±10.6 vs. 62.8%±12.1), TRA-1-81 (61.8%±3.2 vs. 62.8%±12.1) (Fig. 6a). Expression and heterogeneity of the population for OCT4A, SOX2, c-MYC, SSEA4, SSEA3, TRA-1-60, and TRA-1-81 was confirmed by confocal immuno-fluorescence. Both e-CSC and l-CSC also contained a subset of cells expressing KLF4, NANOG and REX1 (Fig. 6b). Positive controls are shown in Fig. S2c.

e-CSC uniquely form embryoid bodies

To investigate if endogenous expression of pluripotency markers translates to an ability to differentiate beyond mesodermal lineages, we tested the capacity of e-CSC and l-CSC to form embryoid bodies (EBs) *in vitro* with differentiation to cells from the three germ layers. We showed that only e-CSC, but not l-CSC, formed EBs when grown in EB-permissive media, which may be linked to lack of *OCT4A* variant 1 expression in l-CSC. The e-CSC EBs began to form at 3 days and by 12 days were larger in size, more complex and compact in morphology and presented a defined peripheral lining (Fig. 7a). EBs from e-CSC expressed differentiation markers representative of all three germ layers: Nestin and *PAX6* (ectoderm), *CK3*, *CK19* and *GATA6* (endoderm), *BMP4* (mesoderm), and *SYCP1* (testis) with down regulation of undifferentiated stem cell marker *NANOG* (Fig.7b). This was confirmed by confocal microscopy for Nestin and MAP2 (ectoderm), GATA4 (mesoderm), AFP (endoderm) and laminin, which was expressed in the peripheral lining of the EBs (Fig. 7c). Positive controls are shown in Fig. S2d.

e-CSC and l-CSC express markers found in primordial germ cells

We next investigated the ontogeny of e-CSC and l-CSC populations. At the transcript level both e-CSC and l-CSC expressed markers found in primordial germ cells (PGC). These included *c-KIT*, a regulator of PGC survival and growth [42,43], which is expressed in PGC, neural crest stem cells and haematopoietic stem cells during embryogenesis [44]. However, e-CSC and l-CSC also expressed other markers which commonly define PGC: Stella, which is thought to maintain pluripotentiality [45,46], Fragilis and Nanos3, which have roles in PGC migration [47,48] as well as SSEA1, an early marker found in PGC and human embryonic germ (hEG) cells but not hESC [49,50] (Fig. 8a). Immuno-fluorescence and flow cytometry supported these findings (positive controls are shown in Fig. S3a). Flow cytometry showed at the protein level a subpopulation of cells in both e-CSC and l-CSC respectively positive for STELLA (29%±14 and 27%±4), FRAGILIS (32%±5 and 38%±12), NANOS3 (25%±9 and 34%±10), SSEA1 (61%±1 and 53%±4), as well as DAZL (30%±10 and 41%±12) and PUM2 (30%±8 and 33%±14), which colocalise in hEG cells [51] (Fig. 8b) Representative flow cytometry images are shown in Fig. S2b. Whereas VASA, a marker uniquely specific to the germ cell lineage [52], was only

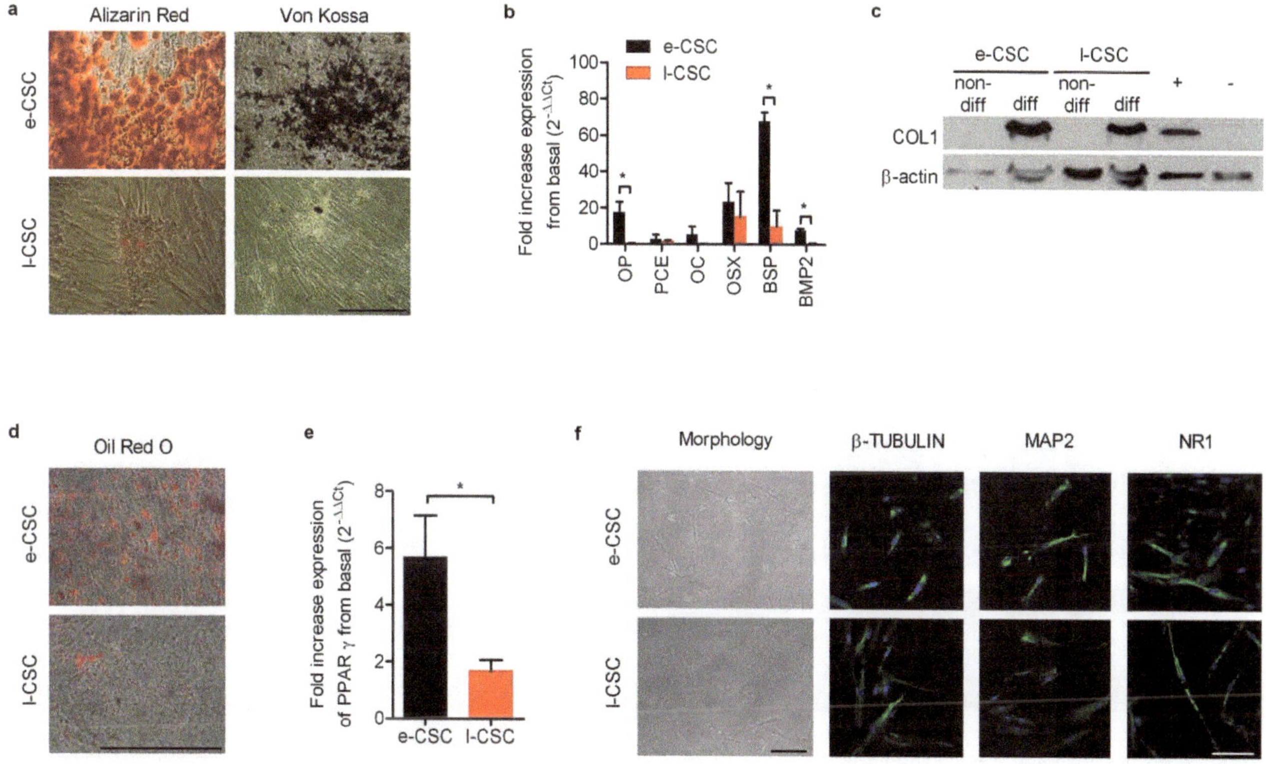

Figure 3. Osteogenic, adipogenic and neurogenic differentiation of e-CSC and l-CSC. (**a**) Alizarin red staining (calcium deposits) and von kossa staining (mineralisation) of cells grown in osteogenic permissive media for 2 weeks. (**b**) Quantitative real time PCR of fold increased expression of osteogenic genes after osteogenic differentiation; osteopontin (OP), procollagen endopeptidase enhancer (PCE), osteocalcin (OC), osterix (OSX), bone sialo-protein-II (BSP) and bone morphogenic protein-2 (BMP2). (**c**) Western blot of COL1 for e-CSC and l-CSC not differentiated (non-diff) and differentiated to bone for 2 weeks (diff). Positive and negative controls shown. Loading control is β-actin. (**d**) Oil Red O staining (lipid droplets) of e-CSC and l-CSC grown in adipogenic permissive media for 3 weeks. (**e**) Quantitative real time PCR of fold increased expression of adipocyte differentiation regulator peroxisome proliferator-activated receptor gamma (PPARγ). (**f**) Morphology and expression of neuronal markers β-TUBULIN, MAP2 and NMDA receptor NR1 shown with FITC (green) of e-CSC and l-CSC after 5 days neurogenic differentiation. Nuclei stained with DAPI (blue). All quantitative real time PCR data normalised to GAPDH and basal expression levels of differentiation genes (i.e. $2^{-\Delta\Delta Ct}$). e-CSC (black) and l-CSC (red). Data. * $P<0.05$; ** $P<0.01$, Student's *t* test, $n=3$ per cell group. Mean ± s.e.m. All scale bars 100 μm.

detected in e-CSC and not l-CSC at the transcript level (Fig. 8b), flow cytometry showed a small proportion of cells expressed VASA in both e-CSC and l-CSC (25%±10 and 17%±5 respectively). Also only a small proportion of e-CSC and l-CSC respectively expressed TNAP, used to denote early PGC [53,54] (20%±12 and 22%±14) and BLIMP1, not found in hEG cells [55], but a key marker in PGC that prevents dedifferentiation into pluripotent stem cells [56] (13%±1 and 15%±1). Down-regulation of Blimp1 in e-CSC and l-CSC may be responsible for the high expression levels of its targets *c-MYC* and *KLF4*, which were at least five times higher than expressed by hESC (Fig.5). PGC reside in extra-embryonic tissues prior to migration along the genital ridge where they undergo epigenetic reprogramming including erasure of methylation on *H19* [57,58]. To determine to which developmental stage of PGC e-CSC and l-CSC are most similar, we investigated the methylation status of *H19*. We found *H19* was both hypermethylated and unmethylated in both e-CSC (50%±0 hypermethylated and 50%±0 unmethylated, n = 3) and l-CSC (70%±8 hypermethylated and 30%±8 unmethylated, n = 3) (Fig. 8c), indicating that e-CSC and l-CSC may contain cells or primordial origin which have been retained in placental tissue prior to the start of genital ridge migration, at a time when imprint erasure has not yet started.

e-CSC showed greater tissue repair capacity *in vivo* compared to l-CSC

We first compared the capacity of e-CSC and l-CSC to improve bone pathology in a mouse model of osteogenesis imperfecta (*oim* mice). *Oim* mice suffer from bone brittleness in response to a mutation in the collagen Type I alpha 2 gene [8,9]; e-CSC and l-CSC (10^6 cells in PBS) were transplanted intraperitoneally into *oim* neonates (n = 15 per transplanted group and n = 30 for non-transplanted *oim*) and culled 8 weeks later for analysis. Transplantation of e-CSC led to an improvement of overall bone quality, i.e. total work (0.90 mJ±0.09 s.e.m for non-transplanted *oim* mice vs. 1.86 mJ±0.42 for e-CSC-transplanted *oim* mice, $P<0.01$) (Fig. 9a) and bone plasticity, i.e. work from yield to fracture (0.62 mJ±0.08 s.e.m for non-transplanted *oim* mice vs. 1.54 mJ±0.40 for e-CSC-transplanted *oim* mice, $P<0.01$) (Fig. 9b), whilst *oim* mice transplanted with l-CSC did not show any significant improvements (1.16 mJ±0.14 for total work, and 0.89 mJ±0.12 for work from yield to fracture) (Fig. 9a and 9b). The differences in bone parameters between *oim* mice transplanted with e-CSC and l-CSC were likely to result from differences in intrinsic donor cell properties, as we observed no difference in donor cell engraftment between e-CSC and l-CSC for both bone marrow (BM) and bone (Fig. S4).

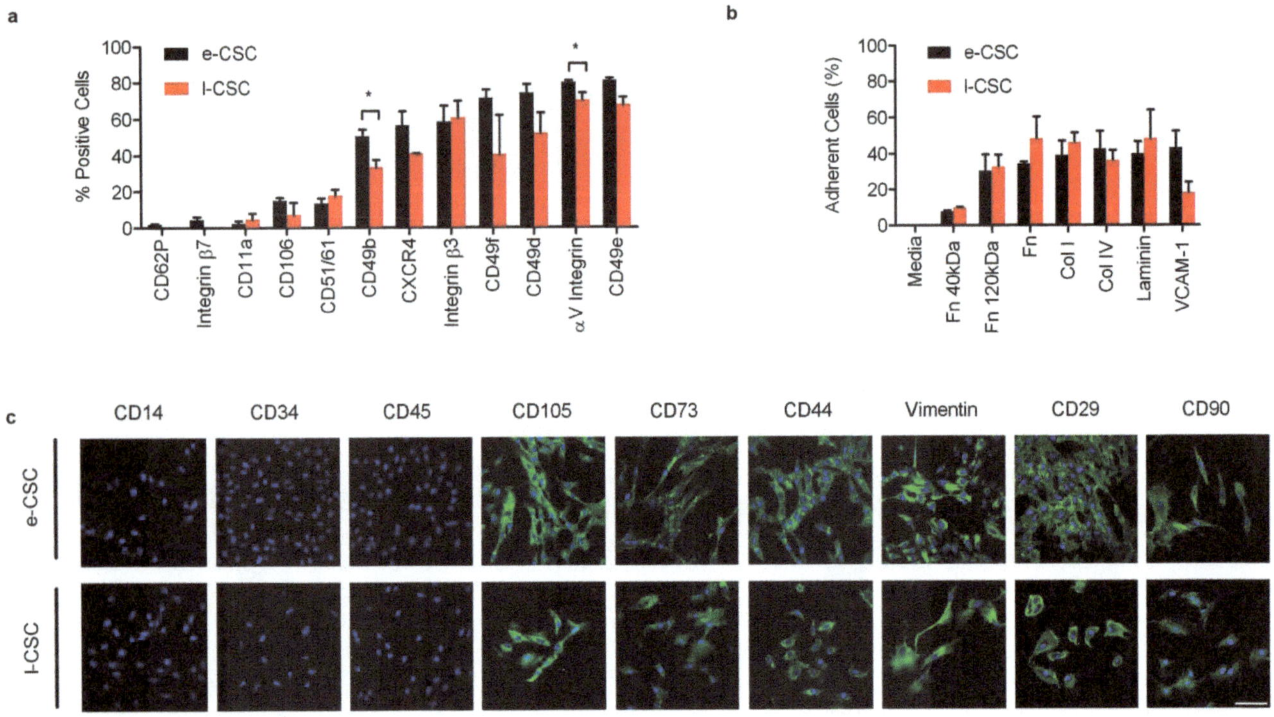

Figure 4. e-CSC and l-CSC have a similar immunophenotype. (**a**) Flow cytometry for percent of e-CSC (black) and l-CSC (red) populations positive for surface adhesion markers; CD62P, Integrin β7, CD11a, CD106, CD51/61, CD49b, CXCR4, Integrin β3, CD49f, CD49d, αV Integrin and CD49e. (**b**) Percentage of e-CSC (black) and l-CSC (red) populations that adhered to fibronectin a-chemotryptic 40 kDa (Fn 40 kDa), fibronectin a-chemotryptic 120 kDa (Fn 120 kDa), whole fibronectin (Fn), collagen I (col I), collagen IV (col IV), human placenta laminin (laminin) and vascular cell adhesion molecule-1 (VCAM-1). Negative control DMEM alone (media). Data. * $P<0.05$, Student's *t* test, $n=3$ per cell group. Mean ± s.e.m. (**c**) Confocal immuno-fluorescence for endothelial marker (CD14), hematopoietic markers (CD34 and CD45), MSC-associated markers (CD105, CD73 and CD44), matrix protein (vimentin) and markers found in pluripotent cells as well as MSC (CD29 and CD90) stained with FITC (green). Nuclei stained with DAPI (blue). Scale bar 100 μm. Positive controls are shown in Fig. S3a.

We next compared the tissue repair capacity of e-CSC and l-CSC during the skin wound healing process. Full thickness dermal excision skin wounds were treated with local application of either e-CSC or l-CSC in PBS or PBS alone ($n=8$ per group). Morphological analysis of H&E sections of the wound beds seven days later showed that the healing process for the e-CSC-treated wounds was more advanced than for either PBS or l-CSC-treated wounds. It was evident by a lower number of infiltrated cells in the e-CSC-treated wounds compared to PBS-treated wounds (53 cells ± 3 s.e.m vs. 66 cells ± 4, respectively, calculated in a 100 μm^2 surface, $P<0.05$). In contrast, l-CSC-treated wounds did not significantly differ from the controls (70 cells ± 6) (Fig. 9c and 9d). Similarly, the epithelial thickness of the wound beds was thinner for the wounds covered with e-CSC compared to controls (46 μm ± 7 s.e.m vs. 76 μm ± 5, respectively, $P<0.01$), whilst l-CSC-treated wounds did not significantly differ from the controls (81 μm±11) (Fig. 9c and 9e).

Discussion

Identifying a safe, suitable and readily obtainable supply of therapeutic cells is a major challenge in regenerative medicine. Our work compares two potential candidates; first trimester and term human placental chorionic stem cells (e-CSC and l-CSC respectively), available from discarded tissues after first trimester chorionic villous sampling or termination of pregnancy, or after delivery. We derived pure populations of cells of fetal not maternal origin to show that e-CSC offer advantages relevant to the clinic when compared with their term counterparts.

Here we showed that e-CSC and l-CSC could differentiate down the osteogenic and adipogenic pathways, as well as neurogenic pathways, as previously reported [31], although they failed to differentiate down the hepatogenic lineage. Both populations shared a common immunophenotype and adhesion molecule profile, both expressing pluripotency markers as well as markers associated with PGC at the transcript and protein levels. However, e-CSC had a smaller size, higher kinetics and longer self-renewal potential, and showed greater osteogenic and adipogenic differentiation qualitatively at the whole population level. e-CSC also expressed pluripotency markers at higher levels than l-CSC, formed embryoid bodies *in vitro*, and expressed the variant 1 transcript of the *OCT4A* isoform, critical for pluripotency [35,36,37,38]. In addition, relevant to cell-based therapy, we showed that e-CSC have higher tissue repair potential *in vivo* when transplanted neonatally into *osteogenesis imperfecta* mice, or applied locally to a dermal excision skin wound. This might be due to superior lineage specific differentiation and/or greater paracrine effects of e-CSC compared to l-CSC. This suggests e-CSC are more primitive than l-CSC and we therefore postulate that the phenotype of stem cells isolated from placental chorionic tissue evolves during pregnancy, with a decrease in plasticity paralleling a down regulation of expression levels of early "stemness" genes.

Our findings parallel our previous work, which showed fetal bone marrow MSC in contrast to their adult counterparts were smaller, grew faster and senesced later, despite sharing a common

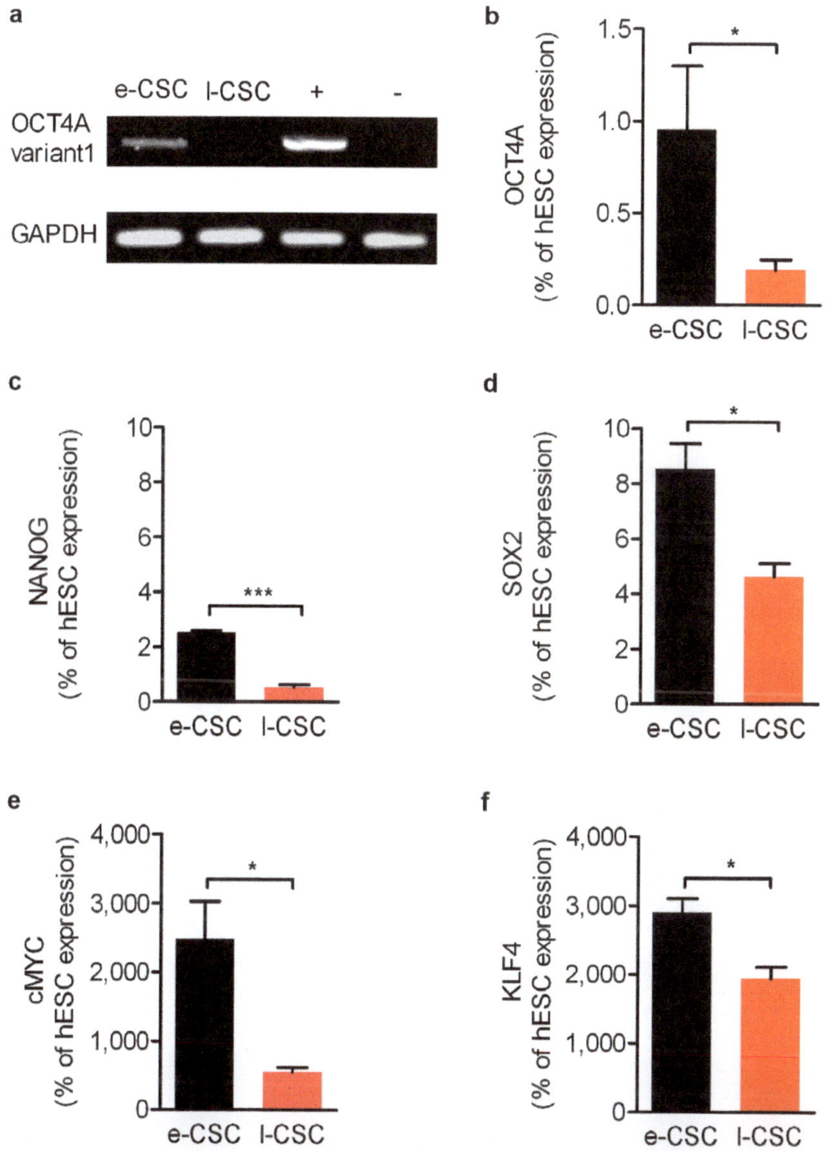

Figure 5. e-CSC have higher expression of pluripotency genes than l-CSC. (**a**) RT-PCR for OCT4A variant 1. GAPDH and positive and negative controls shown. (**b–f**) Quantitative real time PCR for (**b**) OCT4A all variants, (**c**) NANOG, (**d**) SOX2, (**e**) CMYC and (**f**) KLF4 for e-CSC (black) and l-CSC (red) as percentage expression of human embryonic stem cells (hESC). * $P<0.05$; *** $P<0.001$, Student's *t* test, $n=4$ per cell group. Mean ± s.e.m.

immunophenotype with expression of mesenchymal, adhesion and lineage markers in the whole population [4]. Others have previously reported that early placental MSC or e-CSC [13] have increased proliferative potential when compared to adult bone marrow MSC. Our data indicate that human fetal chorionic stem cells retain their immunophenotype with age, but lose associated primitive characteristics, not only from fetal to adult, but also from fetal to neonatal.

The absence of *OCT4A* variant 1 expression in l-CSC at the transcript level and downregulation of other pluripotency markers may be pivotal to the decline of primative characteristics seen in these cells when compared to their early gestational counterparts. *OCT4A* variant 1 transcript is expressed in the nuclei of undifferentiated hESC, embryonal carcinoma (EC) cells and PGC and is responsible for maintaining "stemness", whilst the other pseudogenes are found in the cytoplasm and have no known function [36,37,38]. In particular the lack of *OCT4A* variant 1 in l-CSC may explain the inability of l-CSC to form embryoid bodies (EBs) as *OCT4A* variant 1 is critical to retain pluripotentiality [35,36]. From a technical view, OCT4A was detected at the protein level in the nuclei of both e-CSC and l-CSC using OCT4A-specific antibodies; however the available antibodies may not be specific for variant 1, potentially also binding to variants 2 and 3 [35]. This was confirmed through investigation of the gene expression of all OCT4A variants, which were found expressed in both e-CSC and l-CSC. Therefore the difference between both populations is only at the level of *OCT4A* variant 1 expression.

Although *OCT4A* variant 1 was not detectable in l-CSC, other pluripotency associated markers were found in both e-CSC and l-CSC populations at both the transcript and protein level. Of particular, note both cell populations contained a similar proportion of cells expressing these markers, indicating down regulation with increased gestational age rather than a reduced number of cells expressing these early developmental markers. A

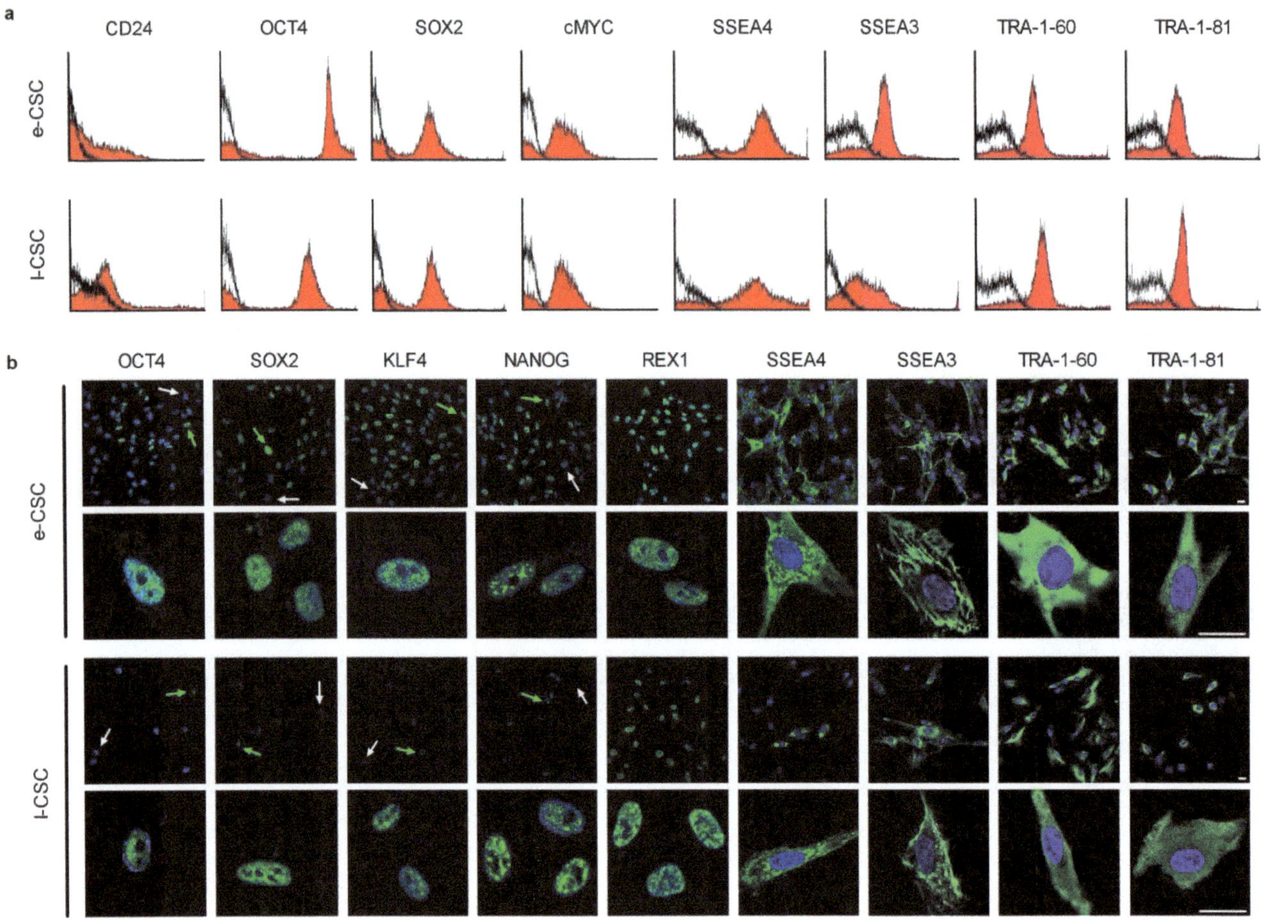

Figure 6. e-CSC and l-CSC contain a subpopulation positive for pluripotency markers. (**a**) Representative flow cytometry ($n=3$) for percent of cells positive for CD24, OCT4, SOX2, CMYC, SSEA4, SSEA3, TRA-1-60 and TRA-1-81 in e-CSC and l-CSC whole populations (isotype control in black). (**b**) Representative confocal immunofluorescence images for OCT4, SOX2, KLF4, NANOG, REX1, SSEA4, SSEA3, TRA-1-60 and TRA-1-81 stained with FITC (green). Positive cells indicated with green arrow, negative cells with white arrow. Nuclei stained with DAPI (blue). Scale bar 25 μm. Positive controls are shown in Fig. S3b.

similar result has been shown in amniotic epithelium stem cells, whereby expression of *NANOG* and *SOX2* was higher in first trimester amnion than term, although levels of *OCT4* were unchanged [6]. While there is little literature on this in human chorionic stem cells [59], work to date on whole chorion tissues indicates a down regulation of telomerase activity over gestation, supporting the decline of primative stem cell characteristics with age in this tissue [60].

Despite both e-CSC and l-CSC expressing early developmental markers, only e-CSC formed EBs, which we validated with typical morphology and expression of markers from the three germ layers. However, the lack of EB formation by l-CSC does not rule out the potential of these cells to differentiate beyond mesodermal lineages. Here we showed both e-CSC and l-CSC differentiated down the neurogenic lineage, but failed to differentiate down the hepatogenic lineage in our study. Others have shown stem cells isolated from both early and term placental tissue can form neural progenitors [13,15,31], cardiomyocytes [61,62,63], hepatocytes [28,64] and pancreatic cells [65]. Results in a number of studies, mainly those on term placental stem cells, may be confounded by contamination with or overgrowth of maternal/decidual cells [23], as verified in some by characterising the cells [21,22,24,25,26] and not tested for in others [14,15,19,20,27,28,29,30,31]. Here we confirmed cells were of fetal origin with no contaminating maternal cells.

The potential of fetal derived MSC for use in bone tissue engineering is promising [3] as is their use for cell therapy in bone related disorders such as brittle bone disease [8,66]. There is however, limited work on placenta-derived stem cells. Surbek's group showed both first trimester and term chorionic stem cells were capable of generating osteogenic grafts [67], indicating the potential of these cells in regenerative medicine. Furthermore, another study showed engraftment of term placental adherent cells inhibited growth of multiple myeloma in bone and stimulated bone formation in an immune-deficient model of medullary myeloma bone disease [68]. Notwithstanding this, neither study excluded a maternal origin for the cell populations used.

This work offers support for Kucia et al's [69,70] hypothesis that cells isolated from adult and fetal tissues that have pluripotency characteristics may share an origin with PGC. This has been shown for murine fetal liver and adult bone marrow derived very small embryonic-like (VSEL) cells [71], and may also be true for umbilical cord blood VSEL cells [72] as well as human fetal bone marrow and lung stem cells [73,74]. During fetal development PGC reside at the junction between embryonic and extra-embryonic tissues, then migrate through the primitive streak

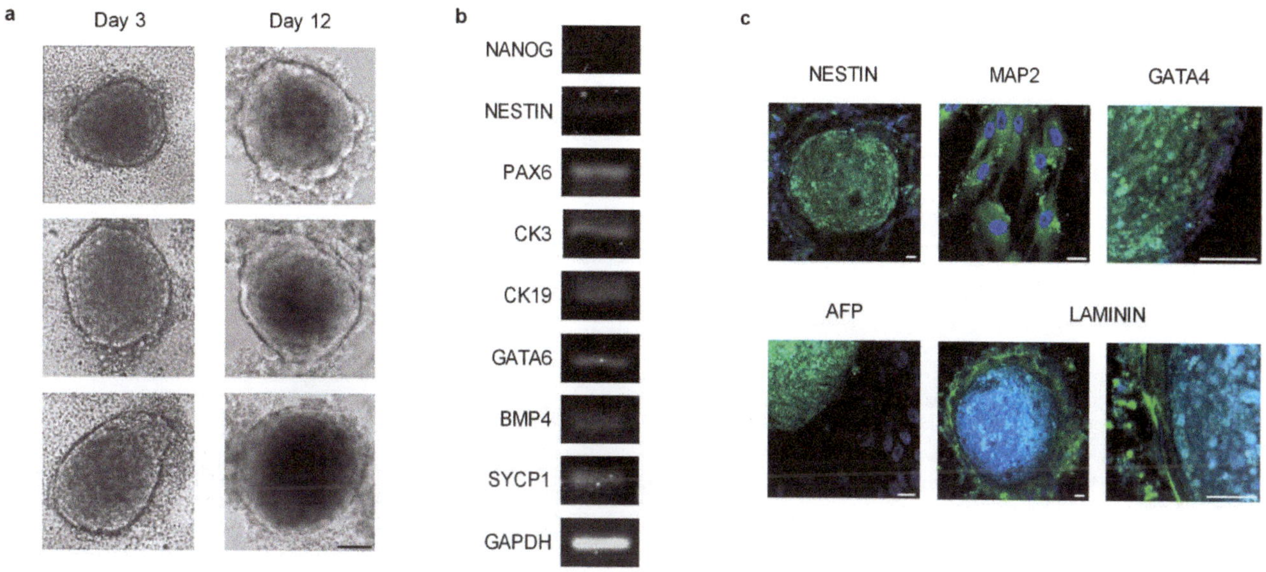

Figure 7. e-CSC uniquely form embryoid bodies (EBs). (**a**) Phase contrast image of e-CSC EBs at day 3 and day 12. Scale bar 100 μm. (**b**) RT-PCR for expression of NANOG, NESTIN, PAX6, CK3, CK19, GATA6, BMP4, SYCP1 and GAPDH in e-CSC EBs. (**c**) Confocal immuno-fluorescence of e-CSC EBs for expression of NESTIN, MAP2, GATA4, AFP and LAMININ stained with FITC (green). Nuclei stained with DAPI (blue). Scale bars 25 μm. Positive controls are shown in Fig. S3c.

to give rise to extra-embryonic mesoderm [75]. Erasure of *H19* methylation occurs during PGC migration. As we have demonstrated e-CSC and l-CSC maintained both methylated and unmethylated regions on *H19* and expressed a wide array of markers associated with pre-migratory PGCs (*OCT4*, *c-KIT*, *STELLA*, *FRAGILIS*, *NANOS3*) [76,77]. The cells may orginate from PGCs deposited in the extra-embryonic tissues before migration and hence before methylation was erased.

Notwithstanding this, we also showed that e-CSC and l-CSC contain a subfraction of cells expressing post-migratory PGC markers (*VASA*, *DAZL*) [76,77], although *VASA* transcripts were not detected in l-CSC. Placental stem cells are known to reside in a vascular niche [19], therefore these cells may have been deposited later in PGC development following their mobilisation into the peripheral blood as shown for VSELs [78].

Human chorionic stem cells are relatively easy to isolate, can be sourced from surplus tissues at routine prenatal diagnostic procedures or after term delivery and thus used without ethical restrictions in allogeneic and autologous applications [79]. Steps towards establishing the safety of e-CSC and l-CSC for use in the clinic are already underway. No adverse clinical signs were seen after administration of allogeneic term placenta derived stem cells in NOD/SCID mice [80] or when delivered by intramuscular injection to mice with critical limb ischemia [81]. Furthermore term placenta cells were safely infused in a patient with acute myeloid leukemia [25], and are currently in Phase III clinical trials by Israel's Pluristem Therapeutics for patients with critical limb ischemia and peripheral artery disease [82], although maternal origin of these cells is possible. Transplantation of e-CSC into immunocompromised mice did not cause tumour formation and cells repopulated depleted hematopoietic tissues [14], while neural

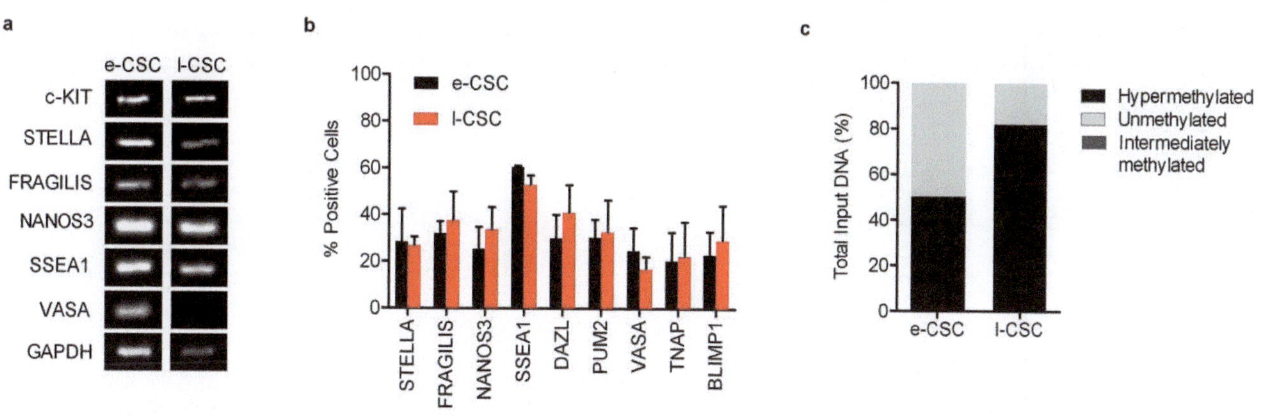

Figure 8. e-CSC and l-CSC express markers found in primordial germ cells. (**a**) RT-PCR for expression of c-KIT, STELLA, FRAGILIS, NANOS3, SSEA1, VASA and GAPDH. (**b**) Flow cytometry for percent of e-CSC (black) and l-CSC (red) populations positive for STELLA, FRAGILIS, NANOS3, SSEA1, DAZL, PUM2, VASA, TNAP and BLIMP1. (**c**) DNA methylation status of imprinted gene *H19* as percent total input DNA hypermethylated (black), unmethylated (light grey) and intermediately methylated (dark grey). $n = 3$ per cell group.

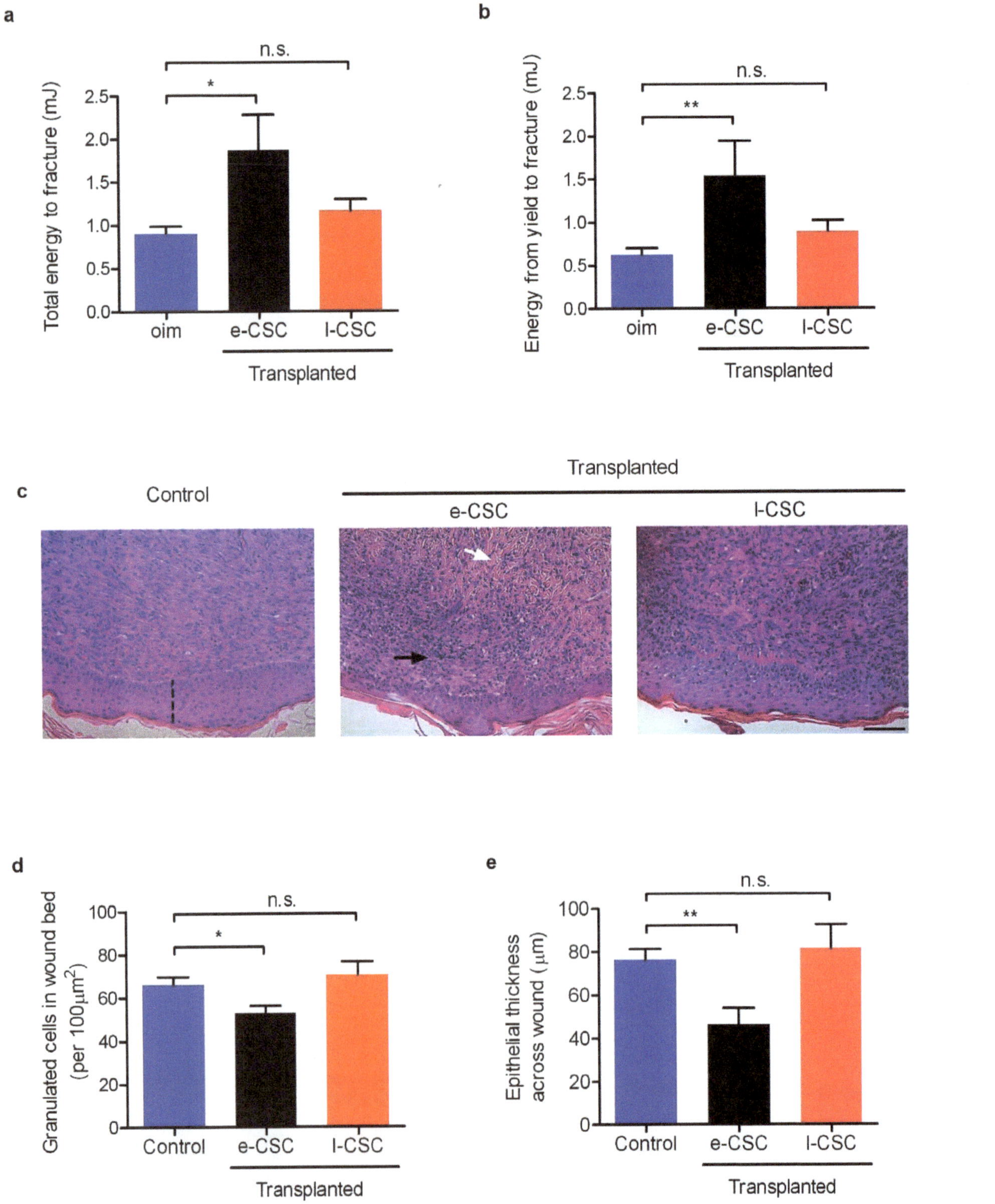

Figure 9. e-CSC showed better tissue repair capacity *in vivo* compared to l-CSC. (**a**) Mechanical 3-point bending data in a mouse model of bone brittleness (*oim*) for overall bone quality shown by total energy input in millijoules to fracture and for (**b**) femoral plasticity shown by energy input in millijoules from yield to fracture. Results shown per mouse for non-transplanted *oim* (blue) and *oim* transplanted with e-CSC (black) and l-CSC (red). (**c**) Cross-section of skin wounds 7 days after 4 mm dermal biopsy in mice treated with either PBS alone (Control) or transplanted with e-CSC or l-CSC; stained with haematoxylin and eosin. Indicated is collagen (white arrow), granulated cells (black arrow) and epithelial thickness (dashed line). In the same model of wound repair, (**d**) number of granulated cells in the wound bed (per 100 μm^2) and (**e**) epithelial thickness across wound in μm. Data. * $P<0.05$; ** $P<0.01$, n.s. (not significant), Student's *t* test. Mean ± s.e.m.

progenitors derived from e-CSC engrafted in the brain of an animal model with hypoxia-ischaemia and improved motor activity [83]. Consequently both e-CSC and l-CSC are emerging as a safe cell source for allogeneic or autologous cell therapy and tissue engineering with potential for cell or tissue banking [13,15].

Here we have shown e-CSC have a comparable immunophenotype and developmental origin to l-CSC, but have advantages for use in regenerative medicine, which may be linked to their earlier state of stemness. Accordingly e-CSC represent a promsing cell choice for experimental and clinical applications.

Materials and Methods

Cell isolation and culture

Human early and term chorionic stem cells (e-CSC and l-CSC respectively) were isolated fresh after first trimester chorionic villous sampling (n = 5, 8^{+3} to 12^{+0} weeks gestational age) and term placenta from caesarean section deliveries (n = 5). The chorionic tissue (5 cm^2 for l-CSC and the totality of the sample for e-CSC) was separated from the amnion, minced mechanically, and trypsinized using 0.05% trypsin EDTA (Invitrogen) for 20 min. Cells were subsequently centrifuged (5 min at 1500 rpm), counted and replated (10^4 cells per cm^2) in Dulbecco's modified Eagle's medium (DMEM-LG) (Invitrogen)+10% fetal bovine serum (BioSera). Isolated cells were plastic adherent and $cKIT^+$. All cells were used at passage 4–8. Ethical approval was given by the Research Ethics Committees of Hammersmith & Queen Charlotte's Hospitals and the joint University College London Hospitals. Committees on the Ethics of Human Research (08/H0714/87).

Bone repair mouse model

All experimental protocols complied with UK Home Office guidelines. Heterozygous male and female (B6C3Fe a/a-Col1a2^{oim}/Col1a2^{oim}) mice (Jackson Laboratory) were housed in individual ventilated cages (IVC) in 12:12-hour light dark cycle (21°C) with water and chow. Offspring were genotyped by sequencing the *oim* fragment then homozygous and wild type colonies established. Homozygous *oim* mice showed multiple fractures and brittle bones [8,9,84]. Progeny were weaned at 30±1 day and culled at 8 weeks of age. Human e-CSC and l-CSC were injected intraperitoneally (10^6 cells in 10 µl PBS) in homozygous *oim* neonates 3–4 days after birth. The mice were weighed and culled at 8 weeks of age for bone analysis. Three-point bending tests were performed on 8 week old unfractured frozen and thawed femurs in a materials testing machine (5866 Instron) at a loading rate of 50 µm/s on two supports 9 mm apart until fracture at mid-diaphysis. Force deflection curves were analyzed with a custom program (Matlab, MathWorks) to measure the bending stiffness (slope of the linear elastic deformation), the yield force (limit between the elastic and plastic deformation), the ultimate force (maximum force sustained), the total work to fracture (total area under the curve) and the plastic work to fracture (area under the curve from the yield point to fracture).

Donor cell engraftment: Femurs from transplanted 8 week old *oim* were dissected and bone marrow flushed. Human donor cell engraftment was determined as previously published [8], RNA was extracted using TRIzol (Invitrogen) followed by cDNA synthesis and qRT-PCR (protocol as above) then amplification of both human specific and human-mouse unspecific sequences of β-actin gene (See Table S2 for primers). Human:mouse chimerism was estimated as a ratio using $2^{-\Delta Ct}$. Samples were considered positive for engraftment with a human-specific β-actin Ct above 35 at a threshold of 0.13 ΔRn (primer specificity was confirmed by absence of amplification of non-transplanted mouse cDNA).

Wound healing model

All experimental protocols complied with UK Home Office guidelines. C57/Bl6 mice were housed in individual ventilated cages (IVC) in 12:12-hour light dark cycle (21°C) with water and chow. Mice were given a dermal excision as previously described by us [85]. After hair removal from the dorsal surface, 4-mm full-thickness excisional dermal wounds were made under anaesthesia. Each mouse was given two dermal excisions on each dorsal side. Stem cells (e-CSC and l-CSC, 10^6 cells in 30 µl PBS) were applied to the left-side wounds and PBS alone applied to the right-side wounds, to avoid inter-mouse variability. Seven days later, the mice were culled and the dermal tissues excised, cut across the middle of the wound and fixed overnight in 10% neutral buffered formalin at 4°C and embedded in paraffin, sectioned and subjected to hematoxylin and eosin staining for histological analysis.

DNA extraction and PCR

Male gender of cells was determined by DNA extraction using TRIzol (Invitrogen), followed by ethanol precipitation, washing and DNA dissolution in 8 mM NaOH. Then PCR was performed in the presence of dNTPs (Promega) and Taq polymerase (Bioline) with *SRY* specific primers (see Supplementary Table S2), followed by gel electrophoresis on 1.5% agarose with ethidium bromide (Sigma).

FISH analysis

No contamination by maternal cells was confirmed at passages 3–9 for *SRY* positive samples using XY FISH analysis. Cells were fixed on slides as previously described [4], then digested with proteinase K before hybridization with human specific centromeric mixture probe (CEP X/Y/18) according to manufacturer's protocol (Vysis). Nuclei were counterstained with DAPI and viewed on a fluorescence confocal laser scanning microscope Leica TCS SP5 (X600 PL APO oil objective). A total of 100 cells per group were analysed.

Cell Size

Size of passage 4–5 cells in suspension was determined using Image J software and calibrated with a haemocytometer.

Growth Kinetics

Cells were plated in triplicate at $1\times10^4/cm^2$ and counted in trypan blue to exclude dead cells in a haemocytometer either twice weekly over 60 days with passaging to maintain sub-confluence or every 2 days for 12 days without passaging. Doubling time (DT) was calculated via the formula: $DT = t/(Log_2[y/m])$, with t = time in culture, y = number of cells at end of culture, m = number of cells at beginning of culture.

Flow cytometry

For intracellular FACS staining cells were detached, blocked with 1% bovine serum albumin (BSA; Sigma) in phosphate-buffered saline (PBS; Invitrogen) then fixed with 0.01% paraformaldehyde (PFA) (Sigma) and permeabilized with 0.5% Triton X-100 (Sigma). For cell surface and intracellular FACS staining detached cells were blocked with 1% BSA and incubated with specific anti-human primary antibodies (see Table S1), either conjugated with PE, FITC or PE-Cy5 or unconjugated. For unconjugated antibodies, cells were subsequently washed with 1%

BSA and incubated with secondary goat anti-murine IgM PE (Santa Cruz). All samples were analyzed in triplicate by FACScalibur flow cytometry (Becton Dickinson). Positive controls were hESC and negative controls were IgG or IgM primary antibody-specific isotype controls.

RNA extraction and cDNA synthesis

Total RNA was extracted from cells using RNeasy mini RNA isolation kit (Qiagen) followed by cDNA synthesis using Pd(N)6 random hexamers and M-MLV reverse transcriptase in the presence of dNTPs (Promega) (10 min, 25°C; 60 min, 42°C and 10 min, 75°C).

RT-PCR

Reverse-transcriptase PCR (RT-PCR) was performed with cDNA using specific primers (see Table S2), dNTPs, ×10 buffer and Taq polymerase for 40 cycles according to the manufacturer's protocol (Qiagen).

qRT-PCR

Quantitative real time polymerase chain reaction (qRT-PCR) was performed using SYBR green dye (Applied Biosystem) and analysed either with MJ-opticon or ABI Prism 7700 Sequence Detection system (Applied Biosystem). Samples were normalized against the internal control (*GAPDH*) and the $2^{-\Delta Ct}$ of each sample plotted ($\log_{10}$ scale) relative to the expression level in human embryonic stem cells (hESC) line H9 (from Wei Cui), arbitrarily set to 100. Primers supplied in Table S2.

Fluorescence immunostaining and confocal microscopy

Cells were fixed and stained as previously described [4] and mounted in VectaShield labelled with DAPI (Vector Labs). Fluorescence confocal laser scanning microscopy images were collected on a Leica TCS SP5 (×400 PL APO oil objective) and transferred to Adobe Photoshop (Adobe Systems). Primary antibodies are listed in Table S1. Secondary antibodies were donkey anti-mouse or anti-rabbit IgG (Jackson ImmunoResearch Laboratories). Positive controls were hESC and negative controls differentiated cells.

Adhesion Assay

Human plasma Fn (Sigma-Aldrich) or 40 and 120 kDa human Fn α-chymotryptic fragments (Chemicon ITL), calf skin collagen type I and human placenta collagen IV (Sigma-Aldrich) were adsorbed at room temperature (2 hr) at 10 μg/cm^2 in PBS. Recombinant human VCAM-1 (ADP5, R&D Systems) and human placenta laminin (Sigma-Aldrich) were adsorbed at 1 μg/cm^2 in PBS at 4°C similarly. Control plates were coated with 2% BSA. Coating solutions were aspirated and plates were incubated with 2% BSA in PBS (30 min) to block non-specific binding. Plates were washed with PBS before 50,000 cells/cm^2 were plated in serum-free medium (DMEM) for 1 hr at 37°C. Non-adherent cells were removed with PBS and adherent cells were trypsinised and counted. Percentage of adhesion was calculated as follows: number of adherent cells/(number of adherent cells+number of non-adherent cells)*100 [86].

Analysis of *H19* methylation

Genomic DNA was extracted using DNeasy Blood and Tissue Kit (Qiagen) and made free from RNA using RNase A (Invitrogen). The EpiTect Methyl qPCR assay (Qiagen) was used to digest DNA with methylation-sensitive and methylation-dependent enzymes to obtain fractions of unmethylated, partially methylated and hypermethylated DNA. The methylation status of *H19* was then quantified using qRT-PCR and specific primers (see Table S2), data were then normalised to the total amount of DNA.

Embryoid body formation and analysis

Undifferentiated e-CSC and l-CSC were induced to differentiate *in vitro* into EBs by incubating confluent cells with differentiation medium containing knockout Dulbecco's modified eagle's medium (Gibco BRL Life Technologies) supplemented with 1 mM L-mercaptoethanol, 1% non-essential amino acids stock (all Gibco BRL Life Technologies) and 20% FBS (Biosera). At confluence, cells were incubated with 1 mg/ml type IV collagenase (Invitrogen) before being dissociated into small clumps on low attachment Petri dish in differentiating medium and allowed to develop into EBs for 12 days. EB suspensions were either pelleted and used for RNA extraction and RT-PCR of lineage specific markers (see Table S2 for primers) or transferred to gelatin-coated coverslips before fixation in 4% PFA and analysis under confocal microscopy for expression of lineage markers from all three germ layers (see Table S1 for primary antibodies used). Positive controls were pluripotent-induced human amniotic fluid stem cell-derived EBs.

Osteogenic Differentiation

Cells were cultured in DMEM supplemented with 10 mM β-glycerophosphate, 0.2 mM ascorbic acid and 10−8 M dexamethasone for 14 days. Then cells were fixed with 10% formalin and stained with von kossa or alizarin red, or pelleted and extracted for RNA or protein for qRT-PCR or western blot analysis respectively. All analysis was compared to undifferentiated cells. qRT-PCR reactions were carried out as described above for osteogenic specific primers (given in Table S2) and normalised to *GAPDH* and compared to basal expression levels in non-differentiated cells.

Adipogenic differentiation

Cells were seeded at a concentration of 5000 cells/cm^2 on tissue culture plastic plates and coverslips and cultured in high glucose DMEM supplemented with 1% penicillin/streptomycin (Invitrogen), 2 mM L-glutamine (Invitrogen) and 10% FBS (Biosera Ltd., East Sussex, UK) for 1 day. The media were then changed to high glucose DMEM supplemented with 0.5 μM hydrocortisone, 0.5 μM isobutyl methylxanthine and 60 μM indomethacin (all Sigma-Aldrich) for 2 weeks. Lipid droplets were subsequently stained with oil-red-O (Sigma-Aldrich). Oil-red-O was prepared at a 1% concentration in isopropanol (BDH Laboratory Supplies, Dorset, UK) then diluted in a 3:2 ratio of stock solution to distilled water. Cells were washed in PBS, fixed with 10% formalin, rinsed with 60% isopropanol (BDH Laboratory Supplies), followed by staining for 10 minutes. The cells were then washed 4 times with distilled water and photographed using an Olympus CK40-SLP microscope and an Olympus C3040-ADL camera (Olympus).

Hepatogenic differentiation

Cells were seeded at a concentration of 5000 cells/cm^2 on tissue culture plastic plates and coverslips coated with Matrigel (BD Biosciences) and cultured in high glucose DMEM supplemented with 1% penicillin/streptomycin (Invitrogen), 2 mM L-Glutamine (Invitrogen) and 10% FBS for 3 days. The media were then changed to high glucose DMEM supplemented with 15% FBS, 1% penicillin/streptomycin, 2 mM L-glutamine, 300 μM monothioglycerol (Sigma), 20 ng/ml hepatocyte growth factor (Peprotech), 10 ng/ml oncostatin M (Peprotech), 10^{-7} dexamethasone (Sigma), 100 ng/ml FGF4 (Peprotech) and 1X ITS (insulin, transferrin, selenium, Sigma). The cells were allowed to differentiate for 21 days and then fixed and stored in PBS for immunofluorescence. The differentiation media were collected and analysed for the presence of urea secreted by the differentiated cells. Urea was subsequently measured using the urea/ammonia determination kit (R-Biopharm AG, Darmstadt, Germany) according to the manufacturers instructions, with positive control being urea (as per manufacturer's instructions) and negative control undifferentiated e-CSC and l-CSC. Cells were subsquently assessed for expression of the hepatocyte markers ALBUMIN and ALPHA FETO-PROTEIN (AFP) using confocal microscopy. Positive control were HepG2 cells (gift from Wei Cui) and negative controls were undifferentiated e-CSC and l-CSC cells.

Neurogenic differentiation

Cells were seeded at a concentration of 5000 cells/cm^2 on tissue culture plastic plates and coverslips and cultured in high glucose DMEM supplemented with 1% penicillin/streptomycin (Invitrogen), 2 mM L-glutamine (Invitrogen) and 10% FBS (Biosera Ltd., East Sussex, UK) for 1 day. The media were then changed to high glucose DMEM supplemented with 0.5% FBS, 1% penicillin/streptomycin, 2 mM L-glutamine and 0.1% baicalin (Sigma), freshly made. C17.2 mouse neural progenitor cells (from Henrik Hagberg and Pierre Gressens) were added in a co-culture setting to promote differentiation, using co-culture membrane inserts (Thermo Scientific Nunc, Loughborough, UK). The cells were allowed to differentiate for 5 days and assessed for expression of the neuronal markers β-tubulin and MAP2 and for the NMDA receptor NR1 using confocal immunostaining. Positive controls were differentiated neural progenitors (gift Henrik Hagberg and from Pierre Gressens) and negative controls were undifferentiated e-CSC and l-CSC cells.

Protein Extraction and Western Blot

Total protein was extracted using RIPAE buffer containing protease inhibitor cocktail and PMSF (Sigma). Protein concentrations were determined using the BCA protein assay (Thermo-scientific), with bovine serum albumin as the standard. Proteins were run on an 8% SDS-PAGE, transferred to nitrocellulose, blocked with milk and incubated with primary anitbody COL1A2 (123 kDa) (Abcam). Secondary HRP-linked anti-rabbit IgG (GE healthcare), followed by enhanced chemiluminescence (Thermo scientific) was used for detection. Loading control used was β-actin (Abcam).

Statistical analysis

Data were expressed as mean ± s.d (standard deviation) or mean ± s.e.m (standard error). Parametric statistics were applied after confirming normal distributions on histograms and unpaired two-tailed Student's *t*-test was used for comparison between groups. $P<0.05$ was considered significant.

Supporting Information

Figure S1 e-CSC and l-CSC hepatogenic differentiation. (**a**) Confocal immuno-fluorescence for ALBUMIN and AFP in e-CSC and l-CSC grown in hepatogenic permissive media for 3 weeks. HepG3 cells used for positive control. All genes stained with FITC (green). Nuclei stained with DAPI (blue). Scale bars 100 μm. (**b**) Presence of urea (mmol/l) in hepatogenic differentiation media (diff) of e-CSC and l-CSC. Positive (urea; according to the manufacturer's instructions) and negative controls; non-differentiated (non-diff) e-CSC and l-CSC are shown. $n = 4$ per cell group. Data. n.s. (not significant), Student's *t* test. Mean ± s.e.m.

Figure S2 Positive controls for confocal immuno-fluorescence. (**a**) Positive controls (differentiated neural progenitors) of neuronal differentiation for expression of β-TUBULIN, MAP2 and NR1. (**b**) Positive controls (human first trimester fetal bone marrow mesenchymal stem cells, hfMSC) CD105, CD73, CD44, CD29 and CD90 (**c**) Positive controls (hES cells line H9) OCT4A, SOX2, KLF4, NANOG, REX1, SSEA4, SSEA3, TRA-1-60, and TRA-1-81 (**d**) Positive controls (pluripotent-induced human amniotic fluid stem cells-derived EBs) NESTIN, MAP2, AFP, GATA4 and LAMININ. All genes stained with FITC (green). Nuclei stained with DAPI (blue). Scale bars 50 μm.

Figure S3 e-CSC and l-CSC express PGC associated markers. (**a**) Representative confocal immuno-fluorescence for protein expression of FRAGILIS, SSEA1, TNAP, NANOS3, BLIMP1, PUM2, STELLA, DAZL and VASA stained with FITC (green). Nuclei stained with DAPI (blue). Scale bar 100 μm. (**b**) Representative flow cytometry for PGC markers; STELLA, FRAGILIS, NANOS3, SSEA1, DAZL, PUM2, VASA, TNAP and BLIMP1 (isotype control in black).

Figure S4 Engraftment of e-CSC and l-CSC in 8 week-old transplanted *oim* mice. Donor cell engraftment in bone marrow and bone was calculated as the $2^{-\Delta Ct}$ of human specific β-actin normalised to human-mouse non-specific β-actin and using quantitative real time PCR in *oim* mice transplanted with e-CSC (black) or l-CSC (red).

Table S1 Antibodies used. List of antibodies used for immuno-fluorescence (IF), flow cytometry (FC) and immuno-histochemistry (IH).

Table S2 Primers used. List of primers used for RT-PCR and quantitative real time PCR and EpiTect Methyl qPCR assay.

Author Contributions

Conceived and designed the experiments: PVG. Performed the experiments: GNJ DM TIPI KK MV GBG. Analyzed the data: GNJ. Contributed reagents/materials/analysis tools: PVG ALD NMF PDC SJS. Wrote the paper: GNJ PVG.

References

1. Phinney DG, Prockop DJ (2007) Concise review: mesenchymal stem/multipotent stromal cells: the state of transdifferentiation and modes of tissue repair–current views. Stem Cells 25: 2896–2902.
2. Chavakis E, Urbich C, Dimmeler S (2008) Homing and engraftment of progenitor cells: a prerequisite for cell therapy. J Mol Cell Cardiol 45: 514–522.
3. Zhang ZY, Teoh SH, Hui JH, Fisk NM, Choolani M, et al. (2012) The potential of human fetal mesenchymal stem cells for off-the-shelf bone tissue engineering application. Biomaterials.
4. Guillot PV, Gotherstrom C, Chan J, Kurata H, Fisk NM (2007) Human first-trimester fetal MSC express pluripotency markers and grow faster and have longer telomeres than adult MSC. Stem Cells 25: 646–654.

5. Zhang ZY, Teoh SH, Chong MS, Schantz JT, Fisk NM, et al. (2009) Superior osteogenic capacity for bone tissue engineering of fetal compared with perinatal and adult mesenchymal stem cells. Stem Cells 27: 126–137.
6. Izumi M, Pazin BJ, Minervini CF, Gerlach J, Ross MA, et al. (2009) Quantitative comparison of stem cell marker-positive cells in fetal and term human amnion. J Reprod Immunol 81: 39–43.
7. Poloni A, Maurizi G, Serrani F, Mancini S, Discepoli G, et al. (2012) Human AB serum for generation of mesenchymal stem cells from human chorionic villi: comparison with other source and other media including platelet lysate. Cell Prolif 45: 66–75.
8. Guillot PV, Abass O, Bassett JH, Shefelbine SJ, Bou-Gharios G, et al. (2008) Intrauterine transplantation of human fetal mesenchymal stem cells from first-trimester blood repairs bone and reduces fractures in osteogenesis imperfecta mice. Blood 111: 1717–1725.
9. Vanleene M, Saldanha Z, Cloyd KL, Jell G, Bou-Gharios G, et al. (2011) Transplantation of human fetal blood stem cells in the osteogenesis imperfecta mouse leads to improvement in multiscale tissue properties. Blood 117: 1053–1060.
10. Guillot PV, Cook HT, Pusey CD, Fisk NM, Harten S, et al. (2008) Transplantation of human fetal mesenchymal stem cells improves glomerulopathy in a collagen type I alpha 2-deficient mouse. J Pathol 214: 627–636.
11. Zhang ZY, Teoh SH, Chong MS, Lee ES, Tan LG, et al. (2010) Neo-vascularization and bone formation mediated by fetal mesenchymal stem cell tissue-engineered bone grafts in critical-size femoral defects. Biomaterials 31: 608–620.
12. Chan J, Kumar S, Fisk NM (2008) First trimester embryo-fetoscopic and ultrasound-guided fetal blood sampling for ex vivo viral transduction of cultured human fetal mesenchymal stem cells. Hum Reprod 23: 2427–2437.
13. Poloni A, Rosini V, Mondini E, Maurizi G, Mancini S, et al. (2008) Characterization and expansion of mesenchymal progenitor cells from first-trimester chorionic villi of human placenta. Cytotherapy 10: 690–697.
14. Spitalieri P, Cortese G, Pietropolli A, Filareto A, Dolci S, et al. (2009) Identification of multipotent cytotrophoblast cells from human first trimester chorionic villi. Cloning Stem Cells 11: 535–556.
15. Portmann-Lanz CB, Schoeberlein A, Huber A, Sager R, Malek A, et al. (2006) Placental mesenchymal stem cells as potential autologous graft for pre- and perinatal neuroregeneration. Am J Obstet Gynecol 194: 664–673.
16. De Coppi P, Bartsch G Jr, Siddiqui MM, Xu T, Santos CC, et al. (2007) Isolation of amniotic stem cell lines with potential for therapy. Nat Biotechnol 25: 100–106.
17. Tsai MS, Lee JL, Chang YJ, Hwang SM (2004) Isolation of human multipotent mesenchymal stem cells from second-trimester amniotic fluid using a novel two-stage culture protocol. Hum Reprod 19: 1450–1456.
18. Prusa AR, Marton E, Rosner M, Bernaschek G, Hengstschlager M (2003) Oct-4-expressing cells in human amniotic fluid: a new source for stem cell research? Hum Reprod 18: 1489–1493.
19. Castrechini NM, Murthi P, Gude NM, Erwich JJ, Gronthos S, et al. (2010) Mesenchymal stem cells in human placental chorionic villi reside in a vascular Niche. Placenta 31: 203–212.
20. Bacenkova D, Rosocha J, Tothova T, Rosocha L, Sarissky M (2011) Isolation and basic characterization of human term amnion and chorion mesenchymal stromal cells. Cytotherapy 13: 1047–1056.
21. Rus Ciuca D, Soritau O, Susman S, Pop VI, Mihu CM (2011) Isolation and characterization of chorionic mesenchyal stem cells from the placenta. Rom J Morphol Embryol 52: 803–808.
22. Soncini M, Vertua E, Gibelli L, Zorzi F, Denegri M, et al. (2007) Isolation and characterization of mesenchymal cells from human fetal membranes. J Tissue Eng Regen Med 1: 296–305.
23. In't Anker PS, Scherjon SA, Kleijburg-van der Keur C, de Groot-Swings GM, Claas FH, et al. (2004) Isolation of mesenchymal stem cells of fetal or maternal origin from human placenta. Stem Cells 22: 1338–1345.
24. Semenov OV, Koestenbauer S, Riegel M, Zech N, Zimmermann R, et al. (2010) Multipotent mesenchymal stem cells from human placenta: critical parameters for isolation and maintenance of stemness after isolation. Am J Obstet Gynecol 202: 193 e191–193 e113.
25. Brooke G, Rossetti T, Pelekanos R, Ilic N, Murray P, et al. (2009) Manufacturing of human placenta-derived mesenchymal stem cells for clinical trials. Br J Haematol 144: 571–579.
26. Barlow S, Brooke G, Chatterjee K, Price G, Pelekanos R, et al. (2008) Comparison of human placenta- and bone marrow-derived multipotent mesenchymal stem cells. Stem Cells Dev 17: 1095–1107.
27. Yen BL, Huang HI, Chien CC, Jui HY, Ko BS, et al. (2005) Isolation of multipotent cells from human term placenta. Stem Cells 23: 3–9.
28. Kim MJ, Shin KS, Jeon JH, Lee DR, Shim SH, et al. (2011) Human chorionic-plate-derived mesenchymal stem cells and Wharton's jelly-derived mesenchymal stem cells: a comparative analysis of their potential as placenta-derived stem cells. Cell Tissue Res 346: 53–64.
29. Fariha MM, Chua KH, Tan GC, Tan AE, Hayati AR (2011) Human chorion-derived stem cells: changes in stem cell properties during serial passage. Cytotherapy 13: 582–593.
30. Miao Z, Jin J, Chen L, Zhu J, Huang W, et al. (2006) Isolation of mesenchymal stem cells from human placenta: comparison with human bone marrow mesenchymal stem cells. Cell Biol Int 30: 681–687.
31. Portmann-Lanz CB, Schoeberlein A, Portmann R, Mohr S, Rollini P, et al. (2010) Turning placenta into brain: placental mesenchymal stem cells differentiate into neurons and oligodendrocytes. Am J Obstet Gynecol 202: 294 e291–294 e211.
32. Dominici M, Le Blanc K, Mueller I, Slaper-Cortenbach I, Marini F, et al. (2006) Minimal criteria for defining multipotent mesenchymal stromal cells. The International Society for Cellular Therapy position statement. Cytotherapy 8: 315–317.
33. Lian Q, Lye E, Suan Yeo K, Khia Way Tan E, Salto-Tellez M, et al. (2007) Derivation of clinically compliant MSCs from CD105+, CD24− differentiated human ESCs. Stem Cells 25: 425–436.
34. Carpenter MK, Rosler ES, Fisk GJ, Brandenberger R, Ares X, et al. (2004) Properties of four human embryonic stem cell lines maintained in a feeder-free culture system. Dev Dyn 229: 243–258.
35. Ryan JM, Pettit AR, Guillot PV, Chan JK, Fisk NM (2011) Unravelling the Pluripotency Paradox in Fetal and Placental Mesenchymal Stem Cells: Oct-4 Expression and the Case of the Emperor's New Clothes. Stem Cell Rev.
36. Atlasi Y, Mowla SJ, Ziaee SA, Gokhale PJ, Andrews PW (2008) OCT4 spliced variants are differentially expressed in human pluripotent and nonpluripotent cells. Stem Cells 26: 3068–3074.
37. Lee J, Kim HK, Rho JY, Han YM, Kim J (2006) The human OCT-4 isoforms differ in their ability to confer self-renewal. J Biol Chem 281: 33554–33565.
38. Pesce M, Scholer HR (2000) Oct-4: control of totipotency and germline determination. Mol Reprod Dev 55: 452–457.
39. Assou S, Le Carrour T, Tondeur S, Strom S, Gabelle A, et al. (2007) A meta-analysis of human embryonic stem cells transcriptome integrated into a web-based expression atlas. Stem Cells 25: 961–973.
40. Adewumi O, Aflatoonian B, Ahrlund-Richter L, Amit M, Andrews PW, et al. (2007) Characterization of human embryonic stem cell lines by the International Stem Cell Initiative. Nat Biotechnol 25: 803–816.
41. Skottman H, Mikkola M, Lundin K, Olsson C, Stromberg AM, et al. (2005) Gene expression signatures of seven individual human embryonic stem cell lines. Stem Cells 23: 1343–1356.
42. De Miguel MP, Cheng L, Holland EC, Federspiel MJ, Donovan PJ (2002) Dissection of the c-Kit signaling pathway in mouse primordial germ cells by retroviral-mediated gene transfer. Proc Natl Acad Sci U S A 99: 10458–10463.
43. Hoyer PE, Byskov AG, Mollgard K (2005) Stem cell factor and c-Kit in human primordial germ cells and fetal ovaries. Mol Cell Endocrinol 234: 1–10.
44. Motohashi T, Aoki H, Chiba K, Yoshimura N, Kunisada T (2007) Multipotent cell fate of neural crest-like cells derived from embryonic stem cells. Stem Cells 25: 402–410.
45. Bowles J, Teasdale RP, James K, Koopman P (2003) Dppa3 is a marker of pluripotency and has a human homologue that is expressed in germ cell tumours. Cytogenet Genome Res 101: 261–265.
46. Payer B, Saitou M, Barton SC, Thresher R, Dixon JP, et al. (2003) Stella is a maternal effect gene required for normal early development in mice. Curr Biol 13: 2110–2117.
47. Tanaka SS, Yamaguchi YL, Tsoi B, Lickert H, Tam PP (2005) IFITM/Mil/fragilis family proteins IFITM1 and IFITM3 play distinct roles in mouse primordial germ cell homing and repulsion. Dev Cell 9: 745–756.
48. Tsuda M, Sasaoka Y, Kiso M, Abe K, Haraguchi S, et al. (2003) Conserved role of nanos proteins in germ cell development. Science 301: 1239–1241.
49. Henderson JK, Draper JS, Baillie HS, Fishel S, Thomson JA, et al. (2002) Preimplantation human embryos and embryonic stem cells show comparable expression of stage-specific embryonic antigens. Stem Cells 20: 329–337.
50. Park JH, Kim SJ, Lee JB, Song JM, Kim CG, et al. (2004) Establishment of a human embryonic germ cell line and comparison with mouse and human embryonic stem cells. Mol Cells 17: 309–315.
51. Moore FL, Jaruzelska J, Fox MS, Urano J, Firpo MT, et al. (2003) Human Pumilio-2 is expressed in embryonic stem cells and germ cells and interacts with DAZ (Deleted in AZoospermia) and DAZ-like proteins. Proc Natl Acad Sci U S A 100: 538–543.
52. Castrillon DH, Quade BJ, Wang TY, Quigley C, Crum CP (2000) The human VASA gene is specifically expressed in the germ cell lineage. Proc Natl Acad Sci U S A 97: 9585–9590.
53. MacGregor GR, Zambrowicz BP, Soriano P (1995) Tissue non-specific alkaline phosphatase is expressed in both embryonic and extraembryonic lineages during mouse embryogenesis but is not required for migration of primordial germ cells. Development 121: 1487–1496.
54. McLaren A (2003) Primordial germ cells in the mouse. Dev Biol 262: 1–15.
55. Durcova-Hills G, Tang F, Doody G, Tooze R, Surani MA (2008) Reprogramming primordial germ cells into pluripotent stem cells. PLoS One 3: e3531.
56. Ohinata Y, Payer B, O'Carroll D, Ancelin K, Ono Y, et al. (2005) Blimp1 is a critical determinant of the germ cell lineage in mice. Nature 436: 207–213.
57. Arnaud P (2010) Genomic imprinting in germ cells: imprints are under control. Reproduction 140: 411–423.
58. Hajkova P, Erhardt S, Lane N, Haaf T, El-Maarri O, et al. (2002) Epigenetic reprogramming in mouse primordial germ cells. Mech Dev 117: 15–23.
59. Sung HJ, Hong SC, Yoo JH, Oh JH, Shin HJ, et al. (2010) Stemness evaluation of mesenchymal stem cells from placentas according to developmental stage: comparison to those from adult bone marrow. J Korean Med Sci 25: 1418–1426.

60. Kyo S, Takakura M, Tanaka M, Kanaya T, Sagawa T, et al. (1997) Expression of telomerase activity in human chorion. Biochem Biophys Res Commun 241: 498–503.
61. Schmidt D, Mol A, Breymann C, Achermann J, Odermatt B, et al. (2006) Living autologous heart valves engineered from human prenatally harvested progenitors. Circulation 114: I125–131.
62. Weber B, Zeisberger SM, Hoerstrup SP (2011) Prenatally harvested cells for cardiovascular tissue engineering: fabrication of autologous implants prior to birth. Placenta 32 Suppl 4: S316–319.
63. Okamoto K, Miyoshi S, Toyoda M, Hida N, Ikegami Y, et al. (2007) 'Working' cardiomyocytes exhibiting plateau action potentials from human placenta-derived extraembryonic mesodermal cells. Exp Cell Res 313: 2550–2562.
64. Shin KS, Lee HJ, Jung J, Cha DH, Kim GJ (2010) Culture and in vitro hepatogenic differentiation of placenta-derived stem cells, using placental extract as an alternative to serum. Cell Prolif 43: 435–444.
65. Susman S, Soritau O, Rus-Ciuca D, Tomuleasa C, Pop VI, et al. (2010) Placental stem cell differentiation into islets of Langerhans-like glucagon-secreting cells. Rom J Morphol Embryol 51: 733–738.
66. Le Blanc K, Gotherstrom C, Ringden O, Hassan M, McMahon R, et al. (2005) Fetal mesenchymal stem-cell engraftment in bone after in utero transplantation in a patient with severe osteogenesis imperfecta. Transplantation 79: 1607–1614.
67. Mohr S, Portmann-Lanz CB, Schoeberlein A, Sager R, Surbek DV (2010) Generation of an osteogenic graft from human placenta and placenta-derived mesenchymal stem cells. Reprod Sci 17: 1006–1015.
68. Li X, Ling W, Pennisi A, Wang Y, Khan S, et al. (2011) Human placenta-derived adherent cells prevent bone loss, stimulate bone formation, and suppress growth of multiple myeloma in bone. Stem Cells 29: 263–273.
69. Kucia M, Machalinski B, Ratajczak MZ (2006) The developmental deposition of epiblast/germ cell-line derived cells in various organs as a hypothetical explanation of stem cell plasticity? Acta Neurobiol Exp (Wars) 66: 331–341.
70. Kucia M, Wu W, Ratajczak MZ (2007) Bone marrow-derived very small embryonic-like stem cells: their developmental origin and biological significance. Dev Dyn 236: 3309–3320.
71. Shin DM, Liu R, Klich I, Wu W, Ratajczak J, et al. (2010) Molecular signature of adult bone marrow-purified very small embryonic-like stem cells supports their developmental epiblast/germ line origin. Leukemia 24: 1450–1461.
72. Kucia M, Halasa M, Wysoczynski M, Baskiewicz-Masiuk M, Moldenhawer S, et al. (2007) Morphological and molecular characterization of novel population of CXCR4+ SSEA-4+ Oct-4+ very small embryonic-like cells purified from human cord blood: preliminary report. Leukemia 21: 297–303.
73. Hua J, Yu H, Dong W, Yang C, Gao Z, et al. (2009) Characterization of mesenchymal stem cells (MSCs) from human fetal lung: potential differentiation of germ cells. Tissue Cell 41: 448–455.
74. Hua J, Pan S, Yang C, Dong W, Dou Z, et al. (2009) Derivation of male germ cell-like lineage from human fetal bone marrow stem cells. Reprod Biomed Online 19: 99–105.
75. Donovan PJ (1998) The germ cell–the mother of all stem cells. Int J Dev Biol 42: 1043–1050.
76. Saiti D, Lacham-Kaplan O (2007) Mouse Germ Cell Development in-vivo and in-vitro. Biomark Insights 2: 241–252.
77. Hayashi K, de Sousa Lopes SM, Surani MA (2007) Germ cell specification in mice. Science 316: 394–396.
78. Ratajczak MZ, Liu R, Marlicz W, Blogowski W, Starzynska T, et al. (2011) Identification of very small embryonic/epiblast-like stem cells (VSELs) circulating in peripheral blood during organ/tissue injuries. Methods Cell Biol 103: 31–54.
79. Abdulrazzak H, Moschidou D, Jones G, Guillot PV (2010) Biological characteristics of stem cells from foetal, cord blood and extraembryonic tissues. J R Soc Interface 7 Suppl 6: S689–706.
80. Ramot Y, Meiron M, Toren A, Steiner M, Nyska A (2009) Safety and biodistribution profile of placental-derived mesenchymal stromal cells (PLX-PAD) following intramuscular delivery. Toxicol Pathol 37: 606–616.
81. Prather WR, Toren A, Meiron M, Ofir R, Tschope C, et al. (2009) The role of placental-derived adherent stromal cell (PLX-PAD) in the treatment of critical limb ischemia. Cytotherapy 11: 427–434.
82. Trounson A, Thakar RG, Lomax G, Gibbons D (2011) Clinical trials for stem cell therapies. BMC Med 9: 52.
83. Park S, Koh SE, Maeng S, Lee WD, Lim J, et al. (2011) Neural progenitors generated from the mesenchymal stem cells of first-trimester human placenta matured in the hypoxic-ischemic rat brain and mediated restoration of locomotor activity. Placenta 32: 269–276.
84. Jones GN, Moschidou D, Lay K, Abdulrazzak H, Vanleene M, et al. (2012) Upregulating CXCR4 in human fetal mesenchymal stem cells enhances engraftment and bone mechanics in a mouse model of osteogenesis imperfecta. Stem Cells Translational Medicine 1: 70–78.
85. Rajkumar VS, Shiwen X, Bostrom M, Leoni P, Muddle J, et al. (2006) Platelet-derived growth factor-beta receptor activation is essential for fibroblast and pericyte recruitment during cutaneous wound healing. Am J Pathol 169: 2254–2265.
86. Huygen S, Giet O, Artisien V, Di Stefano I, Beguin Y, et al. (2002) Adhesion of synchronized human hematopoietic progenitor cells to fibronectin and vascular cell adhesion molecule-1 fluctuates reversibly during cell cycle transit in ex vivo culture. Blood 100: 2744–2752.

A Novel Berbamine Derivative Inhibits Cell Viability and Induces Apoptosis in Cancer Stem-Like Cells of Human Glioblastoma, via Up-Regulation of miRNA-4284 and JNK/AP-1 Signaling

Fan Yang[1], Sangkil Nam[2], Christine E. Brown[3], Robin Zhao[2], Renate Starr[3], David A. Horne[2], Linda H. Malkas[4], Richard Jove[6], Robert J. Hickey[1,5]*

1 Translational Biomarker Discovery Core, Beckman Research Institute, City of Hope Comprehensive Cancer Center, Duarte, California, United States of America, **2** Molecular Medicine, Beckman Research Institute, City of Hope Comprehensive Cancer Center, Duarte, California, United States of America, **3** Cancer Immunotherapeutics & Tumor Immunology, Beckman Research Institute, City of Hope Comprehensive Cancer Center, Duarte, California, United States of America, **4** Molecular and Cellular Biology, Beckman Research Institute, City of Hope Comprehensive Cancer Center, Duarte, California, United States of America, **5** Molecular Pharmacology, Beckman Research Institute, City of Hope Comprehensive Cancer Center, Duarte, California, United States of America, **6** Vaccine & Gene Therapy Institute of Florida, Port St. Lucie, Florida, United States of America

Abstract

Glioblastoma (GBM) is the most common primary brain tumor, accounting for approximately 40% of all central nervous system malignancies. Despite standard treatment consisting of surgical resection, radiotherapy and/or chemotherapy, the prognosis for GBM is poor; with a median survival of 14.6 months. The cancer stem cell or cancer-initiating cell model has provided a new paradigm for understanding development and recurrence of GBM following treatment. Berbamine (BBM) is a natural compound derived from the *Berberis amurensis* plant, and along with its derivatives, has been shown to exhibit antitumor activity in several cancers. Here, we reported that a novel synthetic Berbamine derivative, BBMD3, inhibits cell viability and induces apoptosis of cancer stem-like cells (CSCs) in a time- and dose-dependent manner when the CSCs from four GBM patients (PBT003, PBT008, PBT022, and PBT030) were cultured. These CSCs grew in neurospheres and expressed CD133 and nestin as markers. Treatment with BBMD3 destroyed the neurosphere morphology, and led to the induction of apoptosis in the CSCs. Induction of apoptosis in these CSCs is dependent upon activation of caspase-3 and cleavage of poly (ADP-ribose) polymerase (PARP). MicroRNA-4284 (miR-4284) was shown to be over-expressed about 4-fold in the CSCs following BBMD3 treatment. Furthermore, transfection of synthetic anti-sense oligonucleotide against human miR-4284 partially blocked the anticancer effects of BBMD3 on the GBM derived CSCs. BBMD3 also increased phosphorylation of the c-Jun N-terminal kinase (JNK)/stress-activated protein kinase (SAPK), resulting in an increase expression of phosphorylated c-Jun and total c-Fos; the major components of transcriptional factor AP-1. The JNK-c-Jun/AP-1 signaling pathway plays an important role in the induction of apoptosis in response to UV irradiation and some drug treatments. Targeting glioblastoma stem-like cells with BBMD3 is therefore novel, and may have promise as an effective therapeutic strategy for treating GBM patients.

Editor: Jeffrey K. Harrison, University of Florida, United States of America

Funding: Funding for the study was provided by a grant from the National Institutes of Health/National Cancer Institute #CA-1155674 to RJ and start-up funding to RH from the Beckman Research Institute. Research reported in this publication included work performed in the Translational Biomarker Discovery Core, which is supported in part by the National Cancer Institute of the National Institutes of Health under award number P30CA33572. The funders had no role in study design, data collection and analysis, decision to publish, or preparation of the manuscript.

Competing Interests: The authors have declared that no competing interests exist.

* E-mail: rohickey@coh.org

Introduction

Glioblastoma (GBM) is the most common and lethal primary brain tumor. Despite current advances in multimodality therapy, which include surgery, radiotherapy and chemotherapy, prognosis remains very poor for patients, who typically have a median survival time of less than 15 months [1,2]. The majority of GBM lesions rapidly develop from a less malignant precursor lesion for which there is little or no clinical, radiological, or morphologic evidence, and it has been demonstrated that a highly tumorigenic subpopulation of cancer cells, called GBM stem cells, promotes resistance to chemo- and radio- therapy [3–5]. These cancer stem cells or tumor-initiating cells share some critical characteristics with normal neural stem cells, including expression of several biomarkers, and the ability for self-renewal, differentiation and proliferation. Due to the poor prognosis for GBM patients following currently available therapies, development of more effective protocols for treating GBM is urgently needed. However progress slowing protocol development remains dependent upon further enhancement of our understanding of the processes driving cancer invasion, the onset of resistance to therapeutic interventions and mechanisms driving tumor recurrence in GBM patients.

Thus, the effective treatment of GBM requires directly targeting these GBM stem cells within the tumor mass, since they are the cells that are resistant to standard therapies [6]. In this regard, Brown et al [7] recently provided a rationale for developing an immunotherapeutic approach for eradicating the GBM stem cell population by reporting that human tumor stem/initiating cells from GBM patients could be recognized and killed by $CD8^+$ cytotoxic T lymphocytes.

In addition to this immunological approach, microRNA (miRNA), which is a relatively new class of small non-coding RNA molecule found in eukaryotic cells, has been shown to regulate a wide spectrum of gene expression patterns via a post-transcriptional mechanism [8]. And a considerable body of evidence now indicates that miRNAs play key roles in the pathogenesis of cancer, and can function either as oncogenes or tumor suppressors [9]. It has also been reported that high expression of miR-196 and miR10b in GBM patients correlates with a poor prognosis [10], and that down-regulation of miR-128 leads to reduction in the self-renewal ability of glioma stem cells by inhibiting Bmi1 gene expression. Thus, miRNAs are rapidly emerging as promising targets for the development of novel but highly selective anticancer therapeutic agents.

Several years ago, Berbamine (BBM), a natural bis-benzylisoquinoline alkaloid, was identified from the traditional Chinese medicine *Berberis amurensis*, and along with several of its derivatives, was reported to have potent antitumor activity toward lymphoma, myeloma, hepatocellular carcinoma, lung cancer and breast cancer and low non-specific toxicity [12–16]. BBM was reported to induce apoptosis in human myeloma, and to suppress growth, migration and invasion of highly-metastatic human breast cancer cells through inhibiting NF-kappaB (NF-KB) activity and its downstream targets, such as cyclin D1, Bcl-xL and survivin [13,14]. Thus the low non-specific toxicity of BBM, and some of its derivatives, makes continued development of these novel compounds potentially promising as effective anti-cancer drugs for a variety of types of cancer. Toward this goal, our lab examined thirteen novel BBM derivatives (BBMDs), which were synthesized from natural BBM. We found that BBMD3 is the most potent of this series of novel BBMDs for killing cancer cells. BBMD3 exhibited over a 6-fold increase in anticancer activity toward melanoma and prostate cancer cells as compared to natural BBM. This was attributed to inhibition of the JAK2/STAT3 signaling pathway [17]. Recently, we also reported that BBMD3 inhibits cell viability, and induces apoptosis in human osteosarcoma cells, which, in this case, was attributed to activation of the JNK/AP-1 signaling pathway [18].

The superfamily of mitogen-activated protein kinases (MAPKs) includes: c-Jun N-terminal protein kinase (JNK)/stress-activated protein kinase (SAPK); p38; and extracellular signal-regulated kinase (ERK). In general, JNK and p38 are key mediators of the response to stress and inflammation, while the ERK cascade is most often induced in response to growth factor stimulation [19,20]. The JNK stress pathway participates in many different intracellular processes, including cell growth, differentiation, transformation and apoptosis [21,22]. The JNK protein kinases are encoded by three genes, of which *Jnk1* and *Jnk2* are expressed by all tissues, and the *Jnk3* gene is confined to a more limited pattern of expression such as in brain and heart [22]. JNK was originally identified by its ability to specifically phosphorylate the transcription factor c-Jun on its N-terminal trans-activation domain at Ser63 and Ser73 [23]. c-Jun is a major component of activating protein-1 (AP-1), which is a dimeric transcriptional factor, and is composed of proteins from several gene families [24]. Though the JNK/c-Jun or JNK/AP-1 pathway have dual roles in apoptosis, it is clear that activation of the JNK pathway is involved in the induction of apoptosis by specific cell death stimuli, such as UV irradiation and treatment with certain drugs [25–27].

In our present study, we show that a novel synthetic BBM derivative (BBMD3) induces apoptosis in cancer stem-like cells, which are cultured from GBM patients. We also show that cell killing is associated with up-regulation of miRNA-4284 and the JNK/AP-1 signaling pathway.

Materials and Methods

Reagents and Antibodies

BBMD3 was synthesized by reacting natural BBM with NH_2-containing substitutents. The structure and purity of the products were analyzed using HNMR spectroscopy. BBMD3 displayed over 98% purity by NMR analysis. Human miRNA inhibitors (synthetic oligonucleotides) against hsa-miRNA-4284 and has-miRNA-27a were purchased from GeneCopoeia. RNAiFect Transfection reagent was purchased from Qiagen Inc. Recombinant human epidermal growth factor (EGF) and fibroblast growth factor basic (FGF) were obtained from R&D Systems Inc., and Accutase was purchased from Innovative Cell Technologies. B-27 serum-free supplement (50X) for growth and maintenance of neurons was obtained from Gibco. Heparin (1000 units/ml) was purchased from APP Pharmaceuticals, LLC. Horseradish peroxidase-labeled anti-mouse and anti-rabbit secondary antibodies were purchased from GE Healthcare. Anti-nestin antibody was purchased from EMD Millipore. All other antibodies were purchased from Cell Signaling, Inc.

Cell Cultures

GBM specimens used in this study, (were evaluated by an attending neuropathologist, and assigned a grade IV classification using the World Health Organization (WHO) established guidelines), were obtained from patients undergoing surgical treatment at the City of Hope Medical Center, in accordance with Institutional Review Board (IRB)-approved protocols. Tumor specimens were given unique patient brain tumor (PBT) numbers, (i.e. PBT003, PBT008, PBT022 and PBT030). GBM neurosphere (NS) cultures were cultured and expanded as previously described [7]. Briefly, NS cultures were maintained in neural stem cell media (DMEM/F-12), supplemented with 1:50 B27 serum-free supplement, 20 ng/mL EGF, 20 ng/ml FGF basic and 0.84 unit/ml of heparin. Normal human neurospheres or non-malignant neurospheres were cultured from human fetal brain tissue under the same culture condition as the GBM neurospheres. The established human glioblastoma cell line U87, was purchased from the American Type Culture Collection (ATCC), and both the U87 and the attached or differentiated cells derived from the GBM neurospheres (stem-like cells), were cultured in DMEM media (containing L-glutamine), supplemented with 10% fetal bovine serum (FBS) and 1% Antibiotic-Antimycotic (AA). All cultured cells were grown at 37°C in a humidified atmosphere containing 5% CO_2.

Treatment of BBMD3

BBMD3 was dissolved in dimethyl sulfoxide (DMSO) and diluted with cell culture medium. To obtain single cells from GBM neurospheres, neurospheres were treated with accutase for 10 min at 37°C and passed through a cell strainer (70 μm nylon). Cells were then seeded in cell culture flask or dishes with complete stem cell culture medium according to the protocol outlined in the experimental design section. Five hours later, different concentrations of diluted BBMD3 were added to the cells; while the controls received equal amount of only the vehicle.

Cell Viability Assays

Cell viability assays were performed with the CellTiter 96 Aqueous One Solution Cell Proliferation Assay from Promega, which contains 3-(4,5-dimethylthiazol-2-yl)-5-(3-carboxymethoxyphenyl)-2-(4-sulfophenyl)-2H-tetrazolium (MTS). Each well of a 96-well plate was seeded with 5000 cells suspended in culture medium. Cells were treated with different concentrations of drug. After 24 h or 48 h of treatment, MTS was added to the cells according to the supplier's protocol, and the absorbance of the reaction product formed by the cultured cells was measured at 490 nm using an ELISA plate reader.

Apoptosis Assay

Cells (2×10^5) derived from the neurospheres were seeded in 6-well plates with complete stem cell medium. The following day, the cells were treated for an additional 48 hours with the indicated concentrations of BBMD3. After treatment, all cells were collected, and the apoptotic cells were detected using an Annexin V-FITC Apoptosis Detection Kit (BD Biosciences). The cells were stained with Annexin V-FITC and propidium iodide (PI) according to the supplier's instructions. Viable and apoptotic cells were detected by flow cytometry in our Analytical Cytometry Core here at the City of Hope National Medical Center. Apoptotic cells include both the early apoptotic portion (Annexin V-positive) and the late apoptotic portion (Annexin V and PI-positive).

Photography

Normal NS culture, untreated GBM NS cultures derived from patient brain tumors (i.e., PBT003, PBT008, PBT022, PBT030) or PBT003 NS treated with different concentrations of BBMD3 were photographed at 10X magnification using a Nikon ECLIPSE TE2000-U microscope.

Total Protein Preparation and Protein Assay

Following a 24 hour incubation with 5 μM BBMD3, cells derived from patient brain tumors (i.e., PBT003, PBT008, PBT022 and PBT030) were collected and washed with cold phosphate-buffered saline (PBS). Then, each cell pellet was added to 150 μl of Cell Signaling Lysis buffer (EMD Millipore), into which we had placed a protease inhibitor cocktail tablet (Roche, Inc., Indianapolis, IN). Following three successive cycles of freezing in a dry ice/ethanol bath and thawing in a 37°C water bath, the total protein concentration of the lysate was determined using a Bio-Rad Protein Assay kit.

Immunoblotting Analysis

Twenty micrograms of total protein were resolved through a pre-cast 4–15% gradient Tris-HCl Polyacrylamide gel purchased from BIO-RAD (Hercules, CA). After gel electrophoresis, the proteins were transferred to Hybond-C membranes (Amersham), blocked for 1 hour at room temperature (RT) in 10% non-fat dry milk, which was dissolved in PBST (1X PBS containing 0.1% Tween-20), and incubated overnight at 4°C with primary antibodies (recognizing the proteins outlined in the text) suspended in PBST containing 2% non-fat dry milk. After washing the membranes in PBST, horseradish peroxidase labeled anti-mouse or anti-rabbit secondary antibodies were incubated with the membranes for 1 hour at RT. The location on the Hybond-C membrane of the antigenic proteins specifically bound to the antibodies was detected with SuperSignal West Pico substrate (Pierce).

Quantitative Analysis of microRNAs

Total microRNA was isolated from PBT003, PBT008, PBT022 and PBT030 neurospheres following a 15 hour of treatment with 5 μM BBMD3 or the vehicle control using the miRNeasy Mini kit (Qiagen). Quantification of the relative abundance of each individual microRNA, with or without BBMD3 treatment, was completed in the Integrative Genomics Core at the City of Hope National Medical Center. Briefly, micro-RNA sequencing was performed using the Illumina HiSeq2000 following the manufacturer's protocol (TruSeq Small RNA Sample Prep kit, Illumine, Inc.) with minor optimization. 500 ng of micro-RNA was ligated to the 3′ adapter (5′ TCTCTGTAGGCACCATCAATC) with T4 RNA ligase 2 for 1 hour at 22°C. The unligated 3′ adaptors were blocked by annealing them with RT primer (5′ ATT-GATGGTGCCTACAGAGA) at 75°C for 5 min and 25°C for 15 minutes. The quantified denatured micro-RNA library was loaded in 1 ml of hybridization buffer to a final concentration of 10 pM.

Treatment of micro-RNA Inhibitors

8000 single cells derived from PBT008 and PBT030 neurospheres in complete stem cell culture media were seeded into each well of 96-well plates. Five hours later, 60 nM miR-4284 or miR-27a anti-sense inhibitors were transfected into the cells by RNAiFect transfection reagent. The control cells were transfected with a control micro-RNA inhibitor. 24 hours following transfection with micro-RNA inhibitors, 5 μM BBMD3 was added to the cell culture plates. 24 hours following BBMD3 treatment, cell viability was determined.

Statistics

Student's *t* test was used to evaluate the statistical significance of differences between two groups and $p<0.05$ was considered statistically significant.

Results

Neurospheres Cultured from Specimens of GBM Patients Exhibit Characteristics of Cancer Stem-cells

We cultured and expanded short-term tumor neurospheres in serum-free stem cell media, (supplemented with EGF and FGF), that were derived from four primary GBM patients, (i.e., PBT003, PBT008, PBT022 and PBT030). These GBM neurospheres exhibited a similar morphology to that of normal neural stem cells (Fig. 1A). It is reported that GBM stem-like cells (GBMSCs) and normal neural stem cells (NSCs) share the expression of several markers, such as CD133 and nestin [28]. CD133 is a cell surface marker expressed on normal human NSCs, and its expression is down-regulated in differentiated cell lines. Nestin is an intermediate filament protein produced during development in stem cells originating in the mammalian central nervous system. Co-expression of CD133 and nestin is associated with a poor prognosis for malignant glioma patients [28]. We observed by Western blot analysis that CD133 and nestin were expressed in GBM neurospheres cultured from each of the four GBM patients, while no expression of CD133 and nestin was observed in an established GMB cell line, (i.e., U87 cells), (Fig. 1B). β-actin expression levels were monitored to ensure that equivalent amounts of cell protein were loaded into each lane of the gel. When cultured in

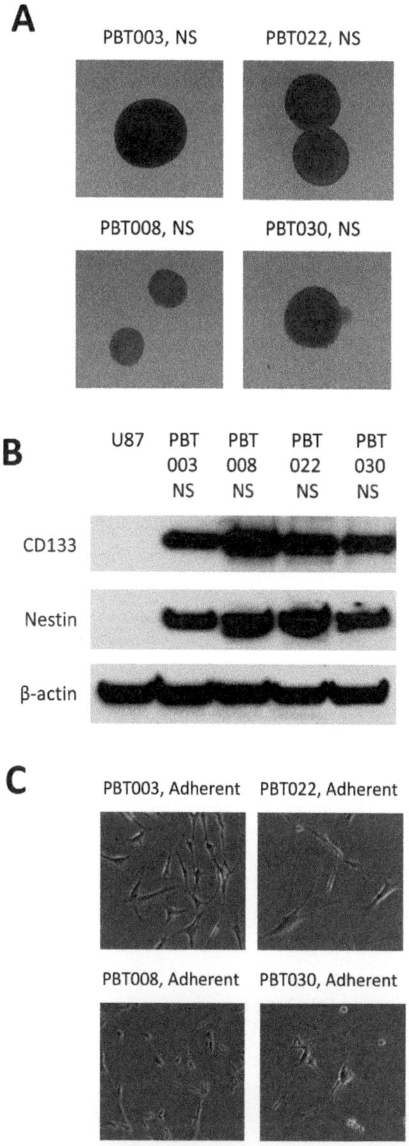

Figure 1. Neurospheres cultured from four GBM patients, PBT003, PBT008, PBT022 and PBT030 exhibited morphological characteristics of neural stem cells. (**A**) Morphology of PBT003, PBT008, PBT022 and PBT030 neurospheres in stem cell culture medium. (**B**) GBM neurospheres expressed specific biomarkers for normal neural stem cells. (**C**) GBM neurospheres exhibit the ability to differentiate.

media with serum, the neurospheres derived from PBT003, PBT008, PBT022 and PBT030 GBM patients have the ability to differentiate into adherent and neuron-like cells, (Fig. 1C). Cell cultures derived from the specimens of the four GBM patients (PBT003, PBT008, PBT022 and PBT030) grew in typical neurospheres, and retained the ability for self-renewal and expressed neural stem cell markers, such as CD133 and nestin when cultured in serum-free media. These neurospheres also had the ability to differentiate in serum-containing growth media. In this report, we therefore defined these neurospheres as either GBM stem-like cells or cancer stem-like cells.

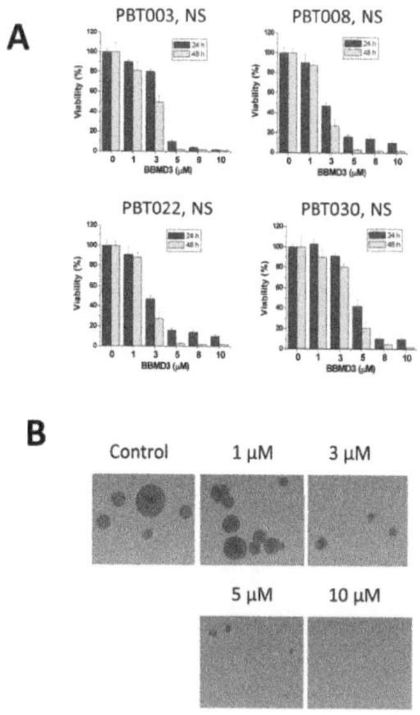

Figure 2. BBMD3 inhibited cell viability and induced cell death in PBT003, PBT008, PBT022 and PBT030 neurospheres. (**A**) Cells from PBT003, PBT008, PBT022 and PBT030 neurospheres were treated with 0, 1, 3, 5, 8, 10 μM BBMD3 for 24 and 48 hours, and viability was determined using the MTS assay, as described in the Methods. Each experiment was performed in triplicate. The top of each bar graph represents the mean of 3 experiments, and the *error bars* represent ± the standard deviation from the mean (SD). (**B**) BBMD3 disrupts the structure of the neurospheres, and induced cell killing in a dose-dependent manner in these PBT003 derived neurospheres, when cells are treated for 48 hours with the drug.

BBMD3 Disrupts the Structure of the Neurospheres and Inhibits the Viability of Cancer Stem-like Cells Cultured from the Four GBM Patients

GBM derived cancer stem-like cells, are less sensitive to, (or become resistant to), conventional chemotherapy [4,5]; highlighting a critical need for finding new and more selectively potent drugs for treating glioblastoma. Berbamine, a natural product derived from the *Berberis amurensis* plant, appears to have cytotoxic activity toward a variety of cancers; prompting the preparation of a series of analogs of Berbamine. These analogs were evaluated to determine whether any of them might be more effective at killing cancer cells than the parent compound. We found that the anticancer efficacy of one of the novel synthetic derivatives, (i.e., BBMD3), on melanoma and prostate cancer cells exceeded that of the other analogs [17]. Thus, we investigated the anticancer effects of BBMD3 on GBM stem-like cells, which were derived from neurospheres cultured from GBM patient's PBT003, PBT008, PBT022 and PBT030. During the study, these cells were treated for 24 or 48 hours in the presence of complete stem cell culture medium containing 1, 3, 5, 8 or 10 μM BBMD3. Control cell cultures were treated with only vehicle (DMSO) in place of BBMD3, and cell viability was determined as described in the Methods. BBMD3 strongly inhibited the viability of each set of the GBM stem-like cancer cells, derived from the individual patient neurospheres, in a time- and dose-dependent manner (Fig. 2A). A 48 hour treatment of the neurospheres with 10 μM BBMD3 inhibited cell viability nearly 100%, and disrupted the structure of neurospheres in a dose-dependent manner. Figure 2B shows that PBT003 derived neurospheres, treated with 0, 1, 3, 5 or 10 μM BBMD3 for 48 hours, disrupted the spherical morphology of the neurospheres following BBMD3 treatment, and that disruption of

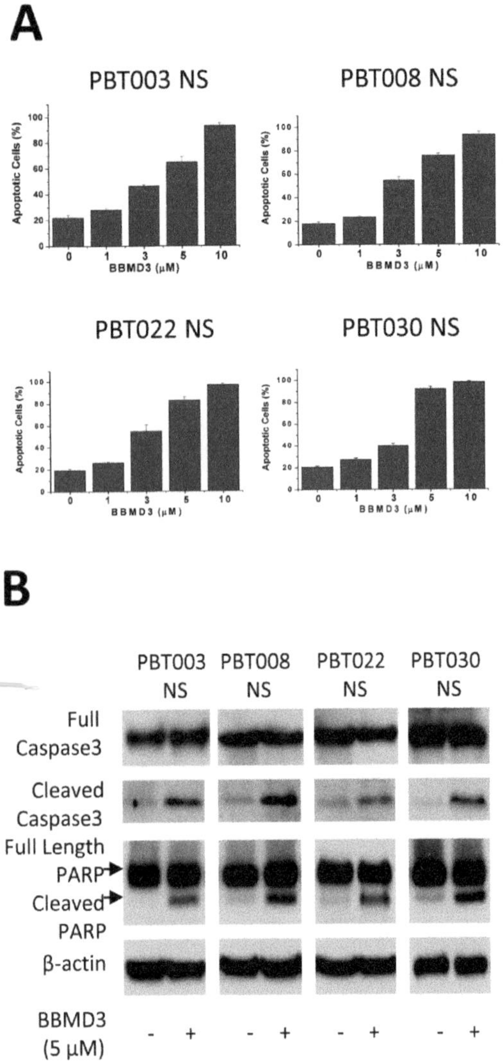

Figure 3. BBMD3 induced apoptosis and increased cleavage of caspase-3 and PARP in GBM neurospheres. (**A**) Apoptotic cells were analyzed for Annexin V-FITC reactivity and PI staining by flow cytometry. Cells from the PBT003, PBT008, PBT022 and PBT030 neurospheres were treated with BBMD3 (0, 1, 3, 5, 10 μM) for 48 hours. Apoptotic cells were identified as being reactive with Annexin V-FITC antibody (early stage of apoptosis) or PI and Annexin V-FITC/PI double-positive (late stage of apoptosis) cells. Each experiment was performed in duplicate or triplicate, and repeated twice as an independent replicate. The top of each bar graph represents the mean of the observed values from these replicates, and the *error bars* represent ± the standard deviation from the mean (SD). (**B**) The extent of cleavage of caspase-3 and PARP was determined following a 24 hour treatment with BBMD3 using Western blotting analyses. β-actin serves as an internal control.

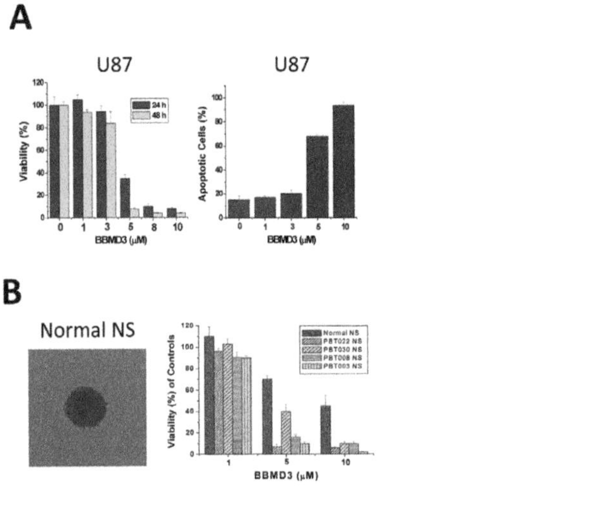

Figure 4. The effects of BBMD3 on non-stem-like GBM tumor cells and normal neurospheres. (**A**) ***Left panel,*** U87 cells were treated with 0, 1, 3, 5, 8, or 10 μM BBMD3 for 24 and 48 hours, and viability was determined using the MTS assay, as described in the Methods. ***Right panel,*** U87 cells were treated with 0, 1, 3, 5, or 10 μM BBMD3 for 48 hours and apoptotic cells were analyzed for Annexin V-FITC reactivity and PI staining by flow cytometry. (**B**) ***Left panel*** shows the morphology of a normal human neurosphere. ***Right panel*** shows the viability of normal fetal brain derived and tumor derived neurospheres following 24 hours of treatment with 1, 5, or 10 μM BBMD3. Each experiment was performed in triplicate, and replicated as an independent experiment at least twice. The top of each bar graph represents the mean of the observed values from these replicates, and the *error bars* denote ± the standard deviation from the mean (SD).

the neurospherical shape was companied by the appearance of floating dead cells.

BBMD3 Induces Caspase-3 Dependent Apoptosis of Cancer Stem-like Cells Cultured from Four GBM Patients

Because BBMD3 treatment killed GBM derived cancer stem-like cells, we examined whether cell killing was mediated by an apoptotic process. Cells (2×10^5) from PBT003, PBT008, PBT022 and PBT030 neurospheres were seeded into six-well tissue culture plates, and treated for 48 hours with different concentrations (0, 1, 3, 5, 10 μM) of BBMD3. All of the cells were collected and analyzed for expression of Annexin V using anti-Annexin V-FITC antibody and propidium iodide staining followed by relative quantification via flow cytometry. Apoptotic cells, shown in Figure 3A, were Annexin V positive (early apoptotic cells), or Annexin V and propidium iodide double-positive (late apoptotic cells). Our results show that BBMD3 induced apoptosis in these GBM stem-like cells in a dose-dependent manner. A 48 hour treatment with 10 μM BBMD3 killed nearly all of the tumor cells. Activation of caspase-3, a critical inducer of the apoptotic process [29], resulted in cleavage of poly (ADP-ribose) polymerase (PARP), which helps cells maintain their viability [30]. To further confirm that the cell killing induced by BBMD3 is an apoptotic process, Western blotting analyses were performed to detect the activation of caspase-3 and the cleavage of PARP in a total cell lysate following a 24 hour treatment with 5 μM BBMD3. BBMD3 increased the cleavage of caspase-3 into its active form, and the cleavage of PARP into its inactive form in cells derived from the four neurosphere cultures (Fig. 3B). These data point toward BBMD3 inducing apoptosis in these GBM stem-like cells via activation of the caspase-3 cascade. Our data also suggest that multiple mechanisms may be involved in the induction of cell killing and the loss in cell viability, since, at the time point assayed, PBT003 NS incubated with 5 μM BBMD3 induced an almost total loss in cell viability, with an apoptosis rate of approximately 60%; while for the PBT030 NS cell viability was approximately 20% after incubation with 5 μM BBMD3, while apoptosis was almost 100% in this culture.

Table 1. Effects of BBMD3 on expression of MiRNAs in PBT003, PBT008, PBT002, and PBT030 neurospheres after 15 h treatment with 5 μM BBMD3.

Downregulation		Upregulation	
Name of miRNA	**Folds**	**Name of miRNA**	**Folds**
hsa-miR-7	−1.06	hsa-miR-7	4.48
hsa-miR-106b	−0.91	hsa- miR-106b	2.25
hsa-miR-25	−0.79	hsa-miR-25	1.26
hsa-miR-20b	−0.77	hsa-miR-20b	1.15
hsa-miR-20a	−0.77	hsa-miR-20a	1.00
hsa-miR-185	−0.64	hsa-miR-185	0.78
hsa-miR-374b	−0.58	hsa-miR-374b	0.71
hsa-miR-503	−0.57	hsa-miR-503	0.70
hsa-miR-148a	−0.56	hsa-miR-148a	0.57

BBMD3 also Reduces the Viability and Induces Apoptosis in Non-stem-like GBM Cells

As we have reported, BBMD3 is cytotoxic to stem-like GBM tumor cells; suggesting that it should also be cytotoxic to non-stem-like GBM cancer cells. To examine this possibility, we used cultured U87 cells, which are an established non-stem-like GBM cancer cell line. We observed that U87 cells did not express the neural stem-cell markers CD133 and nestin (see Fig. 1B for comparison to the stem-like GBM tumor cell lines). As we expected, BBMD3 inhibited cell viability and induced apoptosis in the U87 cells (Fig. 4A) in a similar manner to that seen with the GBM stem-like cells (Fig. 2A and 3A). Our data indicates that BBMD3 can kill both stem-like and non-stem-like GBM tumor cells.

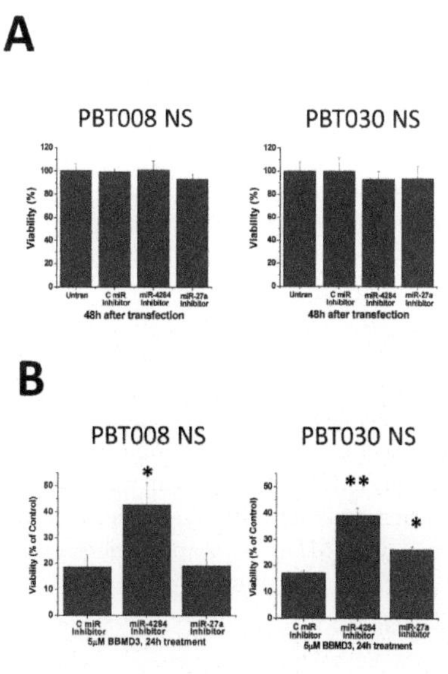

Figure 5. The effect of miR-4284 and miR27a inhibitors on the cytotoxic activity of BBMD3 in PBT003 and PBT030 neurospheres. (**A**) Cells from PBT003 and PBT030 neurospheres were transfected with 60 nM of anti-sense inhibitors against miR-4284, miR-27a and an unrelated control miRNA, and the inhibitor was allowed to interact with the miRNA for 48 hours prior to determining the viability of the cells. (**B**) After cells from PBT003 and PBT030 neurospheres were transfected with 60 nM anti-sense nucleotide inhibitors against miR-4284, miR-27a and an unrelated control anti-sense miRNA, and the inhibitors were allowed to interact with the miRNA for 24 hours; 5 μM of BBMD3 was added to the culture media for another 24 hour. Cell viability was determined at the end of the 48 hour period. Each experiment was performed in triplicate. The top of each bar graph represents the mean of 3 experiments, and the *error bars* represent ± the standard deviation from the mean SD. *, $p<0.01$; **, $p<0.001$.

Normal Human Neurospheres are Less Sensitive to BBMD3 than GBM Derived Neurospheres

Berbamine and its derivatives have also been reported to exert less of a cytotoxic effect toward normal human cells [17] than toward GBM and other types of tumor cells. To further evaluate the differential cytotoxicity of BBMD3 toward normal neurospheres, we tested the effects of BBMD3 on normal human neurospheres derived from the normal fetal brain. Figure 4B shows the morphology of normal neurospheres, and the effects of BBMD3 on normal neurospheres and the neurospheres derived from the four GBM patients following a 24 hour exposure to the drug. In comparison to the tumor derived neurospheres, normal neurosphere viability is reduced by exposure to increasing concentrations of BBMD3, but these normal neurosphere cultures are less sensitive to the cytotoxic effects of BBMD3 than the neurosphere cultures derived from the 4 GBM patients.

BBMD3 Increases the Expression of miR-4284 and miR-27a in PBT003, PBT008, PBT022 and PBT030 Neurospheres

Though miRNAs have been known for only a few years, they have been found to play crucial roles in the processes mediating tumorigenesis, angiogenesis, cell invasion, and apoptosis in a variety of tumor types. We therefore examined whether BBMD3 treatment could alter the expression profile of miRNAs in GBM stem-like cells. Neurospheres, derived from PBT003, PBT008, PBT022 and PBT030 GBM patients, were treated with 5 μM BBMD3 for 16 hours, while control cells were treated with only vehicle (DMSO). Total miRNA was isolated from these neurosphere cultures, and expression of each miRNA was quantified. In comparison to the controls, BBMD3 treatment up-regulated expression of some miRNAs, and down-regulated expression of other miRNAs. Table 1 shows the average miRNA expression profile, relative to the untreated control GBM stem-like cells, for each of the four GBM patients following BBMD3 treatment. miR-7 is the most down-regulated miRNA, with a decrease in

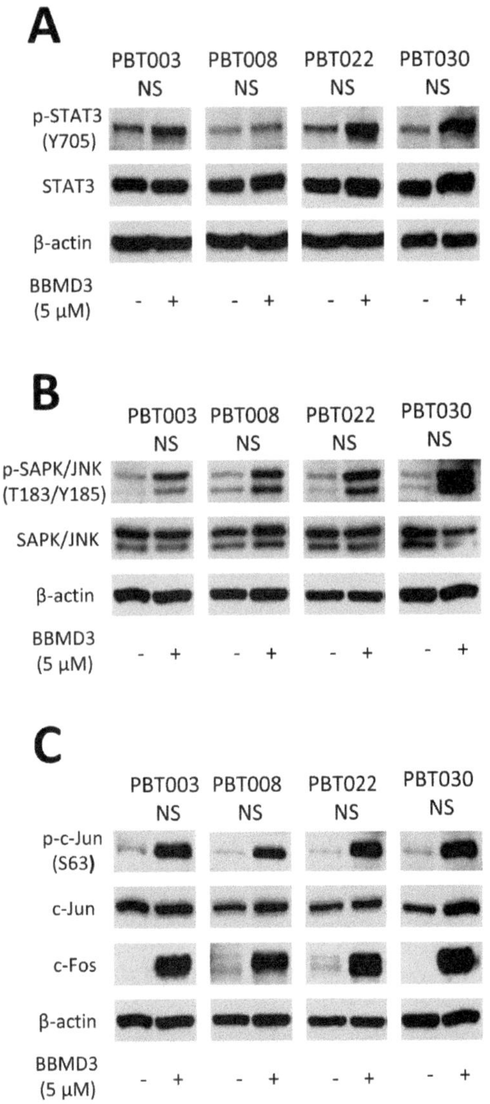

Figure 6. Western blotting analysis of BBMD3 induced phosphorylation of JNK 1 and 2, c-Jun and total c-Fos protein expression in cells derived from human GBM neurospheres. (A) The effect of BBMD3 on the expression of total and phosphorylated STAT3 in cells derived from PBT003, PBT008, PBT022 and PBT030 neurospheres following a 24 hour treatment with BBMD3. (**B**) Western Blotting analysis of cell lysates probing for phosphorylated JNK expression following a 24 hour treatment of cells derived from GBM neurospheres with BBMD3. (**C**) Phosphorylation of c-Jun and quantification of c-Fos total protein expression following BBMD3 treatment.

expression of 1.06 fold. In contrast, the two most up-regulated miRNAs are miR-4284 and miR-27a; increasing 4.48 and 2.25 fold respectively. It was reported that up-regulation of miR-23a, 27a, and 24–2 induces both caspase-dependent and -independent apoptosis in human embryonic kidney cells [31]. To date, we have been unable to find any publication describing the regulatory function(s) of miR-4284.

miR-4284 Inhibitor Blocks the Effects of BBMD3 on the Viability of GBM Neurospheres

To confirm whether the increased expression of miR-27a and miR-4284 induced by BBMD3 treatment correlates with the decrease in GBM neurosphere viability, we examined whether miR-27a and miR-4284 anti-sense sequences could inhibit the cytotoxic activity of BBMD3. Our first step was to examine whether these inhibitors alone inhibited cell viability. 60 nM of human anti-sense miRNA inhibitors (i.e., synthetic oligonucleotides) against hsa-miRNA-4284 and hsa-miRNA-27a were transfected into cells derived from PBT008 and PBT030 neurospheres. Unrelated anti-sense micro-RNA inhibitor was used as a negative control, and 48 hours following transfection with these miRNA inhibitors, cell viability was determined. The results shown in Fig. 5A demonstrated that miR-27a and miR-4284 inhibitors alone did not effectively decrease cell viability relative to that of the untransfected cells (untran). We then tested the effect of miR-27a and miR-4284 inhibitors on the cytotoxic activity of BBMD3 using PBT008 and PBT030 neurospheres. After these GBM neurospheres were transfected with, and allowed to express for 24 hours, the miR-27a and miR-4284 anti-sense inhibitors, or the control miRNA inhibitor sequence, cells were incubated with 5 μM BBMD3 for another 24 hours, at which point cell viability was determined. The results shown in Figure 5B indicated that, in contrast to the control miRNA inhibitor, cells transfected with the anti-sense miR-4284 inhibitor showed greater viability (two fold) following BBMD3 treatment in both the PBT008 and PBT030 neurosphere cultures. However, transfection of the miR-27a anti-sense inhibitor into these neurosphere cultures was not as effective at reducing the cytotoxic effect of BBMD3 treatment on the GBM stem-like cells as transfection with the miR4284 inhibitor. Therefore, BBMD3 mediated inhibition of GBM stem-like cell viability depends, at least in part, on increasing expression of miR-4284.

BBMD3 Increases the Phosphorylation of JNK1 and JNK2 in GBM Stem-like Cells

Since BBMD3 was reported to inhibit the STAT3 signaling pathway in melanoma cells [17], we evaluated whether BBMD3 could also inhibit this pathway in human GBM stem-like cells. Following the treatment of PBT003, PBT008, PBT022, PBT030 derived neurospheres with 5 μM BBMD3 for 24 hours, total cellular protein was isolated and analyzed by Western blotting for phosphorylation of STAT3. Our results indicated that BBMD3 did not inhibit the phosphorylation of STAT3 at Tyr705, the active form of STAT3 in these cells (Fig. 6A). Thus, the anti-cancer potential of BBMD3 on GBM stem-like cells was not a result of inhibiting the STAT3 signaling pathway. We then examined whether BBMD3 could activate the JNK stress-response signaling pathway following BBMD3 treatment. In related work from this laboratory, the JNK pathway was shown to be responsible for the anti-cancer effect of BBMD3 in human osteosarcoma cells [18]. Following a 24 hour incubation of cells derived from PBT003, PBT008, PBT022 and PBT030 neurospheres with 5 μM BBMD3, total protein was isolated and analyzed by Western blotting for expression of the JNK1 (46 kD) and JNK2 (54 kD) isoforms, using antibodies that are specific to these proteins. Surprisingly, BBMD3 dramatically increased the phosphorylation (active form) of both of the JNK isoforms in the neurosphere cultures derived from all four GBM patients (Fig. 6B); suggesting that the up-regulation of the JNK signaling pathway mediates the cytotoxic response exhibited by the neurosphere derived cells following incubation with BBMD3.

BBMD3 Increases the Phosphorylation of c-Jun in GBM Stem-like Cells

Because up-regulation of the JNK/AP-1 signaling pathway was shown to have a direct role in the induction of apoptosis induced by BBMD3 in human osteosarcoma cells [18], this signaling pathway was anticipated to play an important role in the apoptotic process induced by BBMD3 in GBM stem-like cells. c-Jun is a downstream target of the JNK signaling pathway, and is the most extensively studied protein of the activator protein-1 (AP-1) complex. It has an important role in contributing to the regulation of numerous cellular processes, such as cell proliferation, apoptosis, cell survival, tumorigenesis and tissue morphogenesis. Because our data indicates that BBMD3 activates both JNK1 and JNK2, we evaluated whether BBMD3 also activates c-Jun. As expected, BBMD3 contributes to the increased phosphorylation of c-Jun in the neurospheres derived from all four GBM patients (Fig. 6C). BBMD3 also contributed to the increase expression of total c-Fos (Fig. 6C), another major component of transcription factor AP-1.

Discussion

Glioblastoma is incurable by conventional approaches. Though most of the tumor can be removed by surgery, radiotherapy and chemotherapy, recurrence of the tumor eventually occurs. GBM stem cells play a key role in the initiation of the primary tumors and the recurrence of the tumor following therapeutic intervention. Recurrence is attributable to the self-renewal capacity of the GBM stem-like cells and their inherent resistance to traditional therapy [32]. Thus there is an urgent need for new therapeutic approaches which can selectively target the GBM stem cell. In this study, BBMD3, a novel synthetic derivative from the natural product Berbamine, was shown to effectively inhibit the viability and induce apoptosis in GBM stem-like cells. The anti-cancer effects of BBMD3 in these cells was shown to be associated with the up-regulation of miR-4284 expression and activation of the stress-response JNK signaling pathway, as well as the AP-1 transcriptional factor, a JNK downstream target.

Additionally, microRNAs play crucial roles in the processes mediating tumorigenesis, angiogenesis, invasion and apoptosis for various types of tumors. Recent studies have identified dysregulation of specific miRNAs in malignant gliomas [10,11]. In our study, we found that BBMD3 treatment up-regulated the expression of miR-4284 more than four-fold in GBM stem-like cells. An anti-sense inhibitor of miR-4284 activity could partially block the anticancer effects of BBMD3 on those GBM stem-like cells. Thus, BBMD3 inhibits cell viability and induces apoptosis in these stem-like cells, at least in part, through increasing expression of miR-4284. As an aside, it should be noted that all of the functions attributed to miR-4284 expression have not yet been identified. Though miRNA mediated regulation of gene expression remains an emerging field of investigation, expression profiles for a variety of miRNAs have been shown to be useful for predicting GBM patient survival, and these profiles have the potential to identify efficacious therapeutic targets [33].

In this manuscript we reported that BBMD3 reduces GBM stem-like cell viability and induces apoptosis in these cells, in a manner consistent with the reports on the effects of BBMD3 on human melanoma and osteosarcoma tumor cells [17,18]. These cytotoxic effects of BBMD3 are mediated through inhibition of the Jak2/STAT3 signaling pathway in melanoma cells and activation of the stress-response JNK pathway in osteosarcoma cells, respectively. These reported observations lead to the conclusion that BBMD3 may target different signaling pathways in different tumor types. In our present study, BBMD3 inhibited cell viability and induced apoptosis of human GBM stem-like cells with an LD_{50} of less than 5 μM after a 48 hour exposure. However, BBMD3 did not suppress STAT3 activity, but rather strongly activated the JNK signaling pathway in GBM stem-like cells. The JNK signaling pathway contributes to cell proliferation and the apoptotic process; suggesting that the JNK pathway is involve in the induction of apoptosis in the GBM stem-like cells. The involvement of the JNK pathway during the induction of apoptosis comes from the observation that $Jnk1^{-/-}Jnk2^{-/-}$ mice were resistant to apoptosis induction following UV irradiation, exposure to anisomycin or the DNA-alkylating agent methyl methanesulfate [34]. JNK is an intrinsic component of the mitochondrial-dependent death pathway associated with stress-induced apoptosis. Following activation of JNK, BBMD3 increases expression of phosporylated c-Jun protein and the expression of c-Fos. Both of these proteins are downstream targets of the JNK pathway in GBM stem-like cells. c-Jun and c-Fos are themselves major components of AP-1, a transcription factor, which regulates proliferation, inflammation, differentiation, apoptosis and cellular migration processes [24,25]. Our data concerning the GBM stem-like cells along with additional evidence [35] indicates that the JNK-c-Jun/AP-1 signaling pathway plays a key role in mediating the induction of apoptosis in GBM stem-like cells [35].

The GBM stem-like cells we used for this project were cultured from four GBM patients and though derived from tumor tissue, retained some of the properties of normal neural stem cells, such as their spherical morphology, the ability to exhibit self-renewal and the ability to differentiate. BBMD3 not only induced apoptosis in these GBM stem-like cells, it also induced apoptosis in established GBM cancer cell lines, including U87 and U251 cells (data not shown). Our data indicates that BBMD3 exhibits a strong anticancer cell activity toward GBM stem-like cells; while exhibiting a lower cytotoxicity toward normal human cells [17]. We therefore believe that BBMD3 may have the potential to be developed as a promising drug for the treatment of malignant gliomas.

Author Contributions

Conceived and designed the experiments: FY SN CEB RZ RS DAH LHM RJ RJH. Performed the experiments: FY SN RZ. Analyzed the data: FY RJ RJH. Contributed reagents/materials/analysis tools: SN DAH CEB RS. Wrote the paper: FY RJH. Obtained permission for use of cell lines: FY.

References

1. Sathornsumetee S, Reardon DA, Desjardins A, Quinn JA, Vredenburgh JJ, et al. (2007) Molecularly targeted therapy for malignant glioma. Cancer 110: 13–24.
2. Nakada M, Nakada S, Demuth T, Tran NL, Hoelzinger DB, et al. (2007) Molecular targets of glioma invasion. Cell Mol Life Sci 64: 458–478.
3. Galli R, Binda E, Orfanelli U, Cipelletti B, Gritti A, et al. (2004) Isolation and characterization of tumorigenic, stem-like neural precursors from human glioblastoma. Cancer Res 64: 7011–7021.
4. Bao S, Wu Q, McLendon RE, Hao Y, Shi Q, et al. (2006) Glioma stem cells promote radioresistance by preferential activation of the DNA damage response. Nature 444: 756–760.
5. Bao S, Wu Q, Li Z, Sathornsumetee S, Wang H, et al. (2008) Targeting cancer stem cells through L1CAM suppresses glioma growth. Cancer Res 68: 6043–6048.

6. Huang Z, Cheng L, Guryanova OA, Wu Q, Bao S (2010) Cancer stem cells in glioblastoma–molecular signaling and therapeutic targeting. Protein Cell 1: 638–655.
7. Brown CE, Starr R, Martinez C, Aguilar B, D'Apuzzo M, et al. (2009) Recognition and killing of brain tumor stem-like initiating cells by CD8+ cytolytic T cells. Cancer Res 69: 8886–8893.
8. Ambros V (2004) The functions of animal microRNAs. Nature 431: 350–355.
9. Thorsen SB, Obad S, Jensen NF, Stenvang J, Kauppinen S (2012) The therapeutic potential of microRNAs in cancer. Cancer J 18: 275–284.
10. Mizoguchi M, Guan Y, Yoshimoto K, Hata N, Amano T, et al. (2012) MicroRNAs in Human Malignant Gliomas. J Oncol 2012: 732874.
11. Godlewski J, Nowicki MO, Bronisz A, Williams S, Otsuki A, et al. (2008) Targeting of the Bmi-1 oncogene/stem cell renewal factor by microRNA-128 inhibits glioma proliferation and self-renewal. Cancer Res 68: 9125–9130.
12. Du HP, Shen JK, Yang M, Wang YQ, Yuan XQ, et al. 4-Chlorobenzoyl Berbamine induces apoptosis and G2/M cell cycle arrest through the PI3K/Akt and NF-kappaB signal pathway in lymphoma cells. Oncol Rep 23: 709–716.
13. Liang Y, Xu RZ, Zhang L, Zhao XY (2009) Berbamine, a novel nuclear factor kappaB inhibitor, inhibits growth and induces apoptosis in human myeloma cells. Acta Pharmacol Sin 30: 1659–1665.
14. Wang S, Liu Q, Zhang Y, Liu K, Yu P, et al. (2009) Suppression of growth, migration and invasion of highly-metastatic human breast cancer cells by Berbamine and its molecular mechanisms of action. Mol Cancer 8: 81.
15. Duan H, Luan J, Liu Q, Yagasaki K, Zhang G (2010) Suppression of human lung cancer cell growth and migration by Berbamine. Cytotechnology 62: 341–348.
16. Wang GY, Lv QH, Dong Q, Xu RZ, Dong QH (2009) Berbamine induces Fas-mediated apoptosis in human hepatocellular carcinoma HepG2 cells and inhibits its tumor growth in nude mice. J Asian Nat Prod Res 11: 219–228.
17. Nam S, Xie J, Perkins A, Ma Y, Yang F, et al. (2012) Novel synthetic derivatives of the natural product Berbamine inhibit Jak2/Stat3 signaling and induce apoptosis of human melanoma cells. Mol Oncol 6: 484–493.
18. Yang F, Nam S, Zhao R, Tian Y, Liu L, et al. (2013) A novel synthetic derivative of the natural product Berbamine inhibits cell viability and induces apoptosis of human osteosarcoma cells, associated with activation of JNK/AP-1 signaling. Cancer Biol Ther 14.
19. Wada T, Penninger JM (2004) Mitogen-activated protein kinases in apoptosis regulation. Oncogene 23: 2838–2849.
20. Roux PP, Blenis J (2004) ERK and p38 MAPK-activated protein kinases: a family of protein kinases with diverse biological functions. Microbiol Mol Biol Rev 68: 320–344.
21. Seger R, Krebs EG (1995) The MAPK signaling cascade. FASEB J 9: 726–735.
22. Davis RJ (2000) Signal transduction by the JNK group of MAP kinases. Cell 103: 239–252.
23. Hibi M, Lin A, Smeal T, Minden A, Karin M (1993) Identification of an oncoprotein- and UV-responsive protein kinase that binds and potentiates the c-Jun activation domain. Genes Dev 7: 2135–2148.
24. Shaulian E (2010) AP-1–The Jun proteins: Oncogenes or tumor suppressors in disguise? Cell Signal 22: 894–899.
25. Liu J, Lin A (2005) Role of JNK activation in apoptosis: a double-edged sword. Cell Res 15: 36–42.
26. Shieh JM, Huang TF, Hung CF, Chou KH, Tsai YJ, et al. (2010) Activation of c-Jun N-terminal kinase is essential for mitochondrial membrane potential change and apoptosis induced by doxycycline in melanoma cells. Br J Pharmacol 160: 1171–1184.
27. Wei G, Wang M, Carr BI (2010) Sorafenib combined vitamin K induces apoptosis in human pancreatic cancer cell lines through RAF/MEK/ERK and c-Jun NH2-terminal kinase pathways. J Cell Physiol 224: 112–119.
28. Dimov I, Tasic-Dimov D, Conic I, Stefanovic V (2011) Glioblastoma multiforme stem cells. ScientificWorld Journal 11: 930–958.
29. Fernandes-Alnemri T, Litwack G, Alnemri ES (1994) CPP32, a novel human apoptotic protein with homology to Caenorhabditis elegans cell death protein Ced-3 and mammalian interleukin-1 beta-converting enzyme. J Biol Chem 269: 30761–30764.
30. Oliver FJ, de la Rubia G, Rolli V, Ruiz-Ruiz MC, de Murcia G, et al. (1998) Importance of poly(ADP-ribose) polymerase and its cleavage in apoptosis. Lesson from an uncleavable mutant. J Biol Chem 273: 33533–33539.
31. Chhabra R, Adlakha YK, Hariharan M, Scaria V, Saini N (2009) Upregulation of miR-23a-27a-24–2 cluster induces caspase-dependent and -independent apoptosis in human embryonic kidney cells. PLoS One 4: e5848.
32. Fatoo A, Nanaszko MJ, Allen BB, Mok CL, Bukanova EN, et al. (2011) Understanding the role of tumor stem cells in glioblastoma multiforme: a review article. J Neurooncol 103: 397–408.
33. Mizoguchi M, Guan Y, Yoshimoto K, Hata N, Amano T, et al. (2013) Clinical implications of microRNAs in human glioblastoma. Front Oncol 3: 19.
34. Tournier C, Hess P, Yang DD, Xu J, Turner TK, et al. (2000) Requirement of JNK for stress-induced activation of the cytochrome c-mediated death pathway. Science 288: 870–874.
35. Dhanasekaran DN, Reddy EP (2008) JNK signaling in apoptosis. Oncogene 27: 6245–6251.

8

Cell-Based Regenerative Strategies for Treatment of Diabetic Skin Wounds, a Comparative Study between Human Umbilical Cord Blood-Mononuclear Cells and Calves' Blood Haemodialysate

Hala O. El-Mesallamy[1]*, Mohamed R. Diab[2], Nadia M. Hamdy[1], Sarah M. Dardir[2]

1 Biochemistry Department, Faculty of Pharmacy, Ain-Shams University, Cairo, Egypt, **2** Research Department, Holding Company for Biological Products and Vaccines (VACSERA), Giza, Egypt

Abstract

Background: Diabetes-related foot problems are bound to increase. However, medical therapies for wound care are limited; therefore, the need for development of new treatment modalities to improve wound healing in diabetic patients is essential and constitutes an emerging field of investigation.

Methods: Animals were randomly divided into 8 groups (I–VIII) (32 rats/group), all were streptozotocin (STZ)-induced diabetics except groups III and VIII were non-diabetic controls. The study comprised two experiments; the first included 3 groups. Group I injected with mononuclear cells (MNCs) derived from human umbilical cord blood (HUCB), group II a diabetic control group (PBS i.v). The second experiment included 5 groups, groups IV, V, and VI received topical HUCB-haemodialysate (HD), calves' blood HD, and solcoseryl, respectively. Group VII was the diabetic control group (topical saline). Standard circular wounds were created on the back of rats. A sample of each type of HD was analyzed using the high performance liquid chromatography-electrospray ionization-mass spectrometry (HPLC-ESI-MS) system. Wound area measurement and photography were carried out every 4 days. Plasma glucose, catalase (CAT), malondialdehyde (MDA), nitric oxide (NO) and platelets count were assessed. Wound samples were excised for hydroxyproline (HP) and histopathological study.

Results: Treatment with HUCB MNCs or HUCB-HD resulted in wound contraction, increased CAT, NO, platelets count, body weights, and HP content, and decreased MDA and glucose.

Conclusion: Systemic administration of HUCB MNCs and topical application of the newly prepared HUCB-HD or calves' blood HD significantly accelerated the rate of diabetic wound healing and would open the possibility of their future use in regenerative medicine.

Editor: Dong-Gyu Jo, Sungkyunkwan University, Korea, Republic of

Funding: The authors have no support or funding to report.

Competing Interests: Co-authors Mohamed R. Diab and Sarah M. Dardir are employed by Holding Company for Biological Products and Vaccines (VACSERA). There are no patents, products in development or marketed products to declare.

* E-mail: hala.elmesallamy@hotmail.com

Introduction

From the chronic complications of diabetes are neuropathy and variety of connective tissue abnormalities [1]. Foot ulceration affects (15–25%) of all diabetic patients during their lifetime [2]. Only 2/3 of diabetic foot ulcers heal and up to 28% may result in lower extremity amputation [3]. To prevent or reduce surgical intervention, new therapeutic strategies are to be developed to improve diabetic wound healing.

Cell therapy is a promising approach for treating diabetic non-healing wounds [4]. The aim of cell-based regenerative strategies is repair or enhancement of the damaged tissues' biological function, by utilizing cells and/or bioactive molecules [5]. This can be carried out by transplantation, through local delivery or systemic infusion of autologous or allogenic cells [6]. These cells include primary cells, cell lines, and various stem cells [5]. Among the main sources of stem cells that might be used for regeneration of injured skin tissue are adult stem cells and embryonic stem cells (ESCs). ESCs have great capacity for self-renewal and pluripotency, but their clinical applications that are associated with ethical and legal issues have shifted the focus to adult stem cells [7].

A population of adult stem cells especially mesenchymal stem cells (MSC) resides within most of adult mammalian tissues/organs, and the most common sources include bone marrow and umbilical cord blood [5]. Bone marrow (BM) BM-MSC can produce multiple types of skin cell and insulin expressing cells [8], but the source of BM is limited and their capacity for differentiation decline with age [9]. Human umbilical cord blood

(HUCB) as a source of stem cells is readily available, on non-invasive collection and can be routinely harvested without any risk for the donor babies [10]. HUCB contains stem cells in greater number than BM [11]. The incidence of graft-versus host disease (GVHD) is lower in HUCB transplantation than other allogenic cell-based therapies. Therefore, the application of HUCB cell turned out to be an excellent alternative source of haemopoietic stem cells to other allogenic cell-based therapies [12].

Haemodialysate (HD) is another approach that has been developed a few decades ago; to improve situations of impaired healing in both experimental and human approaches [13]. Solcoseryl, referred to as the commercially available HD, is a chemically and biologically standardized, protein free, non pyrogenic, and non antigenic dialysate derived from healthy suckling calves' blood [14]. Solcoseryl activity is ascribed to its constituents, being a broad spectrum of natural low molecular weight substances including electrolytes, amino acids, lipids, phospholipids, essential trace element, and intermediate products of carbohydrate and fat metabolism [15]. Hence, the aim of the present study is to evaluate MNCs derived from HUCB as a cell therapy for diabetic wounds compared to untreated diabetic and normal wounds. In addition, HUCB will be used in a different way to prepare a low molecular weight fraction; "HD" which will be investigated as a potential topical treatment for the same condition in comparison to a HD prepared from calves' blood, and the commercially available product; "solcoseryl". Knowing that this type of HD has never been prepared before from HUCB, and has never been used in an experimental study focusing on diabetic wounds, we believe that this study hopefully, provides an innovation in diabetic wound treatment.

Materials and Methods

1. Ethics statement

Animals were housed in accordance with the Principles of laboratory animal care (NIH publication no. 85–23, revised 1985; http://grants1.nih.gov/grants/olaw/references/phspol.htm). The experimental protocol was approved as well by Faculty of Pharmacy, Ain Shams University/VACSERA ethical committees. An informed written consent was obtained from each mother participating in the study after approval of the General Organization for Teaching Hospitals & Institutes (GOTHI) research ethics committee as well as Faculty of Pharmacy, Ain Shams University/VACSERA ethical committees.

2. Drugs and chemicals

Solcoseryl ampoules were obtained from Misr Co. for Pharm. Ind. S.A.E under license of Valeant Pharmaceuticals Co. Switzerland GMBH. Other biochemical reagents, unless otherwise specified, were purchased from Sigma-Aldrich Chemical Co. (St. Louis, Mo., USA).

3. Animals

The study was performed on 256 adult male Albino rats of Wistar strain weighing (150–200 gm), obtained from the breeding colony of Helwan farm belonging to the Holding company for biological products and vaccines (VACSERA, Egypt).

4. Experimental design

Animals were randomly divided into 8 groups (I–VIII) (32 rats/group), all were diabetic except groups III and VIII were non-diabetic control groups.

5. Experimental induction of diabetes

Rats were made diabetic by a single I.P injection of 50 mg/kg body weight of STZ dissolved in citrate buffer (0.01 mol/l, pH 4.5). Rats serving as controls were given the same volume of sodium citrate. Diabetes was confirmed by determination of fasting blood glucose (FBG) concentration 3-days post STZ injection showing FBG levels above 250 mg/dl [16].

6. Collection of HUCB and isolation of cells

HUCB samples were collected from healthy full-term (36–40 weeks) normal human deliveries [17]; they were obtained from the Department of Obstetrics and Gynecology, El-Galaa hospital, Cairo, Egypt. Each cord blood sample was collected into a 50 ml sterile polypropylene tube containing 5 ml citrate phosphate dextrose (CPD) anticoagulant [18].To isolate MNCs, each UCB unit was carefully layered onto Ficoll separating solution (BIOCHROM AG, Berlin, Germany) (density = 1.077 g/ml) inside a disposable 15 ml centrifuge tube. After density gradient centrifugation at 400×g (~1500 rpm) for exactly 30 min at room temperature, the MNCs layer was removed from the interface, washed twice in PBS and centrifuged for 10 min at 250×g (~1200 rpm) [19]. Cell viability was more than 95% by Trypan-Blue exclusion. The number of cells was optimized to be (400×10^6) viable MNCs suspended in 0.5 ml PBS and injected one time into each animal via the tail vein (slow I.V injection).

7. Preparation of HUCB or calves' blood HD

Blood was applied to an Amicon device (Millipore Corporation-model, 8200, 200 ml, MA, USA) stirred ultra filtration cell containing nitrocellulose membrane of cut-off 10,000 Dalton. The system was surrounded with ice and put on a magnetic stirrer during the whole run (to avoid denaturation or decomposition of the blood components by the action of heat). Ultrafilterated blood containing the very low molecular weight substances was collected in sterile tubes and known as "HD". The produced HD was then filtered through sterile non-pyrogenic filter units (Schleicher & Schuell) (pore size = 0.2 μm) and aliquoted into small tubes, a sample of which was analyzed for its contents using (HPLC-ESI-MS) at the National Research Center (Dokki, Giza, Egypt) [20] and the remaining aliquots were stored inside a deep freezer (−70°C) until used.

8. Acute toxicity (LD50) study

Initially, a dose–response study was executed for selecting HD dose. Tested or prepared drug (calf or human UCB-HD) was injected separately in a group of mice weighing 20–25 g I.P at 0.4, 0.5, 0.6, 0.7, 0.8, 0.9, 1.25, 1.5, 1.75, and 2 ml doses per 20 g of mice, where each dose was given to 6 mice/cage. All mice were subjected to careful observation for any toxic symptoms or deaths for 3 consecutive days (72 h) [21]. This was followed by tissue culture for the determination of the minimum effective therapeutic dose of both types of HD [22].

9. Induction of skin ulcer (Excision wound creation)

All rats (diabetic and non-diabetic) were anesthetized with diethyl ether. The hair of each rat was shaved with an electric clipper. Area of the wound to be created was outlined on the dorsal skin with a marking pen and a template (1.5 cm diameter) just before skin excision [23].The entire wound was left open.

10. Evaluation of wound healing

10.1. Wound area measurement & photography. Starting from the day of wound induction (defined as day 0) and then on days 4, 8, 12, 16, and 20 post-wounding, wounds were photographed using a Canon (Canon PowerShot A2000 IS, Japan) digital camera and wound margins were traced on transparency papers then subsequently transferred to 1 mm^2 graph sheets from which the wound surface area was evaluated. Percentage of wound contraction was calculated using the following formula: [(Initial wound size- specific day wound size)/ Initial wound size]×100 [24]. Period of epithelialization was defined as the number of days taken for the complete healing of wounds so that no raw wound is left behind [25].

10.2. Experimental protocol. The first experiment comprised investigation of the healing effect of undifferentiated MNCs isolated from HUCB on diabetic ulcer. Groups I and II were STZ-induced diabetic. Group I designated as HUCB MNCs group; was treated with (400×10^6) viable MNCs suspended in 0.5 ml PBS (HUCB MNCs) and injected one time slowly into each animal via the tail vein immediately after wound creation. Group II was a diabetic control group and was injected with PBS i.v. The third group served as a normal (non-diabetic) control group, receiving no treatment.

The second experiment was designed to investigate the effect of cord blood and Calf blood HD on diabetic ulcers healing in comparison to solcoseryl (positive control drug). Groups IV, V, VI, and VII were STZ-induced diabetic and were topically treated with 150 μL of the treatment agent. Groups IV and V received prepared HUCB-HD and calves' blood HD, respectively. However, group VI served as a positive control group and was treated with solcoseryl. Group VII was a diabetic control group, which received normal saline, while the last group (VIII) was a non-diabetic control group receiving no treatment.

11. Biochemical analysis

Pre and post- treatment, the animals' body weights and FBG were determined. Blood samples for performing different biochemical and haematological tests were obtained by retro-orbital venous plexus puncture and collected into tubes containing the type of anticoagulant suitable for each test. Fluoride oxalate for determination of FBG [26], EDTA for determination of platelets count using an automated haematology analyzer (Cell-Dyn Emerald, Abbott Diagnostics, France) and plasma NO [27], sodium citrate for determination of MDA by thiobarbituric acid reactive species method [28] and CAT [29] by enzymatic kits supplied by Biodiagnostics (Giza, Egypt). Wound tissue samples were removed for the determination of granulation tissue weight, and stored at −70°C for subsequent biochemical assay of hydroxyproline (HP) [30] as an indicator of total collagen content.

Representative wound tissue from each group were fixed in 10% formol saline for subsequent histopathological examination by hematoxylin and eosin (H&E) [31].

12. Statistical analysis

Data are presented as means ± SD. Multiple group comparisons were carried out using one-way analysis of variance (ANOVA) followed by the Tukey-Kramer test for post hoc analysis. Statistical significance was acceptable to a level of $p\leq0.05$. Data analysis was performed using the Statistical Package for Social Sciences (SPSS) version 19 software package, IBM Corp., USA, 2010).

Results

Only after 4 days of treatment, HUCB MNCs resulted in a significant ($p\leq0.001$) wound closure accounting for about half of the originally opened area of the wound (49.19%) as compared to both diabetic (18.19%) and non diabetic control groups (17.4%). Closure of wounds continued significantly for this group until reaching (98.38%) on day 12 post-wounding and was almost complete (99.46%) by reaching day 16 post-wounding (**Table 1**) (**Figure 1A, B**). HUCB-HD-treated group showed the highest initial closure percentage (50.8%) among all groups. Only (0.02 ± 0.02 cm^2) of the original wound area was still open on day 12 post-wounding, and it was the first group to attain complete healing (100% closure) before day 16. Whereas on day 16, the calves' blood HD group was about to heal (99.46%), followed by the solcoseryl group (96.24%) which was close to the normal control group (96.74%). The diabetic control group had a still open wound by (0.12 ± 0.04 cm^2) (**Table 1**) (**Figure 1C, D, E, F**).

Administration of either HUCB MNCs (i.v) or HUCB-HD (topical) resulted in a significant ($p\leq0.01$) antihyperglycemic effect that was evident from the first week onwards. Moreover, the decrease in blood glucose was significant on the second week in groups treated with calves' blood HD and solcoseryl compared to the diabetic untreated groups, which showed a progressive increase in the blood glucose level throughout the period of wound healing (**Figure 2A, B**).

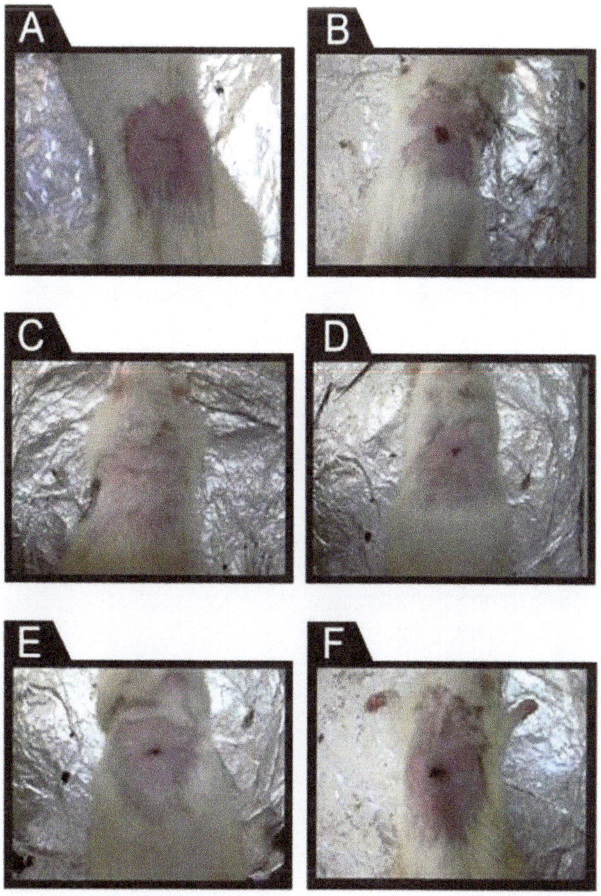

Figure 1. Gross appearance of wound healing on day 16 post-wounding. (a) diabetic wound treated with HUCB MNCs i.v, (b) diabetic control group (PBS i.v) (c) diabetic wound treated topically with HUCB-HD, (d) diabetic wound treated topically with calves' blood HD, (e) positive control group (diabetic wound) treated topically with solcoseryl, (f) diabetic control group (topical saline).

Table 1. Wound area, period of epithelialization, wet and dry granulation wt and hydroxyproline of treated and untreated diabetic and normal rats in the different studied groups.

Experiment	1st Experiment (i.v)			2nd Experiment (topical)				
Parameter	**HUCB MNCs**	**Diabetic Control (PBS)**	**Control (non-diabetic)**	**HUCB-HD**	**Calves' blood HD**	**Solcoseryl**	**Diabetic control (saline)**	**Control (non-diabetic)**
Wound area(cm^2)								
Day 0	1.85±0.2	1.87±0.09	1.85±0.1	1.87±0.1	1.85±0.1	1.85±0.1	1.86±0.1	1.85±0.1
Day 4	0.94±0.2[a]	1.53±0.1	1.52±0.1	0.92±0.3[b]	1.35±0.1	1.47±0.3	1.52±0.3	1.52±0.1
Day 8	0.43±0.1[a]	1.36±0.08	1.15±0.2	0.3±0.1[b]	0.69±0.3[b]	0.96±0.3	1.18±0.2	1.15±0.2
Day 12	0.03±0.03[a]	0.87±0.3	0.56±0.1	0.02±0.02[b]	0.12±0.1[b]	0.4±0.3	0.54±0.4	0.56±0.1
Day 16	0.01±0.01[a]	0.6±0.4	0.04±0.02	0±0	0.01±0.02	0.07±0.1	0.12±0.04	0.04±0.02
Day 20	0±0	0.31±0.4	0.006±0.008	0±0	0±0	0.002±0.004	0.04±0.04	0.006±0.008
Epithelialization(days)	15.7 ±1.2[a]	22.8±2.2	19.7±1.8	13.5±1.4[b]	16.8 ±1.8[b]	18.2±2.3[b]	23.5 ±2.1	19.7±1.8
Wet granulation wt (mg/100 g rat)	163±7.2[a]	110±6.3	120±3.5	168±10[b]	130±2.2[b]	119±2.7	113±2.7	120±3.5
Dry granulation wt (mg/100 g rat)	66±9.1[a]	28±3.1	33±3.7	72±6.7[b]	38±2.6[b]	31±3.5	29±2.9	33±3.7
HP (mg/g tissue)								
Day 0	118.3±1	117.9±2.4	133±1.4	119±1.6	119.2±2.1	117.6±2.7	119.5±2	133±1.4
Day 15	133.6±0.6[a]	38.3±2.1	90.2±0.8[b]	87.6±2[b]	81.1±3[b]	78.5±1.6[b]	37.6±1.1	90.2±0.8[b]

Values are expressed as mean±SD from 6 rats/group.
[a]:significantly different from diabetic control group (PBS i.v) at $p \leq 0.01$,
[b]:significantly different from diabetic control group (topical saline) at $p \leq 0.01$.

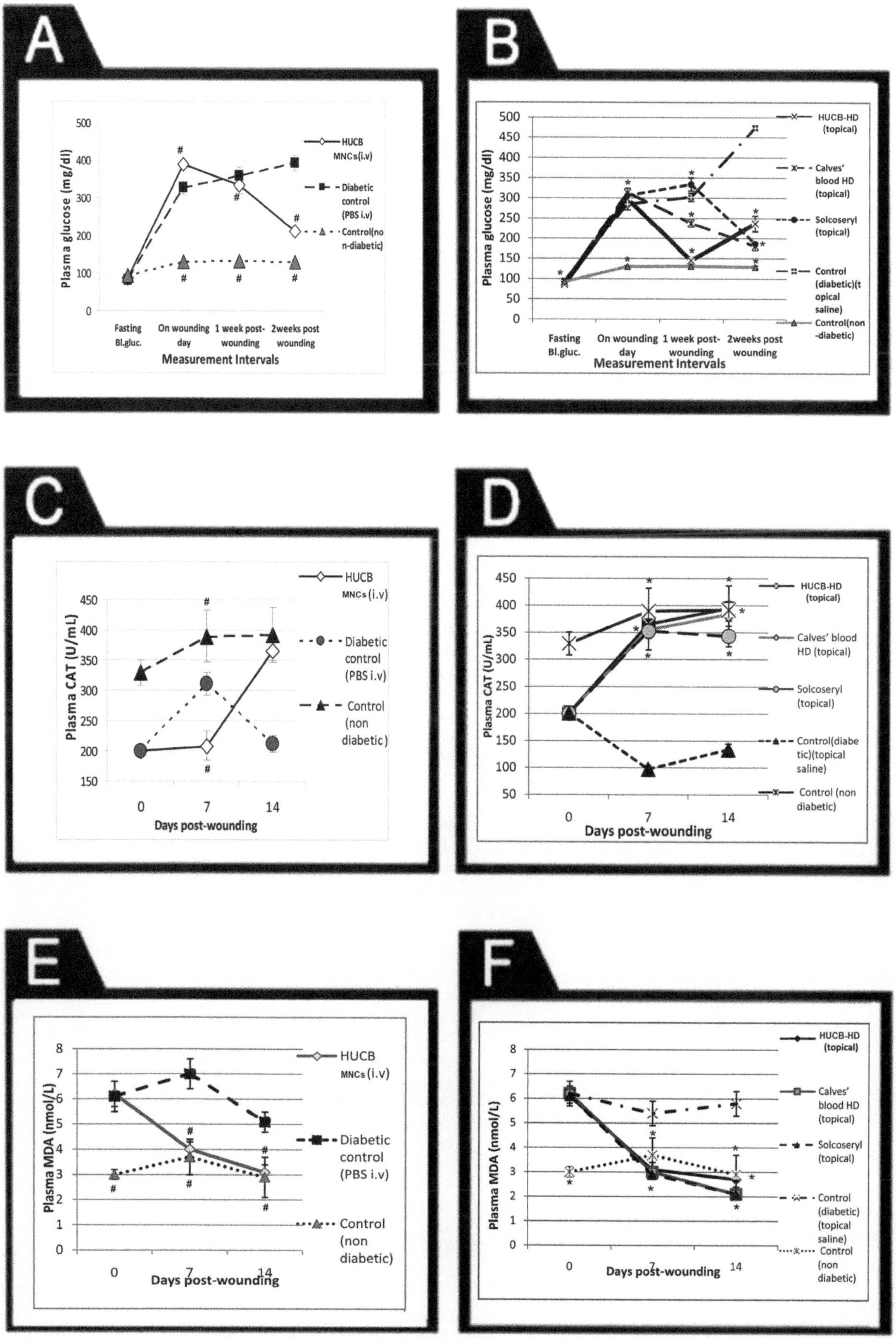
A
Plasma glucose (mg/dl)
HUCB MNCs(i.v)
Diabetic control (PBS i.v)
Control(non-diabetic)
Fasting Bl.gluc.
On wounding day
1 week post-wounding
2weeks post wounding
Measurement Intervals
B
Plasma glucose (mg/dl)
HUCB-HD (topical)
Calves' blood HD (topical)
Solcoseryl (topical)
Control (diabetic)(topical saline)
Control(non-diabetic)
Fasting Bl.gluc.
On wounding day
1 week post-wounding
2weeks post wounding
Measurement Intervals
C
Plasma CAT (U/mL)
HUCB MNCs(i.v)
Diabetic control (PBS i.v)
Control (non diabetic)
Days post-wounding
D
Plasma CAT (U/mL)
HUCB-HD (topical)
Calves' blood HD (topical)
Solcoseryl (topical)
Control(diabetic)(topical saline)
Control (non diabetic)
Days post-wounding
E
Plasma MDA (nmol/L)
HUCB MNCs(i.v)
Diabetic control (PBS i.v)
Control (non diabetic)
Days post-wounding
F
Plasma MDA (nmol/L)
HUCB-HD (topical)
Calves' blood HD (topical)
Solcoseryl (topical)
Control (diabetic) (topical saline)
Control (non diabetic)
Days post-wounding

Figure 2. Plasma glucose (mg/dL), CAT (U/mL), and MDA (nmol/mL) levels in the different studied groups. Left column represents the 1st experiment (i.v); effect of HUCB MNCs i.v and the 2nd is represented in the right column (topical experiment); effect of HUCB-HD, calves' blood HD, and solcoseryl topical treatment. Plasma glucose (A, B) was measured while fasting, on the wounding day, and then on a weekly basis for 2 weeks. CAT (C, D) and MDA (E,F) were measured on the wounding day, on day 7, day14 post-wounding. Values are expressed as mean ± SD from 6 rats/group. # significantly different from diabetic control group (PBS i.v), at $p \leq 0.001$, * significantly different from diabetic control group (topical saline), at $p \leq 0.001$.

Before treatment, on day 0, it was found that all diabetic groups showed oxidative stress markers represented by reduced antioxidant CAT enzyme level, elevated MDA, and reduced NO level as compared to normal non-diabetic control groups of both experiments. Treatment with HUCB MNCs resulted in elevation of CAT on day 7 post-wounding followed by a significant ($p<0.001$) increase on day 14 post-wounding (**Figure 2C**). Likewise, a significant increase in CAT level was observed in all HD-treated groups on days 7 and 14 post-wounding as compared to control diabetic groups (**Figure 2D**), but the extent of increase was higher in HUCB-HD, calves' blood HD, and better than solcoseryl, where the CAT level was significantly reversed to near normal and the significant increase was observed as early as day 7 compared to the HUCB MNCs. Analysis of MDA on day 7 and day 14 post-wounding showed a significant ($p<0.001$) decrease in MDA following treatment with HUCB MNCs and HUCB-HD compared to diabetic untreated control groups of both experiments (**Figure 2E, F**).

NO level at day 10 (32.1±3.5 μmol/L) for the HUCB MNCs group was found to be 2-fold greater than the corresponding diabetic control (PBS i.v) group (14.6±1.3 μmol/L). Production of NO was elevated in solcoseryl and calves' blood HD-treated groups in response to treatment evidenced by high NO levels measured on day 10 post-wounding (22.6±1.8 and 25±1.1 μmol/L), respectively. Moreover, it was highly significantly ($p \leq 0.001$) increased in HUCB-HD-treated group (64.3±5.3 μmol/L) when compared to its diabetic (topical saline) control group (6.1±0.6 μmol/L) and approximately three-folds greater than its day 0 level (20±0.2 μmol/L) (**Figure 3**).

Induction of DM resulted in reduction of HP content (**Table 1**) in diabetic rats' skin granulation tissue as evidenced by low HP content measured on day 0 for all diabetic rats as compared to normal non-diabetic ones. After 15 days of injection with HUCB MNCs, a significant ($p \leq 0.001$) elevation in HP content was observed (133.6±0.6 mg/g tissue) which even exceeded its skin content in normal non-diabetic rats before ulcer induction (133±1.4 mg/g tissue). Although treatment with HUCB-HD, calves' blood HD, or solcoseryl resulted in significant ($p \leq 0.001$) elevation of HP content compared to the diabetic control group, the HP content attained by the HUCB-HD-treated group (87.6±1.9 mg/g tissue) was higher than the other two groups and very close to the content attained by the normal non-diabetic control group (90.2±0.8 mg/g tissue) after 15 days of treatment.

Meanwhile, a significant ($p \leq 0.001$) increase was observed in the dry and wet granulation tissue's weight in groups treated with HUCB MNCs, calves' blood HD and HUCB-HD in comparison to their corresponding diabetic control groups (**Table 1**).

Wounds histopathological changes on day 14 are presented in **Figure 4**. HUCB MNCs and HUCB-HD-treated wounds showed earlier wound contraction and earlier onset of epithelization than calves' blood HD and solcoseryl-treated wounds. This was evidenced by the presence of marked vascular proliferation and fibroplasia. However, diabetic control groups showed failure of wound closure attempts and little evidence of re-epithelialization.

Discussion

Development of new treatment modalities to improve wound healing in diabetic patients is an essential and emerging field of investigation [32], as medical therapies for wound care are limited, the outcome of management is poor and there is uncertainty concerning optimal approaches to management.

Cell-based therapy is an attractive approach for the treatment of wounds with multiple impairments. Healing activity of stem cells is recognized for their ability to separate into the various component cells of injured tissues, as well as to discharge growth factors that may encourage the formation of new blood vessels [33].

Reduction in wound area, in the 1st experiment, was faster in the MNCs-treated group compared to the diabetic and non-diabetic control groups. Similarly, in the 2nd experiment, wound contraction was greater and more prominent in the HUCB-HD-treated group, followed by the calves' blood HD group and finally the solcoseryl group. Therefore, systemic administration of MNCs in the 1st experiment or topical application of HUCB-HD in the 2nd experiment significantly accelerated the rate of wound closure. Despite the comparative results displayed by HUCB MNCs and HUCB-HD, however HUCB-HD is superior for treatment as evidenced by showing an average healing time about 2.2 days shorter than the HUCB MNCs-treated group. These results agree with **Gurdol and colleagues** [34] who observed progressive reductions in wound areas following treatment of diabetic patients foot ulcers using hyperbaric oxygen. A study by **Hamed and colleagues** [35] investigated the effect of topical erythropoietin treatment on defective wound repair in diabetic rats skin. Erythropoietin-containing creams promoted skin wound repair in diabetic rats and the area of open wounds was significantly smaller than that of vehicle-treated wounds. In a human study of chronic non-healing wounds, it was shown that directly applied bone marrow-derived cells can lead to wound closure and possible tissue rebuilding [36].

HUCB MNCs were capable of normalizing blood glucose level in diabetic rats, an effect attributed to the generation of HUCB-derived insulin-producing cells which can be mediated through both fusion-dependent and independent mechanisms. This is in agreement with different *in vitro* and *in vivo* studies which clarified the capability of HUCB-derived cells to generate insulin-producing cells. **Hitherto**, one study carried out by **Pessina and colleagues**, has investigated the expression of endocrine pancreatic progenitor markers in HUCB cells [37]. All the HD compounds administered were also shown to improve blood glucose levels in diabetic rats. However, HUCB-HD was capable of even normalizing blood glucose levels. This may be reclaimed to the clear insulin-like effects of HD compounds on glucose uptake and utilization which have been demonstrated *in vitro* [38] and in different pharmacodynamic studies following the use of a deproteinized HD. Improvement of cellular glucose uptake is based on direct activation of glucose transporters located in the cell membrane. This activation was demonstrated to reach an extent comparable to that of insulin [39]. Moreover, this insulin-like activity of the HD can be explained in the terms of inositolphosphooligosaccharides (IPOs) compounds, which activate glucose

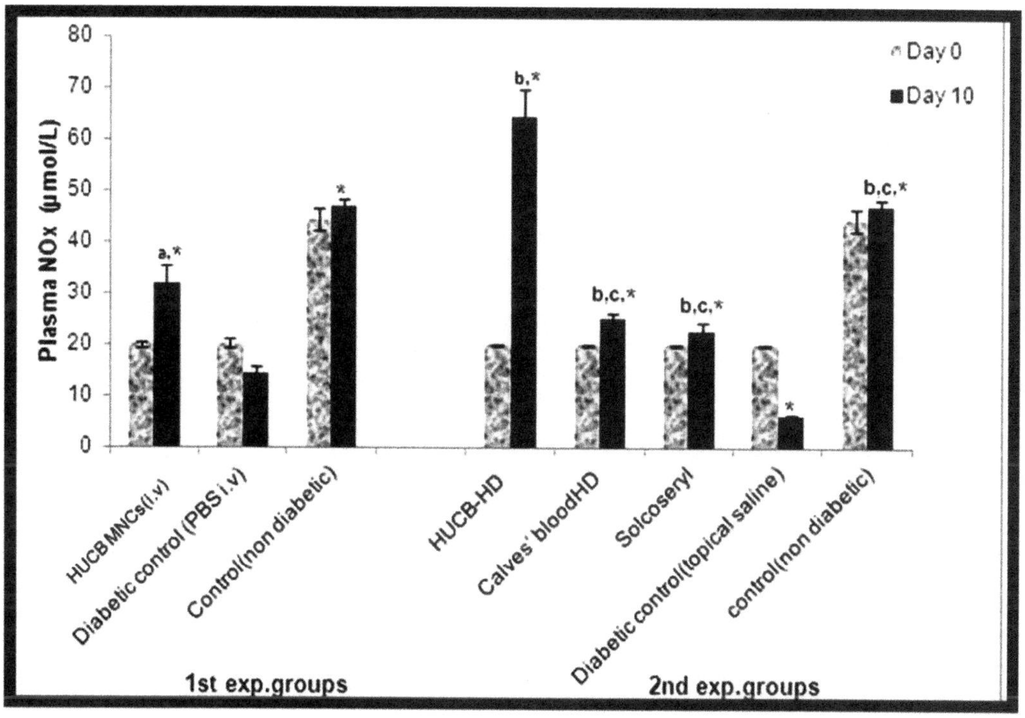

Figure 3. Effect of different therapeutic agents of both experiments on plasma NOx (µmol/L) measured on day 0 and day 10 post-wounding. Values are mean ± SD from 6 rats/group. a, b, c significantly different from diabetic control (PBS i.v), diabetic control (topical saline), and HUCB-HD, respectively, at $p \leq 0.001$. *Significant increase on day 10 compared to day 0 at $p \leq 0.05$.

transporters, with a direct influence on the activities of some enzymes involved in glucose metabolic pathways [40].

Blood glucose levels measured were found to be strongly and inversely related to body weight of rats, where treatment with HUCB MNCs and HUCB-HD resulted in an early and steady increase in body weights (data not shown) till the end of the experiment. In contrast, diabetic control groups exhibited an overall persistent loss of body weight. These results are consistent with the results of [41,42]

Treatment of diabetic wounds with HUCB-derived MNCs, prepared HD of both HUCB and local buffalo calves' blood resulted in alleviation of oxidative stress in wounded STZ-induced diabetic rats. These results are consistent with the results of **Nwanjo and colleagues** [41] and can be explained by the ability of treatment agents to improve activities of the hepatic anti-oxidant enzymes; consequently scavenging toxic reactive oxygen free radicals which are responsible for tissue damage in STZ-induced DM, as well as inhibition of lipid peroxidation and thereby tissues and organs oxidative damage.

Ongoing experimental and clinical wound healing studies have presented NO as a critical mediator of tissue repair [43]. NO is also the most effective antioxidant being a scavenger of superoxide. Moreover, impaired wound healing in diabetics has been associated with reduced NO synthesis. Thus, high level of bioavailable NO in wounds is important for enhancing the healing process [44]. In our study we aimed to determine total NO_2^-/ NO_3^- level, which represent the stable oxidation product of NO as an indicator of NO production in plasma, and to evaluate its relation to healing during foot ulcers' treatment.

Evidence shows that decrease of NO bioavailability might be related to elevation of endogenous nitric oxide synthase (NOS) inhibitors [45]. In diabetes-induced hyperglycemia, there is an increased metabolism of glucose to sorbitol via the polyol pathway. Increased activity of the enzyme aldose reductase requires and may deplete cellular NADPH, which is also a required cofactor for NOS [46]. It is possible that both mechanisms occur during diabetic wound healing and results in drastically reduced NO production [47]. Therefore, treatment with HUCB-derived MNCs and HUCB-HD might have resulted in decreasing NOS inhibitors levels or increasing the availability of cellular NADPH. These effects are probably secondary to glycemic control improvement as well as their antioxidant properties [34]. Nevertheless, the elevation in NO level following treatment with HUCB-HD was remarkable and exceeded that attained by the normal control group. HUCB-HD would be able to elevate NO levels efficiently to an optimum level, contributing to the healing process by reversing the impaired healing condition associated with DM and restoring wound NO levels towards more normal values.

Collagen plays a central role in wound healing, being the principal component of connective tissue providing a structural framework for tissue regeneration [16]. Measurement of HP could be used as an index for collagen turn over [48]. Hence, its estimation in the granulation tissue may throw light on maturation and the healing process [49]. In our study, granulation tissue HP level was significantly increased in both HUCB MNCs group and HUCB-HD group following treatment. However, the substantial rise in HP following MNCs treatment may be attributed to the paracrine effect of stem cells that can promote the wound healing progress [7]. An *in vitro* study indicated that collagen synthesis and levels of bFGF and VEGF were much higher in bone marrow stem cells than those in dermal fibroblasts, so the ability of adult stem cells to alter tissue microenvironment via secretion of soluble factors may contribute to tissue repair even more than their multipotential differentiation [50]. Decreased HP content

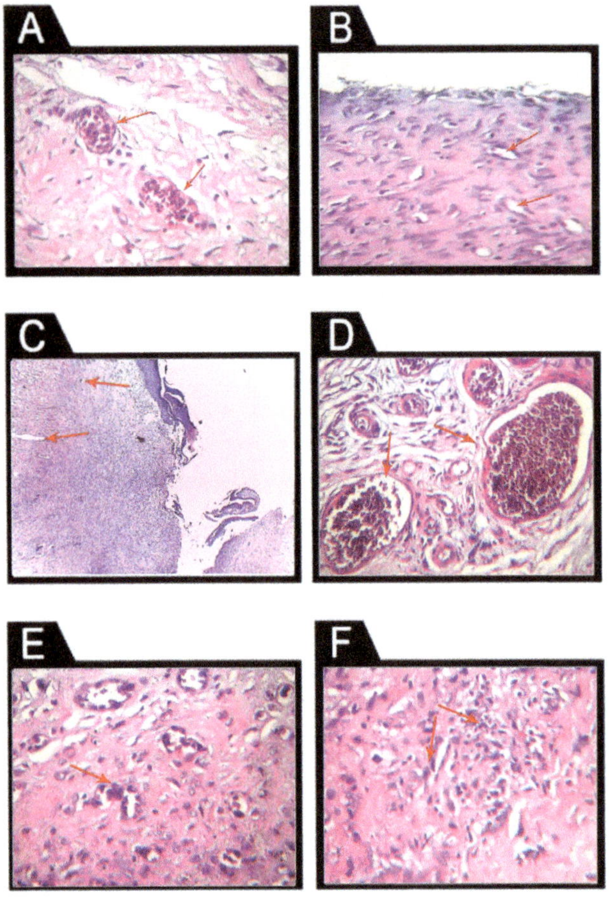

Figure 4. Representative photomicrographs of studied wound sections on day 14 post-wounding stained with hematoxylin-eosin stain (H&E). A: diabetic wounds treated with HUCB MNCs (i.v) showing marked vascular proliferation (arrows) and fibroplasia with increased re-epithelialization (×400). B: diabetic control wounds (PBS i.v) showing failure of wound closure at the upper area of the field with subepithelial exudates of neutrophils, blood vessels were immature (arrows) with minimal proliferation (×200). C: diabetic wound treated topically with solcoseryl showing mild angioplasia encountered, but blood vessels are not yet fully mature, with collapsed lumens. A marked fibro-collagenous reaction (arrows) is encountered and epithelial proliferation is minimal (×100). D: diabetic wound treated topically with HUCB-HD, where marked mature vascular proliferation (arrows) and fibroplasia were noticed, as well as adequate epithelization with subepithelial relatively high angiofibroplasia (×200). E: diabetic wound treated topically with calves' blood HD showing mature vascular proliferation (arrows) and moderate exudates of macrophages, surface epithelization with moderate inflammation, mostly of macrophages (×400). F: diabetic control (topical saline) wound showing failure of wound closure attempts at the upper area of the field, with subepithelial exudates of neutrophils, sustained high neutrophilic count with immature fibrous proliferation. Blood vessels were of a low count appearing irregular and immature (arrows) (×200).

measured in the untreated diabetic rats' wounds parallels decreased weight of granulation tissue in this study, while the observed increase in wet and dry weights of granulation tissue in groups treated with HUCB MNCs and HUCB-HD reflected higher protein content and better healing [48]. This was much confirmed through the histopathological examination of skin tissue samples, where HUCB MNCs were found to decrease wound inflammatory cell infiltration, accelerated new blood vessels formation "angiogenesis", and granulation tissue. This effect is related to the self-renewal and multipotency properties, which stem cells have. Adult stem cells can produce differentiated skin cells and a previous study have found that MSCs, isolated from different tissues including UCB, can contribute to skin reconstitution in cutaneous wounds [51]. The potential of MSCs to differentiate into skin tissue cells has been described [52]. In a study by Korean researchers, transplanting HUCB-derived endothelial progenitor cells (EPCs) resulted in a significant acceleration of wound closure in diabetic mouse models [53]. Adult stem cells can also modulate immune and inflammatory responses to promote wound healing; the transplantation of autologous and allogenic MSCs on the surface of deep burn wounds in rats decreased inflammatory cell infiltration into the wound, and accelerated new blood vessels and granulation tissue formation [54]. The marked vascular proliferation and fibroplasia observed in tissues after 14 days of treatment confirmed near closure. A variety of CB populations have been used including MNCs fractions, with significant beneficial effects. Undoubtedly, the beneficial effects could be attributed to induction of angiogenesis due to the release of a variety of angiogenic and growth factors [55]. Our results concerning HUCB-HD on day 14 post-wounding revealed that the granulation tissue in healed wound contained comparatively few inflammatory cells, more collagen, early efficient epithelization and marked mature vascular proliferation. This is because our HD and fractions isolated from it – consisting exclusively of low molecular weight compounds stimulate the proliferation of coronary endothelial cells which is not a trivial nutritive effect. The ability of endothelial cells to proliferate and to differentiate is a prerequisite of angiogenesis. Thus, agents stimulating the proliferation of EC are of special interest [56]. In addition, HD might strengthen the angiogenic effect of EGF on EC consisting of a proliferation-promoting and a chemoattractive activity, the latter demonstrated in wound fluids [56]. These synergisms would result in faster neovascularization probably accompanied by an improved fibroplasia during formation of granulation tissue. And this synergistic effect is believed to explain the therapeutic efficacy of HD in cases of impaired wound healing, ulcers, burns and occlusive diseases documented by several clinical and experimental studies [57,58]. The present study has also shown that, in rats, HD compounds reduced the inflammatory phase duration in wound healing. This finding is consistent with the observation of **Dri and colleagues** that the HD interacts with polymorphnuclear leucocytes (PMNL) in a dual fashion; higher HD concentrations stimulate PMNL activity, whereas lower concentrations inhibit such activity [59]. HD improves the anabolic phase of healing by providing the wound with sufficient immediately utilizable and necessary low molecular weight starting materials and energy carriers for repair and energy metabolism [60]. We hypothesize that the reason for the enhanced vascularization and the superior positive variables encountered in wounds of HUCB-HD-treated group may be related to its composition. Therefore, a sample of this novel type of HD and another sample of the prepared calves' blood HD were taken for HPLC-ESI-MS analysis in order to find out if there is a difference in the composition. Analysis of the HUCB-HD sample revealed glutamate to be the predominant free amino acid, which is not the case in the Calves' blood HD (**Figure S1**).

Glutamine is a non-essential amino acid that can become a "conditionally essential" amino acid in certain circumstances, including tissue injury/repair [61]. It is used by the inflammatory cells within the wound for proliferation and as a source of energy [62]. Fibroblasts use glutamine for these same purposes, as well as for protein and nucleic acid synthesis. Glutamate is also a migration-promoting factor of neutrophils to the wound site. Rapid migration of neutrophils to the wound site is a prerequisite to the wound healing process. According to a previous study by **Gupta and Chattopadhyay** on placental extract, it was shown that glutamate at an optimum concentration induced phenotypic neutrophil chemotaxis as well as neutrophil chemokinesis [63]. Glutamate-induced chemotaxis was accompanied by polarization of the actin cytoskeleton and by polymerization of F-actin. Moreover, glutamine plays a key role in the immune system [64], so deficiency in this nutrient can significantly slow the healing process. This explains the probable superior wound healing potential of the HUCB-HD over the calves' blood-HD.

Additionally, the present study recorded a relative decrease in platelet count (data not shown) of STZ-induced diabetic control rats similar to that found in human DM, which was corrected only in HUCB MNCs and HUCB-HD-treated groups (data not shown). This finding agree with **Shi and colleagues** who reported a promotion of hematopoietic recovery following treatment of combined radiation and wound injury with a systemic infusion of dermal multipotent cells. They indicated that stem cells through an indirect mechanism could produce a series of cytokines and extracellular matrix molecules, including IL-3, IL-6, IL-15, VEGF, PDGF, HGF, TGF-β, ICAM-1, VCAM-1, and fibronectin, which were closely related to hematopoiesis and wound healing [65].

Increase in platelets count in case of HUCB-HD may be attributed to the indirect mechanism of decreasing the platelets turnover, increasing their survival time or decreasing their aggregability due to improving disturbance in carbohydrate metabolism associated with diabetes. Disturbed carbohydrate metabolism was suggested to be directly or indirectly causing increased platelet aggregation [66] an effect resulting from the insulin-like effect of the HD mentioned previously.

Conclusions

Our results revealed that systemic administration of HUCB-derived MNCs or topical treatment with calves' blood HD and the novel HUCB-derived HD can improve the impaired healing of diabetic wounds as manifested by shortened epithelialization time and rapid wound contraction. These healing promoting effects can be explained on the basis of several mechanisms; antihyperglycemic mechanisms, antioxidant mechanisms, reduction of lipid peroxidation product (MDA) and increase of NO production; a critical mediator of tissue repair, in addition to increasing HP content in the granulation tissue. However, the HUCB-derived HD has turned out to be the best among all treatment agents due to the fastest wound closure and shortest healing time achieved. This could be probably explained by the multifactorial nature of this approach that covers the whole multiphase complex process of wound healing. Results of this study suggest that topical application of HUCB–derived HD should be studied in human diabetic foot ulcers and the mechanism of its action should be further investigated.

Supporting Information

Figure S1 Mass spectra of two HD samples. Mass spectra of two HD samples derived from human umbilical cord blood (a) and Calves's blood (b), respectively are shown. Both samples were subjected to LC-ESI-MS/MS protein analysis. Amino acid content of each type of HD is represented as peaks at their specific m/z ratio, with a relative intensity corresponding to the amount of each amino acid in the sample.

Acknowledgments

We would like to thank Dr. Hossam H. Soliman, Manager of the Labor and Delivery, Dept. of Obstetrics and Gynecology, El-Galaa Hospital, for his assistance in HUCB samples' collection and Dr. Tarek N. El-Bolkainy, Associate Professor of Pathology at the National Cancer Institute, Cairo University, for histological analysis of skin tissue samples.

Author Contributions

Conceived and designed the experiments: HOE MRD. Performed the experiments: HOE MRD NMH SMD. Analyzed the data: HOE MRD NMH SMD. Contributed reagents/materials/analysis tools: SMD. Wrote the paper: HOE MRD NMH SMD.

References

1. Falanga V (2005) Wound healing and its impairment in the diabetic foot. Lancet 366: 1736–1743.
2. Lavery LA, Armstrong DG, Wunderlich RP, Tredwell J,Boulton AJ (2003) Diabetic foot syndrome: evaluating the prevalence and incidence of foot pathology in Mexican Americans and non-Hispanic whites from a diabetes disease management cohort. Diabetes Care 26: 1435–1438.
3. Hinchliffe RJ, Valk GD, Apelqvist J, Armstrong DG, Bakker K, et al. (2008) A systematic review of the effectiveness of interventions to enhance the healing of chronic ulcers of the foot in diabetes. Diabetes Metab Res Rev 24 Suppl 1: S119–144.
4. Bloomgarden ZT (2008) The Diabetic Foot. Diabetes Care 31: 372–376.
5. Fodor WL (2003) Tissue engineering and cell based therapies, from the bench to the clinic: the potential to replace, repair and regenerate. Reprod Biol Endocrinol 1: 102.
6. Shi C, Zhu Y, Su Y,Cheng T (2006) Stem cells and their applications in skin-cell therapy. Trends Biotechnol 24: 48–52.
7. Lu W, Zhang YJ, Jin Y (2009) Potential of stem cells for skin regeneration following burns. Expert Rev Dermatol 4: 97–99.
8. Tang DQ, Cao LZ, Burkhardt BR, Xia CQ, Litherland SA, et al. (2004) In vivo and in vitro characterization of insulin-producing cells obtained from murine bone marrow. Diabetes 53: 1721–1732.
9. Mueller SM, Glowacki J (2001) Age-related decline in the osteogenic potential of human bone marrow cells cultured in three-dimensional collagen sponges. J Cell Biochem 82: 583–590.
10. Koblas T, Harman SM, Saudek F (2005) The application of umbilical cord blood cells in the treatment of diabetes mellitus. Rev Diabet Stud 2: 228–234.
11. Wang JC, Doedens M, Dick JE (1997) Primitive human hematopoietic cells are enriched in cord blood compared with adult bone marrow or mobilized peripheral blood as measured by the quantitative in vivo SCID-repopulating cell assay. Blood 89: 3919–3924.
12. Grewal SS, Barker JN, Davies SM,Wagner JE (2003) Unrelated donor hematopoietic cell transplantation: marrow or umbilical cord blood? Blood 101: 4233–4244.
13. Wilmink JM, Stolk PW, van Weeren PR, Barneveld A (2000) The effectiveness of the haemodialysate Solcoseryl for second-intention wound healing in horses and ponies. J Vet Med A Physiol Pathol Clin Med 47: 311–320.
14. Ghoneimi IG, Bang RL (1997) Use of Solcoseryl in minor burns. Annals of Burns and Fire Disasters 10: 219–222.
15. Markiel Z, Ziaja K, Zejc D, Samorodny J, Simka M, et al. (2002) Efficacy of Solcoseryl in patients with pre-gangrene of lower limbs treated by vascular reconstructive therapy. Pol 4: 167–173.
16. Al Bayaty F, Abdulla M, Abu Hassan MI, Masud M (2010) Wound healing potential by hyaluronate gel in streptozotocin-induced diabetic rats. Sci Res Essays 5: 2756–2760.
17. Bieback K, Kern S, Kluter H, Eichler H (2004) Critical parameters for the isolation of mesenchymal stem cells from umbilical cord blood. Stem Cells 22: 625–634.
18. Ende N, Chen R, Reddi AS (2004) Transplantation of human umbilical cord blood cells improves glycemia and glomerular hypertrophy in type 2 diabetic mice. Biochem Biophys Res Commun 321: 168–171.
19. Boyum A (1968) Separation of leukocytes from blood and bone marrow. Introduction. Scand J Clin Lab Invest Suppl 97: 7.
20. Gu L, Jones AD, Last RL (2007) LC−MS/MS Assay for Protein Amino Acids and Metabolically Related Compounds for Large-Scale Screening of Metabolic Phenotypes. Analytical Chemistry 79: 8067–8075.

21. Sur TK, Biswas TK, Ali L, Mukherjee B (2003) Anti-inflammatory and anti-platelet aggregation activity of human placental extract. Acta Pharmacol Sin 24: 187–192.
22. Al-Watban FA, Andres BL (2001) The effect of He-Ne laser (632.8 nm) and Solcoseryl in vitro. Lasers Med Sci 16: 267–275.
23. Toker S, Gulcan E, Cayc MK, Olgun EG, Erbilen E, et al. (2009) Topical atorvastatin in the treatment of diabetic wounds. Am J Med Sci 338: 201–204.
24. Barua CC, Talukdar A, Begum SA, Sarma DK, Fathak DC, et al. (2009) Wound healing activity of methanolic extract of leaves of Alternanthera brasiliana Kuntz using in vivo and in vitro model. Indian J Exp Biol 47: 1001–1005.
25. Sumitra M, Manikandan P, Gayathri VS,Suguna L (2009) Influence of Honey on Energy Metabolism during Wound Healing in Rats. Scholarly Research Exchange 2009.
26. Trinder P (1969) Determination of blood glucose using an oxidase-peroxidase system with a non-carcinogenic chromogen. J Clin Pathol 22: 158–161.
27. Montgomery H, Dymock J (1961) The determination of nitrite in water. Analyst 86: 414–416.
28. Ohkawa H, Ohishi N,Yagi K (1979) Assay for lipid peroxides in animal tissues by thiobarbituric acid reaction. Anal Biochem 95: 351–358.
29. Aebi H (1984) Catalase in vitro. Methods Enzymol 105: 121–126.
30. Woessner FJ Jr (1961) The determination of hydroxyproline in tissue and protein samples containing small proportions of this imino acid. Archives of Biochemistry and Biophysics 93: 440–447.
31. Banchroft J, Stevens A, Dawson MP (1996) Theory and practice Of Histological Techniques.: Churchil Livingstone, New York, London, San Francisco, Tokyo.
32. Trousdale RK, Jacobs S, Simhaee DA, Wu JK, Lustbader JW (2009) Wound closure and metabolic parameter variability in a db/db mouse model for diabetic ulcers. J Surg Res 151: 100–107.
33. Barcelos LS, Duplaa C, Krankel N, Graiani G, Invernici G, et al. (2009) Human CD133+ progenitor cells promote the healing of diabetic ischemic ulcers by paracrine stimulation of angiogenesis and activation of Wnt signaling. Circ Res 104: 1095–1102.
34. Gurdol F, Cimsit M, Oner-Iyidogan Y, Kocak H, Sengun S, et al. (2010) Collagen synthesis, nitric oxide and asymmetric dimethylarginine in diabetic subjects undergoing hyperbaric oxygen therapy. Physiol Res 59: 423–429.
35. Hamed S, Ullmann Y, Masoud M, Hellou E, Khamaysi Z, et al. (2010) Topical erythropoietin promotes wound repair in diabetic rats. J Invest Dermatol 130: 287–294.
36. Badiavas EV, Falanga V (2003) Treatment of chronic wounds with bone marrow-derived cells. Arch Dermatol 139: 510–516.
37. Pessina A, Eletti B, Croera C, Savalli N, Diodovich C, et al. (2004) Pancreas developing markers expressed on human mononucleated umbilical cord blood cells. Biochem Biophys Res Commun 323: 315–322.
38. Machicao F, Mushack J, Seffer E, Ermel B, Haring HU (1990) Mannose, glucosamine and inositol monophosphate inhibit the effects of insulin on lipogenesis. Further evidence for a role for inositol phosphate-oligosaccharides in insulin action. Biochem J 266: 909–916.
39. Obermaier-Kusser B, Muhlbacher C, Mushack J, Seffer E, Ermel B, et al. (1989) Further evidence for a two-step model of glucose-transport regulation. Inositol phosphate-oligosaccharides regulate glucose-carrier activity. Biochem J 261: 699–705.
40. Gottschalk WK, Jarett L (1988) The insulinomimetic effects of the polar head group of an insulin-sensitive glycophospholipid on pyruvate dehydrogenase in both subcellular and whole cell assays. Arch Biochem Biophys 261: 175–185.
41. Nwanjo H, Oze G, Okafor M, Nwosu D, Nwankpa P (2007) Protective role of phyllantus niuri extract on serum lipid profiles and oxidative stress in hepatocytes of diabetic rats. Afr J Biotech 6: 1744–1749.
42. Phuc P, Buu Truc P, Thuy D, Nhung T, Chung D, et al. (2010) Regeneration of Pancreatic B Cells of Type 1 Diabetic Mouse by Stem Cell Transplatation. In: V . Toi and T . Khoa, editors. The Third International Conference on the Development of Biomedical Engineering in Vietnam. Springer Berlin Heidelberg. pp.163–166.
43. Schwentker A, Billiar TR (2003) Nitric oxide and wound repair. Surg Clin North Am 83: 521–530.
44. Soneja A, Drews M, Malinski T (2005) Role of nitric oxide, nitroxidative and oxidative stress in wound healing. Pharmacol Rep 57 Suppl: 108–119.
45. Lin KY, Ito A, Asagami T, Tsao PS, Adimoolam S, et al. (2002) Impaired nitric oxide synthase pathway in diabetes mellitus: role of asymmetric dimethylarginine and dimethylarginine dimethylaminohydrolase. Circulation 106: 987–992.
46. Sima AA, O'Neill SJ, Naimark D, Yagihashi S, Klass D (1988) Bacterial phagocytosis and intracellular killing by alveolar macrophages in BB rats. Diabetes 37: 544–549.
47. Dashti N, Ansari M, Shabani M, Vardasbi S (2004) Nitric oxide donor (deta nonoate) enhance experimental wound healing in streptozotocin-induced diabetic rats. Pak J Med Sci 20: 211–214.
48. Nayak S, Nalabothu P, Sandiford S, Bhogadi V, Adogwa A (2006) Evaluation of wound healing activity of Allamanda cathartica. L. and Laurus nobilis. L. extracts on rats. BMC Complement Altern Med 6: 12.
49. Peacock EE Jr (1984) Wound repair: structure, synthesis and interaction of fibrous protein matrix. 56–101 p.
50. Phinney DG, Prockop DJ (2007) Concise review: mesenchymal stem/multipotent stromal cells: the state of transdifferentiation and modes of tissue repair–current views. Stem Cells 25: 2896–2902.
51. Yoshikawa T, Mitsuno H, Nonaka I, Sen Y, Kawanishi K, et al. (2008) Wound therapy by marrow mesenchymal cell transplantation. Plast Reconstr Surg 121: 860–877.
52. Sasaki M, Abe R, Fujita Y, Ando S, Inokuma D, et al. (2008) Mesenchymal stem cells are recruited into wounded skin and contribute to wound repair by transdifferentiation into multiple skin cell type. J Immunol 180: 2581–2587.
53. Kim JY, Song SH, Kim KL, Ko JJ, Im JE, et al. (2010) Human cord blood-derived endothelial progenitor cells and their conditioned media exhibit therapeutic equivalence for diabetic wound healing. Cell Transplant 19: 1635–1644.
54. Shumakov VI, Onishchenko NA, Rasulov MF, Krasheninnikov ME, Zaidenov VA (2003) Mesenchymal bone marrow stem cells more effectively stimulate regeneration of deep burn wounds than embryonic fibroblasts. Bull Exp Biol Med 136: 192–195.
55. Chen HK, Hung HF, Shyu KG, Wang BW, Sheu JR, et al. (2005) Combined cord blood stem cells and gene therapy enhances angiogenesis and improves cardiac performance in mouse after acute myocardial infarction. Eur J Clin Invest 35: 677–686.
56. Schönwald D, Sixt B, Machicao F, Marx E, Haedenkamp G, et al. (1991) Enhanced proliferation of coronary endothelial cells in response to growth factors is synergized by hemodialysate compounds in vitro. Research in Experimental Medicine 191: 259–272.
57. Niinikoski J, Renvall S (1979) Effect of a deproteinized blood extract on experimental granulation tissue. Acta Chir Scand 145: 287–291.
58. Konturek SJ, Brzozowski T, Dembinski A, Warzecha A, Drozdowicz D (1988) Comparison of solcoseryl and epidermal growth factors (EGF) in healing of chronic gastroduodenal ulcerations and mucosal growth in rats. Hepatogastroenterology 35: 25–29.
59. Dri P, Cramer R, Mittenzwei H, Patriarca P (1989) Dual effect of a protein-free hemodialysate on the oxygen uptake of phagocytosing human polymorphonuclear leucocytes. Arzneimittelforschung 39: 1565–1567.
60. Isler H, Bauen A, Baschong W (1991) Topical treatment of standardized burns with a protein-free haemodialysate. Burns 17: 93–97.
61. Wilmore DW (2001) The effect of glutamine supplementation in patients following elective surgery and accidental injury. J Nutr 131: 2543S–2549S; discussion 2550S–2541S.
62. Newsholme P (2001) Why is L-glutamine metabolism important to cells of the immune system in health, postinjury, surgery or infection? J Nutr 131: 2515S–2522S; discussion 2523S–2514S.
63. Gupta R, Chattopadhyay D (2009) Glutamate is the chemotaxis-inducing factor in placental extracts. Amino Acids 37: 271–277.
64. Calder PC, Yaqoob P (1999) Glutamine and the immune system. Amino Acids 17: 227–241.
65. Shi C, Cheng T, Su Y, Mai Y, Qu J, et al. (2004) Transplantation of dermal multipotent cells promotes survival and wound healing in rats with combined radiation and wound injury. Radiat Res 162: 56–63.
66. Eriksson U, Ewald U, Tuvemo T (1983) Increased platelet volume in manifest diabetic rats. Ups J Med Sci 88: 17–23.

9

Influence of Delivery Method on Neuroprotection by Bone Marrow Mononuclear Cell Therapy following Ventral Root Reimplantation with Fibrin Sealant

Roberta Barbizan[1], Mateus V. Castro[1], Benedito Barraviera[2], Rui S. Ferreira Jr.[2], Alexandre L. R. Oliveira[1]*

1 Laboratory of Nerve Regeneration, Department of Structural and Functional Biology, University of Campinas - UNICAMP, Campinas, São Paulo, Brazil, **2** Center for the Study of Venoms and Venomous Animals (CEVAP), São Paulo State University (UNESP – Univ Estadual Paulista), Botucatu, São Paulo, Brazil

Abstract

The present work compared the local injection of mononuclear cells to the spinal cord lateral funiculus with the alternative approach of local delivery with fibrin sealant after ventral root avulsion (VRA) and reimplantation. For that, female adult Lewis rats were divided into the following groups: avulsion only, reimplantation with fibrin sealant; root repair with fibrin sealant associated with mononuclear cells; and repair with fibrin sealant and injected mononuclear cells. Cell therapy resulted in greater survival of spinal motoneurons up to four weeks post-surgery, especially when mononuclear cells were added to the fibrin glue. Injection of mononuclear cells to the lateral funiculus yield similar results to the reimplantation alone. Additionally, mononuclear cells added to the fibrin glue increased neurotrophic factor gene transcript levels in the spinal cord ventral horn. Regarding the motor recovery, evaluated by the functional peroneal index, as well as the paw print pressure, cell treated rats performed equally well as compared to reimplanted only animals, and significantly better than the avulsion only subjects. The results herein demonstrate that mononuclear cells therapy is neuroprotective by increasing levels of brain derived neurotrophic factor (BDNF) and glial derived neurotrophic factor (GDNF). Moreover, the use of fibrin sealant mononuclear cells delivery approach gave the best and more long lasting results.

Editor: Graca Almeida-Porada, Wake Forest Institute for Regenerative Medicine, United States of America

Funding: The present work was supported by a grant from Fundação de Amparo a Pesquisa do Estado de São Paulo - FAPESP, Brazil (2010/0986-5). Barbizan R. was supported by Fundação de Amparo a Pesquisa do Estado de Sao Paulo - FAPESP (process number: 2010/00729-2). This work was funded by grants from Coordenação de Aperfeiçoamento de Pessoal de Nível Superior (CAPES), Conselho Nacional de Desenvolvimento Cientifico e Tecnológico (CNPq) and Fundação de Amparo à Pesquisa do Estado de São Paulo (FAPESP). The funders had no role in study design, data collection and analysis, decision to publish, or preparation of the manuscript.

Competing Interests: The authors have declared that no competing interests exist.

* Email: alroliv@unicamp.br

Introduction

In order to enhance the success of adult stem cell (SC) translational medicine efforts, the source as well as the most effective delivery method has to be considered. The bone marrow contains endothelial progenitor cells and mononuclear cells (MC). The MC fraction corresponds to the totality of hematopoietic and mesenchymal stem cells. MC present clinical advantages over other stem cells, based on the minimally invasive harvesting procedures, which are fast and cost-effective. Also, the possibility of autografting avoids the use of immunosuppressants, present low oncogenic potential and does not raise ethical issues [1] as compared to other SC. Moreover, MCs have similar potential therapeutic outcome for nerve regeneration in comparison to mesenchymal cells [2]. The peripheral nerve regeneration after MC has been connected to the local production of neurotrophic factors [1,3,4]. Relevantly, stem cell therapy may also present an immunomodulatory effect, reducing pro-inflammatory events as well as glial reaction following lesion.

Ventral root avulsion in rats has been used as a model for brachial plexus lesion (BPL). BPL is frequently caused by motorbike accidents in young adults as well as following complicated child-birth delivery [5]. It causes paralysis in the corresponding muscle groups and loss of sensory functions [6]. The degenerative impact on motoneurons is well characterized and is potentiated by pulling out the ventral roots from the CNS/PNS interface at the spinal cord surface [6]. Similarly to BPL, VRA results in extensive loss of neurons in the first weeks after injury [7,8].

Reimplantation of avulsed roots can rescue motoneurons from degeneration, increasing the regenerative capacity of axonal regrowth [9,10]. As a result, anatomical and functional reinnervation of denervated muscles can be obtained [11–13]. As seen in a previous work [10], a snake venom derived fibrin sealant allowed successful and stable ventral root implantation. Nevertheless, additional therapeutic approaches need to be developed, since root reimplantation alone, although neuroprotective, results in insufficient functional sensory-motor recovery [12,14–16].

In order to improve the outcome following VRA, regarding neuronal survival, several attempts have been made to provide neurotrophic molecules at the site of injury. In this regard, the association of the root reimplantation with BDNF and CNTF resulted in rescue of injured motoneurons after avulsion in rabbits [17]. Therefore, the use of neurotrophic factors in combination with root reimplantation is a potential therapy to be used in patients.

The use of recombinant neurotrophic factors, however, present important drawbacks. One of them is the need of relatively large amounts of the purified substance, to reach the target lesioned area. Due to the short biological activity window of such substances, there is also need of constant perfusion, what may contribute to infection and further lesion of the affected spinal cord area. Additionally, it is improbable that a single neurothrophic molecule will be sufficient to provide the necessary conditions for optimal regeneration.

Based on such facts, the advent of stem cell technology brought new insights on cell therapy and local delivery of trophic substances. To date, however, there is not sufficient data on the delivery method to the nervous system, especially following VRA. So far, it is known that mesenchymal stem cells synthesize and possibly release BDNF and GDNF, when grafted to the VRA lesion area [18]. No data, however, indicates that MC exhibit the same properties.

Therefore, the present study investigated two delivery strategies of MC, comparing the local injection to the spinal cord with the possibility of mixing MC with fibrin sealant on the interface of the CNS/PNS. Local production of BDNF and GDNF were evaluated in both situations.

The results herein demonstrate that MC therapy is neroprotective and increases the transcript and protein levels of BDNF and GDNF in the lesioned spinal cord area. Moreover, the administration method significantly influenced treatment outcome, so that the use of fibrin sealant MC delivery approach gave the best and more long lasting results.

Material and Methods

2.1- Experimental animals

Adult female Lewis (LEW/HsdUnib) rats, 7 weeks old, were obtained from the Multidisciplinary Center for Biological Investigation (CEMIB/UNICAMP) and housed under a 12-hour light/dark cycle with free access to food and water. The study was approved by the Institutional Committee for Ethics in Animal Experimentation (Committee for Ethics in Animal Use – Institute of Biology - CEUA/IB/UNICAMP, proc. n° 2073-1). All experiments were performed in accordance with the guidelines of the Brazilian College for Animal Experimentation. The animals were subjected to unilateral avulsion of the L4-L6 lumbar ventral roots and divided into 4 groups: 1) VRA without reimplantation (AV 1 week, n = 5; AV 4 weeks, n = 15 and AV 8 weeks, n = 10); 2) VRA followed by lesioned roots reimplantation with fibrin sealant (AV+S 1 week n = 5, AV+S 4 weeks, n = 15 and AV+S 8 weeks, n = 10); 3) VRA followed by lesioned roots reimplantation with fibrin sealant and MC homogenized to the sealant (AV+S+HC 1 week, n = 5, AV+S+HC 4 weeks, n = 15 and AV+S+HC 8 weeks, n = 10) and 4) VRA followed by lesioned roots reimplantation with fibrin sealant and injection of MC (AV+S+IC 4 weeks, n = 15 and AV+S+IC 8 weeks, n = 10).

The peroneal functional index was calculated weekly up to 8 weeks after injury (n = 10 for each group). Animals were killed 1 week and 4 weeks after injury and their lumbar spinal cords were processed for PCR (n = 5 for each group). The animals were killed after 4 and 8 weeks after avulsion and the lumbar spinal cords were processed for immunohistochemistry (n = 5 for each group) and neuronal survival counting (n = 5 for each group). Each animal's unlesioned, contralateral spinal cord side, served as an internal control.

2.2- Bone Marrow Mononuclear Cells Extraction

MC were extracted from transgenic Lewis rats (LEW-Tg EGFP F455/Rrrc), with the EGFP (Enhanced green fluorescent protein) gene under Ubiquitin C promoter control. The animals were imported from Missouri University (EUA) and were provided by Dr. Alfredo Miranda Góes, Federal University of Minas Gerais – UFMG, Brazil.

The EGFP rats were killed with lethal dose of halothane (Tanohalo, Cristália Chemicals and Pharmaceuticals, Brazil), and the femur and tibia were dissected out from the muscular and connective tissue. The cells were isolated by density gradient centrifugation using Histopaque 1077 following the separation of mononuclear cells methods of Sigma-Aldrich n° 1119 protocol. The cell suspension consists of heterogeneous cell populations (13.2% CD11b, 1.52% CD3, 92.2% CD45, 20.8% CD34 of CD45). After 2 washing steps, cells were immediately transplanted.

2.3- Ventral root avulsion (VRA)

The rats were anesthetized with 50 mg/Kg of Ketamine (Fort Dodge) and 10 mg/Kg of xylasine (Köning) and subjected to unilateral avulsion of the lumbar ventral roots as previously described [19]. Right side avulsion was performed at the L4, L5 and L6 lumbar ventral root after laminectomy. A longitudinal incision was made to open the dural sac, the denticulate ligament was dissected and the ventral roots associated with the lumbar intumescence could be identified and avulsed with fine forceps (№ 4). Finally, the musculature, fascia and skin were sutured in layers. Chlorhydrate of tramadol was administrated by gavage after the surgical procedures (20 mg/kg) and 2.5 mg/day soluble in water during 5 days.

2.4- Reimplantation of the motor roots

In AV+S, AV+S+MC and AV+S+IC groups, the roots were replaced at the exact point of detachment, on the ventral surface of the lumbar spinal cord at the avulsion site with the aid of a snake venom fibrin sealant [10]. The fibrin sealant was kindly provided by CEVAP/UNESP and is under the scope of the Brazilian Patents BR 10 2014 011432 7 and BR 10 2014 011436 0 [20–22]. The sealant used herein is composed of three separate solutions (1-fibrinogen, 2-calcium chloride and 3-thrombin-like fraction). During surgical repair of the avulsed roots, the first two components were applied and the avulsed roots were returned to their original sites. The third component was then added for polymerization. The reimplanted roots were then gently pulled from the spinal cord, and the stability of the fixation was observed to evaluate the success of the repair.

2.5- Mononuclear cells transplantation

In group 3 (AV+S+HC), 3×10^5 MCs were added to the fibrin sealant in avulsed roots reimplantation at the moment of implantation. In this case, the cells are grouped in the right ventrolateral surface of the spinal cord (Figure 1 A and B). In group 4, 3×10^5 MCs were placed directly into the lateral funiculus on the lesioned side (ipsilateral) in the segments L4–L6 of the spinal cord with the aid of a thick capillary Pasteur pipette coupled to a Hamilton syringe and then the avulsed roots were repaired with sealant. In this group, the cells are at the site of injection (Figure 1 C and D).

2.6- Specimen preparation

The animals were anaesthetized with an overdose of mixture of xylasine and Ketamine, and the vascular system was transcardially perfused with phosphate buffer 0.1 M (pH 7.4). For PCR, the rats

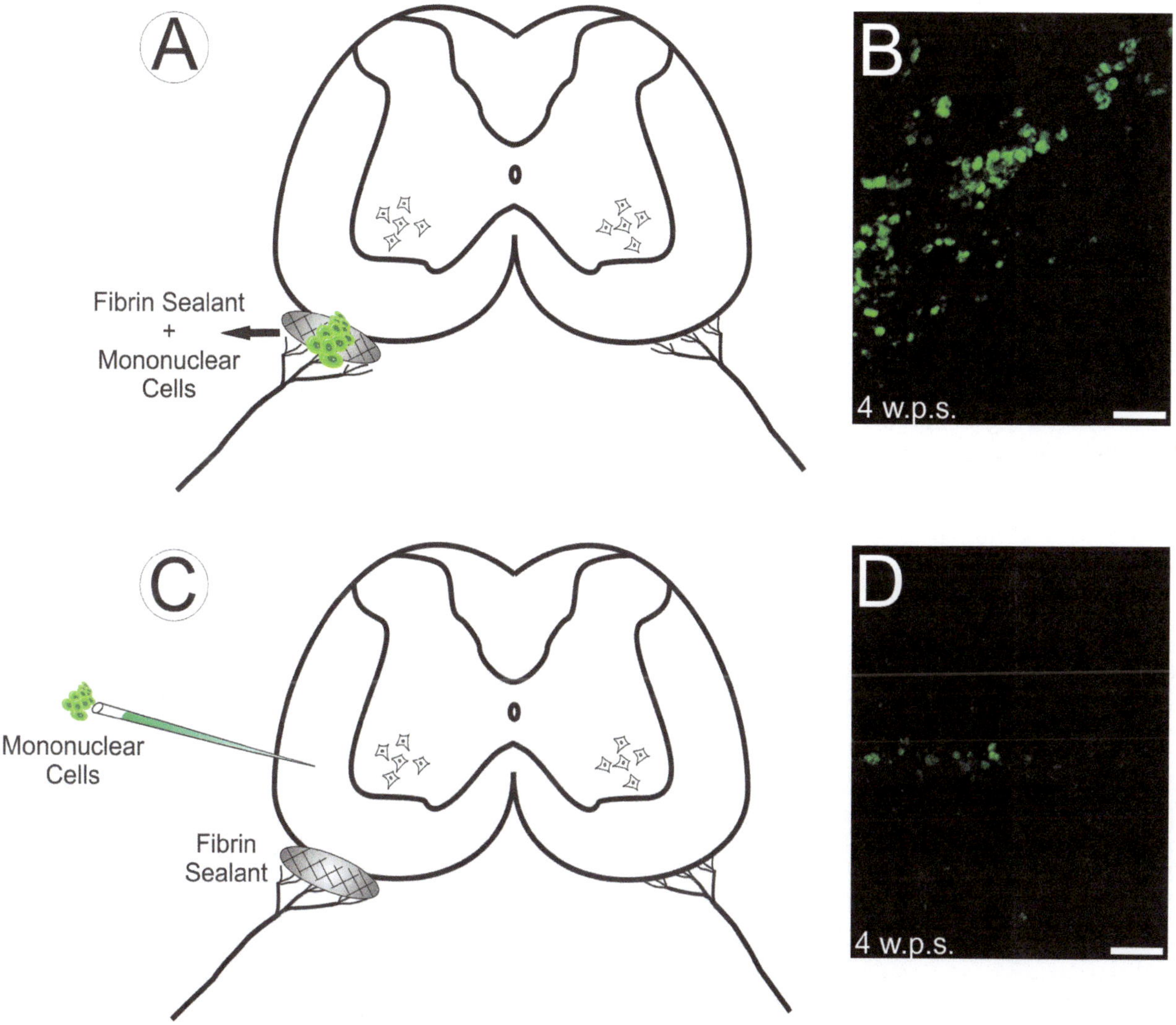

Figure 1. Scheme of the spinal cord subjected to motor root avulsion and reimplantation, with the two different cell delivery locations. A) Scheme of transverse spinal cord section subjected to motor avulsion followed by fibrin sealant reimplantation plus MC. B) MC are grouped in the right ventrolateral surface of the spinal cord (AV+S+HC), four weeks post surgery (4 w.p.s.). C) Scheme of the transverse spinal cord section subjected to motor root avulsion followed by fibrin sealant reimplantation and cell injection in the lateral funiculus. D) The cells are at the site of injection (AV+S+IC), 4 w.p.s. Such cells were E-GFP fluorescent donors. Scale bar = 50 μm.

were killed 1 and 4 weeks after VRA and their lumbar intumescences were frozen in liquid nitrogen. For neuron survival counting and immunohistochemistry the rats were killed 4 and 8 weeks after VRA and fixed by vascular perfusion of 10% formaldehyde in phosphate buffer (pH 7.4). The lumbar intumescence was dissected, post-fixed overnight and then washed in phosphate buffer and stored overnight in sucrose 20% before freezing. Transverse cryostat 12 μm thick sections of spinal cords were obtained and transferred to gelatin-coated slides and dried at room temperature for 30 min before being stored at −20°C.

2.7- Counting of motoneurons survival

Cell counts were performed on sections from the lumbar enlargement. Transverse cryostat sections of the spinal cords were stained for 3 min in aqueous 1% cresyl fast violet solution (Sigma-Aldrich, USA). The sections were then washed in distilled water, dehydrated and mounted with Entellan (Merck, USA). The motoneurons were identified (based on their morphology, size, and location in the dorsolateral lamina IX) and cells with a visible nucleus and nucleolus were counted. The absolute number of motoneurons on the lesioned and non-lesioned side of each section was used to calculate the percentage of surviving cells in each specimen. This percentage was calculated by dividing the number of motoneurons in the ipsilateral (lesioned) side by the number of neurons in the contralateral (non-lesioned) side and multiplying the result by 100. Abercrombie's formula was used to correct for the duplicate counting of neurons [23].

$$N = nt/(t+d)$$

where N is the corrected number of counted neurons, n is the counted number of cells, t is the thickness of the sections (12 μm) and d is the average diameter of the cells. Because differences in cell size can significantly affect cell counts, the value of d was calculated specifically for each experimental group and for both ipsilateral and contralateral neurons. The diameters of 15 randomly chosen neurons from each group were measured by using the ImageTool software (version 3.00, The University of Texas Health Science Center, San Antonio, USA).

2.8- Immunohistochemistry

Transverse sections of spinal cord were incubated with mouse anti-synaptophysin (Dako, 1:250), goat anti-GFAP (Dako, 1:900), and rabbit anti-Iba1 (Wako, 1:800) diluted in a solution containing 1% BSA in TBS-T (Tris-Buffered Saline and Tween) and 2% Triton X-100 in PB 0.1 M (phosphate buffer). All sections were incubated for 6 hours at room temperature in a moist chamber. After rinsing in TBS-T, the sections were incubated according to the primary host antibody (CY-3, Jackson Immunoresearch; 1:250) for 45 minutes in a moist chamber at room temperature. After 3 times washing in TBS-T, the slides were mounted in a mixture of glycerol/PBS (3:1) and observed with a Nikon eclipse TS100 inverted microscope (Nikon, Japan). For quantitative measurements, 3 representative images (with 2 MNs) of the spinal cord (L4–L6 at lamina IX, ventral horn) from each animal were captured at a final magnification of ×200. Double blind quantification was performed in IMAGEJ software (version 1.33 u, National Institute of Health, USA) using the enhanced contrast and density slicing two features [24]. The integrated density of pixels was systematically measured in six representative areas of the motor nucleus from each section, according to [19]. The integrated pixel density was calculated for each section of spinal cord, and then a mean value for each spinal cord was calculated. The data are represented as the mean ± standard error (SE).

2.9- Real time polymerase chain reaction (PCR)

Total RNA was extracted from the ipsilateral and contralateral sides of the frozen lumbar intumescences, 1 and 4 weeks after lesion, using the RNeasy Lipid Tissue Kit (cat n° 74804, Quiagen), according to the manufacturer's recommendations. The RNA was quantified using a NanoDrop Spectrophotometer (A260/280; model 2000, Thermo Scientific). The RNA (1 µg) obtained from five samples was reverse-transcribed using a commercial kit (AffinityScripts QPCR cDNA Synthesis Kit - Agilent Technologies, La Jolla, CA, USA) to achieve a final reaction volume of 20 µL. Real time quantitative PCR was performed on Mx3005P qPCR System (Agilent Technologies, La Jolla, CA, USA), after an initial denaturation for 10 minutes at 95°C, followed by 35 cycles of amplification (95°C for 30 seconds followed by 72°C for one minute). The reactions were carried out with 12.5 µL 2×SYBR Green PCR master mix (Agilent Technologies), 0.2 µM of each forward and reverse primer 100 ng cDNA template, in a final reaction volume of 20 µL. All quantifications were normalized to the house keeping gene β-actin. A non-template control with non-genetic material was included to eliminate contamination or nonspecific reactions. Each sample (n = 5) was tested in triplicate and then used for the analysis of the relative transcription data using the $2^{-\Delta\Delta CT}$ method [25]. The following forward (F) and reverse (R) primers were used: BDNF: (F) 5′-CCACAATGTTC-CACCAGGTG-3′, (R) 5′-TGGGCGCAGCCTTCAT-3′; GDNF: (F) 5′-CCACCATCAAAAGACTGAAAAG-3′, (R) 5′-CGGTTCCTCTCTCTTCGAGGA-3′; Iba-1 (F) 5′- CCC-CACCTAAGGCCACCAGC-3′, (R) 5′-TCCTGTTGGCTTT-CAGCAGTCC-3′; GFAP (F) 5′-TGCTGGAGGGCGAA-GAAAACCG-3′ 5′-CCAGGCTGGTTTCTCGGATCTGG-3′; β-actin (F) 5′-GGAGATTACTGCCCTGGCTCCTA-3′ (R) 5′-GACTCAICGTACTCCTGCTTGCTG-3′. The BDNF and GDNF were based on [18] and the β-actin was based on [26].

2.10- Functional Analysis

For the gait recovery analysis, the CatWalk system (Noldus Inc., The Netherlands; was used. In this set up, the animal crosses a walkway with an illuminated glass floor. A high-speed video camera Gevicam (GP-3360, USA) equipped with a wide-angle lens (8.5 mm, Fujicon Corp., China) is positioned underneath the walkway and the paw prints are automatically recorded and classified by the software. The paw prints from each animal were obtained before and after the VRA. Post-operative CatWalk data were collected twice a week for 12 weeks. The peroneal functional index (PFI) was calculated as the distance between the third toe and hind limb pads (print length) and the distance between the first and fifth toes (print width). Measurements of these parameters were obtained from the right (lesioned) and left (unlesioned) paw prints, and the values were calculated using the following formula [27].

$$PFI = 174.9\ x\ ((EPL - NPL)/NPL)) \\ + 80.3x\ ((ETS - NTS)/NTS)) - 13.4$$

Where N: normal, or non-operated side; E: experimental, or operated; PL: print length; TS: total toe spread, or distance between first to fifth toe.

The pressure exerted on the platform by individual paws was also evaluated. The Catwalk data from each day were expressed as an ipsi-/contralateral ratio.

2.11- Statistical analysis

Statistical analysis was performed with Graphpad Prisma 4.0 software. The neuronal survival, immunohistochemistry, and PCR were firstly evaluated with one-way ANOVA. Data from the functional analysis was evaluated with two-way ANOVA. Bonferroni post-test was used to identify intergroup differences. The data are presented as the mean ± SE and the differences between groups were considered significant when the P-value was <0.05 (*), <0.01 (**) and <0.001 (***).

Results

3.1- Neuroprotection after root reimplantation plus MC treatment after VRA

Neuronal survival was assessed as the ipsi-/contralateral ratio of motoneurons present in the lamina IX of the ventral horn. No significant differences between the numbers of motoneurons on the contralateral side in the different experimental conditions were observed. After four weeks, there was severe degeneration of affected motorneurons in the AV group (Fig. 2A). In implanted groups (Fig 2 C, E and G) a higher number of surviving neurons was observed. Such neuroprotection was even more evident in the AV+S+HC group (Fig. 2I) (AV 36.09%±3.67%; AV+S 65.19%±6.93%; AV+S+HC 89.44%±1.89%; AV+S+IC 64.60%±1.33% percentage of survival ipsi-/contralateral ± SEM; p<0.001).

Neuroprotection in the implanted groups remained superior to the avulsion only throughout the time course of the study, i.e. up to 8 weeks post lesion (Fig. 2 D, E and F). Nevertheless, the AV+S+HC group was similar to AV+S. Figure 2I shows the ipsi-/contralateral ratio for the groups where a statistically significant neuroprotective effect was observed (AV 34.58%±4.12%; AV+S 66.46%±3.88%; AV+S+HC 67.11%±4.83%; AV+S+IC 46.81%±5.87% percentage of survival ipsi-/contralateral ± SEM; p<0.001).

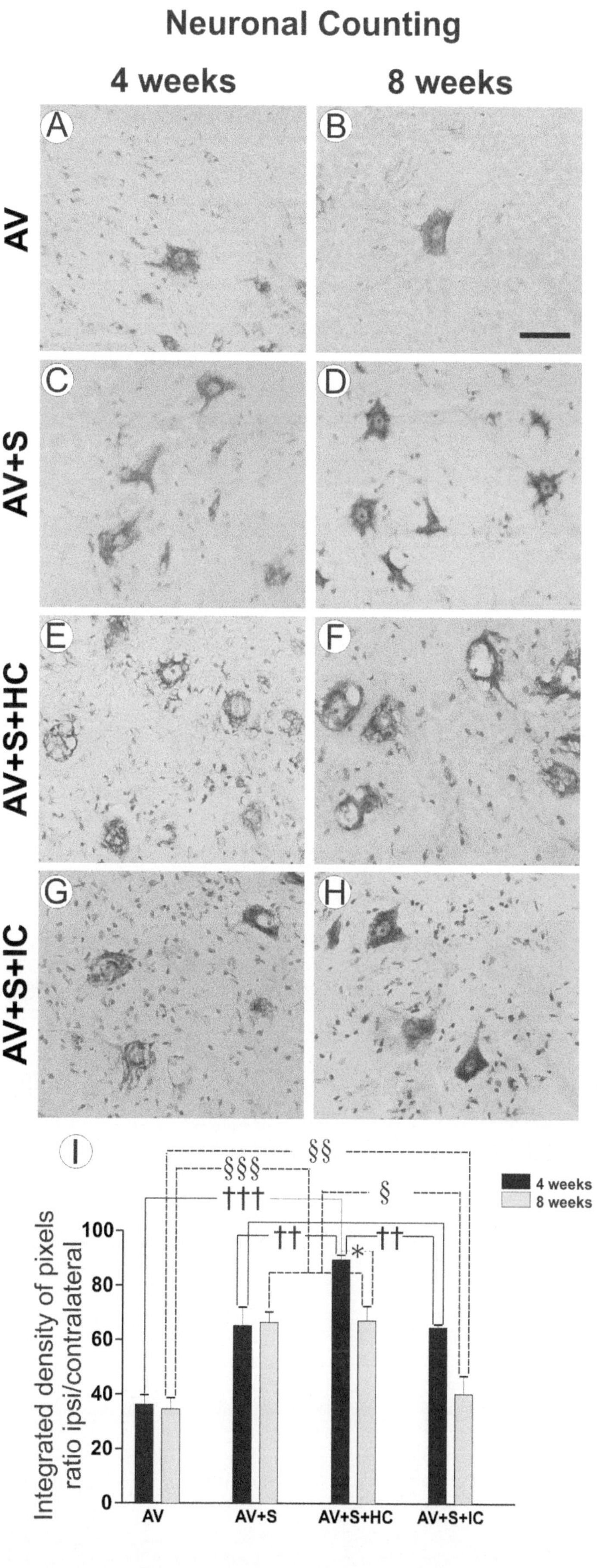
Neuronal Counting
4 weeks
8 weeks
AV
AV+S
AV+S+HC
AV+S+IC
A
B
C
D
E
F
G
H
I
Integrated density of pixels ratio ipsi/contralateral
100
80
60
40
20
0
AV
AV+S
AV+S+HC
AV+S+IC
4 weeks
8 weeks
§§
§§§
†††
§
††
††
*

Figure 2. Nissl-stained spinal cord transverse sections at lamina IX illustrating the neuroprotective effects of root reimplantation and MC treatment on motoneurons 4 and 8 weeks after VRA. Motoneuron cell bodies of the ipsilateral side of VRA after 4 weeks (A, C, E and G) and after 8 weeks (B, D, F and H). (A and B) AV, (C and D) AV+S, (E and F) AV+S+HC and (G and H) AV+S+IC. Scale bar = 50 μm. (I) Percentage of neuronal survival after ventral root avulsion, reimplantation and reimplantantion with MC. Note a significant rescue of lesioned neurons in the implanted groups with and without cells in two different survival times (4 and 8 weeks). This neuroprotection was even more intense in AV+S+HC 4 weeks after avulsion. (†† $p<0.01$ and ††† $p<0.001$ comparing all groups 4 weeks after avulsion, § $p<0.05$, §§ $p<0.01$ and §§§ $p<0.001$ comparing all groups 8 weeks after AV and *$p<0.05$ comparing AV+S+HC at two different times after avulsion, n = 5).

3.2- Decreased synaptic elimination after VRA followed by implantation and cell treatment

Synaptic network changes after root avulsion were evaluated in the ventral horn by immunohistochemistry with an antibody against synaptophysin. Quantitative measurements of synaptophysin immunoreactivity in the sciatic motor nuclei after avulsion (AV) and after avulsion followed by ventral root implantation (AV+S), implantation with cells homogenized on the sealant (AV+S+HC) and implantation with cell injection (AV+S+IC) were carried out. As shown in Fig. 3, AV only led to a significant decrease in synaptophysin expression four weeks after avulsion, which remained until 8 weeks. Such results indicate a significant decrease of complexity of intraspinal networks following lesion. In contrast, in implanted groups with or without cells, the repair resulted in preservation of synaptophysin immunoreactivity, in the immediate vicinity of the motoneurons. In AV+S+IC group the synaptic inputs decreased when comparing 4 and 8 weeks. Four weeks after avulsion: (AV 0.43±0.01; AV+S 0.66±0.09; AV+S+HC 0.65±0.07; AV+S+IC 0.60±0.01 mean ratio ipsi-/contralateral ± SE with $p<0.01$) and 8 weeks after avulsion: (AV 0.40±0.05; AV+S 0.72±0.05; AV+S+HC 0.84±0.07; AV+S+IC 0.855±0.03 mean ratio ipsi-/contralateral ± SE, $p<0.05$).

3.3- Astroglial reactivity is not further enhanced by mononuclear cells around motoneuron vicinity

Immunoreactivity against GFAP was used to analyze the degree of astroglial reactivity after lesion. This demonstrates the presence of GFAP-positive astrocytic processes in the vicinity of the avulsed motoneurons. Figure 4 shows that the astroglial reactivity was not significantly further increased after implantation and cell treatment in both experimental times: Four weeks after avulsion (AV 2.65±0.48; AV+S 2.49±0.60; AV+S+HC 2.62±0.43; AV+S+IC 2.17±0.12 mean ratio ipsi-/contralateral ± SE); eight weeks after avulsion (AV 3.45±0.61; AV+S 2.09±0.26; AV+S+HC 2.58±0.37; AV+S+IC 2.30±0.32 mean ratio ipsi-/contralateral ± SE).

The real-time PCR analysis was performed to measure GFAPmRNA in the RNA in all ventral horn after avulsion. Figure 4 demonstrated that the levels of GFAPmRNA were similar in all groups, 1 week (Fig 4 J) after lesion (AV 1.51±0.38; AV+S 1.20±0.24; AV+S+HC 1.65±0.26 mean ratio ipsi-/contralateral ± SE). However, 4 weeks (Fig 4 K) after avulsion, AV+S group presented a decreased number of transcripts to GFAP (AV 2.64±0.45; AV+S 1.35±0.14; AV+S+HC 2.11±0.12; AV+S+IC 2.00±0.12 mean ratio ipsi-/contralateral ± SE).

3.4- Microglial reactivity is enhanced by mononuclear cells one week after tranplantation

To detect possible changes in the microglial cells close to large motoneurons cell bodies after avulsion, the immunoreactivity against Iba-1 was evaluated in the ventral horn in the different experimental groups. Figure 5 shows that the microglial reactivity was not significantly further increased after implantation and cell treatment in both experimental times. However, the AV+S and AV+S+IC groups showed decreased microglia reaction in the course of experimental time. Four weeks after avulsion: (AV 4.71±0.41; AV+S 4.73±0.94; AV+S+HC 3.02±0.64; AV+S+IC 5.67±1.34; mean ratio ipsi-/contralateral ±SE with $p<0.01$) and 8 weeks after avulsion: (AV 2.70±0.51; AV+S 2.18±0.57; AV+S+HC 2.32±0.59; AV+S+IC 2.25±0.35 mean ratio ipsi-/contralateral ± SE, $p<0.05$).

The real-time PCR analysis was performed to measure Iba-1 mRNA in the in ventral horn RNA after avulsion. Figure 5 demonstrated that the levels of Iba-1 mRNA in AV+S+HC were significantly greater than in the AV and AV+S groups (AV 0.95±0.30; AV+S 1.63±0.60; AV+S+HC 4.15±0.48 mean ratio ipsi-/contralateral ± SE). Nevertheless, 4 weeks (Fig 5 K) after avulsion, Iba-1 mRNA were similar in all groups (AV 2.07±0.26; AV+S 1.07±0.18; AV+S+HC 1.64±0.26; AV+S+IC 1.73±0.03 mean ratio ipsi-/contralateral ± SE).

3.5- Mononuclear cells enhance BDNF and GDNF expression one week after avulsion

Real time PCR was used to detect BDNF and GDNF mRNAs in the lumbar spinal cord after avulsion (Figure 6). The results demonstrated the levels of both neurotrophic factors in the AV+S+HC group were significantly greater, when compared to the other groups, one week after surgery (A and C) (BDNF - AV 1.20±0.29; AV+S 1.33±0.18; AV+S+HC 2.48±0.10 mean ratio ipsi-/contralateral ± SE *$p<0.05$) and (GDNF - AV 1.71±0.56; AV+S 1.24±0.23; AV+S+HC 2.86±0.34; mean ratio ipsi-/contralateral ± SE *$p<0.05$). Four weeks after avulsion, BDNF and GDNF mRNA levels were similar in all groups (B and D) (BDNF - AV 2.04±0.18; AV+S 1.45±0.35; AV+S+HC 1.53±0.16; AV+S+IC 1.70±0.17 mean ratio ipsi-/contralateral ± SE) and (GDNF - AV 2.25±0.10; AV+S 1.92±0.43; AV+S+HC 1.72±0.30; AV+S+IC 1.77±0.01 mean ratio ipsi-/contralateral ± SE).

3.6- Functional motor recovery

The recovery of motor function was analysed by the CatWalk System. Post-operative assessments of peroneal function were performed for eight consecutive weeks. The preoperative peroneal functional index mean values (Figure 7A) did not significantly differ between groups. The peroneal functional index drastically declined after avulsion. The three implanted groups presented significantly better peroneal index performance as compared to AV group (AV −228.89±23.0; AV+S −139.75±12.5; AV+S+HC −155.50±17.2; AV+S+IC −205.14±18.6, mean ratio ipsi-/contralateral ± SE, *$p<0.05$, **$p<0.01$ and ***$p<0.001$. These results are consistent with the footprint paw pressure data (Figure 7B), indicating that all three implanted groups presented a better performance as compared to avulsion group (AV 0.20±0.07; AV+S 0.81±0.09726454; AV+S+HC 0.66±0.078; AV+S+IC 0.40±0.11 mean ratio ipsi-/contralateral ± SE with *$p<0.05$ and ***$p<0.001$).

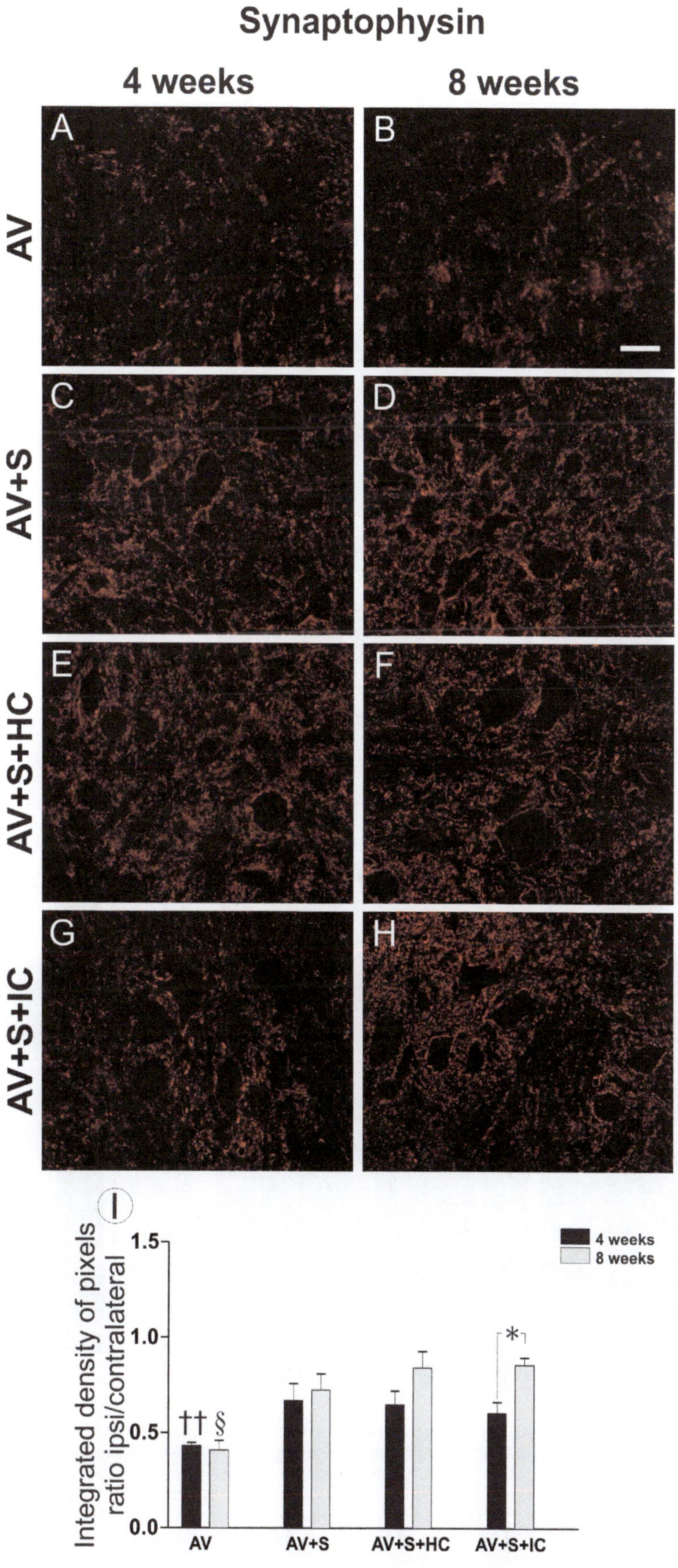
Synaptophysin
4 weeks
8 weeks
A
B
C
D
E
F
G
H
AV
AV+S
AV+S+HC
AV+S+IC
I
Integrated density of pixels ratio ipsi/contralateral
1.5
1.0
0.5
0.0
4 weeks
8 weeks
††
§
*
AV
AV+S
AV+S+HC
AV+S+IC

Figure 3. Immunohistochemical analysis of the spinal cord ventral horn stained with anti-synaptophysin 4 and 8 weeks after VRA. Observe the preservation of synaptophysin labeling, especially at the surface of the lesioned motoneurons in implanted group and both implanted with mononuclear cells treatment. After 4 weeks post lesion (A, C, E and G) and after 8 weeks (B, D, F and H). (A and B) AV, (C and D) AV+S, (E and F) AV+S+HC and (G and H) AV+S+IC. Scale bar =50 μm. (I) Quantification of synaptic covering obtained by the ratio ipsi/contralateral sides of the integrated density of pixels at lamina IX four and eight weeks after injury. (†† $p<0.01$ comparing all groups 4 weeks after avulsion, § $p<0.05$ comparing all groups 8 weeks after AV and * $p<0.05$ comparing AV+S+IC at two different times after avulsion, n=5).

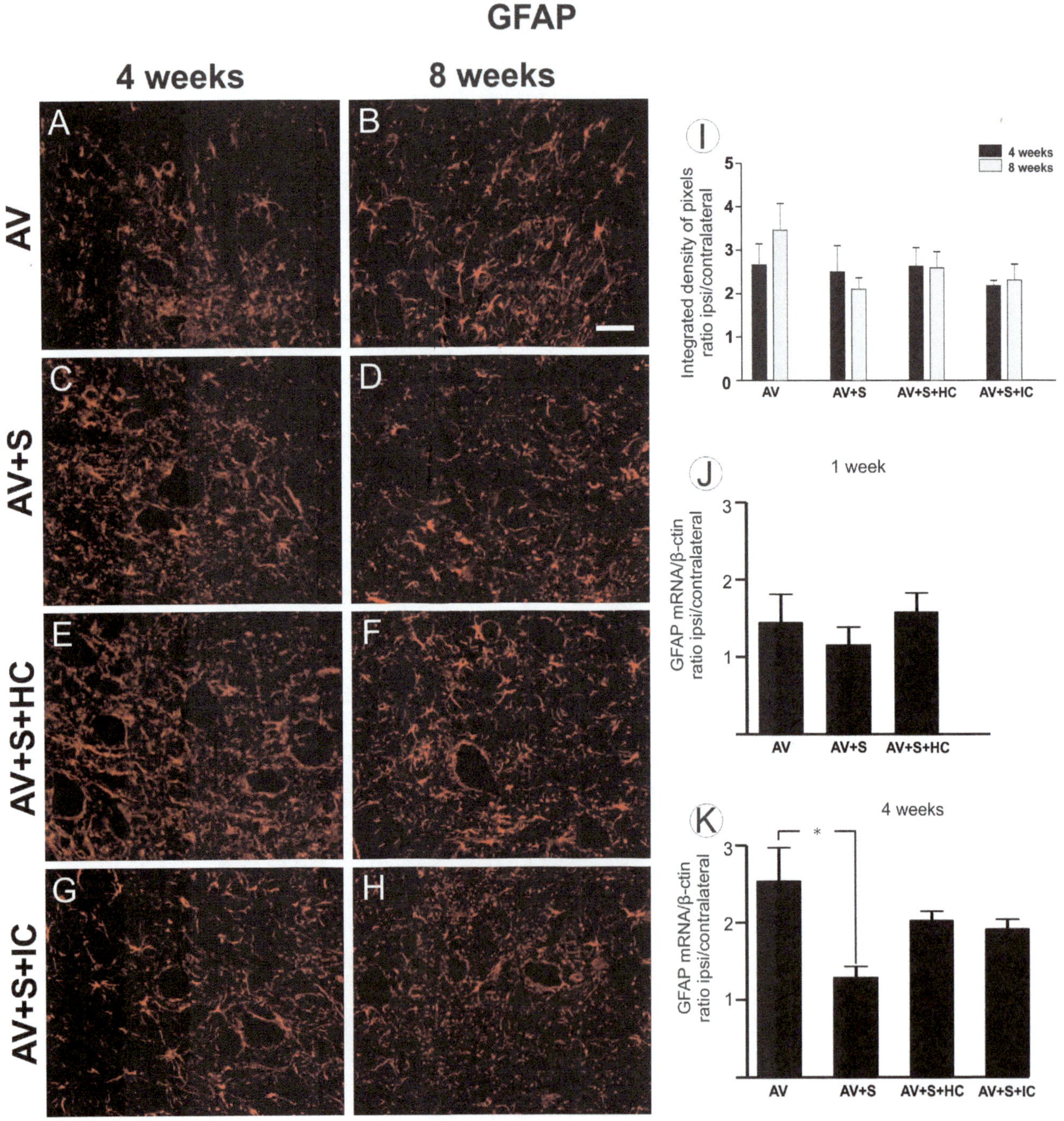

Figure 4. Glial fibrillary acidic protein (GFAP) in the spinal cord ventral horn. Immunohistochemical analysis of the anterior horn of the spinal cord was labeled with anti-GFAP, 4 and 8 weeks after injury to assess the degree of astroglial reactivity after root avulsion (A–H). Representative images of AV, AV+S, AV+S+HC and AV+S+IC. Scale bar =50 μm. Observe that ventral root implantation and cell treatment did not increase astroglial reaction. (I) The mean ratio of the ipsi-/contralateral integrated intensity of pixels of the ipsilateral and contralateral sides in all groups. (* $p<0.05$, n=5). (J) One week after avulsion, there were no differences between groups by GFAPmRNA analysis. (K) The RT-qPCR performed 4 weeks after avulsion demonstrated significant decrease in the synthesis of GFAP mRNA in AV+S group compared with AV (*$p<0.05$, n=5).

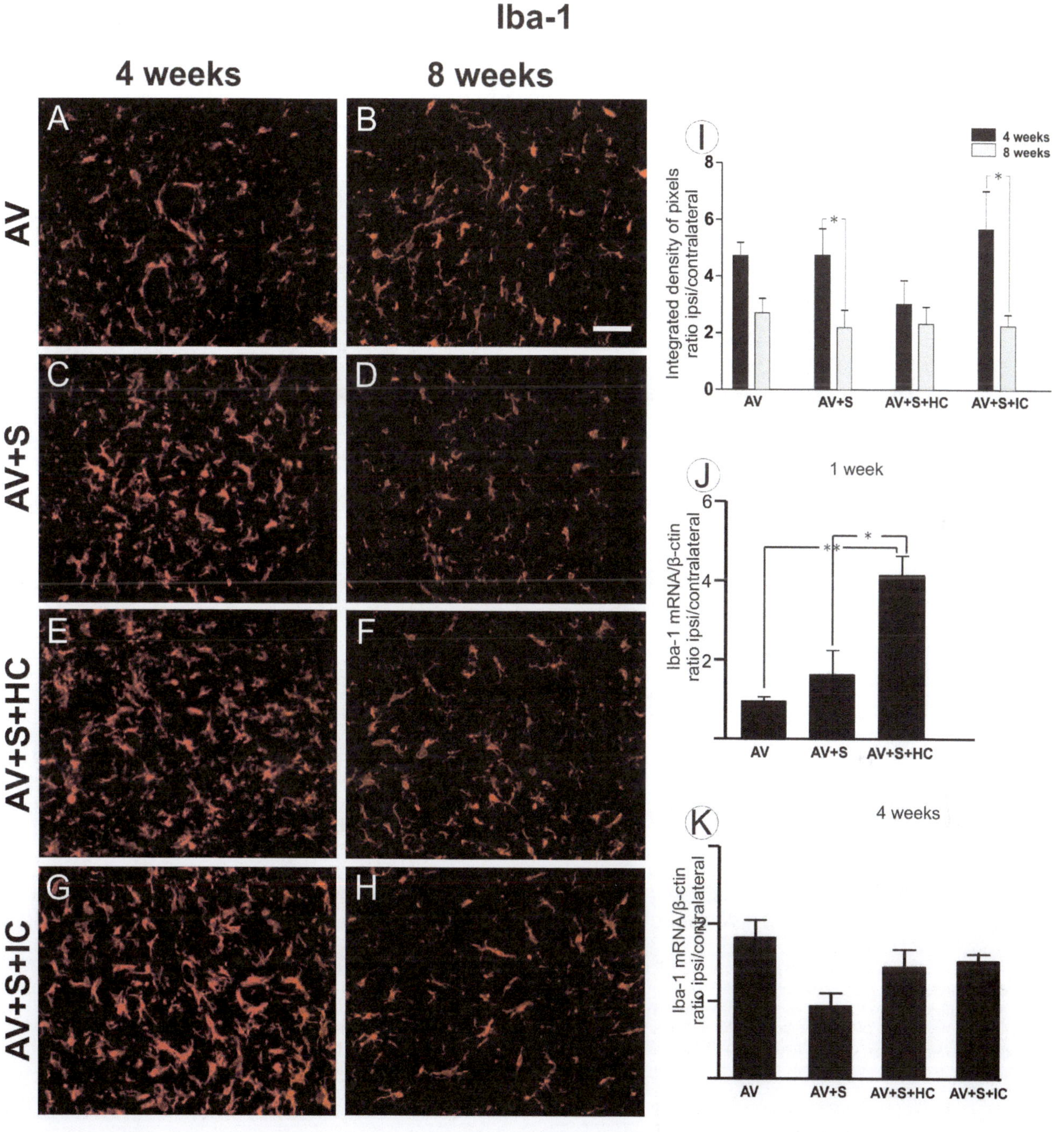

Figure 5. Microglial analysis of the spinal cord ventral horn 4 and 8 weeks after VRA. The quantification analysis on motoneuron cell bodies stained with anti- Iba1 of the ipsilateral side of VRA after 4 weeks (A, C, E and G) and after 8 weeks (B, D, F and H). (A and B) AV, (C and D) AV+S, (E and F) AV+S+HC and (G and H) AV+S+IC. Scale bar $=50$ μm. (I) The mean ratio of the ipsil-/contralateral integrated intensity of pixels of the ipsilateral and contralateral sides in both groups. Note the decrease in microglial reactivity in AV+S and AV+S+IC as compared 4 and 8 weeks after injury. (J) There were significant increase in the synthesis of Iba-1 mRNA in the lumbar spinal cord one week after avulsion ($^{*}p<0.05$ and $^{**}p<0.01$, $n=5$). (K) The RT-qPCR, performed 4 weeks after avulsion, did not demonstrate significant differences between groups in Iba-1 mRNA.

Discussion

Brachial plexus lesions are particularly debilitating and usually affect young adults. To date, new technological refinements aiming at improving the repair of such traumatic injury are necessary. Although nerve reimplantation has been proposed [7,10] the relatively poor clinical outcome requires further attention. Treatment with neurotrophic factors, following root reimplantation, has proven to be efficient [17]. Importantly, it has been previously shown that mesenchymal stem cells naturally produce BDNF and GDNF when grafted to the VRA lesion area [18]. Considering it, the present study main objective was to compare the root reimplantation regenerative outcome by either injecting or engrafting the mononuclear cells at the wounded area,

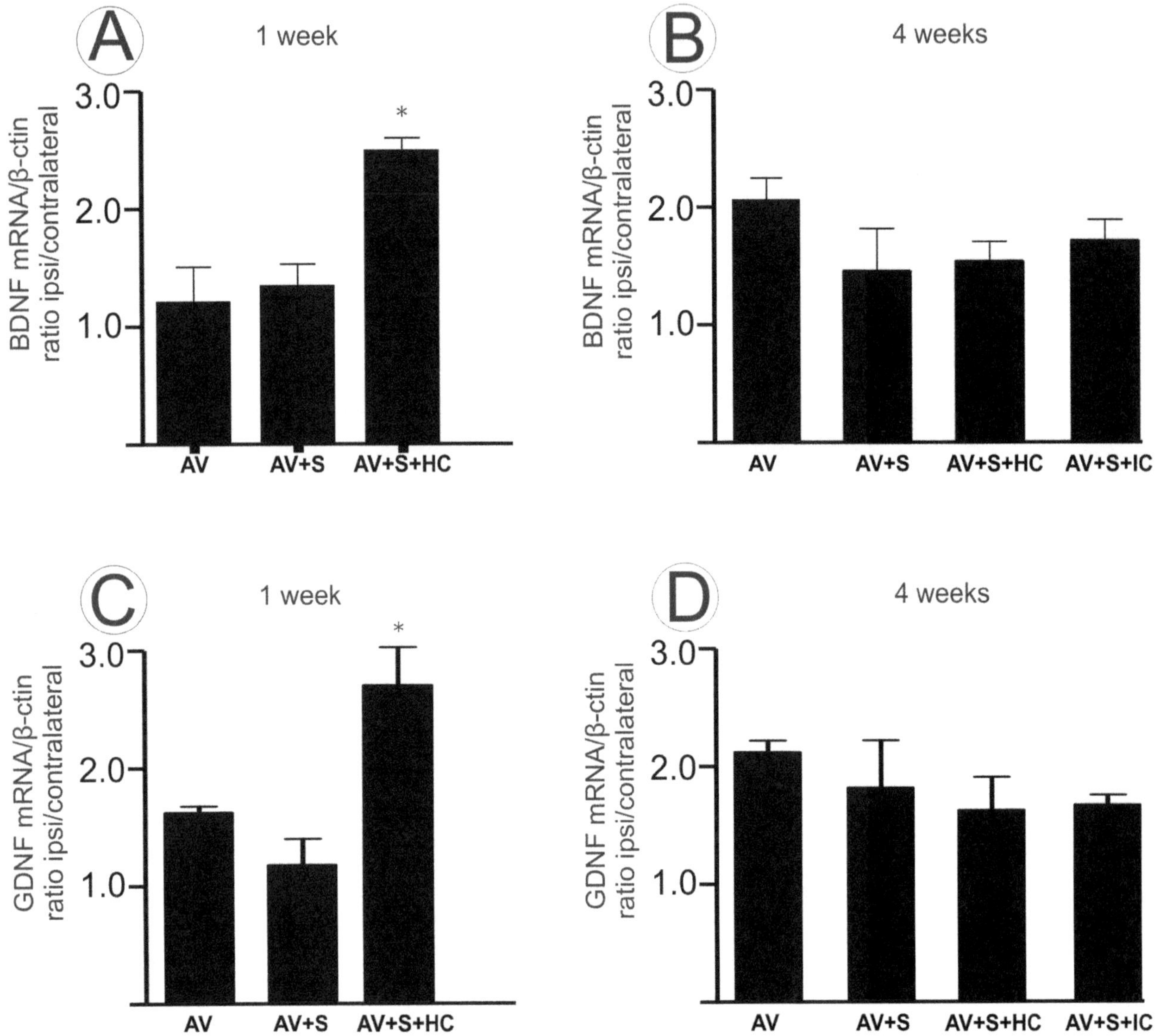

Figure 6. Neurotrophic factor expression (BDNF and GDNF) by RT-qPCR in the ventral horn spinal cord 1 and 4 weeks after avulsion. Expression of mean ratio of the ipsi-/contralateral mRNA for BDNF obtained by RT-qPCR in the lumbar spinal cord one (A) and four (B) weeks after avulsion. Note that one week after avulsion AV+S+HC enhanced the BDNF mRNA production compared to others groups (*$p<0.05$, $n=5$). Expression of mean ratio of the ipsi-/contralateral mRNA for GDNF obtained by RT-qPCR in the lumbar spinal cord one (C) and four (D) weeks after avulsion. Note that, one week after avulsion in the AV+S+HC group, there is enhanced GDNF mRNA production as compared to others groups (*$p<0.05$, $n=5$).

with aid of a fibrin sealant scaffold. However, evaluating the perspective of translational studies, the use of bone marrow MC has been preferred instead of mesenchymal stem cells.

The use of MC resulted in neuroprotection following peripheral nervous system lesion [1,3] and following spinal cord injury [28]. Furthermore, the mononuclear fraction is easy to obtain, implicating in lower costs and less processing time, as compared with other cell types [1,29].

An important point to be observed is the best moment for performing the cell therapy. Therefore, in some instances, it has been proposed a late cell transplantation, following the acute phase after injury [30]. However, immediate cell engrafting increases the survival of motoneurons after root avulsion [17].

Based on such observations, the present work was based on cells treatment immediately after avulsion, in order to obtain the best neuroprotective results possible. This is particularly important following CNS/PNS injuries that cause extensive degeneration of adult motoneurons [31–33]. Such neuronal loss reaches up to 80% of motoneuron degeneration in the first two weeks post injury [7]. Our results showed that both intraspinal administration and fibrin sealant implantation of MC preserved a significant number of motoneurons when compared to the untreated group (avulsion). The neuroprotective effect observed herein reinforces that the combination of root reimplantatio and cell therapy leads to enhancement of axotomized motoneurons. On the contrary, most of the motoneurons degenerated following AV, which reinforces the need of acute neuroprotective treatments following VRA.

Neuronal rescue was highest following root reimplantation and fibrin sealant cell engrafting, 4 weeks after injury. Such effect may be the result of stem cell neurotrophic factor release in the site of

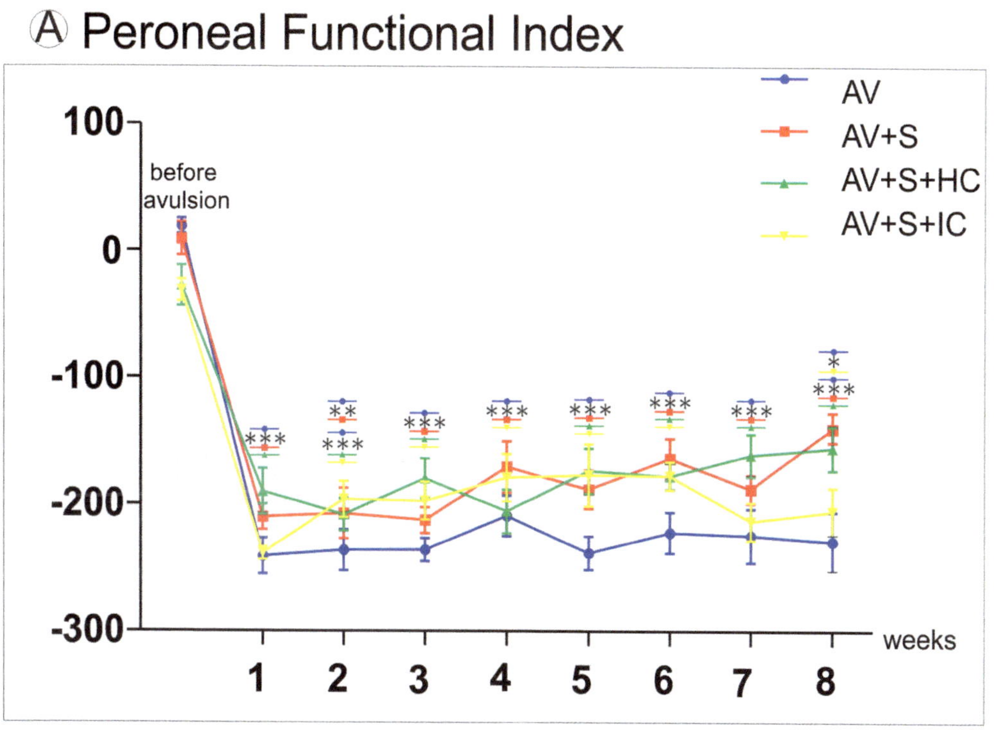

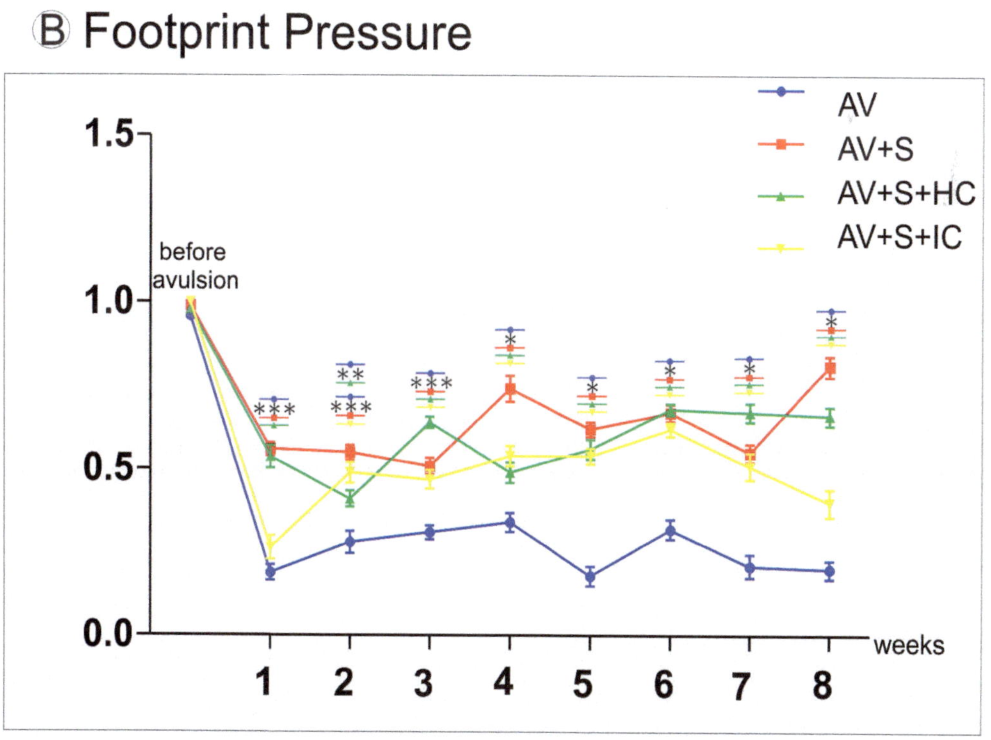

Figure 7. Motor function recovery after ventral root treatment. (A) Graph of the peroneal nerve functional index up to 8 weeks after avulsion. There is a significantly better performance of the three implanted groups compared to AV from the first week post lesion until the eighth week (***$p<0.001$, **$p<0.01$ and *$p<0.05$ n = 10). (B) Restoration of weight-bearing capacity following avulsion. There is also a restoration of weight-bearing capacity following avulsion in implanted groups with or without cells from the first up to the eight week after injury. Values are expressed as the ratio of ipsi-/contralateral pressure exerted by the paw on the catwalk platform (***$p<0.001$ and *$p<0.05$, n = 10).

injury [17,18]. Nevertheless, the production of extracellular matrix molecules has also been suggested as neuroprotective [34]. Neuronal survival decreased at 8 weeks post injury, indicating that the therapeutic window for MC is relatively short and further cell treatments may be necessary to maintain the axotomized neurons.

Cell injection to the spinal cord gave similar results to the reimplantation alone, 4 weeks post injury. However, neuronal degeneration increased up to 8 weeks, reaching the level of avulsion only. We believe that the injection into the CNS/PNS interface result in a late inflammatory response that cause delayed cell death. In this sense, using a similar reimplantation approach, followed by mesenchymal stem cell injection, neuron preservation has been shown without significant functional recovery [35]. Overall, the neuronal survival results indicate that fibrin sealant homogenized MC result in acutely better results as compared to intraspinal engrafting.

According to prior studies, MC express BDNF and GDNF [3] in the PNS following sciatic nerve injury. BDNF is also produced by MCs in the CNS, 3 and 7 days after MC therapy [4]. In line with this, several studies have demonstrated neuroprotective effects of such neurotrophic factors [7,32,36], especially after spinal cord injury [37]. Neuroprotection is also achieved in neonatal animals [38], where the effects of axotomy are more drastic than in adults. The results herein are in line with the literature and show BDNF and GDNF production by the MC, 1 week post injury.

Interestingly, BDNF, which is exacerbated in the implant group, promotes axonal sprouting and elongation after peripheral nerve injury and during development [39]. This is in line with the motor recovery observed in all root reimplanted groups, suggesting regeneration of axons after avulsion.

Functional analysis of the peroneal nerve, as well as the pressure exerted by the ipsilateral paw, provides reliable data on functional recovery of animals with motor problems [40,41]. The motor improvement observed herein is noteworthy because gait restoration is necessarily related to motor units recovery, together with cerebral cortex control [42]. Importantly, the preservation of pre-synaptic inputs to the injured motoneurons after implantation, observed by immunohistochemistry, indicates preservation of supraspinal contacts.

An important occurrence following VRA is the reduction of pre-synaptic terminals to motoneurons [42,43]. This synaptic decrease become irreversible if the regrowing axons within the root do not reach their target muscles [43]. Importantly, the groups subjected to root reimplantation showed a significant preservation of synapses as compared to the avulsion only group. In this case, cell treatment gave no further benefit, indicating that the restoration of CNS/PNS connection is sufficient to preserve spinal synaptic networks. Coupled with such synaptic changes, glial cells became reactive following VRA. Such reactive gliosis is characterized by hypertrophy of the cell body and processes of astrocytes and microglial hyperplasia [44–46].

Although the microglial reaction was morphologically decreased by MC treatment four weeks post lesion, at one week, an enhancement of Iba-1 gene transcripts was obtained. One possibility is that MCs increased or accelerated acute microglial response [47]. Regarding the astrogliosis, no differences were perceived between groups by immunohistochemical analysis. qPCR data, on the other hand, indicated a decreased number of GFAP gene transcripts following root reimplantation alone. This is in line with what has been observed 12 weeks after root reimplantation [10].

Conclusions

Taken together, the results of the present study indicate that MCs therapy further preserves injured motoneurons up to 4 weeks after implantation, in comparison to root reimplantation alone. Such fact is possibly related to the production of neurotrophic factors, such as BDNF and GDNF. Nonetheless, such difference was not seen at 4 weeks post injury. This may in turn indicate that restoration of the CNS/PNS connection must be carried out in a short period of time after avulsion. Importantly, the present data indicate that cell injection to the spinal cord does not result in long lasting neuroprotection. On the other hand, engrafting of MC at the site of injury with the aid of a fibrin scaffold is effective and may represent a more practical approach, with regard to translational medicine.

Acknowledgments

The authors are thankful to Prof. Dr. Alfredo M. Goes for providing the EGFP transgenic rats used in the present study. To Prof. Dr. Antonio C. Rodrigues for the fruitful discussions and to Prof. Dr. Dawidson A. Gomes and Prof. Dr. Ana Leda Longhini for expert flow cytometry analysis and comments.

Author Contributions

Conceived and designed the experiments: ALRO. Performed the experiments: ALRO MVC RB. Analyzed the data: ALRO MVC RB BB RSF. Contributed reagents/materials/analysis tools: ALRO BB RSF. Contributed to the writing of the manuscript: ALRO MVC RB BB RSF.

References

1. Goel RK, Suri V, Suri A, Sarkar C, Mohanty S, et al. (2009) Effect of bone marrow-derived mononuclear cells on nerve regeneration in the transection model of the rat sciatic nerve. J Clin Neurosci 16: 1211–1217.
2. de Freitas HT, da Silva VG, Giraldi-Guimaraes A (2012) Comparative study between bone marrow mononuclear fraction and mesenchymal stem cells treatment in sensorimotor recovery after focal cortical ablation in rats. Behav Brain Funct 8: 58.
3. Lopes-Filho JD, Caldas HC, Santos FC, Mazzer N, Simoes GF, et al. (2011) Microscopic evidences that bone marrow mononuclear cell treatment improves sciatic nerve regeneration after neurorrhaphy. Microsc Res Tech 74: 355–363.
4. Gubert F, Zaverucha-do-Valle C, Figueiredo FR, Bargas-Rega M, Paredes BD, et al. (2013) Bone-marrow cell therapy induces differentiation of radial glia-like cells and rescues the number of oligodendrocyte progenitors in the subventricular zone after global cerebral ischemia. Stem Cell Res 10: 241–256.
5. Thatte MR, Babhulkar S, Hiremath A (2013) Brachial plexus injury in adults: Diagnosis and surgical treatment strategies. Ann Indian Acad Neurol 16: 26–33.
6. Carlstedt T (2009) Nerve root replantation. Neurosurg Clin N Am 20: 39–50, vi.
7. Koliatsos VE, Price WL, Pardo CA, Price DL (1994) Ventral root avulsion: an experimental model of death of adult motor neurons. J Comp Neurol 342: 35–44.
8. Li L, Houenou LJ, Wu W, Lei M, Prevette DM, et al. (1998) Characterization of spinal motoneuron degeneration following different types of peripheral nerve injury in neonatal and adult mice. J Comp Neurol 396: 158–168.
9. Penas C, Casas C, Robert I, Fores J, Navarro X (2009) Cytoskeletal and activity-related changes in spinal motoneurons after root avulsion. J Neurotrauma 26: 763–779.
10. Barbizan R, Castro MV, Rodrigues AC, Barraviera B, Ferreira RS, et al. (2013) Motor recovery and synaptic preservation after ventral root avulsion and repair with a fibrin sealant derived from snake venom. PLoS One 8: e63260.
11. Cullheim S, Carlstedt T, Linda H, Risling M, Ulfhake B (1989) Motoneurons reinnervate skeletal muscle after ventral root implantation into the spinal cord of the cat. Neuroscience 29: 725–733.
12. Hoang TX, Nieto JH, Dobkin BH, Tillakaratne NJ, Havton LA (2006) Acute implantation of an avulsed lumbosacral ventral root into the rat conus medullaris promotes neuroprotection and graft reinnervation by autonomic and motor neurons. Neuroscience 138: 1149–1160.
13. Chang HY, Havton LA (2008) Surgical implantation of avulsed lumbosacral ventral roots promotes restoration of bladder morphology in rats. Exp Neurol 214: 117–124.

14. Carlstedt T (2008) Root repair review: basic science background and clinical outcome. Restor Neurol Neurosci 26: 225–241.
15. Eggers R, Tannemaat MR, Ehlert EM, Verhaagen J (2010) A spatio-temporal analysis of motoneuron survival, axonal regeneration and neurotrophic factor expression after lumbar ventral root avulsion and implantation. Exp Neurol 223: 207–220.
16. Giuffre JL, Kakar S, Bishop AT, Spinner RJ, Shin AY (2010) Current concepts of the treatment of adult brachial plexus injuries. J Hand Surg Am 35: 678–688; quiz 688.
17. Lang EM, Asan E, Plesnila N, Hofmann GO, Sendtner M (2005) Motoneuron survival after C7 nerve root avulsion and replantation in the adult rabbit: effects of local ciliary neurotrophic factor and brain-derived neurotrophic factor application. Plast Reconstr Surg 115: 2042–2050.
18. Rodrigues Hell RC, Silva Costa MM, Goes AM, Oliveira AL (2009) Local injection of BDNF producing mesenchymal stem cells increases neuronal survival and synaptic stability following ventral root avulsion. Neurobiol Dis 33: 290–300.
19. Oliveira AL, Thams S, Lidman O, Piehl F, Hokfelt T, et al. (2004) A role for MHC class I molecules in synaptic plasticity and regeneration of neurons after axotomy. Proc Natl Acad Sci U S A 101: 17843–17848.
20. Gasparotto VP, Landim-Alvarenga FC, Oliveira AL, Simoes GF, Lima-Neto JF, et al. (2014) A new fibrin sealant as a three-dimensional scaffold candidate for mesenchymal stem cells. Stem Cell Res Ther 5: 78.
21. Barros LC, Soares AM, Costa FL, Rodrigues VM, Fuly AL, et al. (2011) Biochemical and biological evaluation of gyroxin isolated from Crotalus durissus terrificus venom. The Journal of Venomous Animals and Toxins including Tropical Diseases 17: 23–33.
22. Barros LC, Ferreira RS, Jr., Barraviera SR, Stolf HO, Thomazini-Santos IA, et al. (2009) A new fibrin sealant from Crotalus durissus terrificus venom: applications in medicine. J Toxicol Environ Health B Crit Rev 12: 553–571.
23. Abercrombie M, Johnson ML (1946) Quantitative histology of Wallerian degeneration; nuclear population in rabbit sciatic nerve. J Anat 80: 37–50.
24. Freria CM, Velloso LA, Oliveira AL (2012) Opposing effects of Toll-like receptors 2 and 4 on synaptic stability in the spinal cord after peripheral nerve injury. J Neuroinflammation 9: 240.
25. Livak KJ, Schmittgen TD (2001) Analysis of relative gene expression data using real-time quantitative PCR and the 2(-Delta Delta C(T)) Method. Methods 25: 402–408.
26. Yao L, Chen X, Tian Y, Lu H, Zhang P, et al. (2012) Selection of housekeeping genes for normalization of RT-PCR in hypoxic neural stem cells of rat in vitro. Mol Biol Rep 39: 569–576.
27. Bain JR, Mackinnon SE, Hunter DA (1989) Functional evaluation of complete sciatic, peroneal, and posterior tibial nerve lesions in the rat. Plast Reconstr Surg 83: 129–138.
28. Yoshihara T, Ohta M, Itokazu Y, Matsumoto N, Dezawa M, et al. (2007) Neuroprotective effect of bone marrow-derived mononuclear cells promoting functional recovery from spinal cord injury. J Neurotrauma 24: 1026–1036.
29. Fernandes M, Valente SG, Fernandes MJ, Felix EP, Mazzacoratti Mda G, et al. (2008) Bone marrow cells are able to increase vessels number during repair of sciatic nerve lesion. J Neurosci Methods 170: 16–24.
30. Okada S, Ishii K, Yamane J, Iwanami A, Ikegami T, et al. (2005) In vivo imaging of engrafted neural stem cells: its application in evaluating the optimal timing of transplantation for spinal cord injury. FASEB J 19: 1839–1841.
31. Novikov L, Novikova L, Kellerth JO (1995) Brain-derived neurotrophic factor promotes survival and blocks nitric oxide synthase expression in adult rat spinal motoneurons after ventral root avulsion. Neurosci Lett 200: 45–48.
32. Kishino A, Ishige Y, Tatsuno T, Nakayama C, Noguchi H (1997) BDNF prevents and reverses adult rat motor neuron degeneration and induces axonal outgrowth. Exp Neurol 144: 273–286.
33. Oliveira AL, Langone F (2000) GM-1 ganglioside treatment reduces motoneuron death after ventral root avulsion in adult rats. Neurosci Lett 293: 131–134.
34. Ide C (1996) Peripheral nerve regeneration. Neurosci Res 25: 101–121.
35. Torres-Espin A, Corona-Quintanilla DL, Fores J, Allodi I, Gonzalez F, et al. (2013) Neuroprotection and axonal regeneration after lumbar ventral root avulsion by re-implantation and mesenchymal stem cells transplant combined therapy. Neurotherapeutics 10: 354–368.
36. Yan Q, Elliott J, Snider WD (1992) Brain-derived neurotrophic factor rescues spinal motor neurons from axotomy-induced cell death. Nature 360: 753–755.
37. Bregman BS, Coumans JV, Dai HN, Kuhn PL, Lynskey J, et al. (2002) Transplants and neurotrophic factors increase regeneration and recovery of function after spinal cord injury. Prog Brain Res 137: 257–273.
38. Sendtner M, Holtmann B, Kolbeck R, Thoenen H, Barde YA (1992) Brain-derived neurotrophic factor prevents the death of motoneurons in newborn rats after nerve section. Nature 360: 757–759.
39. Mendell LM, Munson JB, Arvanian VL (2001) Neurotrophins and synaptic plasticity in the mammalian spinal cord. J Physiol 533: 91–97.
40. Dijkstra JR, Meek MF, Robinson PH, Gramsbergen A (2000) Methods to evaluate functional nerve recovery in adult rats: walking track analysis, video analysis and the withdrawal reflex. J Neurosci Methods 96: 89–96.
41. Varejao AS, Meek MF, Ferreira AJ, Patricio JA, Cabrita AM (2001) Functional evaluation of peripheral nerve regeneration in the rat: walking track analysis. J Neurosci Methods 108: 1–9.
42. Purves D, Lichtman JW (1978) Formation and maintenance of synaptic connections in autonomic ganglia. Physiol Rev 58: 821–862.
43. Brannstrom T, Kellerth JO (1998) Changes in synaptology of adult cat spinal alpha-motoneurons after axotomy. Exp Brain Res 118: 1–13.
44. Privat A, Valat J, Fulcrand J (1981) Proliferation of neuroglial cell lines in the degenerating optic nerve of young rats. A radioautographic study. J Neuropathol Exp Neurol 40: 46–60.
45. Norton WT, Aquino DA, Hozumi I, Chiu FC, Brosnan CF (1992) Quantitative aspects of reactive gliosis: a review. Neurochem Res 17: 877–885.
46. Eng LF, Ghirnikar RS, Lee YL (2000) Glial fibrillary acidic protein: GFAP-thirty-one years (1969–2000). Neurochem Res 25: 1439–1451.
47. Grove JE, Bruscia E, Krause DS (2004) Plasticity of bone marrow-derived stem cells. Stem Cells 22: 487–500.

10

Baculovirus as an Ideal Radionuclide Reporter Gene Vector: A New Strategy for Monitoring the Fate of Human Stem Cells *In Vivo*

Yu Pan[1]ʘ, Shuai Liu[1]ʘ, Haifei Wu[1], Jing Lv[1], Xiaoqian Xu[2], Yifan Zhang[1]*

1 Department of Nuclear Medicine, Ruijin Hospital, School of Medicine, Shanghai Jiao Tong University, Shanghai, China, 2 Department of Hematology, Changhai Hospital, Secondary Military Medical University, Shanghai, China

Abstract

Purpose: Radionuclide reporter gene imaging holds promise for non-invasive monitoring of transplanted stem cells. Thus, the feasibility of utilizing recombinant baculoviruses carrying the sodium iodide symporter (NIS) reporter gene in monitoring stem cell therapy by radionuclide imaging was explored in this study.

Methods: Recombinant baculoviruses carrying NIS and green fluorescent protein (GFP) reporter genes (Bac-NIS and Bac-GFP) were constructed and used to infect human induced pluripotent stem cells (hiPSCs), human embryonic stem cells (hESCs) and human umbilical cord blood mesenchymal stem cells (hUCB-MSCs). Infection efficiency, total fluorescence intensity and duration of transgene expression were determined by flow cytometry. Cytotoxicity/proliferative effects of baculovirus on hUCB-MSCs were assessed using CCK-8 assays. ^{125}I uptake and perchlorate inhibition assays were performed on Bac-NIS-infected hUCB-MSCs. Radionuclide imaging of mice transplanted with Bac-NIS-infected hUCB-MSCs was performed by NanoSPECT/CT imaging.

Results: Infection efficiencies of recombinant baculovirus in hESCs, hiPSCs and hUCB-MSCs increased with increasing MOIs (27.3%, 35.8% and 95.6%, respectively, at MOI = 800). Almost no cytotoxicity and only slight effects on hUCB-MSCs proliferation were observed. Obvious GFP expression (40.6%) remained at 8 days post-infection. The radioiodide was functionally accumulated by NIS gene products and specifically inhibited by perchlorate (ClO_4^-). Radioiodide uptake, peaking at 30 min and gradually decreasing over time, significantly correlated with hUCB-MSCs cell number ($R^2 = 0.994$). Finally, radionuclide imaging showed Bac-NIS-infected hUCB-MSCs effectively accumulated radioiodide *in vivo*, which gradually weakened over time.

Conclusion: Baculovirus as transgenic vector of radionuclide reporter gene imaging technology is a promising strategy for monitoring stem cell transplantation therapy.

Editor: Katriina Aalto-Setala, University of Tampere, Finland

Funding: This study was supported by grants from the National Natural Science Foundation of China (Nos. 81171367, http://isisn.nsfc.gov.cn/egrantweb/) and the Innovation Program of Shanghai Municipal Education Commission (Nos. 12Y041, http://www.shmec.gov.cn/). The funders had no role in study design, data collection and analysis, decision to publish, or preparation of the manuscript.

Competing Interests: The authors have declared that no competing interests exist.

* E-mail: zhangyifan1992@yahoo.com.cn

ʘ These authors contributed equally to this work.

Introduction

Stem cells are important in basic research as well as clinical applications in cell therapy and drug screening due to their capabilities of self-renewal and multilineage differentiation. In recent years, human stem cell transplantation therapy has been explored in clinical trials for numerous degenerative diseases, such as cardiovascular system diseases [1], nerve system diseases [2,3] and diabetes [4,5], which have obtained significant achievements and shown prospects of wide applicability. There are many potential sources of human stem cells for therapy. The three main types are human embryonic stem cells (hESCs), human induced pluripotent stem cells (hiPSCs) and human somatic stem cells.

hESCs are capable of totipotent differentiation [6] and unlimited self-renewal [7]; however, some vital hurdles must be overcome. Aside from the ethical and political controversies associated with hESCs, issues such as potential immunogenicity [8,9] and tumorigenicity [10,11] are ever-present safety concerns. In 2006, hiPSCs were reported as a significant breakthrough and a major milestone in life science research [12,13]. Unlike hESCs, the use of hiPSCs would avoid ethical and immunogenicity problems because they are derived from the patient's autologous cells [14]. However, critical concerns and challenges remain with this promising stem cell technology [15,16].

At present, the most clinically applicable stem cell type is human mesenchymal stem cells/mesenchymal stromal cells (hMSCs).Umbilical cord blood (UCB), which is readily available as it is usually discarded after the delivery of a baby, is an accepted source of hMSCs because hUCB-MSCs are less mature, less immunogenic and have a much higher capability of proliferation and expansion, compared to hMSCs from other sources [17–20]. More importantly, unrelated donor UCB units can be obtained and preserved in public or private banks for instant access. Recently, increasing number of preclinical studies using hUCB-MSCs as material for stem cell therapy have showed great potential in the treatment of human degenerative diseases [21–25]. Therefore, hUCB-MSCs may be a more practical material for stem cell transplantation therapy but will require further validation in long-term clinical trials.

Despite the potential of stem cell transplantation therapy as a novel and promising treatment, there are still numerous problems need to be solved [26], especially the ability to monitor the long-term fate of transplanted stem cells *in vivo* using non-invasive and sensitive methods, determining how stem cells integrate, proliferate and differentiate would be of great value for understanding their biology and for optimizing stem cell transplantation techniques to gain the maximum therapeutic benefits [27,28].

Reporter gene imaging is a noninvasive, sensitive and repetitive method that has been developed rapidly in recent years for monitoring cells *in vivo*, especially the radionuclide reporter gene imaging system which has specific advantages of high sensitivity ($<10^{-9}$ M) and specificity, quantitative measurement, and the ability to image multiple target sites and employ various biological tracers for functional assessment [29–31]. Moreover, the radionuclide reporter gene imaging also overcomes the shortages of other traditional imaging modalities (e.g. inability of optical imaging to detect cells deeper in the body and the inability of radionuclide labeling imaging or magnetic resonance imaging to distinguish viable cells from nonviable cells) [29].

In radionuclide reporter gene imaging systems, a transgenic vector is required for delivering the radionuclide reporter gene into target stem cells. Among the vectors (e.g., adenovirus, lentivirus, retrovirus and baculovirus) which have been commonly utilized for transgene delivery, baculovirus has several advantages compared with other vectors [32,33] and have captured growing attention as a versatile and powerful vector system for production of proteins, vaccine development, *in vitro* and *in vivo* gene delivery, surface display of eukaryotic proteins, cell-based assays for drug development and cancer therapy [32,34].

The sodium iodide symporter (NIS) can effectively participate in the uptake of radiounclides such as ^{131}I (scintigraphic imaging), ^{123}I (single photon emission computed tomography, SPECT), ^{125}I (SPECT), ^{124}I (positron emission tomography, PET) $^{94m}TcO_4^-$ (PET) and $^{99m}TcO_4^-$ (SPECT), and was considered as an excellent reporter gene for imaging [35–37]. Therefore, in this study, we infected hUCB-MSCs, hESCs and hiPSCs with a recombinant baculovirus carrying the GFP or NIS reporter gene to investigate the feasibility of baculovirus mediated radionuclide reporter gene imaging as a new strategy in monitoring human stem cells *in vivo*.

Materials and Methods

Construction and Preparation of Recombinant Baculovirus Vectors

The baculovirus vector pFBGFPR was a gift from the Institute of Molecular Biology (Hong Kong University, Hong Kong, China), and pcDNA-NIS was obtained from Sissy Jhiang (Ohio State University, Columbus, OH, USA). Recombinant baculovirus vectors carrying NIS or GFP reporter genes (Bac-NIS and Bac-GFP) were constructed and prepared as described previously [38]. These baculovirus vectors were stored in phosphate-buffered saline (PBS; pH 7.4) at 4°C, and the titer (pfu/ml) was determined by plaque assay.

Stem Cell Cultures

hiPSCs and hESCs (X-01 cell line) were graciously provided by Lei Xiao [39,40] (Institute of Biochemistry and Cell Biology, Chinese Academy of Sciences, Shanghai, China). hESCs were approved by the Howard Hughes Medical Institute (Harvard University, Boston, MA, USA; Harvard Agreement Number: A13080). All hESCs experiments were conducted in accordance with the guidelines for research on human embryonic stem cells, jointly issued by the Ministry of Science and Technology and the Ministry of Health of China, and approved by the Ethical Committee of Shanghai Institutes for Biological Sciences. hUCB-MSCs were supplied by the Hematology Department of Changhai Hospital (Secondary Military Medical University, Shanghai, China) after collection (the collection of tissue samples was approved by the Changhai Hospital Ethical Committee, and all patients provided written informed consent) and identification by cell-specific markers [CD14(−), CD34(−), CD106(−), CD45(−), HLA-DR(−), CD29(+), CD44(+), CD90(+), CD105(+), HLA-ABC(+)] and osteogenic and adipogenic differentiation (data not shown).

hiPSCs and hESCs were maintained in an undifferentiated, pluripotent state with 1000 IU/mL leukemia inhibitory factor (LIF; Millipore, Billerica, MA, USA) and grown over murine embryonic fibroblast feeder layers which had been inactivated by 10 μg/mL mitomycin C (Sigma, St. Louis, MO, USA). The stem cells were cultured in serum-free medium, which was composed of DMEM/F12 supplemented with 20% KnockOut Serum Replacement, 0.1 mM non-essential amino acids, 1 mM L-glutamine and 0.1 mM β-mercaptoethanol (all from Invitrogen, Carlsbad, CA, USA). The cells were passaged with collagenase type IV (Invitrogen) at a ratio of 1:10 every 3 or 4 days. The culture conditions were determined as described by Amit *et al.* [41].

hUCB-MSCs were cultured in medium, which was composed of basal medium for human MSCs supplemented with hMSC stimulatory supplements (StemCell, Vancouver, BC, Canada), and routinely passaged with 0.05% trypsin-EDTA (Invitrogen) when reaching ~80% confluency. Cells at the 5~10th passages were used for the present study.

Infection Efficiency of Stem Cells with Bac-GFP

hUCB-MSCs were seeded in 12-well plates at a density of 1×10^5 cells per well. The medium was aspirated after 24 h, and cells were washed twice with PBS. Bac-GFP virus stock supernatant was adjusted with PBS to the final volume of 500 μl/well for different multiplicities of infection (MOI) of 0, 20, 50, 100, 200, 400, 600 or 800, then the hUCB-MSCs were incubated in above virus solution at 25~27°C for 4 h, as described elsewhere [42]. The virus solution was aspirated after the infection, and cells were washed with PBS twice and replenished with fresh medium. The efficiency of GFP expression was observed using a fluorescent inverted phase contrast microscope after 24 h, and the percentage of GFP-positive cells (GFP^+ %) and the mean fluorescence intensity (MFI) were analyzed by flow cytometry (blue light excitation wave-length: 488 nm; detection wave length: 520 nm) at the same time. The total fluorescence intensity (TFI) representing the total transgene expression level was calculated as follows: $TFI = GFP^+ \% \times MFI \times \text{cell number } (100{,}000)$.

hiPSCs and hESCs were seeded in 12-well plates at a density of 1×10^5 per well (cell density was determined after dissociating the cell clusters into single cells with trypsin-EDTA). The cell clusters were mechanically reduced to a size of about 200~300 cells and replated onto Matrigel (BD Biosciences, Bedford, MA, USA) coated plates. The medium was aspirated after 24 h, and the cells were washed twice with PBS. Opti-MEM (Invitrogen) containing Bac-GFP at the MOI of 0, 50, 200 or 800 in a final volume of 500 μl was added to the cells and incubated at 25~27°C for 4 h. Subsequently, the virus solution was aspirated, and cells were washed with PBS twice and replenished with fresh medium. The efficiency of GFP expression was observed using a fluorescent inverted phase contrast microscope after 24 h, and the GFP^+ % and MFI were analyzed by flow cytometry at the same time after dissociating the cell clusters into single cells by trypsin-EDTA.

Cell Viability and Proliferation Assays

hUCB-MSCs were seeded into 96-well plates at a density of 4×10^3 cells per well. After 24 h, 100 μl PBS containing the virus supernatant at the MOI of 20, 50, 100, 200 or 400 was added into each well, respectively and incubated for 4 h. The control group was mock-infected hUCB-MSCs (MOI = 0), and the blank group contained only medium without cells. At 1, 3, 5, 7, 9, 11, 13 and 15 days post-infection (dpi), the Cell Counting Kit-8 reagent (CCK-8; Dojindo, Mashikimachi, kamimashiki gun Kumamoto, Japan) was added at 10 μl per well of each group after changing the medium with PBS, and the absorbance was read using a microplate spectrofluorometer (detection wavelength: 450 nm; reference wavelength: 650 nm). The cells were reseeded at 4×10^3 per well when reaching ~80% confluency.

Thereafter, the cell viability (%) was calculated using the following formula: cell viability (%) = $(A_{test}-A_{blank})/(A_{control}-A_{blank})\times100$. Here, A_{test} represents absorbance of each experimental group, including Bac-NIS-infected cells, PBS and CCK-8 reagent; A_{blank} represents absorbance of the blank group, including only PBS and CCK-8 reagent; $A_{control}$ represents absorbance of control group, including mock-infected cells, PBS and CCK-8 reagent.

Durability of Bac-GFP Expression in Infected hUCB-MSCs

hUCB-MSCs were seeded in 12-well plates at a cell density of 1×10^5 per well. After 24 h, the cells were infected with Bac-GFP virus at MOI = 200 (or mock-infected, MOI = 0) as above. The GFP^+ % and MFI of infected hUCB-MSCs were analyzed by flow cytometry at 1, 2, 3, 4, 8, 12, 16 and 20 dpi until the GFP^+ % was lower than 10%. Cell cultures were maintained by reseeding at 1×10^5 per well when reaching ~80% confluency.

Iodide Uptake of Bac-NIS-infected hUCB-MSCs

hUCB-MSCs were seeded in 24-well plates at 2×10^4 per well for 24 h. After washing the cells with PBS, 250 μl PBS containing the Bac-NIS virus supernatant at the MOI of 0, 20, 50, 100, 200, 400, 600 or 800 was added to each well for 4 h at 25~27°C. Thereafter, the virus solution was removed, and cells were incubated for another 24 h at 37°C. Before the radioiodide uptake assay, the medium was aspirated, and cells were washed with 500 μl Hank's balanced salt solution (HBSS), which was adjusted with Hepes buffer (Invitrogen) to pH 7.3. The iodide uptake assay was performed by adding 500 μl buffered HBSS containing 0.1 μCi (3.7 kBq) $Na^{125}I$ and 10 μM NaI to each well [43]. The cells were incubated at 37°C for 30 min, and then radioiodide uptake was terminated by aspirating the radioactive solution and washing the cells rapidly with 500 μl ice-cold buffered HBSS. To determine the intracellular amount of ^{125}I, 500 μl 1 M NaOH was added to each well for 20 min, and then samples were transferred into vials for determination of counts per minute (cpm) with a γ counter. In dynamic iodide uptake assay, the cells in MOI = 0 and 200 groups were incubated with radioactive solution for various time points (5, 15, 30, 60, 90 and 120 min) before terminating the radioiodide uptake.

$NaClO_4$ Inhibition of Iodide Uptake

hUCB-MSCs were seeded into 24-well plates at 2×10^4 per well for 24 h. The cells were then infected with Bac-NIS at MOI = 200. After infection, 500 μl buffered HBSS containing 0.1 μCi (3.7 kBq) $Na^{125}I$, 10 μM NaI and various concentrations of $NaClO_4$ (0 μM, 30 μM and 300 μM groups) were added. The cells were incubated at 37°C for 30 min, and then iodide uptake was terminated and determined with a γ counter as above.

Correlation Analysis between ^{125}I Uptake and Number of Bac-NIS-infected hUCB-MSCs *in vitro*

hUCB-MSCs were seeded in 24-well plates at densities of 5×10^3, 1×10^4, 2×10^4, 4×10^4 and 8×10^4 per well for 24 h. The cells were then infected with Bac-NIS at MOI = 200 or 0 (as control group). The virus solution was removed after 4 h infection, and cells were incubated at 37°C for another 24 h. The radioiodide uptake assay of each group was then performed as above.

Nano-single-photon Emission Computed Tomography/Computed Tomography (NanoSPECT/CT) Imaging of Nude Mice Transplanted with Bac-NIS-infected hUCB-MSCs

hUCB-MSCs were infected with Bac-NIS at MOI of 200. At 24 h post-infection (hpi), 1×10^7 infected and mock-infected cells were respectively harvested, centrifuged, resuspended in 100 μl PBS and then transplanted subcutaneously into each axilla of male nude mice (BALB/c nu/nu, 4 weeks old) with a 1-inch needle. After administration with 300 μCi (11.1 MBq) $Na^{125}I$ via the tail vein, the mice were anesthetized with 4% isoflurane and then placed in a spread prone position on the scanner bed of NanoSPECT/CT (Bioscan, Washington, DC, USA). The mice were imaged at 30, 60 and 120 min under inhalational isoflurane anesthesia (1.5% in oxygen at flow rate of 0.6 L/min). NanoSPECT/CT images were reconstructed by using Bioscan InVivoScope® 1.43 software (Bioscan) and displayed in coronal, sagittal and horizontal planes for visual evaluation. All mice used in this study were purchased from Shanghai Slaccas Experiment Animal Corporation (Shanghai Institutes for Biological Science, Shanghai, China), and all procedures were conducted according to the Animal Care and Use guideline and were approved by the Institutional Animal Care and Use Committee of the Ruijin Hospital.

Statistical Analysis

All data were processed using SAS ver. 8.02 software (SAS Institute Inc., Cary, NC, USA), and all graphs were drawn using GraphPad Prism ver. 5.01 software (GraphPad Software Inc., La Jolla, CA, USA). Data were expressed as means ± standard deviation (SD), and continuous variables were compared by the paired Student's t-test. Differences between groups were compared by ANOVA using Dunnett's Multiple Comparison Test. *P*-values less than 0.05 (two-tailed) were considered to be significant.

Results

Amplification and Determination of Recombinant Bac-GFP and Bac-NIS

The recombinant baculoviruses carrying GFP and NIS under the control of a cytomegalovirus immediate-early (CMV-IE) promoter were constructed successfully. The viruses were amplified, and titers determined by plaque assay reached as high as $\sim 1.4 \times 10^9$ pfu/ml.

Infection Efficiency of Bac-GFP in Human Stem Cells

The hUCB-MSCs with a fusiform shape were distributed uniformly at 24 h after infection with Bac-GFP, and the majority expressed GFP (Fig. 1A and B). The infection efficiency and total transgene expression level of infected hUCB-MSCs, expressed as GFP^+ % (Fig. 1C) and TFI (Fig. 1D), respectively, were elevated with increasing MOI. The use of MOI = 800 yielded the highest GFP^+ % and TFI, which were as high as 95.6% and 3.35×10^9, respectively, while the GFP^+ % reached 76.7% at the MOI of 200. Though the GFP^+ % gradually approached saturation levels with the increase of MOI, the TFI still increased and seemed to positively correlate with the MOI. These findings clearly indicated that a higher infection efficiency and transgene expression level could be achieved by increasing MOI.

Observations by fluorescence microscopy revealed that the scattered cells and the cells surrounding the hESCs and hiPSCs cell clusters obviously expressed GFP, while little fluorescence was found in the center of the clusters (Fig. 2A and B). The GFP^+ % of these two types stem cells were, respectively, 27.3% and 35.8% at MOI = 800, and only 8.6% and 17.7% at MOI = 200 (Fig. 2C).

Effects of Recombinant Baculovirus on Viability and Proliferation of hUCB-MSCs

At 24 h after infection, cell viabilities of each MOI group were determined (Fig. 3A). There were no statistically significant differences between each experimental and control group. However, proliferation rates of infected cells were slightly lower than that of mock-infected cells, and the effects were more obvious with the increase of MOI. The proliferation rates of infected cells returned to normal levels and approached that of the mock-infected group after passaging (Fig. 3B).

Durability of GFP Expression in Bac-GFP-infected hUCB-MSCs

As quantitatively determined by flow cytometry, the GFP expression of infected hUCB-MSCs decreased gradually over time after infection. The GFP^+ % at 1, 2, 3, 4, 8 and 12 dpi were 76.7%, 72.7%, 71.2%, 62.7%, 40.6% and 7.4%, respectively (Fig. 3C), indicating that the Bac-GFP expression was especially stable in the first 4 days and remained at a high level for at least 1 week before declining with cell division.

Radioiodide Uptake Assays of Bac-NIS-infected hUCB-MSCs *in vitro*

The amount of ^{125}I uptake in Bac-NIS-infected hUCB-MSCs increased with increasing MOIs (Fig. 4A), and this was consistent with the TFI of Bac-GFP-infected cells at different MOIs, which also represented the total transgene expression level of a given number of cells (Fig. 1D).

Iodide uptake in thyroid cells can be specifically inhibited by ClO_4^-. In the iodide uptake inhibition assay, the ^{125}I uptake of Bac-NIS-infected hUCB-MSCs was also inhibited by $NaClO_4$ effectively (Fig. 4B). Even at the $NaClO_4$ concentration of only 30 μM, the ^{125}I uptake could be inhibited significantly. Compared

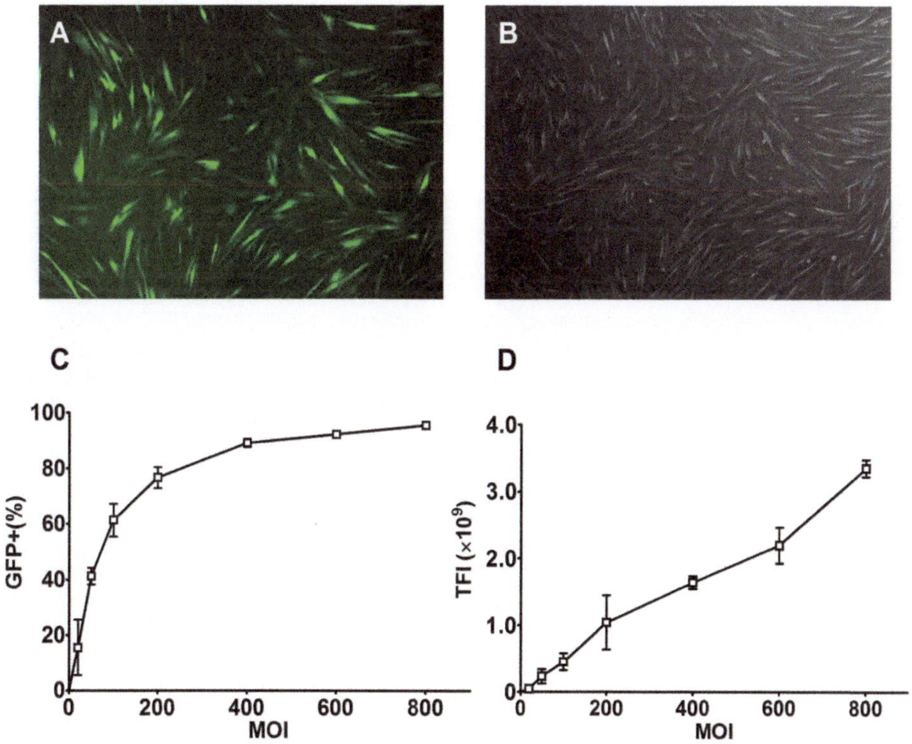

Figure 1. Infection efficiency of recombinant baculovirus in hUCB-MSCs. A and **B**: Fluorescence and bright field microscopy images of hUCB-MSCs, respectively, which were seeded at 1×10^5 cells per well and infected with Bac-GFP at MOI = 200 (40×). **C** and **D**: GFP^+ % and TFI of Bac-GFP-infected hUCB-MSCs at 24 hpi with different MOIs. Results are means ± SD (n = 3). **Abbreviations**: Bac-GFP, recombinant baculovirus carrying green fluorescence protein reporter gene; hUCB-MSCs, human umbilical cord blood mesenchymal stem cells; MOI, multiplicity of infection; GFP+%, percentage of GFP positive cells; TFI, total fluorescence intensity; SD, standard deviation.

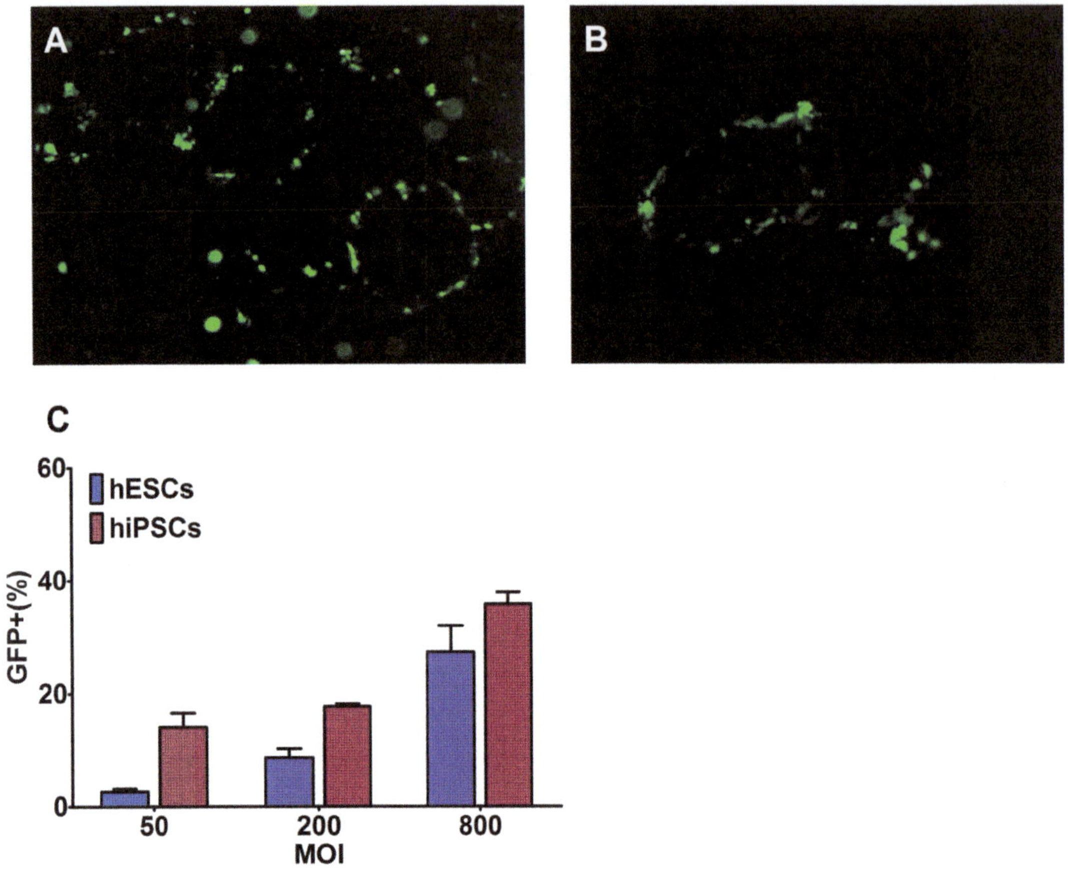

Figure 2. Infection efficiency of recombinant baculovirus in hESCs and hiPSCs. A and **B**: Fluorescence microscopy images of hESCs and hiPSCs, respectively, which were seeded at 1×10^5 cells per well and infected with Bac-GFP at MOI = 200 (40×). **C**: GFP^+ % of Bac-GFP-infected hESCs and hiPSCs at 24 hpi with different MOIs. Results are means ± SD (n = 3). **Abbreviations**: hESCs, human embryonic stem cells; hiPSCs, human induced pluripotent stem cells.

with the 0 μM control group, the inhibition ratios of the $NaClO_4$ 30 μM and 300 μM groups reached 89.3% and 90.9%, respectively, and there was no statistically significant difference between these two experimental groups.

In the dynamic radioiodide uptake assay, the amount of intracellular ^{125}I uptake increased dramatically within 30 min and peaked at 30 min. Thereafter, the level of intracellular ^{125}I in Bac-NIS-infected hUCB-MSCs decreased gradually; meanwhile the ^{125}I export increased. And there was no functional iodide uptake observed in the control group (Fig. 4C). These results also indicated that 30 min after administration with radioiodide may be the optimal time point for imaging of transplanted stem cells *in vivo*.

Correlation Analysis between ^{125}I Uptake and Number of Bac-NIS-infected hUCB-MSCs *in vitro*

To evaluate the correlation between the amount of ^{125}I uptake and cell number of Bac-NIS-infected hUCB-MSCs, the cells were serially diluted from 5×10^3 to 8×10^4 per well. There was no statistically significant difference between each mock-infected group. By contrast, Bac-NIS-infected hUCB-MSCs showed a significant increase of ^{125}I uptake along with increasing cell number (Fig. 5A). A rather high correlation ($R^2 = 0.994$, $P<0.001$) was observed between the ^{125}I uptake amount and the cell number of infected cells (Fig. 5B), which provided a possibility of indirectly and quantitatively determining the viable cell number *in vivo* by measuring intracellular radionuclides.

NanoSPECT/CT Imaging of Transplanted hUCB-MSCs *in vivo*

At 30 min after administration with $Na^{125}I$, the right axilla of mice, where was transplanted with Bac-NIS-infected hUCB-MSCs (MOI = 200, 1×10^7 cells) 24 h before, was obviously visible by SPECT imaging, whereas the left axilla transplanted with mock-infected hUCB-MSCs as a control was not. At the same time, radioactivity signals were observed in the pharynx oralis, salivary gland, thyroid, heart, stomach and intestines area. The bladder was also visible as a result of ^{125}I excretion in the urine (Fig. 6A). By observation at 60 min and 120 min after administration with $Na^{125}I$, radioactivity in the right axilla area appeared to decrease over time, although it was still clearly visible at 120 min. While radioactivity in the heart and intestines area gradually reduced and eventually vanished, the thyroid and stomach sustained high radioactivity levels, as did the bladder as the excretion pathway (Fig. 6B and C).

Discussion

In recent years, molecular imaging based on radionuclide technology, which enable non-invasive, repetitive and quantitative visualization of various cellular events and exogenous/endogenous

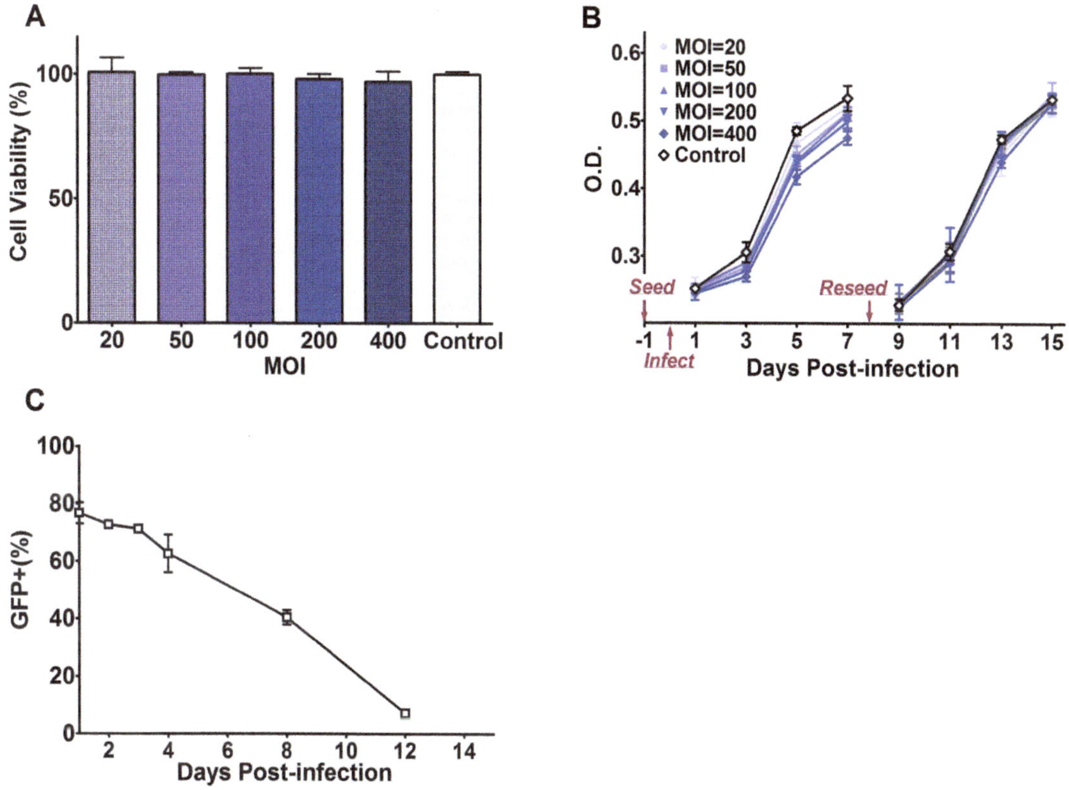

Figure 3. Effects and duration of transgene expression in baculovirus-infected hUCB-MSCs. A: Cell viability of hUCB-MSCs infected with Bac-NIS at different MOIs at 24 hpi as detected by the CCK-8 assay. **B**: Cell proliferation of hUCB-MSCs infected by Bac-NIS at different MOIs over time (1, 3, 5, 7, 9, 11, 13 and 15 dpi), as detected by the CCK-8 assay. The cells were seeded 1 day before infection with Bac-NIS and reseeded at 8 dpi when the cells reached ~80% confluency. **C**: GFP^+ % of Bac-GFP-infected hUCB-MSCs over time (1, 2, 3, 4, 8 and 12 dpi). The cells were reseeded at 8 dpi when reached ~80% confluency. Results are means ± SD (n = 3). **Abbreviations**: Bac-NIS, recombinant baculovirus carrying sodium-iodide symporter reporter gene; dpi, days post-infection; CCK-8, Cell Counting Kit-8.

gene expression in living organisms, has been rapidly developed and widely used in the biomedical research field. The most widely used radionuclide reporter gene imaging strategy for monitoring and evaluating stem cell transplantation therapy currently is indirect method of using reporter genes and their "radionuclide reporter probes". For this strategy, it is obvious that an ideal transgenic vector is crucially important for transducing the radionuclide reporter genes into target stem cells.

In this study, we constructed a recombinant baculovirus containing the CMV-IE promoter to transduce the GFP reporter gene into three types of stem cells. The recombinant baculovirus was found to infect hUCB-MSCs efficiently and reach a remarkable 76.7% at the MOI of 200 without assistance of any reagent like butyrate. However, the infection efficiencies in hESCs and hiPSCs were much lower (8.6% and 17.7% respectively) at MOI = 200, and improved but were still not ideal at MOI = 800 (27.3% and 35.8%, respectively). The main reason for this phenomenon may be due to the promoter of the recombinant baculovirus. Zeng *et al.* [44] and Du *et al.* [45] found that the woodchuck hepatitis post-transcriptional regulatory element (WPRE) and the human elongation factor 1-α (EF-1α) promoter could significantly enhance the baculovirus infection efficiency in hESCs. In addition, the formation and growth of stem cell clusters, the infection surrounding solution medium, and the infection time and temperature before virus removal have also been found to contribute to the infection efficiency [46–48]. Thus, improving the infection conditions for hESCs and hiPSCs with baculovirus vector will require further studies.

In order to address biosafety concerns, we assessed the effects of recombinant baculovirus on the viability and proliferation of hUCB-MSCs in this study. The recombinant baculovirus did not compromise the stem cell viability but somewhat impeded the cell proliferation rate, which could be restored to normal levels after passaging. These transient effects may be explained by the perturbation of host cell global gene expression and triggering of innate immune responses by the baculovirus [49]. Of note, baculoviruses are not known to either replicate inside stem cells or integrate its genome into host chromosomes in the absence of selective pressure [49]. It has also been reported that recombinant baculoviruses compromise neither the viability, phenotype and pluripotency/differentiation of stem cells [44,50], nor the expression of human stem cell markers and teratoma formation [45].

Long-term transgene expression is also an important attribute of an ideal vector, therefore we examined the duration of reporter gene expression in baculovirus infected hUCB-MSCs and found that the transgene expression level remained at a relative high level for at least 8 dpi but nearly extinguished at 12 dpi. For baculovirus cannot replicate in mammalian cells, its viral genome and the transgene will dilute with cell division, which may be responsible for the gradual decrease of transgene expression. Recent studies on the development of baculoviruses for sustained long-term transgene expression may overcome this flaw and thus optimize this vector system for long-term monitoring of stem cells [51,52].

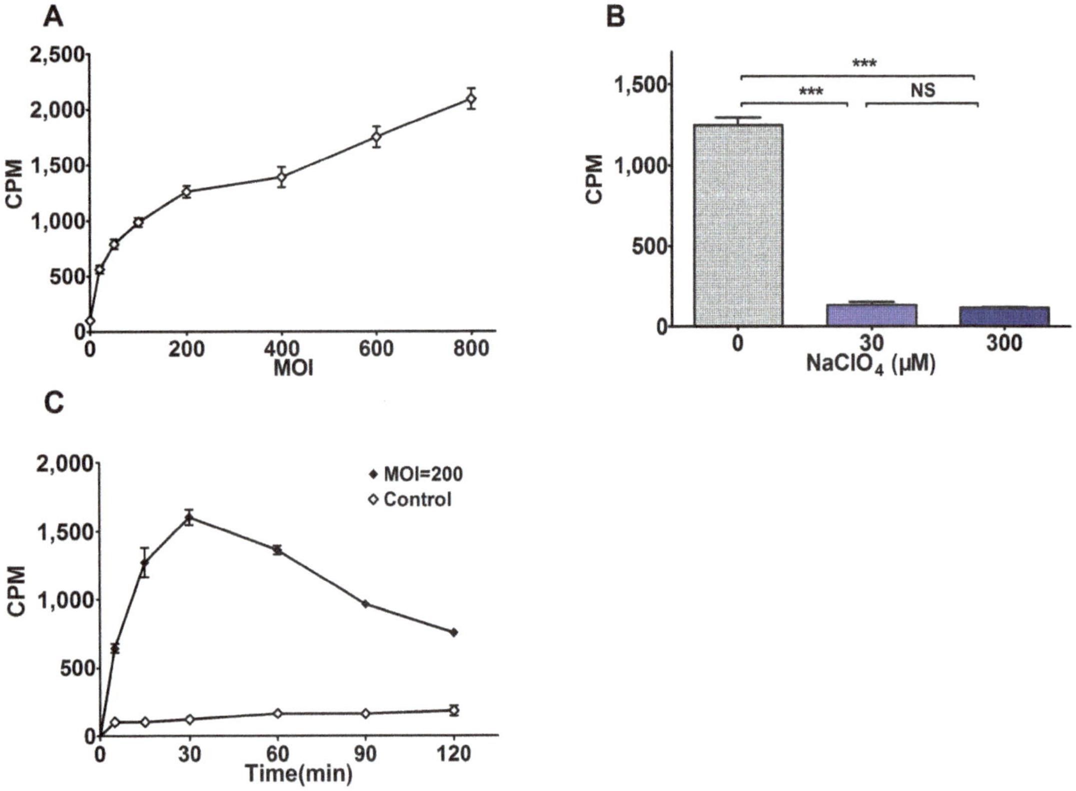

Figure 4. Radioiodide uptake of Bac-NIS-infected hUCB-MSCs *in vitro*. A: The cpm of intracellular ^{125}I in Bac-NIS-infected hUCB-MSCs with different MOIs at 24 hpi. **B**: Inhibitory effects of $NaClO_4$ on ^{125}I uptake in Bac-NIS-infected hUCB-MSCs at MOI = 200. **C**: Dynamic ^{125}I uptake curve of Bac-NIS-infected hUCB-MSCs at MOI = 200. Results are means ± SD (n = 3). **Abbreviations**: cpm, counts per minute; ***, $P<0.001$; NS, no statistically significant difference.

This novel baculovirus transgenic vector has other advantageous features as well, such as neither replicating in mammalian cells nor randomly integrating into the host genome (randomly integration may lead to mutations) [53,54]. It also has a packaging capacity of more than 100 kb [55], thus allowing for insertion of multiple genes and regulatory elements. Moreover, recombinant baculovirus construction is easy and can be propagated to high titers by infecting its natural host insect cells [56]. Furthermore, it can infect mammalian cells in the stationary phase as well as the mitotic phase. Taken above together, baculovirus may be a suitable transgenic vector of choice for infecting human stem cells in reporter gene imaging.

However, thus far, studies on radionuclide reporter gene imaging for hUCB-MSCs mediated by a baculovirus vector have not been reported. Therefore, in this study, we also successfully constructed and prepared a recombinant baculovirus carrying the

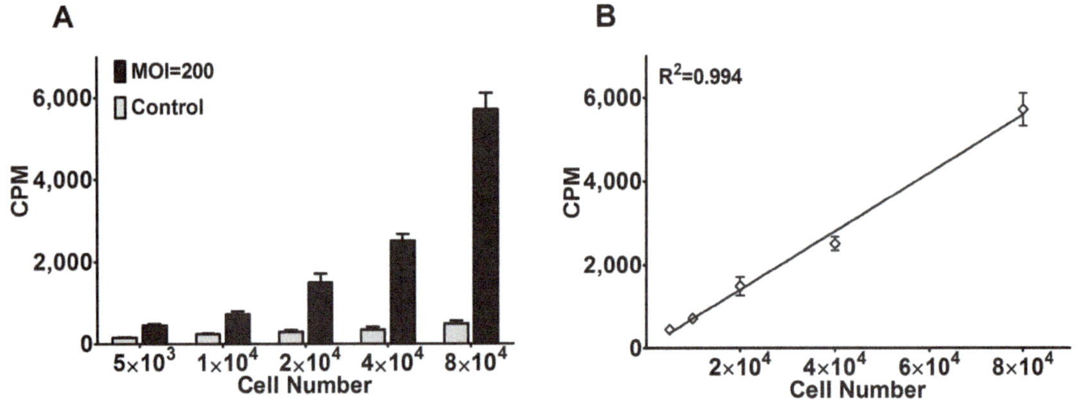

Figure 5. Correlation between ^{125}I uptake in Bac-NIS-infected hUCB-MSCs and cell number *in vitro*. A: ^{125}I uptake (cpm) in Bac-NIS-infected and mock-infected hUCB-MSCs at MOI = 200 in serially diluted cells (ranging from 5×10^3 to 8×10^4 per well). **B**: Correlation analysis between cpm and cell number of Bac-NIS-infected hUCB-MSCs. Results are means ± SD (n = 3).

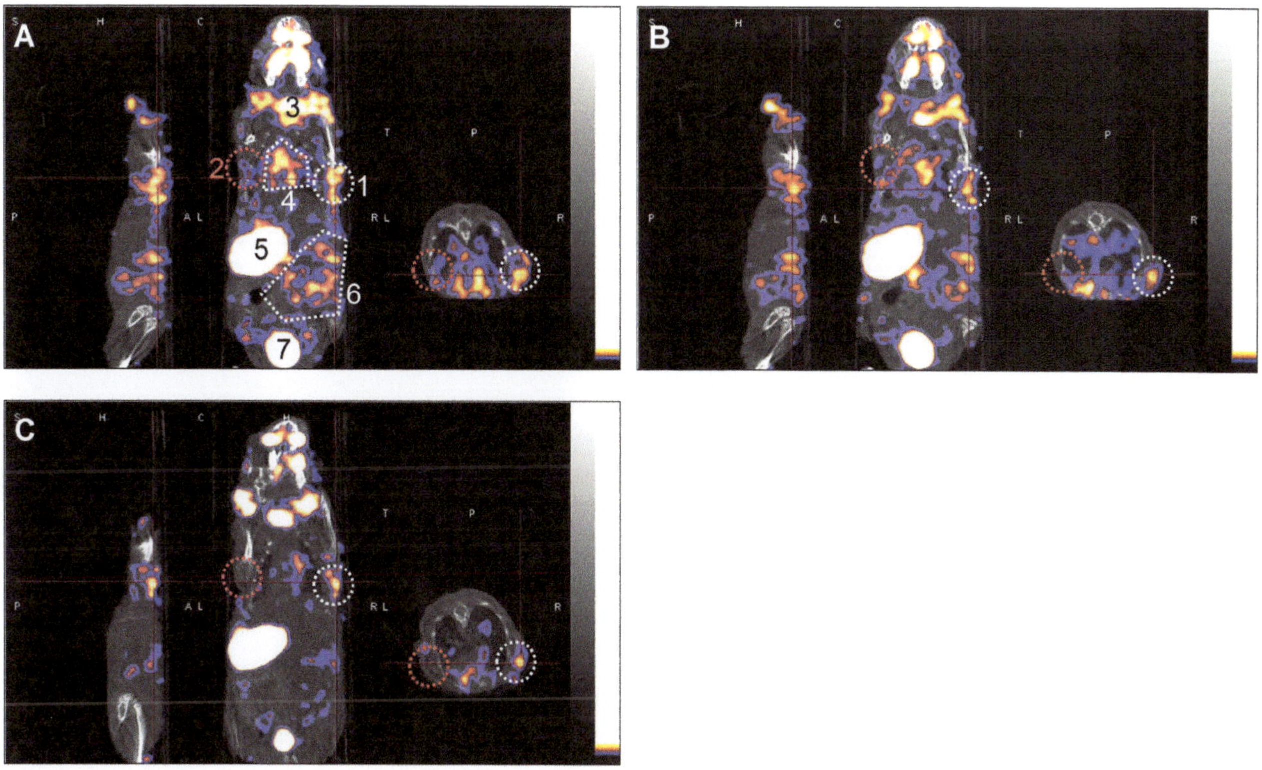

Figure 6. *In vivo* imaging of Bac-NIS-infected hUCB-MSCs transplantation with NanoSPECT/CT. A: Overlapping SPECT and CT images at 30 min after administration of 300 μCi (11.1 MBq) Na^{125}I. The right axilla (white circular area) was transplanted with Bac-NIS-infected hUCB-MSCs (MOI = 200, 1×10^7 cells) and showed a high radioiodide uptake, while the left axilla (red circular area) was transplanted with mock-infected cells as a control and showed no obvious radioiodide uptake. From left to right side are, respectively, the coronal, sagittal and horizontal sections. All CT images are shown with a grey palette, and all SPECT images are shown with a warm palette. **B** and **C**: SPECT/CT images at 60 min and 120 min after radioiodide administration. **Abbreviations**: CT, computed tomography; SPECT, single photon emission computed tomography; **1**, Bac-NIS-infected hUCB-MSCs; **2**, mock-infected hUCB-MSCs; **3**, thyroid; **4**, heart; **5**, stomach; **6**, intestinal area; **7**, bladder.

NIS reporter gene for exploring the feasibility of utilizing the baculovirus as radionuclide reporter gene vector in monitoring hUCB-MSCs transplantation. The *in vitro* results showed that, the NIS proteins expressed in Bac-NIS-infected hUCB-MSCs allowed iodide uptake effectively, which could be specifically inhibited by perchlorate (ClO_4^-). The dynamic changes in radioiodide uptake, with a rapid increase and a slow decrease, provided a peak time point for monitoring transplanted hUCB-MSCs and also minimized radioactive damage on the stem cells. This study also demonstrated a significantly high correlation between the cell number of hUCB-MSCs and the level of radioiodide uptake, supporting the further possibility of this method for quantitatively monitoring transplanted stem cells *in vivo*.

In *in vivo* experiment, the hUCB-MSCs were infected with Bac-NIS *in vitro* before transplantation, which also avoided the limitation of baculovirus being inactivated by serum complement proteins, as would occur with the use of this vector directly *in vivo*. Then Bac-NIS-infected hUCB-MSCs were then transplanted subcutaneously into axillae of nude mice to preliminarily evaluate the feasibility of monitoring hUCB-MSCs *in vivo* by radionuclide reporter gene imaging. The Bac-NIS-infected hUCB-MSCs successfully expressed NIS proteins, which effectively mediated the accumulation of radioiodide *in vivo*, as detected by SPECT/CT, while no radioactivity was observed with mock-infected transplanted cells. The radioactivity of Bac-NIS-infected hUCB-MSCs *in vivo* was most obvious at 30 min after Na^{125}I administration and gradually decreased over time. Therefore, imaging of transplanted hUCB-MSCs was determined to be most optimal at 30 min after radioiodide administration.

However, the radioactivity level of transplanted stem cells mainly depended on the transient radioiodide uptake from the flowing blood. The tissues expressing endogenous NIS (such as thyroid gland, salivary glands, stomach, lactating mammary glands, lachrymal glands, small intestine, rectum, heart and lung [57–59]), could accumulate radioiodide from the blood and then could be clearly observed in SPECT/CT imaging at an early period after administration (Fig. 6A). These non-specific radioiodide accumulations may interfere the imaging of transplanted stem cells or even lead to false results when the transplantation sites are located in or near these tissues, as would be the case, for example, if the stem cells are targeted for therapy of myocardial infarction. By contrast, stem cell transplantation in tissues such as liver, muscle or spinal cord, which have low endogenous NIS expression, will be ideal for NIS imaging. The feasible solution to this problem of endogenous NIS interference may be delaying imaging to a time point at which the intracellular radioiodide in these tissues is nearly eliminated, while transplanted cells in the same area are still clearly visible. Alternatively, the animal model can be pre-treated with thyroid blocking agents before stem cell transplantation, which can decrease the radioiodide uptake of thyroid and other tissues expressing endogenous NIS and thus increase the transient uptake of Bac-NIS-infected stem cells.

With rapid advances in human stem cell research and with high demands for testing their regenerative potential in preclinical models, we can foresee that non-invasive and repetitive stem cell *in vivo* imaging will play a critical role, and the strategy in our study of using baculovirus as a transgenic vector in radionuclide reporter gene imaging system may provide valuable data to support further studies of hUCB-MSCs (or other types of stem cells) transplantation therapy for specific diseases.

Conclusion

In view of the results that have been obtained, this study showed that the baculovirus vector with significantly high transduction efficiency and low cytotoxicity is suitable and potential for transducing reporter genes into hUCB-MSCs *in vitro*. More importantly, this study also supported the feasibility of using the baculovirus vector expressing the NIS radionuclide reporter gene for non-invasively monitoring hUCB-MSCs transplantation therapy *in vivo*.

Acknowledgments

We wish to thank Hong Zhou and Xiangqin Weng for excellent technical assistances.

Author Contributions

Conceived and designed the experiments: YFZ YP SL. Performed the experiments: YP SL. Analyzed the data: HFW JL. Contributed reagents/materials/analysis tools: XQX. Wrote the paper: YP YFZ.

References

1. Strauer BE, Brehm M, Zeus T, Kostering M, Hernandez A, et al. (2002) Repair of infarcted myocardium by autologous intracoronary mononuclear bone marrow cell transplantation in humans. Circulation 106: 1913–1918. DOI: 10.1161/01.cir.0000034046.87607.1c.
2. Fassas A, Kimiskidis VK, Sakellari I, Kapinas K, Anagnostopoulos A, et al. (2011) Long-term results of stem cell transplantation for MS A single-center experience. Neurology 76: 1066–1070. DOI: 10.1212/WNL.0b013e318211.
3. Morizane A, Darsalia V, Guloglu MO, Hjalt T, Carta M, et al. (2010) A simple method for large-scale generation of dopamine neurons from human embryonic stem cells. J Neurosci Res 88: 3467–3478. DOI: 10.1002/jnr.22515.
4. Voltarelli JC, Couri CEB, Stracieri ABPL, Oliveira MC, Moraes DA, et al. (2007) Autologous nonmyeloablative hematopoietic stem cell transplantation in newly diagnosed type 1 diabetes mellitus. JAMA-J Am Med Assoc 297: 1568–1576. DOI: 10.1001/jama.297.14.1568.
5. D'Amour KA, Bang AG, Eliazer S, Kelly OG, Agulnick AD, et al. (2006) Production of pancreatic hormone-expressing endocrine cells from human embryonic stem cells. Nat Biotechnol 24: 1392–1401. DOI: 10.1038/nbt1259.
6. van der Bogt KEA, Swijnenburg RJ, Cao F, Wu JC (2006) Molecular imaging of human embryonic stem cells - Keeping an eye on differentiation, tumorigenicity and immunogenicity. Cell Cycle 5: 2748–2752. DOI: 10.4161/cc.5.23.3533.
7. Hayflick L (1961) The establishment of a line (WISH) of human amnion cells in continuous cultivation. Exp Cell Res 23: 14–20. DOI: 10.1016/0014-4827(61)90059-3.
8. Swijnenburg RJ, Schrepfer S, Cao F, Pearl JI, Xie X, et al. (2008) In vivo imaging of embryonic stem cells reveals patterns of survival and immune rejection following transplantation. Stem Cells Dev 17: 1023–1029. DOI: 10.1089/scd.2008.0091.
9. Swijnenburg RJ, Schrepfer S, Govaert JA, Cao F, Ransohoff K, et al. (2008) Immunosuppressive therapy mitigates immunological rejection of human embryonic stem cell xenografts. Proc Natl Acad Sci U S A 105: 12991–12996. DOI: 10.1073/pnas.0805802105.
10. Cao F, van der Bogt KEA, Sadrzadeh A, Xie X, Sheikh AY, et al. (2007) Spatial and temporal kinetics, of teratoma formation from murine embryonic stem cell transplantation. Stem Cells Dev 16: 883–891. DOI: 10.1089/scd.2007.0160.
11. Lee AS, Tang C, Cao F, Xie X, van der Bogt K, et al. (2009) Effects of cell number on teratoma formation by human embryonic stem cells. Cell Cycle 8: 2608–2612. DOI: 10.4161/cc.8.16.9353.
12. Takahashi K, Tanabe K, Ohnuki M, Narita M, Ichisaka T, et al. (2007) Induction of pluripotent stem cells from adult human fibroblasts by defined factors. Cell 131: 861–872. DOI: 10.1016/j.cell.2007.11.019.
13. Yu J, Vodyanik MA, Smuga-Otto K, Antosiewicz-Bourget J, Frane JL, et al. (2007) Induced pluripotent stem cell lines derived from human somatic cells. Science 318: 1917–1920. DOI: 10.1126/science.1151526.
14. Sun N, Longaker MT, Wu JC (2010) Human iPS cell-based therapy Considerations before clinical applications. Cell Cycle 9: 880–885. DOI: 10.4161/cc.9.5.10827.
15. Pournasr B, Khaloughi K, Salekdeh GH, Totonchi M, Shahbazi E, et al. (2011) Concise review: alchemy of biology: generating desired cell types from abundant and accessible cells. Stem Cells 29: 1933–1941. DOI: 10.1002/stem.760.
16. Okita K, Yamanaka S (2011) Induced pluripotent stem cells: opportunities and challenges. Philos Trans R Soc B-Biol Sci 366: 2198–2207. DOI: 10.1098/rstb.2011.0016.
17. Yang SE, Ha CW, Jung MH, Jin HJ, Lee MK, et al. (2004) Mesenchymal stem/progenitor cells developed in cultures from UC blood. Cytotherapy 6: 476–486. DOI: 10.1080/14653240410005041.
18. Kern S, Eichler H, Stoeve J, Kluter H, Bieback K (2006) Comparative analysis of mesenchymal stem cells from bone marrow, umbilical cord blood, or adipose tissue. Stem Cells 24: 1294–1301. DOI: 10.1634/stemcells.2005-0342.
19. Barry FP, Murphy JM (2004) Mesenchymal stem cells: clinical applications and biological characterization. Int J Biochem Cell Biol 36: 568–584. DOI: 10.1016/j.biocel.2003.11.001.
20. Watt SM, Contreras M (2005) Stem cell medicine: Umbilical cord blood and its stem cell potential. Semin Fetal Neonatal Med 10: 209–220. DOI: 10.1016/j.siny.2005.02.001.
21. Pierro M, Ionescu L, Montemurro T, Vadivel A, Weissmann G, et al. (2012) Short-term, long-term and paracrine effect of human umbilical cord-derived stem cells in lung injury prevention and repair in experimental bronchopulmonary dysplasia. Thorax. In press. DOI: 10.1136/thoraxjnl-2012-202323.
22. Zhang X, Zhang L, Xu W, Qian H, Ye S, et al. (2013) Experimental therapy for lung cancer: umbilical cord-derived mesenchymal stem cell-mediated interleukin-24 delivery. Curr Cancer Drug Targets 13: 92–102. DOI: http://dx.doi.org/10.2174/1568009113804486665.
23. Jin JL, Liu Z, Lu ZJ, Guan DN, Wang C, et al. (2013) Safety and efficacy of umbilical cord mesenchymal stem cell therapy in hereditary spinocerebellar ataxia. Curr Neurovasc Res 10: 11–20. DOI: http://dx.doi.org/10.2174/156720213804805936.
24. Liu R, Zhang Z, Lu Z, Borlongan CV, Pan J, et al. (2012) Human umbilical cord stem cells ameliorate experimental autoimmune encephalomyelitis by regulating immunoinflammation and remyelination. Stem Cells Dev. In press. DOI: 10.1089/scd.2012.0463.
25. Yan C, Li S, Li Z, Peng H, Yuan X, et al. (2013) Human umbilical cord mesenchymal stem cells as vehicles of cd20-specific trail fusion protein delivery: a double-target therapy against non-hodgkin's lymphoma. Mol Pharm 10: 142–151. DOI: 10.1021/mp300261e.
26. Dove A (2002) Cell-based therapies go live. Nat Biotechnol 20: 339–343. DOI: 10.1038/nbt0402-339.
27. Sykova E, Jendelova P (2007) Migration, fate and in vivo imaging of adult stem cells in the CNS. Cell Death Differ 14: 1336–1342. DOI: 10.1038/sj.cdd.4402140.
28. Wu JC, Abraham MR, Kraitchman DL (2010) Current perspectives on imaging cardiac stem cell therapy. J Nucl Med 51 Suppl 1: 128S–136S. DOI: 10.2967/jnumed.109.068239.
29. Higuchi T, Anton M, Dumler K, Seidl S, Pelisek J, et al. (2009) Combined reporter gene pet and iron oxide mri for monitoring survival and localization of transplanted cells in the rat heart. J Nucl Med 50: 1088–1094. DOI: 10.2967/jnumed.108.060665.
30. Bai JZ, Ding W, Yu MJ, Du J, Liu ZJ, et al. (2004) Radionuclide imaging of mesenchymal stem cells transplanted into spinal cord. Neuroreport 15: 1117–1120. DOI: 10.1097/01.wnr.0000126505.97607.1b.
31. Acton PD, Zhou R (2005) Imaging reporter genes for cell tracking with PET and SPECT. Q J Nucl Med Mol Imaging 49: 349–360.
32. Chen CY, Lin CY, Chen GY, Hu YC (2011) Baculovirus as a gene delivery vector: Recent understandings of molecular alterations in transduced cells and latest applications. Biotechnol Adv 29: 618–631. DOI: 10.1016/j.biotechadv.2011.04.004.
33. Kost TA, Condreay JP, Jarvis DL (2005) Baculovirus as versatile vectors for protein expression in insect and mammalian cells. Nat Biotechnol 23: 567–575. DOI: 10.1038/nbt1095.
34. Lin SY, Chen GY, Hu YC (2011) Recent patents on the baculovirus systems. Recent Pat Biotechnol 5: 1–11. DOI: http://dx.doi.org/10.2174/187220811795655904.
35. Kim YH, Lee DS, Kang JH, Lee YJ, Chung JK, et al. (2005) Reversing the silencing of reporter sodium/iodide symporter transgene for stem cell tracking. J Nucl Med 46: 305–311.
36. Hwang DW, Jang SJ, Kim YH, Kim HJ, Shim IK, et al. (2008) Real-time in vivo monitoring of viable stem cells implanted on biocompatible scaffolds. Eur J Nucl Med Mol Imaging 35: 1887–1898. DOI: 10.1007/s00259-008-0751-z.
37. Chung JK (2002) Sodium iodide symporter: Its role in nuclear medicine. J Nucl Med 43: 1188–1200.
38. Zhou X, Li B, Wang J, Yin H, Zhang Y (2010) The feasibility of using a baculovirus vector to deliver the sodium-iodide symporter gene as a reporter. Nucl Med Biol 37: 299–308. DOI: 10.1016/j.nucmedbio.2009.12.007.

39. Liao J, Wu Z, Wang Y, Cheng L, Cui C, et al. (2008) Enhanced efficiency of generating induced pluripotent stem (iPS) cells from human somatic cells by a combination of six transcription factors. Cell Res 18: 600–603. DOI: 10.1038/cr.2008.51.
40. Wu Z, Li H, Rao LJ, He LXZ, Bao L, et al. (2011) Derivation and characterization of human embryonic stem cell lines from the Chinese population. J Genet Genomics 38: 13–20. DOI: 10.1016/j.jgg.2011.12.006.
41. Amit M, Carpenter MK, Inokuma MS, Chiu CP, Harris CP, et al. (2000) Clonally derived human embryonic stem cell lines maintain pluripotency and proliferative potential for prolonged periods of culture. Dev Biol 227: 271–278. DOI: 10.1006/dbio.2000.9912.
42. Ho YC, Chung YC, Hwang SM, Wang KC, Hu YC (2005) Transgene expression and differentiation of baculovirus-transduced human mesenchymal stem cells. J Gene Med 7: 860–868. DOI: 10.1002/jgm.729.
43. Weiss SJ, Philp NJ, Grollman EF (1984) Iodide transport in a continuous line of cultured cells from rat thyroid. Endocrinology 114: 1090–1098. DOI: 10.1210/endo-114-4-1090.
44. Zeng J, Du J, Zhao Y, Palanisamy N, Wang S (2007) Baculoviral vector-mediated transient and stable transgene expression in human embryonic stem cells. Stem Cells 25: 1055–1061. DOI: 10.1634/stemcells.2006-0616.
45. Du J, Zeng JM, Zhao Y, Boulaire J, Wang S (2010) The combined use of viral transcriptional and post-transcriptional regulatory elements to improve baculovirus-mediated transient gene expression in human embryonic stem cells. J Biosci Bioeng 109: 1–8. DOI: 10.1016/j.jbiosc.2009.06.017.
46. Shen HC, Lee HP, Lo WH, Yang DG, Hu YC (2007) Baculovirus-mediated gene transfer is attenuated by sodium bicarbonate. J Gene Med 9: 470–478. DOI: 10.1002/jgm.1037.
47. Mahonen AJ, Makkonen KE, Laakkonen JP, Ihalainen TO, Kukkonen SP, et al. (2010) Culture medium induced vimentin reorganization associates with enhanced baculovirus-mediated gene delivery. J Biotechnol 145: 111–119. DOI: 10.1016/j.jbiotec.2009.11.003.
48. Hsu CS, Ho YC, Wang KC, Hu YC (2004) Investigation of optimal transduction conditions for baculovirus-mediated gene delivery into mammalian cells. Biotechnol Bioeng 88: 42–51. DOI: 10.1002/bit.20213.
49. Chen CY, Wu HH, Chen CP, Chern SR, Hwang SM, et al. (2011) Biosafety assessment of human mesenchymal stem cells engineered by hybrid baculovirus vectors. Mol Pharm 8: 1505–1514. DOI: 10.1021/mp100368d.
50. Ho YC, Lee HP, Hwang SM, Lo WH, Chen HC, et al. (2006) Baculovirus transduction of human mesenchymal stem cell-derived progenitor cells: variation of transgene expression with cellular differentiation states. Gene Ther 13: 1471–1479. DOI: 10.1038/sj.gt.3302796.
51. Lo WH, Hwang SM, Chuang CK, Chen CY, Hu YC (2009) Development of a hybrid baculoviral vector for sustained transgene expression. Mol Ther 17: 658–666. DOI: 10.1038/mt.2009.13.
52. Lin CY, Lin KJ, Kao CY, Chen MC, Lo WH, et al. (2011) The role of adipose-derived stem cells engineered with the persistently expressing hybrid baculovirus in the healing of massive bone defects. Biomaterials 32: 6505–6514. DOI: 10.1016/j.biomaterials.2011.05.059.
53. Hu YC (2008) Baculoviral vectors for gene delivery: A review. Curr Gene Ther 8: 54–65. DOI: 10.2174/156652308783688509.
54. Zhang XJ, Godbey WT (2006) Viral vectors for gene delivery in tissue engineering. Adv Drug Deliv Rev 58: 515–534. DOI: 10.1016/j.addr.2006.03.006.
55. Hartley JL (2006) Cloning technologies for protein expression and purification. Curr Opin Biotechnol 17: 359–366. DOI: 10.1016/j.copbio.2006.06.011.
56. Hu YC (2006) Baculovirus vectors for gene therapy. Adv Virus Res 68: 287–320. DOI: 10.1016/s0065-3527(06)68008-1.
57. Nicola JP, Basquin C, Portulano C, Reyna-Neyra A, Paroder M, et al. (2009) The Na+/I− symporter mediates active iodide uptake in the intestine. Am J Physiol-Cell Physiol 296: C654–C662. DOI: 10.1152/ajpcell.00509.2008.
58. Spitzweg C, Joba W, Eisenmenger W, Heufelder AE (1998) Analysis of human sodium iodide symporter gene expression in extrathyroidal tissues and cloning of its complementary deoxyribonucleic acids from salivary gland, mammary gland, and gastric mucosa. J Clin Endocrinol Metab 83: 1746–1751. DOI: 10.1210/jc.83.5.1746.
59. Vayre L, Sabourin JC, Caillou B, Ducreux M, Schlumberger M, et al. (1999) Immunohistochemical analysis of Na+/I- symporter distribution in human extra-thyroidal tissues. Eur J Endocrinol 141: 382–386. DOI: 10.1530/eje.0.1410382.

11

Oxygen Transport and Stem Cell Aggregation in Stirred-Suspension Bioreactor Cultures

Jincheng Wu[1], Mahboubeh Rahmati Rostami[1], Diana P. Cadavid Olaya[2], Emmanuel S. Tzanakakis[1]*

1 Department of Chemical and Biological Engineering, Tufts University, Medford, Massachusetts, United States of America, **2** Department of Chemical and Biological Engineering, State University of New York at Buffalo, Buffalo, New York, United States of America

Abstract

Stirred-suspension bioreactors are a promising modality for large-scale culture of 3D aggregates of pluripotent stem cells and their progeny. Yet, cells within these clusters experience limitations in the transfer of factors and particularly O_2 which is characterized by low solubility in aqueous media. Cultured stem cells under different O_2 levels may exhibit significantly different proliferation, viability and differentiation potential. Here, a transient diffusion-reaction model was built encompassing the size distribution and ultrastructural characteristics of embryonic stem cell (ESC) aggregates. The model was coupled to experimental data from bioreactor and static cultures for extracting the effective diffusivity and kinetics of consumption of O_2 within mouse (mESC) and human ESC (hESC) clusters. Under agitation, mESC aggregates exhibited a higher maximum consumption rate than hESC aggregates. Moreover, the reaction-diffusion model was integrated with a population balance equation (PBE) for the temporal distribution of ESC clusters changing due to aggregation and cell proliferation. Hypoxia was found to be negligible for ESCs with a smaller radius than 100 μm but became appreciable for aggregates larger than 300 μm. The integrated model not only captured the O_2 profile both in the bioreactor bulk and inside ESC aggregates but also led to the calculation of the duration that fractions of cells experience a certain range of O_2 concentrations. The approach described in this study can be employed for gaining a deeper understanding of the effects of O_2 on the physiology of stem cells organized in 3D structures. Such frameworks can be extended to encompass the spatial and temporal availability of nutrients and differentiation factors and facilitate the design and control of relevant bioprocesses for the production of stem cell therapeutics.

Editor: Francesco Pappalardo, University of Catania, Italy

Funding: Funding support has been provided by the National Institutes of Health (NHLBI, R01HL103709) and the New York Stem Cell Science Trust (NYSTEM, contracts C024355 and C026714) to EST. JW is the recipient of a Mark Diamond Research Foundation award. The funders had no role in study design, data collection and analysis, decision to publish, or preparation of the manuscript.

Competing Interests: The authors have declared that no competing interests exist.

* Email: Emmanuel.Tzanakakis@tufts.edu

Introduction

Realization of the therapeutic potential of pluripotent stem cells (PSCs) including embryonic stem cells (ESCs) and induced pluripotent stem cells (iPSCs) hinges on the development of platforms for large-scale PSC expansion and directed differentiation. The culture of PSCs has been demonstrated in laboratory scale stirred-suspension bioreactors (SSBs) as cell aggregates [1] or on various scaffold types [2,3]. Compared to static culture, SSBs afford higher cell densities and tighter control of process variables leading to more efficient utilization of media and growth factors [1].

Despite the advantages of culturing PSCs as aggregates in stirred-suspension vessels, less attention has been devoted to limitations in the transfer of medium components to and within PSC clusters. Concentration gradients in the bioreactor environment contribute to stem cell population heterogeneity leading to variable responses to self-renewal or differentiation stimuli. Such unevenness is more pronounced for oxygen characterized by low solubility in aqueous solutions and increased rate of consumption by metabolically active stem cells. Oxygen availability affects directly the viability, proliferation and differentiation propensity of stem cells in vivo and in vitro [4]. Cultivated human ESCs (hESCs) are less prone to spontaneous differentiation and chromosomal aberrations at hypoxic (2–5%) than normoxic (21%) O_2 level without significantly reduced proliferation [5,6]. Lower O_2 tension (pO_2) in culture also predisposes stem cells to commit along particular lineages including endothelial cells [7] and chondrocytes [8] similar to in vivo processes [4]. More details emerge about O_2 modulation of the activity of stem cell-fate controlling pathways such as the canonical Wnt/β-catenin [9] and Notch cascades [10]. These effects are largely mediated by transcriptional regulators such as the hypoxia-inducible factors (HIF) interacting with a wide gamut of genes including the pluripotency marker POU5F1 (Oct4) [11]. Hence, knowledge of O_2 profile among stem or progenitor cells is important for cell fate prediction and control.

The distribution of O_2 was reported for single hESC aggregates under static conditions [12,13]. The findings from these studies however may not apply directly to aggregate cultures of ESCs in SSBs where the transfer of O_2 through multiple interfaces complicates the analysis of the O_2 tension (pO_2) that each cell experiences. In most small-scale setups, O_2 is transferred via the gas/liquid interface (headspace aeration) to the culture medium under agitation. From the medium bulk O_2 fluxes to cells via a boundary layer surrounding each aggregate and a pore network within each cluster. These ultrastructural characteristics of aggregates have also not been considered explicitly to date. And although a steady state assumption simplifies the mathematical framework for O_2 transfer in stem cell aggregate culture, it may not proper in the case of SSB cultures. This is because the continuous cell proliferation and dynamics of agitation-induced aggregation result in the temporal modulation of O_2 levels. Furthermore, eliminating time as a variable in the analysis of mass transfer does not allow the calculation of the duration that a particular cell (or fraction) experiences a pO_2 below a certain threshold. Differences in the "residence time" at a certain O_2 concentration among cells may be consequential for the differentiation and/or self-renewal of the stem cell population.

Here, the distribution of O_2 was calculated from experimental data linked with a mathematical model for mouse ESC (mESC) and hESC aggregates cultivated in dishes and spinner flasks. Aggregates cultured in spinner flasks under different agitation rates were analyzed at different time points and ultrastructural attributes such as porosity and tortuosity were determined for the first time. The effective diffusivity of O_2 within the aggregates and parameters associated with the specific rate of O_2 consumption were also computed by coupling measurements with a transient diffusion-reaction model. This model was subsequently paired to a population balance equation (PBE) depicting the time evolution of aggregate size distribution of proliferating ESCs under agitation. The multiscale PBE/diffusion-reaction model facilitated the calculation of the O_2 profile in the medium bulk and inside the aggregates in stirred suspension. Moreover, the fraction of cells experiencing hypoxia was predicted. The computational model described in this study can be utilized in the development of differentiation strategies and the design and control of relevant bioprocesses for the production of stem cell therapeutics.

Material and Methods

Experimental methods

Embryonic stem cell culture. Human ESCs (H9 cells, WiCell, Madison, WI; passages 30–60) were maintained on dishes coated with Matrigel (BD Biosciences, San Jose, CA) and grown in 5% CO_2/95% air at 37°C with mTESR1 medium (Stem Cell Technologies, Vancouver, BC) replaced daily. Four hours before harvesting, cells were treated with the Rho-associated kinase inhibitor (ROCKi) Y-27632 (10 μM, EMD Chemicals, Gibbstown, NJ). Colonies were dissociated with Accutase (Innovative Cell Technologies, San Diego, CA) into single hESCs. For static aggregate culture, cells were transferred to poly HEMA-treated (Sigma-Aldrich, St. Louis, MO) Petri dishes with mTESR1. Dispersed hESCs (10^5/ml) were seeded in 125-ml ProCulture spinner flasks (Corning, Corning, NY) with mTESR1 medium and 10 μM ROCKi. Agitation was kept constant during each experiment and medium was replaced daily thereafter without the addition of ROCKi.

Mouse E14Tg2a ESCs (Mutant Mouse Regional Resource Centers (MMRRC), University of California-Davis, CA) were maintained in defined serum-free medium (DSFM) as described [14] (see *Materials S1*). For stirred-suspension culture [14], mESCs were transferred in DSFM to 125-ml ProCulture spinner flasks at 5×10^4 cells/ml. The agitation was kept constant (60–100 rpm) throughout each run.

Oxygen consumption. Cell aggregate samples of known volumes were transferred to 15 ml-centrifuge tubes and allowed to settle. Without removing aggregates, the medium was replaced by PBS (pH 7.4) containing 0.35 g/L HEPES, 0.5 g/L bovine serum albumin, and 300 mg/dL glucose (Sigma-Aldrich, St. Louis, MO) with a Bunsen solubility coefficient of $\alpha_m = 1.27\times10^{-9}$ mol/(cm^3 mm Hg) at 37°C [15,16]. The cell suspension was transferred to a 30-ml cylindrical container (2.3 cm wide) with a magnetic cylindrical stirring bar (1.59 cm×0.79 cm). The container was set on a submersible magnetic stirrer plate in a water bath at 37°C bath and agitation was set to 60 rpm. A dissolved O_2 (DO) polarographic probe (VWR, Bridgeport, NJ) was inserted in the container through the cap with a rubber seal ensuring airtight fitting. DO data (mg/l) were logged on a computer connected to the DO probe/meter. Then, aggregates were transferred to a 96-well plate and images were taken and processed (see below) to obtain the total number of aggregates and pertinent size distributions [17].

Stem cell aggregate micrograph acquisition and analysis. Samples of known volume from spinner flasks, dishes or the DO chamber were transferred to a 96-well plate and images from each well were acquired at 4x magnification with a camera (Moticam 2300, Motic, Richmond, BC) connected to the microscope. Raw images were imported into ImageJ (U.S. National Institutes of Health, Bethesda, Maryland, USA, http://imagej.nih.gov/ij/) and for counting total numbers of aggregates. After background subtraction (rolling ball method), the diameter of each aggregate was calculated as the mean of two perpendicular diameters and size distributions were generated [17].

Porosity ε and tortuosity τ were determined in aggregates transferred to culture medium containing 10 mg/ml fluorescein isothiocyanate (FITC)-dextran (MW: 4.4 kDa, Sigma). The aggregate suspension was then transferred to a microscopy slide with a well, sealed with a coverslip and visualized by confocal laser scanning microscopy (CLSM) with a Zeiss LSM 510 Metal NLO system (Zeiss, Thornwood, NY). Aggregates were optically sectioned (10–12 sections per aggregate) at 5-μm steps and the micrographs were processed in MATLAB (Mathworks, Natick, MA) to obtain porosity and tortuosity values (**Fig. 1A**; see *Materials S1*).

Statistical Analysis. Data were expressed as mean±st.dev. unless stated otherwise. ANOVA and the *post hoc* Tukey test were performed using Minitab (Minitab Inc, State College, PA) with $p < 0.05$ considered as significant.

Modeling methods

Transient reaction-diffusion model. For the transient reaction-diffusion model employed in this study, the aggregates were assumed to be spherical and thus spherical coordinates were employed to describe the O_2 concentration (C) profile in each aggregate:

$$\frac{\partial C}{\partial t} = \frac{D_t^*}{r^2}\frac{\partial}{\partial r}\left(r^2\frac{\partial C}{\partial r}\right) - V_{O_2} \tag{1}$$

where r is the aggregate radius, t is the time and D_t^* is the effective diffusion coefficient. Taking into account the porosity ε and tortuosity τ, D_t^* is calculated as

$$D_t^* = \frac{\varepsilon D_t}{\tau} \tag{2}$$

where D_t is the diffusion of O_2 in the tissue (aggregate).

The consumption of O_2 (V_{O_2}in mol/(cm^3·s)) was described by Michaelis-Menten (M-M)-type kinetics:

$$V_{O_2} = \frac{V_{max}C}{K_M + C} \tag{3}$$

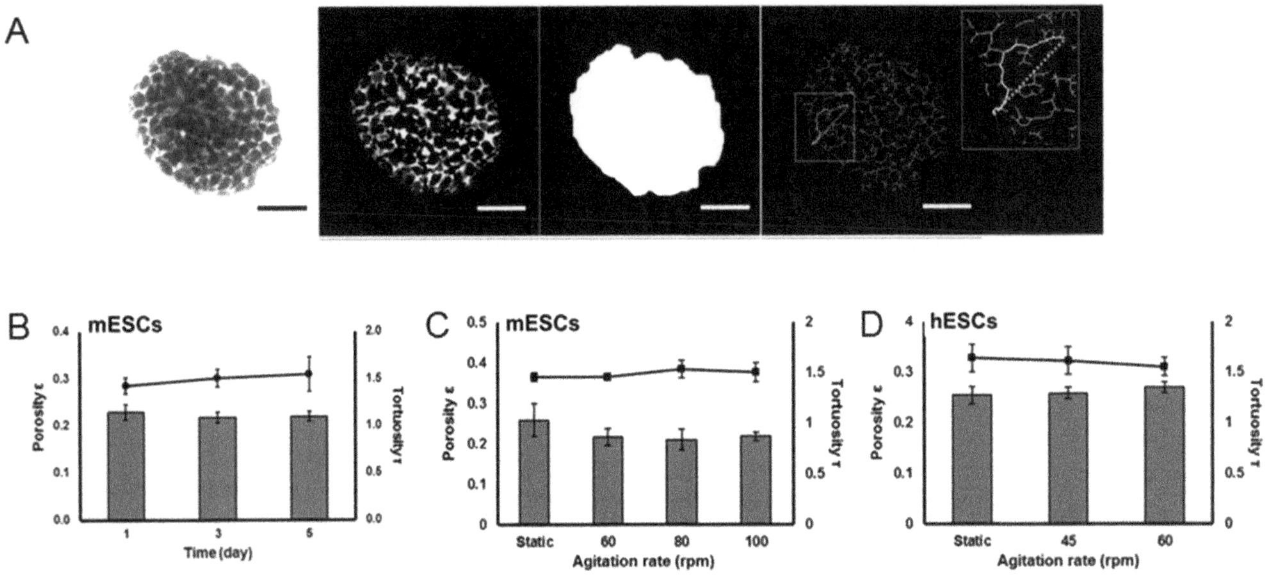

Figure 1. Structural characteristics (tortuosity and porosity) of mESC and hESC aggregates under different conditions. (A) Optical sections were acquired for each aggregate via CLSM and processed to obtained tortuosity and porosity values. For each condition, at least 10 aggregates were imaged by CLSM (30 sections/aggregate) for data analysis. Tortuosity (lines) and porosity (bars) data are shown for (B) mESC aggregates at different time points at 100 rpm, (C) mESC aggregates at different agitation rates on day 3, and (D) hESC aggregates at different agitation rates on day 4. The values in (B–D) are shown as mean±st. dev.

where V_{max} is the maximum O_2 consumption rate and K_M is the M-M constant. Because the system comprises aggregates of multiple sizes, cell clusters were classified into L groups or bins. Here, the utilization of 7 bins provided sufficient resolution of the aggregate size distribution. Equation 1 was solved for each group [18] (see *Materials S1*) subject to appropriate initial and boundary conditions.

$$C = C(r,0) \tag{4}$$

$$\left.\frac{\partial C}{\partial r}\right|_{r=0} = 0 \text{ (symmetry about the aggregate center)} \tag{5}$$

$$D_t^* \left.\frac{\partial C}{\partial r}\right|_{r=R_l} = k_l(C_M - C(R_l)) \tag{6}$$

where R_l is the radius for the aggregates in the lth bin. Calculation of the value of the O_2 transfer coefficient k_l from the bulk liquid to the aggregates was based on the Frössling equation [19] relating the particle Sherwood (*Sh*) and Schmidt (*Sc*) numbers shown in the *Materials S1* section.

The total rate of O_2 transfer from the bulk to all aggregates (L groups) is:

$$-V_M \frac{dC_M(t)}{dt} = \sum_{l=1}^{L} f_l z_{total} \int_0^{R_l} \frac{V_{max} C(r,t)}{K_M + C(r,t)} 4\pi r^2 dr \tag{7}$$

where f_l and z_{total} are the fraction of aggregates with average diameter $2R_l$ and the total number of aggregates, respectively. Integrals were calculated via the trapezoidal rule using C values from solving Equation 1 and the Euler method was implemented yield estimates of the O_2 concentration $C_M^{t+1}|_{calc}$ in the medium bulk at the $(t+1)$-time step in Equation 7.

Values for D_t^*, V_{max} and K_M were obtained by minimizing the objective function [20] comparing the calculated values $C_M^{t+1}|_{calc}$ with the experimental ones $C_M^{t+1}|_{exp}$ at each time point:

$$\Delta_C = \left(\frac{C_M^{t+1}|_{exp} - C_M^{t+1}|_{calc}}{C_M^{t+1}|_{exp}} \right)^2 \tag{8}$$

Development of PBE model for the bioreactor culture of ESC aggregates. A size-structured PBE model was employed to simulate the temporal evolution of size of ESC aggregates cultured in spinner flasks [1] considering the contributions of ESC proliferation and collisions between particles to aggregate size growth:

$$\frac{\partial n(x,t)}{\partial t} + \frac{\partial \left[n(x,t) \frac{\partial x}{\partial t} \right]}{\partial x} = \frac{1}{2} \int_{x_0}^{x} n(x_C,t) n(x',t) K(x_C|x') dx' - \int_{x_0}^{\infty} n(x,t) n(x',t) K(x|x') dx' \tag{9}$$

The density function $n(x,t)$ is defined such that $n(x,t)dx$ is the fraction of aggregates with sizes between x and $x+dx$ at time t per unit volume of the culture. The size x corresponds either to the aggregate volume or mass since the buoyant density of cells does not vary significantly during the cell cycle [21,22]. The mass of individual cells was taken as x_o and $x_C = x - x'$. Aggregate breakage was not observed particularly after the first day of culture and therefore was not considered.

The rate of aggregate size change, $\frac{\partial x}{\partial t}$, due to cell proliferation was modeled after the Gompertz equation [23]:

$$\frac{\partial x}{\partial t} = \alpha_G x ln\left(\frac{M}{x}\right) \tag{10}$$

where M is the aggregate mass reached as $t \rightarrow \infty$ and α_G is a constant characteristic of the cell proliferation. Both parameters were estimated from temporal data of aggregates in static cultures.

Different aggregation kernels $K(x|x')$ have been reported in various systems [24,25]. Given the non-rigid nature of cells/aggregates, we applied here an aggregation kernel (Eq. 11) for liquid-liquid dispersions assuming dynamic deformation of the colliding bodies [26]. The coalescence efficiency (exponential term) decreases with increasing average size of the colliding aggregates. The parameters k, k_1 and a are obtained from ESC aggregate size distributions measured in spinner flask cultures.

$$K(x|x') = kexp(-k_1\bar{x}^a) \left(x^{\frac{1}{3}} + x'^{\frac{1}{3}}\right)^{\frac{7}{3}} k_1, a > 0 \tag{11}$$

The PBE was expressed with respect to the aggregate radius, which was measured experimentally, assuming spherical aggregates. However, the density function was transformed in terms of the logarithm of the aggregate radius normalized to that of a single cell (~7.5 μm) to allow for an extended range of aggregate sizes to be considered (see [17] and *Materials S1*).

The integro-partial differential PBE was solved numerically by finite (backward) differences over equally spaced nodes in the *lnr* domain. Kernel parameters for different agitation rates were obtained from aggregate size data. The total number of aggregates and biomass were calculated as the zeroth and first moments of the cell distribution.

Integration of PBE and reaction-diffusion models. The PBE and reaction-diffusion models were combined for predicting the O_2 profile in stirred-suspension cultures of ESC aggregates (**Fig. 1**). The total number of aggregates z_{total}, size distribution of aggregates f_l, oxygen concentration in the medium bulk C_M, working volume of bioreactor V_M (assumed equal to that of the medium) and agitation rate N_{RPM} were used for initialization of the integrated model (see *Materials S1*). Cell growth and aggregation were modeled (Eq. 10–11) and the consumption of O_2 by all the aggregates in the bioreactor was determined (Eq. 7). This required knowledge of the O_2 concentration distribution within individual aggregates (Eq. 1). Subsequently, the bulk O_2 level C_M was updated based on the net effect of O_2 consumption and supply:

$$V_M\frac{dC_M(t)}{dt} = k_l^* a\cdot(C^* - C_M(t)) - \sum_{l=1}^{L} f_l z_{total} \int_0^{R_l} \frac{V_{max}C(r,t)}{K_M + C(r,t)} 4\pi r^2 dr f_l \tag{12}$$

The rate of O_2 supply through headspace aeration was:

$$V_M\frac{dC_M(t)}{dt} = k_l^* a\cdot(C^* - C_M(t)) \tag{13}$$

where C^* is the O_2 partial pressure in the gas phase (95% air/5% CO_2, 37°C, 760 mm Hg), C_M is the oxygen partial pressure in bulk medium, $k_l{}^*a$ is the mass transfer coefficient at the gas-liquid interface. Values of $k_l{}^*a$ were determined experimentally (see *Materials S1*).

Results

Porosity and tortuosity of ESC aggregates

The transport of O_2, nutrients and pluripotency-maintaining or differentiation-inducing factors to and within stem cell aggregates is greatly influenced by the aggregate ultrastructure. Moreover, whether the intra-aggregate ultrastructure is strongly affected by culture conditions in SSBs -particularly agitation- remains unknown. Thus, before proceeding with the solution of the diffusion-reaction model for ESC aggregates, we used CLSM and image analysis to determine two ultrastructural descriptors – porosity (ε) and tortuosity (τ)– for mESC and hESC aggregates formed in dishes or spinner flasks at different stirring rates (**Fig. 1A**). The porosity of mESC aggregates in spinner flasks at 100 rpm was 0.229±0.016 at 24 hours (**Fig. 1B**) and decreased slightly after 3 (0.218±0.011) and 5 days (0.220±0.011) of culture suggesting that aggregates were undergoing compaction. For 3-day cultures the ε values for 60 and 80 rpm were 0.217±0.02 and 0.210±0.025, respectively (**Fig. 1C**). When cultured in dishes, mESCs formed looser aggregates with ε = 0.259±0.041 (day 3). Yet, no statistically significant differences were noted in the porosity between aggregates cultured in dishes and spinner flasks at different agitation rates. Human ESCs in static culture formed aggregates with a porosity of 0.254±0.017 (**Fig. 1D**) whereas hESC aggregates from spinner flasks at 45 and 60 rpm displayed porosities of 0.258±0.011 and 0.270±0.007, respectively.

The tortuosity of mESC aggregates at 100 rpm was 1.427±0.083 at 24 hours and did not change significantly after 3 (1.504±0.093) and 5 days (1.547±0.181) of culture (**Fig. 1B**). Moreover, clusters formed at 60 rpm (1.460±0.035), 80 rpm (1.534±0.086) and in dishes (1.457±0.422) exhibited similar tortuosity. Human ESC aggregates were characterized by slightly higher average τ values than mESC clusters (**Fig. 1D**). After 4 days in stirred suspension an average τ of 1.612±0.135 (45 rpm) or 1.551±0.086 (60 rpm) was determined whereas the tortuosity of dish aggregates was 1.638±0.136.

Our findings provide a first account of porosity and tortuosity values determined directly for mESC and hESC aggregates formed in dish and stirred-suspension cultures at different times and under typical agitation rates. The data did not show strong dependence of the ultrastructural descriptors considered here on the cultivation mode, i.e. dish vs. stirred suspension nor on the stirring speeds tested.

Effective diffusion and consumption of O_2

A transient reaction-diffusion model was employed to determine the values of the diffusion coefficient D_l^* and consumption (V_{max}, K_M) of O_2 in the ESC aggregates from experimentally measured O_2 concentration values in the medium. First, the mean size was determined of aggregates in dishes and stirred-suspension cultures at different agitation rates (**Fig. 2A**). At the end of the culture, mESC aggregates at 100 rpm were smaller (mean diameter: 113.04±15.64 μm, $p<0.05$) than those at 60 rpm

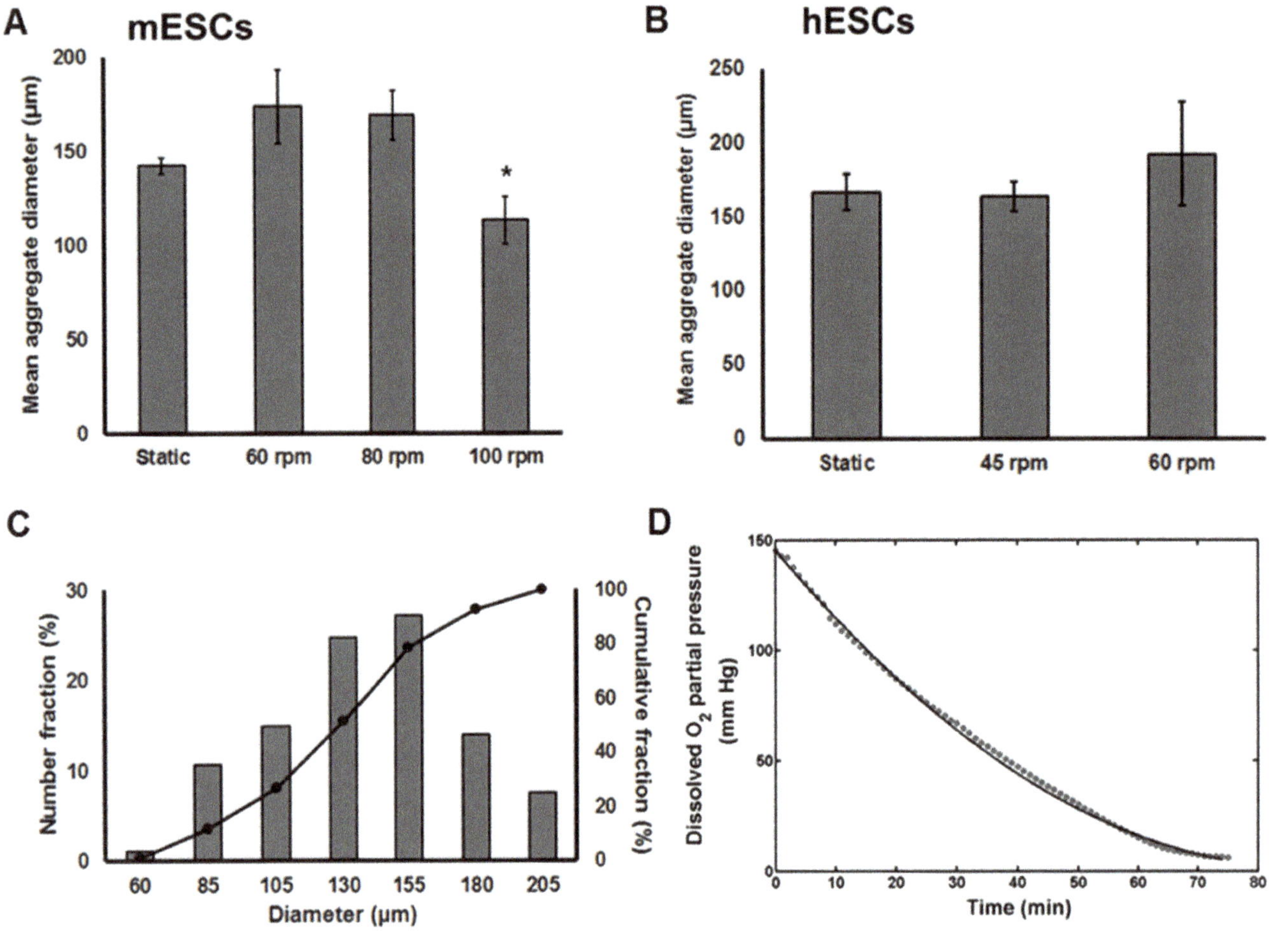

Figure 2. Experimental measurement of ESC aggregate size and O_2 consumption profiles for static and spinner flask cultures. Mean aggregate diameter data are shown for dish and bioreactor cultures at different agitation rates for (A) mESCs (*p<0.05 vs. 60 and 80 rpm) and (B) hESCs. (C) A representative ESC aggregate size distribution is shown (mESC culture at 100 rpm). (D) Experimental (solid line) and simulation (dotted line) results for the consumption of O_2 over time. Parameter (V_{max}, D, and K_M) values were acquired by fitting the transient reaction-diffusion model to the experimental data for different agitation rates (Table 1).

(173.8±23.9 µm) and 80 rpm (169.01±16.13 µm). The mean diameter of hESC aggregates at 45 and 60 rpm at day 4 was 163.8 and 192.45 µm, respectively and similar to those in static cultures (166.61±12.4 µm) (**Fig. 2B**). For each sample, the number fraction f_l distribution was determined (**Fig. 2C**).

Cells within aggregates consumed O_2 resulting in a gradual decrease of the DO level in the medium (**Fig. 2D**). Typically, DO measurements were performed until the O_2 concentration in the medium dropped below 10 mm Hg. Solution of the model coupled with experimental data on DO, yielded values for D_t^*, $V_{\max}$ and K_M (**Table 1**). It should be noted that parameter values were calculated for a minimum (critical) O_2 concentration at the aggregate center of either (i) 0 mm Hg, or (ii) 8 mm Hg [27]. Irrespective of the minimum O_2 level considered, the values of D_t (see Eq. 2) were lower than that of the diffusion coefficient D_{O_2} of O_2 in pure water at 37°C (2.78×10^{-5} cm^2/s). As a measure of maximum O_2 consumption rate, $V_{\max}$ was higher for mESCs in suspension culture at 60 rpm than for hESCs ($p<0.05$). Mouse ESCs also displayed greater $V_{\max}$ at 100 vs. 60 rpm suggesting an effect of the agitation on O_2 metabolism ($p<0.05$). The mean $V_{\max}$ for hESCs at 45 and 60 rpm did not differ significantly.

Oxygen profile within ESC aggregates in static and stirred-suspension cultures

Once the values of D_t^*, V_{max} and K_M were determined, the reaction-diffusion model was implemented to calculate the O_2 profile for ESC aggregates of different size. The diffusion of O_2 becomes limited as the aggregate radius increases leading to reduced O_2 levels for cells residing closer to the core. To better illustrate this, the O_2 distribution was calculated in hESC aggregates sampled from spinner flask cultures assuming a constant O_2 concentration in the bulk of 148 mm Hg (saturation) (**Fig. 3A**). Because hypoxia studies typically employ pO_2 of ~30 mm Hg (4–5% O_2) [5], we calculated the fractions of ESCs exposed to O_2 below this level for different aggregate sizes. Human ESC aggregates smaller than 200 µm were normoxic throughout their volume. In contrast, almost 23% of the hESCs in aggregates larger than 400 µm were under hypoxia. This fraction increased to 70% for aggregates with 1,000 µm diameter (**Fig. 3A**). The O_2 profiles in hESC aggregates of various sizes maintained in suspension (60 rpm) or static culture are shown in **Fig. 3B**. Similarly, the O_2 distributions within mESC aggregates cultured at 60 rpm or statically are also shown (**Fig. 3C**). It should be noted that the results shown in **Figs. 3B–C** were obtained assuming that

Table 1. Parameter values with a critical concentration of 8 mm Hg.

Culture condition	D_t (×10⁵ cm²/s)	V_{max} (×10⁸ mol/cm³·s)	K_M (×10⁸ mol/cm³)
mESCs			
Static culture (aggregates)	1.631±0.19 [0.242±0.024]	2.615±0.997 [4.732±0.318]	2.81±1.904 [0.518±0.165]
60 rpm	1.716±0.223 [0.364±0.02]	2.667±0.378*,# [5.055±0.163]	2.721±0.385 [0.782±0.329]
80 rpm	1.283±0.864 [0.418±0.05]	3.103±1.03 [3.602±1.06]	4.575±1.915 [0.392±0.233]
100 rpm	2.375±0.199 [0.386±0.114]	4.831±0.593 [7.322±0.83]	4.063±0.693 [0.563±0.06]
hESCs			
Static culture (aggregates)	1.20±0.170 [0.217±0.022]	0.995±0.464 [2.029±0.478]	3.473±0.232 [0.062±0.033]
45 rpm	1.536±0.322 [0.547±0.093]	1.16±0.355 [1.591±0.963]	3.23±0.257 [1.564±0.155]
60 rpm	1.163±0.454 [0.69±0.223]	1.36±0.427 [1.266±0.542]	4.126±1.138 [0.876±0.443]

Values in brackets are calculated with a 0 mm Hg critical concentration.
*p = 0.005 [0.019] for V_{max} at 60 vs. 100 rpm,
#p = 0.032 [0.001] for V_{max} for mESCs vs. hESCs at 60 rpm.

all aggregates in culture are the same size. Simulation results for additional agitation rates are shown in **Figure S1**.

In general, the fractions of cells experiencing an O_2 concentration below a particular threshold were lower in stirred suspension compared to dish cultures for given size clusters. For instance, the hypoxic cell fractions are larger in the plots from dishes compared to spinner flask cultures for the same size aggregates (**Figs. 3B–C**). This is most likely due to the faster transfer (k_l) of O_2 from the medium bulk to each aggregate in agitated vessels compared to static cultures (Eq. S2). Practically however, we rarely observed aggregates with a radius over 300 μm and only in static cultures. The higher hindrance for O_2 (and nutrient) transport to large clusters possibly impacts ESC viability and reduces proliferation by stimulating differentiation. This and the agitation-induced shear in spinner flask cultures may limit further growth of bigger aggregates.

Aggregates from stirred-suspension cultures showed similar O_2 profiles across all the agitation rates we tested (**Fig. S1**). As the aggregate size increased, cells positioned over 100 μm from the aggregate surface experienced less than 10 mm Hg of O_2. Compared to hESCs, the fractions of mESCs under the same O_2 concentration threshold were greater for equal size aggregates owing most likely to the higher V_{max} of the latter. The difference was most significant for aggregates with a radius between 100–250 μm.

Nevertheless, the expansion of mESCs and hESCs in spinner flasks gave rise to aggregate diameters mostly below 200–300 μm (**Fig. 2**). Thus, if a normoxic concentration of O_2 can be maintained in the bulk only 2.2% of the cells (located in the aggregate center) experience O_2 tension below 30 mm Hg (**Fig. 3A**). In fact, when we analyzed cells by qPCR for HIF1α expression (**Fig. S2**) there was no notable differences between static and stirred-suspension cultures. Also, compared to monolayer cultures at normoxia, HIF1α expression increased only slightly for mESCs (<2.0-fold) and hESCs (<1.5-fold) within aggregates.

PBE model for ESC aggregation in stirred-suspension culture

For the O_2 profiles presented above, the O_2 concentration in the medium was assumed to be fixed at saturation (~148 mm Hg). This assumption is valid if O_2 can be supplied efficiently and continuously to the cultivation system or the culture is sparse so that O_2 consumption is trivial. However, stirred-suspension cultures accommodate high cell densities resulting in fluctuating O_2 levels in the bioreactor. Agitation and proliferation also drive the formation and growth of aggregates over time with concomitant changes in the intra- and inter-aggregate concentrations of O_2. Given the multitude of effects of O_2 on stem cell physiology, predicting the time-variant distribution of O_2 in scalable ESC aggregate cultures is highly desirable.

To that end, a PBE model was developed for the temporal evolution of the ESC cluster size distribution in an SSB. The growth rate of ESC aggregates due to cell proliferation was described by the Gompertz equation. For this purpose, the growth of sparsely cultured mESC aggregates in dishes was monitored and the Gompertz parameters were evaluated ($M = 9.71\times10^6$, $\alpha_G = 5.72\times10^{-3}\ hr^{-1}$) based on the recorded size changes (**Fig. 4A**). The Gompertz model accurately described the growth of mESC aggregates ($R^2 = 0.99$, $n = 20$). The maximum attainable size for mESC aggregates was calculated at 1600 μm (lnr_{max}~5.4).

Parameters for the aggregation kernel were determined from size data of mESC aggregates cultured in spinner flasks (**Fig. 4B**). It should be noted that the kernel parameters were calculated from data of cell cultures at 0–72 hours (**Table 2**). At longer times, aggregation was negligible and cell proliferation drove the increase in cluster size. With increasing agitation rate, the parameter value trends are such that the coalescence frequencies (i.e. collisions leading to aggregate formation) become lower. At 60 rpm, the mean aggregate radius increased to ~105 μm on day 2 (mainly due to collision-driven aggregation) in line with the model predictions (**Fig. 4C**). Subsequent increase in aggregate size was less pronounced as the mean radius was approximately 150 μm (day 4). The corresponding average size at higher agitation rates was lower (e.g. ~90 μm at 100 rpm, **Fig. S3**). The simulation results were corroborated by observations that a lower agitation rate generally promotes formation of larger aggregates. It should also be noted that the cell viability was not affected significantly by agitation (>90%).

Taken together, the PBE model developed here allowed the prediction of the temporal evolution of the ESC aggregate size distribution (and the total aggregate number and biomass; see below). Simulation results were in good agreement with the

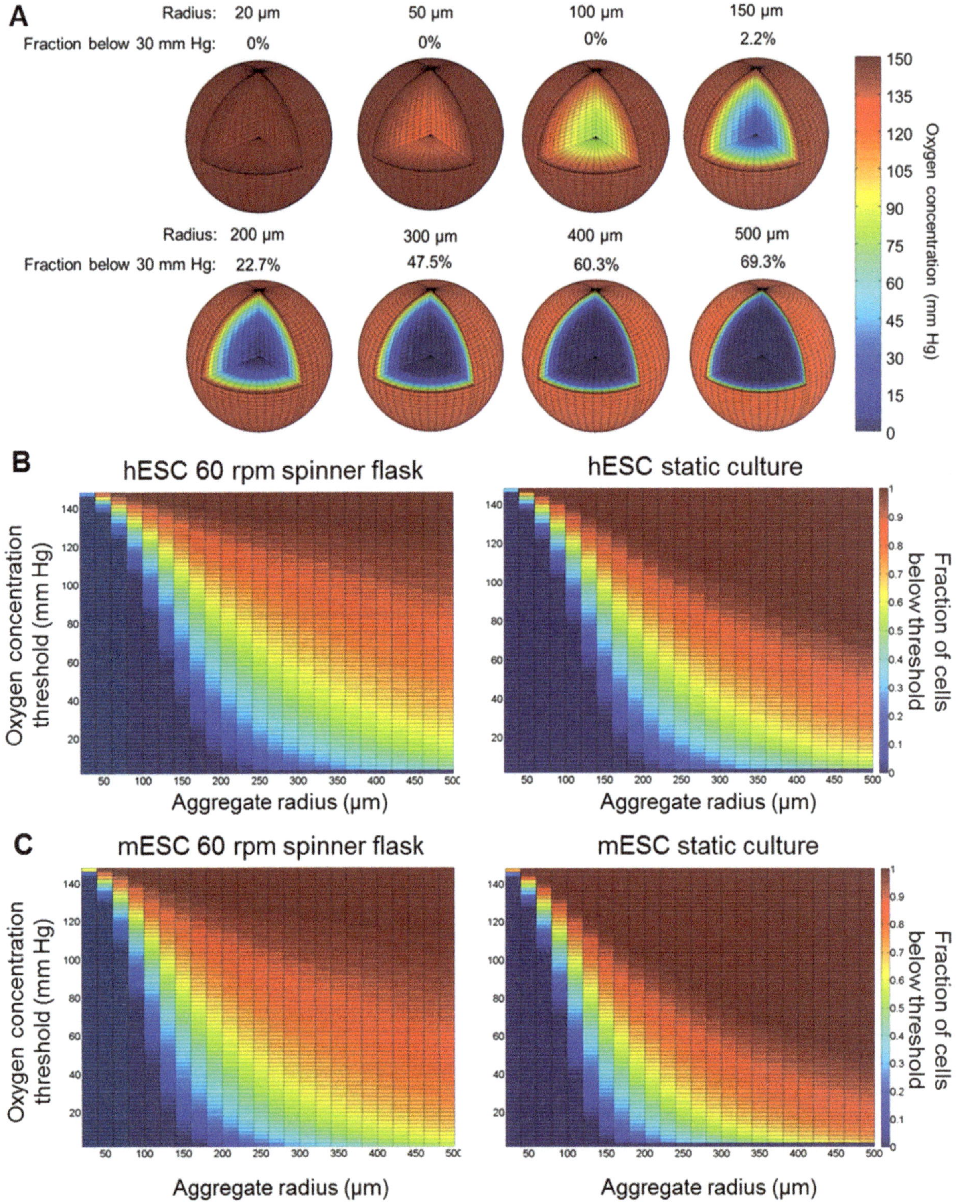

Figure 3. Steady-state O_2 profile within ESC aggregates of varying sizes. (A) Three-dimensional distribution of O_2 and hypoxic cell fraction of hESC aggregates with radius ranging from 20 µm to 500 µm. The bulk oxygen level was taken as 148 mm Hg. Fractions of ESCs exposed to O_2 below a particular threshold (y-axis) for different size aggregates cultured at 60 rpm or in dishes (static) shown for (B) hESCs and (C) mESCs. Results were obtained assuming that all cultured aggregates were the same size. The minimum intra-aggregate O_2 level was fixed at 8 mm Hg (critical concentration).

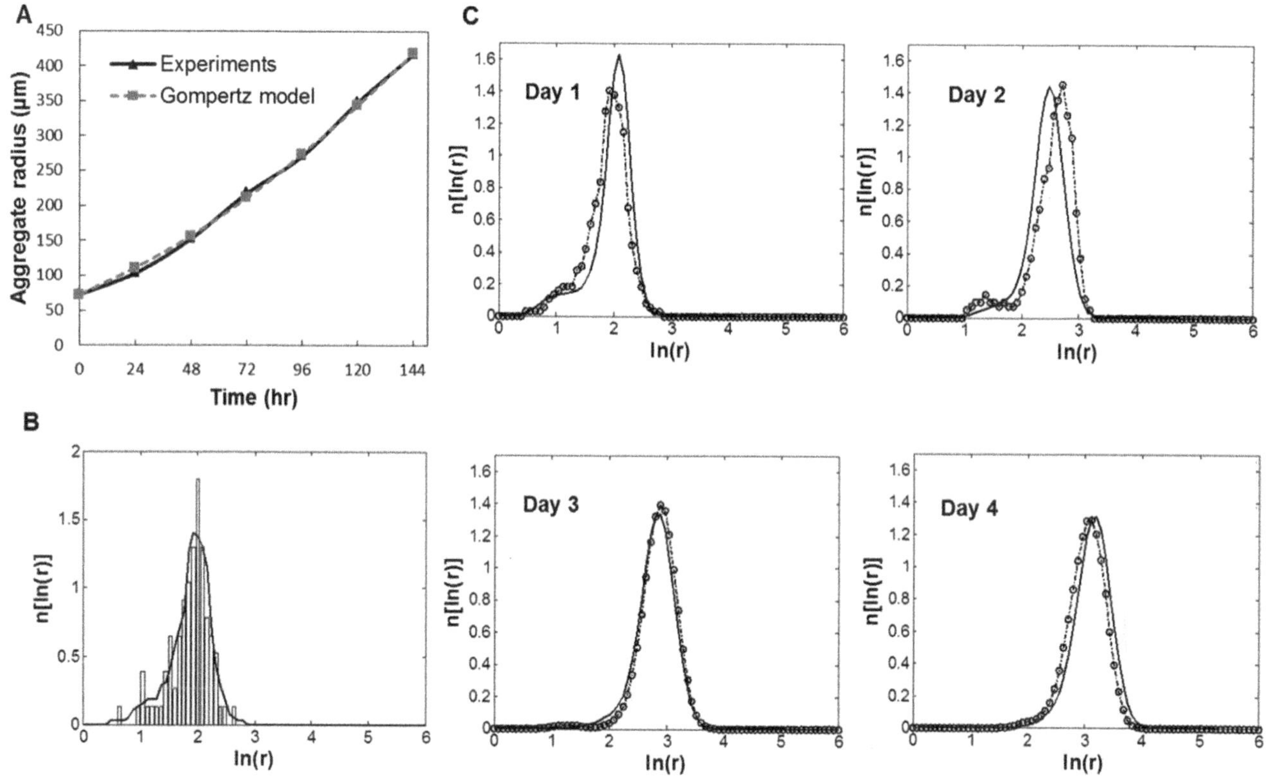

Figure 4. PBE model for the ESC aggregate culture in spinner flasks. (A) Mouse ESC aggregates in dishes were monitored and their size was measured over several days. The experimental data for aggregate growth due to mESC proliferation were compared with the Gompertz model. Relevant parameters were determined: $M = 9.71\times10^{6}$, $\alpha_G = 5.72\times10^{-3}$ hr^{-1}. (B) Representative data of size (radius) distribution for mESC aggregates cultured at 60 rpm (day 1). Histogram and a continuous distribution curve are shown. (C) Model size distributions (lines) shown for a 96-hour culture at 60 rpm. The original distributions (dash-circle lines) for each time point are shown for comparison.

experimental observations from stirred-suspension ESC culture at different agitation rates.

Prediction of O_2 profile and hypoxia fraction in ESC aggregate stirred-suspension culture

With the PBE model of ESC aggregation in the SSB in place, the O_2 distribution in the medium and within ESC aggregates was computed by coupling the diffusion reaction and PBE models. For this purpose, the coefficient k_l^* was measured for the O_2 transfer via headspace aeration to the bulk medium (air-liquid interface, **Fig. S4**). The slope ($k_l^*\cdot a/V_M$) was 4.87 ± 0.13 hr^{-1} yielding a k_l^* of 16.57 ± 0.43 cm/hr for our cultivation system. Simulations were run for spinner flask cultures of mESCs at 60 rpm and a seeding density of 5×10^{4} cells/ml. The temporal size distribution of mESC aggregates is shown in **Fig. 5A**. During the first day, there was a significant increase in mean cluster size mainly due to aggregation. By day 3, mESC aggregates had a mean diameter of around 200 µm in agreement with our cell culture data. The total number of aggregates dropped substantially during the first day due to aggregation but remained relatively steady afterwards (**Fig. 5B**). At day 4, the calculated total number of aggregates was around 2.7×10^{4}/50 ml of culture similar to our experiments and the corresponding biomass had increased 160-fold.

The fractions of cells exposed to various levels of O_2 in spinner flask cultures were also predicted. Here, the fractions were computed of cells experiencing O_2 tension below 30 mm Hg, which may be favorable for maintenance of ESC pluripotency (**Fig. 5C**). We also calculated the fractions for extremely low pO_2 (<10 mm Hg), which hampers ESC self-renewal and induces death [5]. The pO_2 in the medium decreased to 82 mm Hg after 4

Table 2. Kernel parameter values at different agitation rates.

Agitation rate (rpm)	k (hr^{-1} $\mu m^{-7/3}$)	k_1 ($\mu m^{-3\alpha}$)	α (unitless)
60	$1.26\pm0.77\times10^{-3}$*	$1.94\pm1.12\times10^{-4}$	$8.06\pm0.79\times10^{-1}$*
80	$6.36\pm4.47\times10^{-6}$	$4.62\pm3.25\times10^{-4}$	$5.96\pm0.085\times10^{-1}$*
100	$6.53\pm6.23\times10^{-6}$	$4.11\pm0.63\times10^{-5}$	$9.91\pm0.65\times10^{-1}$*

The parameter values were shown as mean ± st.dev. calculated by model simulation to experimental data from mESC cultures (at least 3 independent runs per agitation rate).
*$p<0.05$ when compared to the mean values obtained for the other agitation rates.

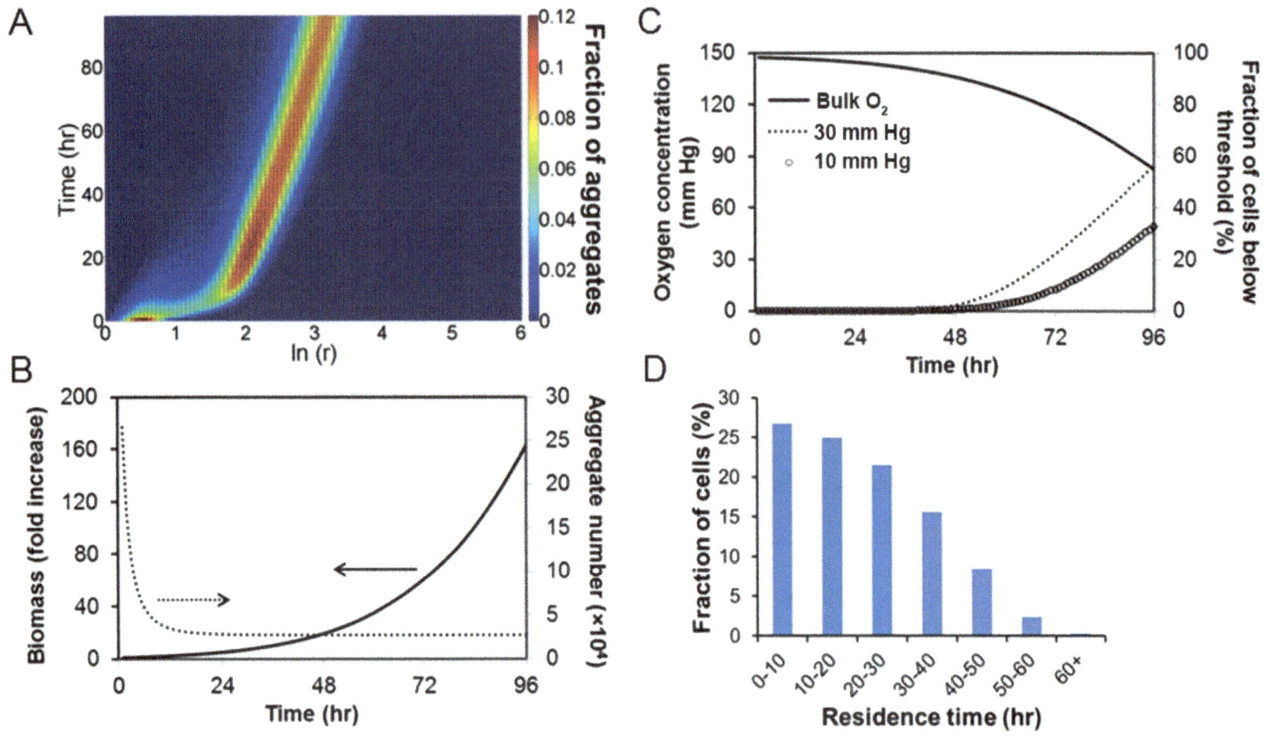

Figure 5. Model prediction of cluster size and O_2 concentration distribution in a stirred suspension bioreactor of cultured mESC aggregates. (A) Density plot of mESC aggregate size distribution during a 4-day culture. (B) Total aggregate number and biomass of mESC aggregates over 4 days of culture. (C) Oxygen concentration in the medium bulk and fraction of ESCs experiencing pO_2 below 30 mm Hg. The model was run for a 50-ml working volume and 60 rpm with mESCs seeded at 5×10^4 cells/ml. Specific model parameters are from Tables 1–2. (D) Distribution of residence time of ESCs exposed to $\leq$30 mm Hg pO_2 during a 4-day culture.

days with 58% of cells experiencing O_2 less than 30 mm Hg. The corresponding fraction of ESCs at 10 mm Hg or less was 32%. If the O_2 concentration in the culture bulk was fixed at 148 mm Hg throughout the cultivation period, only 33% of cells would be at pO_2<30 mm Hg by day 4 (data not shown).

In addition, the residence time distribution (**Fig. 5D**) shows that mESCs entered low O_2 level (<30 mm Hg) at different time points. This variation is attributed to the distributed size of cell clusters and random aggregation under stirring. The data clearly illustrate that although a group of cells is exposed to a particular O_2 range (e.g. below 30 mm Hg), not all cells within the group experience the O_2 level for the same amount of time. Therefore, information about the fraction of cells at different pO_2 levels should be coupled to residence times provided by the model presented here. Our findings show that stem cell aggregate size heterogeneity is linked to population heterogeneity with respect to the duration of exposure at different levels of O_2 concentration despite the fairly homogeneous environment induced by mixing in SSBs.

Discussion

The transport of factors among cells within 3D structures such as aggregates is an important aspect of the performance of cultivation systems. The importance becomes even more pronounced for stem cell culture modalities given the central role of O_2 in pluripotency and lineage specification. We employed a transient reaction-diffusion model to predict the O_2 profile inside ESC aggregates of varying sizes. The effective diffusivity and consumption rate parameters of O_2 in hESC and mESC aggregates were determined for the first time based on experiments without a steady state constraint. A PBE model was further developed for the SSB culture of ESC aggregates. The combined diffusion-reaction/PBE framework was utilized to predict the aggregation outcome and the O_2 profile inside ESC clusters and the medium bulk in stirred-suspension over time. As a result, the fractions of cells exposed to particular levels of O_2 and pertinent residence times were determined allowing to draw conclusions about the ensuing population heterogeneity and state of the cells (e.g. viability, differentiation potential).

The oxygen level to which ESCs are exposed to significantly affects cell fate decisions. Hypoxia (~30 mm Hg) does not modulate adversely the proliferation rate of hESCs, which exhibit reduced spontaneous differentiation and chromosomal abnormalities [5,6,28]. Human ESCs and iPSCs exhibit differentiation efficiencies along particular lineages dependent on O_2 availability. For instance, low pO_2 enhances hESC differentiation to chondrogenic, endothelial and cardiac cells [7,8,29]. Mouse ESC EBs generate significantly more myeloid and erythroid progenitors when cultured at 3% O_2 instead of 21% O_2 [30]. However, extremely low pO_2 can be detrimental for cells. Thus, delicate pO_2 control is essential in culture systems to avoid hampering stem cell growth, viability and self-renewal or commitment. This is particularly relevant in scaling up stem cell cultivation where local concentration gradients and transfer of factors are more pronounced. The model presented in this study permits the prediction of the actual O_2 level facilitating informed decisions about dynamic pO_2 control and optimal aggregate size in systems for stem cell expansion.

For the results reported here, an implicit assumption was made that the O_2 consumption rate (i.e. K_M and V_{max}) does not vary with time or the location of cells within each cluster. Although this is a premise that is common with previous reports, we cannot exclude the possibility of ESC metabolism adaptation to available O_2 levels. For example, V_{max} was estimated to vary with pO_2 albeit within the same order of magnitude [31]. Human ESCs reportedly lower their O_2 uptake rate when the dissolved O_2 decreases from 21% to at 5% and 1% O_2 [32]. Such adjustment may aid the cells to maintain the concentration of available O_2 above zero at the core of aggregates with a diameter >200 μm thereby reducing the likelihood for necrosis (**Fig. S1** in the absence of the constraint of 8 mm Hg critical pO_2). However, further studies are warranted on the adaptation of O_2 consumption by pluripotent stem cells in response to microenvironment O_2 levels. With more information becoming available, our model can accommodate spatially (e.g. radial distance from the aggregate surface) and temporally changing O_2 consumption rates.

Our study also raises the issue of 'residence time' (duration of exposure) of cells especially to low O_2 levels. To our knowledge, this has not been investigated previously and even in reports of stem cell aggregate cultivation in stirred-suspension vessels, emphasis is placed on the actual pO_2 the cells experience. It is important to note that even after the first 5 minutes of mESC culture in 2.2% O_2, increases are observed in HIF-1α expression and reactive O_2 species [33]. At 12–24 hours of hypoxia, cell-cycle regulatory proteins such as cyclins D1 and E, CDK2 and CDK4 along with retinoblastoma phosphorylation increase yielding a larger fraction of mESCs in the S phase and higher overall cell numbers. Accelerated cell proliferation in aggregate regions of hypoxia will lead to increased O_2 consumption further depressing pO_2 levels. As our model indicates, an increasing fraction of cells will reside in the hypoxic region of the aggregates which grow continuously. The 'cut-off' time of cell exposure to low pO_2 before hypoxia-induced changes (e.g. cell differentiation, apoptosis etc.) become irreversible is unknown. Such information will be essential in formulating strategies (e.g. by modulating agitation (**Table 2**)) to achieve ESC aggregate sizes so that a high cell fraction remains pluripotent for downstream differentiation to a desired phenotype.

The effects of O_2 on gene expression are mediated largely by HIF transcription factors and as noted, mESCs upregulate HIF-1α within 5 minutes of exposure to 2.2% O_2. Cells in our cultures did not exhibit significant differences in HIF-1α expression under different agitation rates (**Fig. S2**). This may be attributed to the average cluster diameter, which generally ranged between 200–300 μm (**Fig. 3**), leading to rather limited fractions of cells exposed to 2.2% O_2 or less. Yet, we acknowledge that HIF-1α transcripts (probed by qPCR) may not be representative of the corresponding protein amount and activity levels. Moreover, the qPCR results are population 'averages' and HIF expression variation at locales with different pO_2 cannot be discerned. Lastly, changes due to fluctuating O_2 can be brought about by HIF-independent mechanisms such as the environmental sensing mammalian target of rapamycin (mTOR) [34].

Although this study focused on the transport of O_2 in ESC aggregates in dish and stirred-suspension cultures, the concurrent availability and consumption of other components such as nutrients and factors for self-renewal or differentiation should be considered. A steady state analysis of the diffusion of glucose and (generic) cytokines (in addition to O_2) in human EBs was recently presented [12]. The diffusivities of glucose and cytokine were lower than that of O_2 suggesting that their transport places further limits on ESC aggregates to avoid cell starvation and death. Such limitations on transport may actually hinder the growth of aggregates and cause their size to stabilize. Interestingly enough, Cameron et al. [35] reported that over 21 days hESC aggregates displayed a maximum size at day 10 ranging between 400 and 500 μm without further increase. These observations support the choice of the Gompertz equation for modeling aggregate growth due to cell proliferation. The Gompertz equation in our study suggested a maximum radius larger than our experimental observations. This discrepancy may be explained by the fact that experimental data were obtained from static cultures with relatively constant bulk concentration of substrates (oxygen, glucose, etc.) due to daily medium changes whereas realistically these concentrations should decrease markedly over time. The maximum radius attained by ESC aggregates may be lower if substrate-dependent growth kinetics are considered. Additional mass balance equations for nutrients and growth factors can be incorporated in the model presented here to provide a more accurate representation of the bioreactor stem cell culture.

The diffusion and consumption of O_2 in mESC and hESC cultures has been investigated in various studies [12,13,31,36,37] although only in a subset of these reports parameters were determined from experiments. The values reported for V_{max} and K_M are the same order of magnitude (10^{-8} mol·cm^{-3}·s^{-1} and 10^{-8} mol·cm^{-3}, respectively) as those presented here. The O_2 consumption rate for mESCs was previously [38] measured at 4×10^{-17} (mol/cell·s), i.e. close to our observed rate assuming cells are spherical with a 15 μm diameter. Values for the diffusion coefficient based on experimentally determined ultrastructural ESC characteristics of aggregates such as porosity and tortuosity were not available to date in the literature for comparison. In fact, the ε and τ values did not vary significantly with the culture modality, i.e. dish or stirred suspension and the agitation speeds tested. This could suggest that cells adopt particular arrangements in the aggregate interior without significant influence from external mechanotransduction but this warrants further investigation. Such organizations may be mediated by cytoskeletal actions, which are central to the assembly of other cell types into aggregates [39].

It should be noted that no overt changes were observed in the values of V_{max}, K_M or D_t at different agitation rates. In part this can be explained by the fact that the majority of cells are within aggregates not experiencing the different levels of shear induced by the stirring rates tested here. Moreover, the DO measurements were performed in PBS-based solution with a known Bunsen solubility coefficient. This facilitates the comparison with findings from previous studies and eliminates potential discrepancies due to differences in culture media utilized in those reports. Even though we cannot rule out potential changes in parameter values if the measurements were performed in culture medium, these changes are expected to be minimal as shown for other systems [36,37,40].

A multiscale approach was described here with the coupling of the transient diffusion-reaction model for individual clusters with the PBE for the temporal evolution of size for ESC aggregate population in the bioreactor due to cell proliferation and collisions. The kernel used in this study comprises two parts: (i) The sum of cubic roots of volumes of the aggregates colliding raised to the power of 7/3 is applicable for a shear field with non-linear velocity profile [25,26] as in the case of bioreactors with impellers and baffles. This is in contrast to the Smoluchowski kernel [41] used for cell aggregation (e.g. of platelets [42]) assuming a linear shear flow. (ii) A multiplicative factor dependent on operating conditions including the agitation rate represents the coalescence frequency written as the product of the collision frequency and the coalescence efficiency. Use of the kernel assumes that the aggregates are not rigid spheres but deform during collision.

More importantly, the coalescence efficiency (collisions leading to new aggregate formation) decreases with increasing aggregate size in line with our experiments. Because of these attributes, the kernel is deemed suitable for describing the aggregation of stem cells in stirred-suspension vessels despite its original derivation for droplet coalescence in liquid-liquid dispersions. Explicit functions of the kernel parameters dependent on agitation rates can be estimated from the data (**Table 2**), and additional operating variables may be included (e.g. cell seeding density which was kept constant for all experiments here). Since transient size distributions of ESC aggregates cultivated in spinner flasks are available, a proper approach to determining the kernel for the PBE model is to solve the inverse problem [43]. Work is underway for deriving kernel functions via solving the inverse problem and comparing the solution to the functional utilized in this study.

The approach described here is certainly not without limitations but amenable to improvements. As already mentioned, the transport and consumption or secretion of factors other than O_2 were not considered. Fluctuations in the consumption of nutrients and/or accumulation of waste will almost certainly affect the microenvironment within aggregates. These can be considered by expanding the model, for instance, to include reaction-diffusion equations (see Equation 1) for additional species. Moreover, O_2 consumption was assumed invariant with the position of each cell within aggregates. Pertinent variations can be introduced by considering different expressions for V_{max} and K_M based on spatial criteria or other conditions (e.g. concentration of particular nutrients). In the context of envisioned stem cell process designs, the model in its current or expanded form can be used in conjunction with investigating relevant bounds in culture variables.

Nonetheless, the framework presented here can be applied widely to address both basic questions of stem cell physiology and issues surrounding relevant bioprocesses utilizing SSBs. The model can be employed to generate maps of concentrations of O_2 (and other factors) expediting research on the role of hypoxia on stem cell self-renewal or commitment along particular lineages. Gaining a better understanding of the effects of O_2 on fate selection is critical in developing optimal strategies for stem cell differentiation. The design and control of stem cell bioprocesses will also benefit from quantitative multiscale approaches such as the one described here, thereby bringing us closer to the realization of the potential of stem cell therapeutics.

Supporting Information

Figure S1 Steady-state O_2 profile for ESC aggregates of different sizes cultured under different conditions. Oxygen concentration profile for (A) hESC aggregates cultured in static or stirred suspension (45 rpm and 60 rpm) culture, and (B) mESC aggregates cultured in static or stirred suspension (60 rpm, 80 rpm and 100 rpm) culture. The distance in the horizontal axis is measured from the aggregate center. The O_2 level in the medium bulk was kept fixed at 148 mm Hg for all conditions. The aggregate radius ranged from 30 μm to 500 μm. Values for V_{max}, K_M and D_i^* were taken from Table 1.

Figure S2 HIF1α gene expression in (A) mESCs and (B) hESCs cultured as aggregates in dishes and spinner flasks.

Figure S3 Comparison of experimental data (dash-circle lines) and model (solid lines) results for different agitation rates at day 4 of culture: (A) 60 rpm, (B) 80 rpm, (C) 100 rpm.

Figure S4 Measurement of O_2 transfer to the medium through headspace aeration in spinner flasks. (A) Oxygen concentration during the experiment. The dissolved O_2 in the medium bulk was first depleted by gassing with nitrogen. After the depletion phase, air was allowed to fill the space above the medium and the O_2 level was measured. (B) Linear regression of the recovery phase for O_2 marked in (A). The standard deviation was calculated from at least three independent experiments. The O_2 transfer coefficient was determined from the slope of the curve.

Materials S1 Supplemental information on cell culture, image analysis, modeling methods and measurements of bioreactor O_2 transfer coefficient and HIF1a gene expression levels.

Acknowledgments

We are thankful to Prof. Johannes Nitsche for critical discussions during the preparation of the manuscript. This work was performed in part at the University at Buffalo's Center for Computational Research.

Author Contributions

Conceived and designed the experiments: JW MRR EST. Performed the experiments: JW MRR DCO. Analyzed the data: JW EST. Contributed reagents/materials/analysis tools: JW MRR EST DCO. Wrote the paper: JW EST.

References

1. Kehoe DE, Jing D, Lock LT, Tzanakakis EM (2010) Scalable Stirred-suspension Bioreactor Culture of Human Pluripotent Stem Cells. Tissue Eng Part A 16: 405–421.
2. Jing D, Parikh A, Tzanakakis ES (2010) Cardiac cell generation from encapsulated embryonic stem cells in static and scalable culture systems. Cell Transplant 19: 1397–1412.
3. Lock LT, Tzanakakis ES (2009) Expansion and differentiation of human embryonic stem cells to endoderm progeny in a microcarrier stirred-suspension culture. Tissue Eng Part A 15: 2051–2063.
4. Simon MC, Keith B (2008) The role of oxygen availability in embryonic development and stem cell function. Nat Rev Mol Cell Biol 9: 285–296.
5. Ezashi T, Das P, Roberts RM (2005) Low O2 tensions and the prevention of differentiation of hES cells. Proc Natl Acad Sci U S A 102: 4783–4788.
6. Forsyth NR, Musio A, Vezzoni P, Simpson AH, Noble BS, et al. (2006) Physiologic oxygen enhances human embryonic stem cell clonal recovery and reduces chromosomal abnormalities. Cloning Stem Cells 8: 16–23.
7. Prado-Lopez S, Conesa A, Arminan A, Martinez-Losa M, Escobedo-Lucea C, et al. (2010) Hypoxia promotes efficient differentiation of human embryonic stem cells to functional endothelium. Stem Cells 28: 407–418.
8. Koay EJ, Athanasiou KA (2008) Hypoxic chondrogenic differentiation of human embryonic stem cells enhances cartilage protein synthesis and biomechanical functionality. Osteoarthritis Cartilage 16: 1450–1456.
9. Mazumdar J, O'Brien WT, Johnson RS, LaManna JC, Chavez JC, et al. (2010) O2 regulates stem cells through Wnt/beta-catenin signalling. Nat Cell Biol 12: 1007–1013.
10. Gustafsson MV, Zheng X, Pereira T, Gradin K, Jin S, et al. (2005) Hypoxia requires notch signaling to maintain the undifferentiated cell state. Dev Cell 9: 617–628.
11. Covello KL, Kehler J, Yu H, Gordan JD, Arsham AM, et al. (2006) HIF-2alpha regulates Oct-4: effects of hypoxia on stem cell function, embryonic development, and tumor growth. Genes Dev 20: 557–570.
12. Van Winkle AP, Gates ID, Kallos MS (2012) Mass transfer limitations in embryoid bodies during human embryonic stem cell differentiation. Cells Tissues Organs 196: 34–47.

13. Khoury M, Bransky A, Korin N, Konak LC, Enikolopov G, et al. (2010) A microfluidic traps system supporting prolonged culture of human embryonic stem cells aggregates. Biomed Microdevices 12: 1001–1008.
14. Kehoe DE, Lock LT, Parikh A, Tzanakakis ES (2008) Propagation of Embryonic Stem Cells in Stirred Suspension without Serum. Biotechnol Prog 24: 1342–1352.
15. Panten U, Klein H (1982) O2 consumption by isolated pancreatic islets, as measured in a microincubation system with a Clark-type electrode. Endocrinology 111: 1595–1600.
16. Groebe K, Thews G (1989) Effects of red cell spacing and red cell movement upon oxygen release under conditions of maximally working skeletal muscle. Adv Exp Med Biol 248: 175–185.
17. Cadavid Olaya DP (2012) Bioreactor culture of embryonic stem cell aggregates: Oxygen transfer dynamics and population balance modeling of aggregation [M.S.]. Ann Arbor: State University of New York at Buffalo. 88 p.
18. Tosaka N, Miyake S (1982) Analysis of a nonlinear diffusion problem with Michaelis-Menten kinetics by an integral equation method. Bulletin of mathematical biology 44: 841–849.
19. Frössling N (1938) Uber die Verdunstung fallender Tropfen. Gerlands Gerlands Beitrage zur Geophysik 52: 170–175.
20. Nelder JA, Mead R (1965) A Simplex-Method for Function Minimization. Computer Journal 7: 308–313.
21. Anderson EC, Petersen DF, Tobey RA (1970) Density invariance of cultured Chinese hamster cells with stage of the mitotic cycle. Biophys J 10: 630–645.
22. Loken MR, Kubitschek HE (1984) Constancy of cell buoyant density for cultured murine cells. Journal of Cellular Physiology 118: 22–26.
23. Brunton GF, Wheldon TE (1980) The Gompertz equation and the construction of tumour growth curves. Cell Tissue Kinet 13: 455–460.
24. Aldous DJ (1999) Deterministic and stochastic models for coalescence (aggregation and coagulation): a review of the mean-field theory for probabilists. Bernoulli 5: 3–48.
25. Smit DJ, Hounslow MJ, Paterson WR (1994) Aggregation and gelation- I. Analytical solutions for CST and batch operation. Chemical Engineering Science 49: 1025–1035.
26. Tobin T, Muralidhar R, Wright H, Ramkrishna D (1990) Determination of coalescence frequencies in liquid-liquid dispersions - Effect of drop size dependence. Chemical Engineering Science 45: 3491–3504.
27. Wartenberg M, Donmez F, Ling FC, Acker H, Hescheler J, et al. (2001) Tumor-induced angiogenesis studied in confrontation cultures of multicellular tumor spheroids and embryoid bodies grown from pluripotent embryonic stem cells. FASEB J 15: 995–1005.
28. Ludwig TE, Levenstein ME, Jones JM, Berggren WT, Mitchen ER, et al. (2006) Derivation of human embryonic stem cells in defined conditions. Nat Biotechnol 24: 185–187.
29. Burridge PW, Thompson S, Millrod MA, Weinberg S, Yuan X, et al. (2011) A universal system for highly efficient cardiac differentiation of human induced pluripotent stem cells that eliminates interline variability. PLoS One 6: e18293.
30. Adelman DM, Maltepe E, Simon MC (1999) Multilineage embryonic hematopoiesis requires hypoxic ARNT activity. Genes Dev 13: 2478–2483.
31. Powers DE, Millman JR, Huang RB, Colton CK (2008) Effects of oxygen on mouse embryonic stem cell growth, phenotype retention, and cellular energetics. Biotechnol Bioeng 101: 241–254.
32. Abaci HE, Truitt R, Luong E, Drazer G, Gerecht S (2010) Adaptation to oxygen deprivation in cultures of human pluripotent stem cells, endothelial progenitor cells, and umbilical vein endothelial cells. Am J Physiol Cell Physiol 298: C1527–1537.
33. Lee SH, Lee MY, Han HJ (2008) Short-period hypoxia increases mouse embryonic stem cell proliferation through cooperation of arachidonic acid and PI3K/Akt signalling pathways. Cell Prolif 41: 230–247.
34. Arsham AM, Howell JJ, Simon MC (2003) A novel hypoxia-inducible factor-independent hypoxic response regulating mammalian target of rapamycin and its targets. J Biol Chem 278: 29655–29660.
35. Cameron CM, Hu WS, Kaufman DS (2006) Improved development of human embryonic stem cell-derived embryoid bodies by stirred vessel cultivation. Biotechnol Bioeng 94: 938–948.
36. Gassmann M, Fandrey J, Bichet S, Wartenberg M, Marti HH, et al. (1996) Oxygen supply and oxygen-dependent gene expression in differentiating embryonic stem cells. Proc Natl Acad Sci U S A 93: 2867–2872.
37. Cochran DM, Fukumura D, Ancukiewicz M, Carmeliet P, Jain RK (2006) Evolution of oxygen and glucose concentration profiles in a tissue-mimetic culture system of embryonic stem cells. Ann Biomed Eng 34: 1247–1258.
38. Zur Nieden NI, Cormier JT, Rancourt DE, Kallos MS (2007) Embryonic stem cells remain highly pluripotent following long term expansion as aggregates in suspension bioreactors. J Biotechnol 129: 421–432.
39. Tzanakakis ES, Hansen LK, Hu WS (2001) The role of actin filaments and microtubules in hepatocyte spheroid self-assembly. Cell Motil Cytoskeleton 48: 175–189.
40. Powers DE, Millman JR, Bonner-Weir S, Rappel MJ, Colton CK (2010) Accurate control of oxygen level in cells during culture on silicone rubber membranes with application to stem cell differentiation. Biotechnol Prog 26: 805–818.
41. von Smoluchowski M (1917) Versuch einer mathematischen Theorie der Koagulationskinetik kolloider Lösungen. Zeitschrift f Physik Chemie 92: 129–168.
42. Belval TK, Hellums JD (1986) Analysis of shear-induced platelet aggregation with population balance mathematics. Biophys J 50: 479–487.
43. Wright H, Muralidhar R, Tobin T, Ramkrishna D (1990) Inverse Problems of Aggregation Processes. Journal of Statistical Physics 61: 843–863.

12

Loss of Sirt3 Limits Bone Marrow Cell-Mediated Angiogenesis and Cardiac Repair in Post-Myocardial Infarction

Heng Zeng, Lanfang Li, Jian-Xiong Chen*

Department of Pharmacology and Toxicology, University of Mississippi Medical Center, Jackson, Mississippi, United States of America

Abstract

Sirtuin-3 (Sirt3) has a critical role in the regulation of human aging and reactive oxygen species (ROS) formation. A recent study has identified Sirt3 as an essential regulator of stem cell aging. This study investigated whether Sirt3 is necessary for bone marrow cell (BMC)-mediated cardiac repair in post-myocardial infarction (MI). *In vitro*, BMC-derived endothelial progenitor cells (EPCs) from wild type (WT) and Sirt3KO mice were cultured. EPC angiogenesis, ROS formation and apoptosis were assessed. *In vivo*, WT and Sirt3 KO mice were subjected to MI and BMCs from WT and Sirt3 KO mice were injected into ischemic area immediately. The expression of VEGF and VEGFR2 was reduced in Sirt3KO-EPCs. Angiogenic capacities and colony formation were significantly impaired in Sirt3KO-EPCs compared to WT-EPCs. Loss of Sirt3 further enhanced ROS formation and apoptosis in EPCs. Overexpression of Sirt3 or treatment with NADPH oxidase inhibitor apocynin (Apo, 200 and 400 microM) rescued these abnormalities. In post-MI mice, BMC treatment increased number of $Sca1^+$/c-kit^+ cells; enhanced VEGF expression and angiogenesis whereas Sirt3KO-BMC treatment had little effects. BMC treatment also attenuated NADPH oxidase subunits $p47^{phox}$ and $gp91^{phox}$ expression, and significantly reduced ROS formation, apoptosis, fibrosis and hypertrophy in post-MI mice. Sirt3KO-BMC treatment did not display these beneficial effects. In contrast, Sirt3KO mice treated with BMCs from WT mice attenuated myocardial apoptosis, fibrosis and improved cardiac function. Our data demonstrate that Sirt3 is essential for BMC therapy; and loss of Sirt3 limits BMC-mediated angiogenesis and cardiac repair in post-MI.

Editor: Guo-Chang Fan, University of Cincinnati, College of Medicine, United States of America

Funding: This study was supported by grants from the National Institutes of Health (NIH grant HL102042 to J.X. Chen). The funders had no role in study design, data collection and analysis, decision to publish, or preparation of the manuscript.

Competing Interests: The authors have declared that no competing interests exist.

* Email: JChen3@umc.edu

Introduction

Sirtuins belong to a highly conserved family (Sirtuin 1–7) of histone/protein deacetylases and its activity is closely associated with the prolong lifespan of organisms such as yeast, worms and flies as well as mammalian [1]. Sirtuins mediate histone protein post-translational modification by coupling lysine deacetylation to NAD^+ hydrolysis [2,3]. Sirtuins have critical roles in the regulation of various cell functions, including cardiomyocytes and endothelial cells [4–7]. Sirtuins have been shown to involve in biological functions related to cell growth, aging, stress tolerance as well as cell metabolism [1,8]. Accumulating evidence indicates an important role of Sirt3 in the genetic control of human aging. Older individuals have about 40% reduced Sirt3 levels when compared with younger subjects, and the health benefits of older patients are accompanied by the elevation of Sirt3 levels [9]. Further, increased levels of Sirt3 are associated with an extended lifespan of man [10,11]. Calorie restriction has been shown to increase Sirt3 expression and improve cardiovascular function [12]. Moreover, Sirt3 KO mice are resistant to the protective effects of calorie restriction against oxidative damage [13]. In contrast, overexpression of Sirt3 blocks cardiac hypertrophy by suppressing reactive oxygen species (ROS) formation. Knockout of Sirt3 in mice promotes angiotensin II-induced cardiac hypertrophy [14,15].

Myocardial infarction (MI) has been shown to induce rapid mobilization of bone marrow derived cells (BMCs) such as endothelial progenitor cells (EPCs), mesenchymal stromal cells and pluripotent very small embryonic-like cells into circulation and home to sites of ischemic area, and promote neovascularization [16,17]. BMC treatment has been shown to lead to a reduction of infarct size and improvement of cardiac function in post-MI [18]. BMCs have been identified as promising candidates for use as cellular therapies for cardiac regeneration and repair in post-MI. Sirt3 is a key regulator of mitochondria ROS formation and telomere length [14,19]. Recent studies highlight the important role of Sirt3 in BMC-derived stem cells [20,21]. Sirt3 expression is significantly reduced in aged hematopoietic stem cells (HSCs). Moreover, upregulation of Sirt3 expression improves their regenerative capacity in aged HSCs [21]. Consistent with these findings, our recent study also shows that loss of Sirt3 attenuates apelin-overexpressing BMC-mediated improvement of cardiac repair and function in post-MI [20]. However, the direct roles of

Sirt3 in the stem cell therapy-mediated cardiac repair and functional recovery in post-MI remain undefined. In this study, we tested our hypothesis that Sirt3 in the BMCs is essential for the stem cell therapy-mediated angiogenesis and cardiac repair in post-MI.

Using BMCs and EPCs from wild type (WT) mice and Sirt3 knockout (Sirt3KO) mice, this study was to determine: (1) whether loss of Sirt3 in EPCs reduces angiogenic growth factor expression and blunts their proangiogenic and anti-apoptotic capacities; (2) whether loss of Sirt3 in BMCs dampens BMC-mediated angiogenesis and cardiac repair in post-MI mice.

Materials and Methods

Ethics Statement

All procedures conformed to the Institute for Laboratory Animal Research Guide for the Care and Use of Laboratory Animals and were approved by the University of Mississippi Medical Center Animal Care and Use Committee (Protocol ID: 1280). The investigation conforms to the Guide for the Care and Use of Laboratory Animals published by the US National Institutes of Health (NIH Publication No. 85–23, revised 1996).

Surgical procedures

Global Sirt3 knockout mice and wild type control of Sirt3 mice (WT) was purchased from Jackson laboratory (Bar Harbor, ME) and breeding by our laboratory. Male Sirt3KO and WT mice at age 12 weeks were used for the experiments. Male mice were anesthetized with ketamine (100 mg/kg) plus xylazine (15 mg/kg), intubated, and artificially ventilated with room air. Adequate anesthesia was monitored by toe pinch. Myocardial infarction was achieved by ligation of the left anterior descending coronary artery (LAD). Sham controls underwent surgery without the LAD [22,23,24]. After induction of myocardial ischemia (IS), mice were intramyocardial injected with fresh donor bone marrow–derived mononuclear cells (1×10^7cells) immediately [20]. Two weeks after myocardial infarction, mice were sacrificed by cervical dislocation under anesthesia with isoflurane.

Analysis of APJ$^+$/Sca1$^+$/c-kit$^+$ cells in the heart

Heart tissue sections (8 μm) from injected area of ischemia were incubated with Sca1 and c-kit (1:200 Santa Cruz, CA) antibodies overnight. Sca1 was visualized using FITC labeled goat anti-mouse IgG antibodies; c-kit was visualized with Fluorolink Cy3 labeled goat anti-mouse IgG antibodies. Myocardial Sca1$^+$/c-kit$^+$ cells in the injected area were assessed by counting the number of positive cells per 100 nuclei [22,25].

Western analysis of Sirt3, VEGF, VEGFR2, eNOS, Akt, p47^{phox}, gp91^{phox}, CXCR4, beclin-1 and LC3-I/II expression

The hearts or EPCs were harvested and homogenized in lysis buffer for Western blot analysis. Total protein concentrations were determined using a BCA protein assay kit (Pierce Co, IL). Fifteen μg of protein were subjected to SDS-PAGE on 10% polyacrylamide gels and transferred to a nitrocellulose membrane. The blot was probed with Sirt3, VEGFR2, Akt and eNOS (1:1000, Cell Signaling, MA), VEGF, CXCR-4, gp91^{phox}, p47^{phox}, LC3-I/II and beclin-1 (1:1000, Santa Cruz, CA) antibodies. The membranes were then washed and incubated with a secondary antibody coupled to horseradish peroxidase and densitometric analysis was carried out using image acquisition and analysis software (TINA 2.0).

Analysis of myocardial capillary and arteriole densities

Eight-micrometer sections were cut and incubated with fluorescerin-labeled Isolectin B4 (1:200; IB4, Molecular Probe, Invitrogen, OR) and Cy3-conjugated anti-α smooth muscle actin (SMA) (1:100; Sigma). The number of capillaries (IB4-positive EC) in the border zone area was counted and expressed as capillary density per square millimeter (mm^2) of tissue. Myocardial arteriole density was measured using image analysis software (Image J, NIH, MD) [22,24].

Myocardial apoptosis and ROS formation

Heart tissue sections were stained with transferase deoxyuridine nick end labeling (TUNEL) following the manufacturer's instructions (Promega, WI). Apoptosis was indexed by counting TUNEL positive cells per 100 nuclei in the infarcted area [22,23]. ROS formation in the infarcted area was measured and quantified by staining with DHE as previously described [26].

Hemodynamic measurements

Experimental mice were anesthetized with ketamine (100 mg/kg) plus xylazine (15 mg/kg), intubated and artificially ventilated with room air. A 1.4-Fr pressure–conductance catheter (SPR-839, Millar Instrument, TX) was inserted into the left ventricle (LV) to record baseline cardiac hemodynamics of the hearts [22,25].

Heart weight to body weight ratio (HW/BW) and fibrosis

Cardiac hypertrophy was assessed by measuring heart-to-body weight ratio at 14 days post-myocardial ischemia. Cardiac β-myosin heavy chain (β-MHC) (1:1000, abcam, MA) and atrial natriuretic peptide (ANP) (1:1000, Santa Cruz, CA) expression were examined by western blot analysis. Cardiac fibrosis was stained with Masson's trichrome (MT, Sigma, MO) and quantified by measuring the blue fibrotic area [24].

Cultured EPC proliferation, tuber formation and apoptosis

Wild type (WT) mice and Sirt3 knockout mice were sacrificed by cervical dislocation under anesthesia with isoflurane. BM–derived EPC were obtained by flushing the tibias and femurs with 10% FBS EGM. EPC was isolated and cultured from femur and tibia bone marrow of WT and Sirt3KO mice as described previously [22,23]. Two EPC markers, IB4 (1:50 dilute) and CD34 (1:200 dilute), were used for EPC identification by immunohistochemistry. Deficiency of Sirt3 in the EPCs was verified by western blot analysis. For the cell proliferation measurement, EPCs were cultured in 10%FBS EGM for 72 hours. The proliferative capacity of cultured EPCs was assayed using a cell proliferation (MTT) kit according to the manufacturer's instructions (Roche Diagnostic Corp., IN, USA) [27,28]. In the apoptosis study, EPCs apoptosis was induced by exposure of cultured EPCs to serum-free medium for 48 hours. The number of apoptotic cells was then examined by counting TUNEL positive cells per 100 nuclei in cultured EPCs.

Measurement of intracellular ROS formation in cultured EPCs

Intracellular ROS were determined by oxidative conversion of cell permeable chloromethyl-2′,7′-dichlorodihydrofluorescein diacetate (CM-H_2DCFDA, Molecular Probes, OR) to fluorescent dichlorofluorescein (DCF). Briefly, BMCs cultured in 2 well chamber slides were incubated with 10 μM CM-H_2DCFDA in PBS for 30 minutes. DCF fluorescence was measured over the whole field of vision using an EVOS fluorescence microscope connected to an imaging system as previously described [29,30].

BM colony-forming cell assay

BM-derived mononuclear cells were isolated from WT and Sirt3 KO mice. BMCs (105 cells per dish) were then seeded in 2% methylcellulose medium. After 7 days of incubation, BMC colony formation and colony number were scored under phase-contrast microscopy [31].

Statistical analysis

The results were expressed as the mean ± SD. Statistical analysis was performed using one way ANOVA followed by post hoc multiple comparisons test. Significance was set at $P<0.05$.

Results

Loss of Sirt3 in BMCs reduces c-kit$^+$/Sca1$^+$ cells in post-MI mice

We first examined whether Sirt3 expression is altered in the hearts of post-MI mice. As shown in Fig 1A, there was a significant reduction of Sirt3 expression in the hearts of post-MI mice. Interestingly, BMC treatment led to a significant increase in Sirt3 expression in post-MI mice when compared with control post-MI mice (**Fig 1A**). Treatment with Sirt3 KO-BMCs also increased Sirt3 expression in the ischemic hearts of WT mice, but it was significantly less than WT-BMC treatment (Fig 1A).

We next examined whether BMC treatment increases vascular progenitor cells in the infarcted hearts. The number of c-kit$^+$/Sca1$^+$ cells was evaluated at 14 days after BMC intramyocardial injection. As shown in **Fig 1 B and C**, the number of c-kit$^+$/Sca1$^+$ cells was increased in the mouse infarcted heart after 14 days of MI, however, no c-kit$^+$/Sca1$^+$ cell was detected in the non-ischemic sham control mice. BMC treatment significantly increased the number of c-kit$^+$/Sca1$^+$ cells compared to saline treatment (**Fig 1 B and C**). Injection with Sirt3KO-BMCs had a significant lower number of c-kit$^+$/Sca1$^+$ cells in infarcted hearts in comparison with BMC treated mice (**Fig 1 B and C**). To determine whether the increased c-kit$^+$/Sca1$^+$ cells came from the donor or the recipient, mice were intramyocardial injected with GFP$^+$-BMCs or GFP$^+$-Sirt3KO-BMCs. No GFP$^+$-BMCs or GFP$^+$-Sirt3KO-BMCs were found in hearts of post-MI mice after 14 and 28 days of BMC treatment (data not shown), indicating these cells may came from recipient but not from donor.

Loss of Sirt3 increases ROS formation and apoptosis in EPCs

Knockout of Sirt3 in BM-derived HSCs has been reported to lead to a 50% reduction in self- renewal compared to WT mice after serial transplantations [21]. We then examined whether loss of Sirt3 affected EPCs function *in vitro*. Our western blot analysis confirmed that Sirt3 expression was absent in EPCs isolated from Sirt3KO mice (**Fig 2A**). In cultured EPCs, there was a significant increase in ROS formation in Sirt3KO-EPCs when compared with WT-EPCs (**Fig 2 B and C**). Moreover, knockout of Sirt3 in EPCs resulted in a significant increase in stress-induced cell apoptosis. Overexpression of Sirt3 significantly reduced stress-induced EPC apoptosis (**Fig 2 D**). Moreover, treatment of Sirt3KO-EPCs with NADPH oxidase inhibitor apocynin (Apo, 200 and 400 microM) attenuated EPC apoptosis in a dose-dependent manner (**Fig 2E**). Interestingly, autophagy marker LC3-I/II was reduced in Sirt3KO-EPCs. Treatment of Sirt3KO-EPCs with Apo (200 and 400 microM) resulted in an increase in LC3-II levels. Furthermore, overexpression of Sirt3 rescued impairment of LC3-II expression in Sirt3KO-EPCs (**Fig 2F**).

Loss of Sirt3 reduces angiogenic growth factor expression and angiogenesis in EPCs

Similarly, the proliferation of EPCs was significantly reduced in Sirt3KO-EPCs compared with WT-EPCs (**Fig 3A**). Loss of Sirt3 in EPCs resulted in a significant decrease in tube formation when compared with WT-EPCs. In contrast, overexpression of Sirt3 in WT-EPCs significantly enhanced EPC tube formation (**Fig 3B**). EPC colony formation was significantly lower in Sirt3KO-EPCs than WT-EPCs (**Fig 3C**). In addition, VEGF and VEGFR2 levels

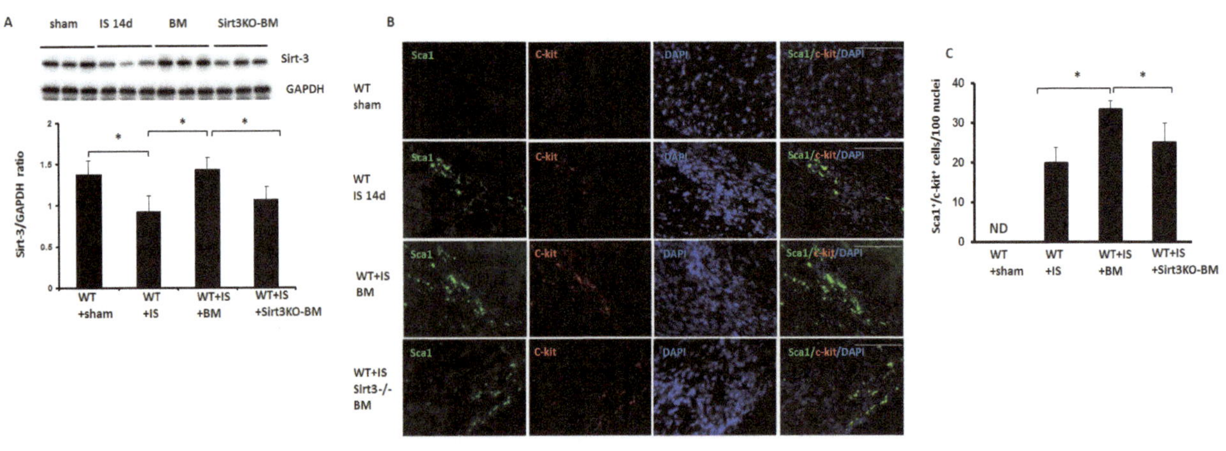

Figure 1. Loss of Sirt3 reduces c-kit$^+$/Sca1$^+$ cell in post MI mice. A. Western blot analysis revealing that Sirt3 expression was significantly reduced in post-MI mice. BMC treatment significantly increased Sirt3 expression compared to post-MI mice. Loss of Sirt3 blunted BMC-mediated upregulation of Sirt3. n = 6 mice, *$p<0.05$. **B.** Immunofluorescence images showing co-localization of Sca1 and c-kit in the border zone of ischemic mouse hearts. Sca1 was stained with mouse Sca1 antibody (green, 10X). c-kit was stained with rabbit c-kit antibody (red, 10X) and nuclei were stained with DAPI (blue, 10X). **C.** Quantitative analysis of Sca1$^+$/c-kit$^+$ cells demonstrating that the number of Sca1$^+$/c-kit$^+$ cells was increased at 14 days of post-MI mice. BMCs significantly increased Sca1$^+$/c-kit$^+$ cells compared to saline treatment. Treatment with Sirt3KO-BMC had a significant lower number Sca1$^+$/c-kit$^+$ cells in infracted heart compared to the BMC treated mice. All data represent mean ± SD; n = 5, *$p<0.05$. ND = not detected.

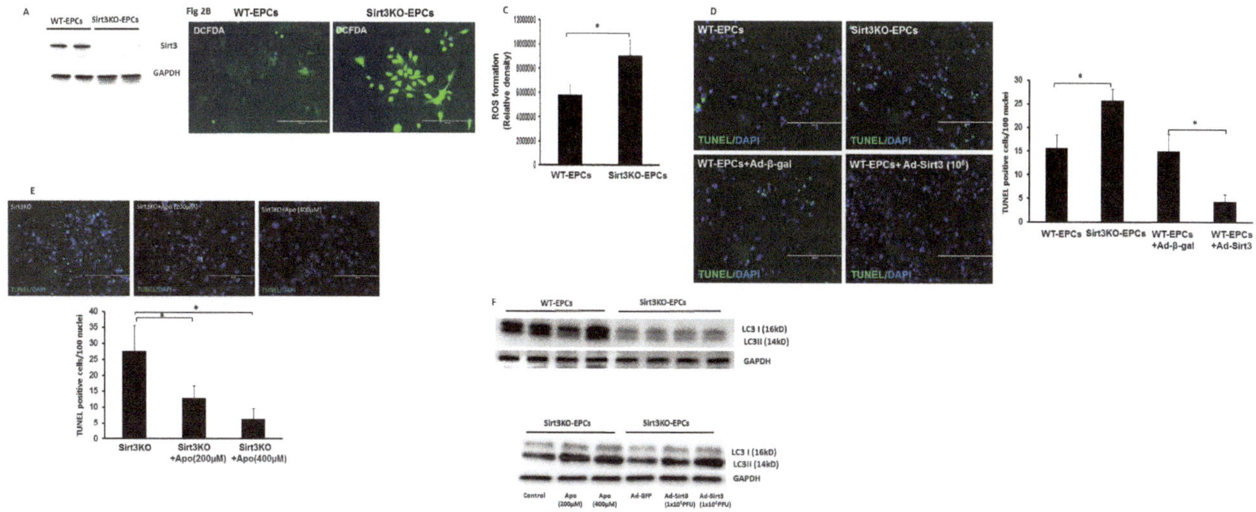

Figure 2. Loss of Sirt3 increases ROS formation and apoptosis in EPCs. A. Western blot analysis showing that Sirt3 expression was absent in EPCs isolated from Sirt3KO mice. n = 2 mice. **B and C.** Representative images and quantification of intracellular ROS formation measured by CM-H_2DCFDA staining in cultured EPCs. ROS formation was significantly increased in cultured EPCs of Sirt3KO mice when compared with that of WT mice (n = 4 mice, *p<0.05). **D and E.** Representative images and quantification of TUNEL positive cell in cultured EPCs. Apoptotic cells were identified by TUNEL staining (green, 20x). Cell apoptosis was significantly increased in cultured EPCs of Sirt3KO mice compared to that of WT mice. Infection of WT-EPC with Ad-Sirt3 (10^6 PFU) significantly reduced EPC apoptosis (n = 6 mice, *p<0.05). Treatment of Sirt3KO-EPC with NADPH oxidase inhibitor apocynin (Apo, 200 and 400 μM significantly reduced EPC apoptosis (n = 6 mice, *p<0.05). **F.** Western blot analysis showing that the basal levels of autophagy gene LC3 II expression were dramatic reduced in the Sirt3KO-EPCs when compared with WT-EPCs. Treatment of Sirt3KO-EPCs with Apo 200 and 400 μM or infection of Sirt3KO-EPCs with Ad-Sirt3 resulted in an increase in LC3 II expression (n = 3 mice).

were reduced in Sirt3KO-EPCs; whereas treatment with Apo (200 and 400 microM) or overexpression of Sirt3 rescued impaired VEGF/VEGFR2 expression in Sirt3KO-EPCs (**Fig 3D and E**). Intriguingly, basal levels of CXCR-4, an EPC recruitment factor, were significantly decreased in Sirt3KO-EPCs when compared with WT-EPCs (**Fig 3F**).

Loss of Sirt3 in BMCs increases $p47^{phox}$ and $gp91^{phox}$ expression in the heart of post-MI

To determine whether Sirt3 is involved in BMC-mediated suppression of ROS formation, NADPH oxidase subunits $p47^{phox}$ and $gp91^{phox}$ expression was examined in the hearts of post-MI mice. BMC treatment led to a significant reduction of NADPH oxidase subunits $p47^{phox}$ and $gp91^{phox}$ expression in ischemic hearts (**Fig 4 A and B**). This was accompanied by a significant reduction of ROS formation in post-MI mice (**Fig 4 C**). In contrast, treatment with Sirt3KO-BMCs did not inhibit $p47^{phox}$ and $gp91^{phox}$ expression in post-MI mice (**Fig 4 A and B**). ROS formation was significantly elevated in Sirt3KO-BMCs + MI mice when compared to WT-BMCs + MI mice (**Fig 4C**).

Loss of Sirt3 in BMCs impairs angiogenesis in post-MI

To elucidate whether Sirt3 in the BMCs is necessary for the myocardial angiogenesis, VEGF, capillary and arteriole densities were examined in post-MI mice. BMC treatment significantly increased VEGF expression in the hearts of post-MI mice. VEGF expression was significantly decreased in Sirt3KO-BMCs + MI mice compared with WT-BMCs + MI mice (**Fig 5A**). BMC treatment further significantly increased capillary and arteriole densities in the border zone of ischemic hearts. However, myocardial capillary and arteriole densities were not significantly increased in Sirt3KO-BMCs + MI mice when compared with post-MI mice (**Fig 5 B-E**).

Sirt3 is necessary for BMC-mediated anti-apoptosis in ischemic hearts

To further elucidate whether Sirt3 in the BMCs is also necessary for the cardiac repair of BMC therapy, myocardial autophagy gene expression, apoptosis, fibrosis and hypertrophy were examined in post-MI mice. Treatment with BMCs led to a significant increase in autophagy gene beclin-1 expression and elevation of LC3-II/I ratio in post-MI mice (**Fig 6A and B**). Beclin-1 expression and LC3-II/I ratio was significantly decreased in Sirt3KO-BMCs+MI mice when compared with WT-BMCs + MI mice. BMC treatment significantly elevated phosphorylation levels of Akt and eNOS in post-MI mice (**Fig 6C and D**). The phosphorylation levels of Akt and eNOS were significantly reduced in Sirt3KO-BMCs + MI mice compared to BMCs+ MI mice. BMC treatment furthermore significantly reduced cardiac apoptosis, but Sirt3KO-BMC treatment did not affect cell apoptosis in post-MI mice (**Fig 6 E and F**).

Loss of Sirt3 in BMCs limits BMC-mediated cardiac repair and functional recovery

In comparison with control post-MI mice, the HW/BW ratio and expression of hypertrophic gene β-MHC and ANP were significantly reduced in the BMCs + MI mice (**Fig 7 A, B, C**). Sirt3KO-BMC treatment did not suppress cardiac hypertrophy compared with BMC treatment (**Fig 7 A, B, C**). BMC treatment also significantly suppressed cardiac fibrosis formation whereas Sirt3KO-BMC treatment had little effects (**Fig 7D**).

After 28 days of myocardial ischemia, post-MI mice exhibited a significant elevation of end-systolic volume (ESV) and reduction of end-systolic pressure (ESP) (**Fig 7E**). The post-MI mice also showed a significant decline in +dp/dtmax and -dp/dtmin pressures compared to non-ischemic sham controls (**Fig 7F**). BMC treatment resulted in a significant decrease in ESV and a dramatic improvement of ESP, +dp/dtmax and -dp/dtmin

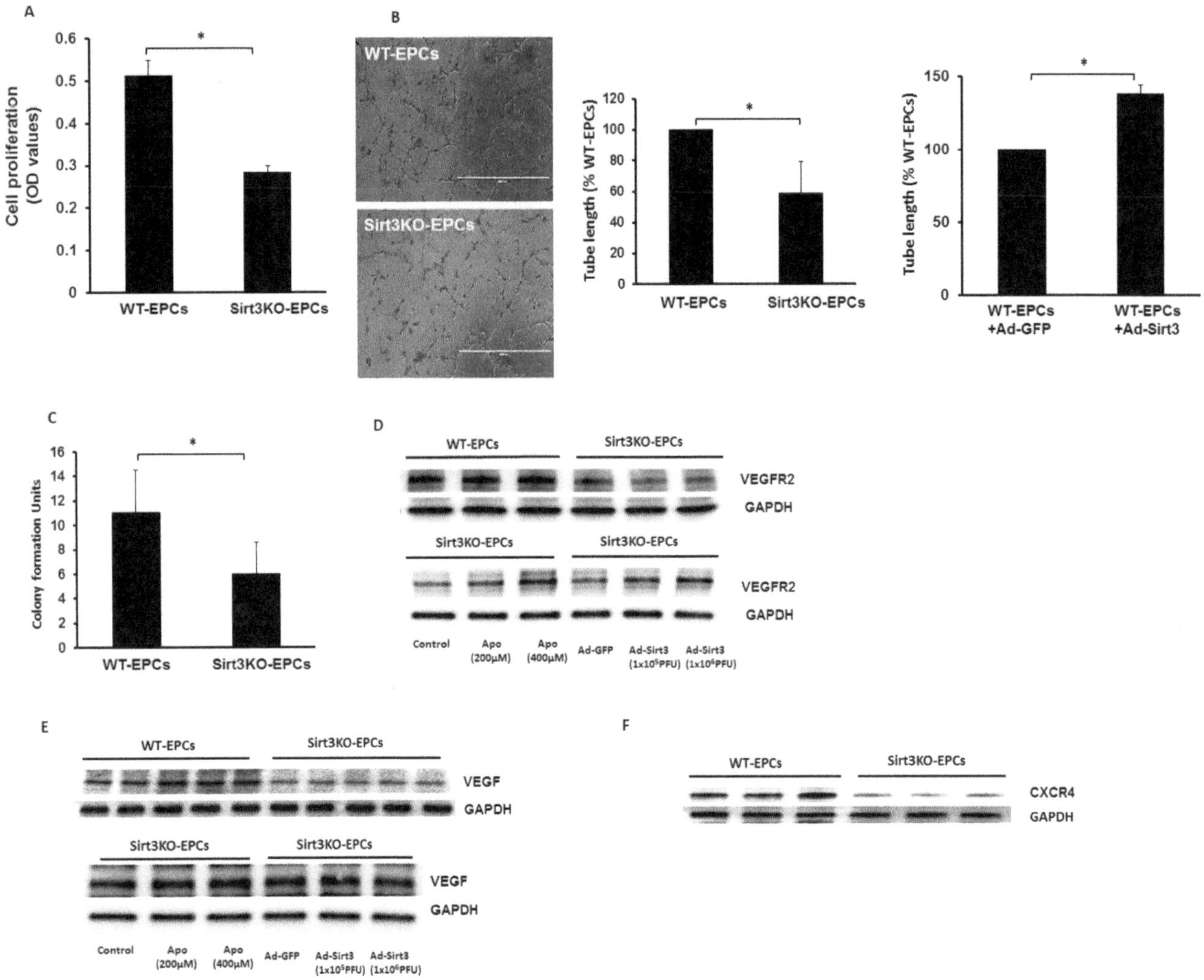

Figure 3. Loss of Sirt3 impairs VEGF/VEGFR2 expression and angiogenesis in EPCs. A. Cell proliferation was measured by MTT assay. The proliferative rate of EPCs was significantly reduced in cultured EPCs of Sirt3KO mice compared to that of WT mice (n = 4 mice, *$p < 0.05$). **B**. EPC tube formation was significantly reduced in EPCs lack of Sirt3 when compared with control EPCs. Overexpression of Sirt3 significantly increased EPC tube formation (n = 4–6 mice, *$p < 0.05$). **C**. BMC colony formation units. EPC colony formation was significantly reduced in Sirt3KO-EPCs when compared with WT-EPCs (n = 6 mice, *$p < 0.05$). **D and E**. Western blot analysis showing that the basal levels of VEGF and VEGFR2 were significantly reduced in Sirt3KO-EPCs (n = 3 mice). Treatment of Sirt3KO-EPCs with NADPH oxidase inhibitor apocynin 200 and 400 μM or infection of Sirt3KO-EPCs with Ad-Sirt3 increased levels of VEGF and VEGFR2 expression (n = 3 mice). **F**. Western blot analysis showing that the basal levels of CXCR-4 expression were dramatic reduced in the Sirt3KO-EPCs (n = 3 mice).

pressures in post-MI mice. Sirt3KO-BMC treatment had little effects on the improvement of these parameters when compared with WT-BMC treatment (**Fig 7E and F**). In contrast, treatment of Sirt3KO post-MI mice with WT-BMCs resulted in a significant reduction of cardiac apoptosis and cardiac fibrosis formation (**Fig 7 G and H**). This was accompanied by a significant improvement of cardiac function in Sirt3KO post-MI mice (**Fig 7I**).

Discussion

Our present study demonstrates that loss of Sirt3 in EPCs reduced angiogenic growth factor expression and angiogenic capacity. Loss of Sirt3 in EPCs increased ROS formation and promoted cell apoptosis *in vitro*. Furthermore, loss of Sirt3 in BMCs abolished BMC therapy mediated protective effects and limited cardiac repair in post-MI mice. Our study suggests that Sirt3 in BMCs is necessary for the protective effects of stem cell therapy in post-MI.

Sirt3 has been reported to be a major mitochondrial deacetylase in human [2,32,33]. Previous studies show that Sirt3 exists in the mitochondria of the heart [34,35]. Our recent study indicates a critical role of Sirt3 in apelin-overexpressing BMC-mediated improvement of angiogenesis and cardiac function in post-MI mice [20]. In present study, we show that treatment with BMCs resulted in a significant increase in Sirt3 expression in post-MI mice. We then further investigated if BMC treatment improved $Sca1^+/c\text{-}kit^+$ progenitor cells in ischemic hearts. Our data demonstrated that the number of $Sca1^+/c\text{-}kit^+$ progenitor cells in ischemic hearts was increased at 14 days of post-MI. Injection of BMCs significantly increased the number of $Sca1^+/c\text{-}kit^+$ cells and promoted cardiac repair at ischemic area in post-MI mice. Intriguingly, the number of $Sca1^+/c\text{-}kit^+$ cells was significantly reduced in Sirt3KO-BMC treatment. This was accompanied by a

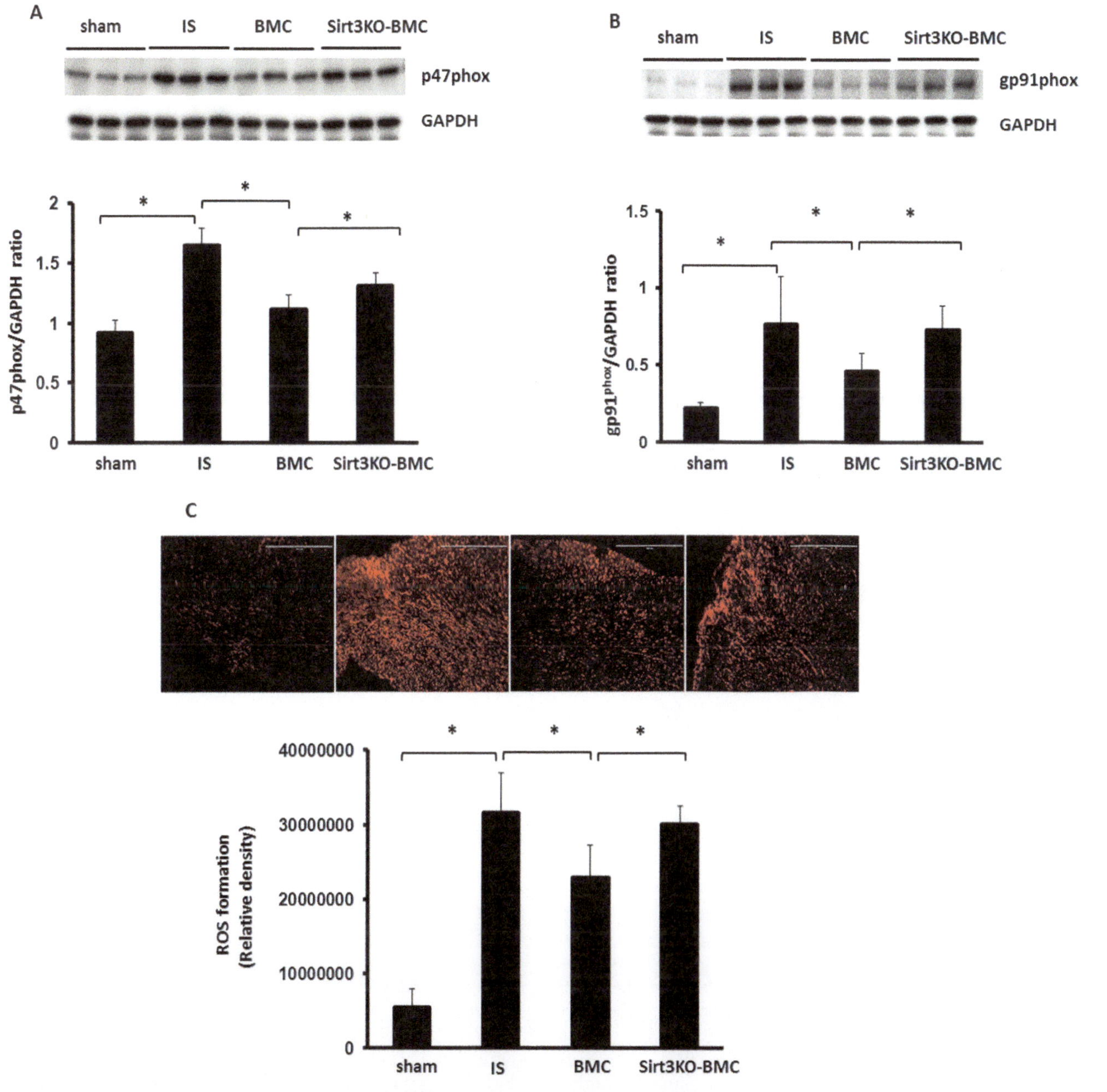

Figure 4. Loss of Sirt3 in BMCs increases ROS formation in post-MI mice. A and B. Western blot analysis revealing that BMC treatment resulted in a significant downregulation of p47^{phox} (A) and gp91^{phox} (B) expression compared to post-MI mice. Sirt3KO-BMC treatment did not inhibit expression of p47^{phox} and gp91^{phox} compared to BMC treatment. n = 6 mice, *p<0.05. **C.** Quantitative analysis of ROS formation by DHE staining revealing that BMC treatment resulted in a significant suppression of ROS formation compared to post-MI mice. Myocardial ROS formation was significantly higher in Sirt3KO-BMC treated mice than BMC treated mice. n = 5 mice; *p<0.05.

significant decline of cardiac function in post-MI mice. A recent study demonstrated that there was no difference in the number of BM derived HSCs (Lin$^-$, c-kit$^+$, sca1$^+$) between WT and Sirt3 KO mice [21]. Consistent with this study, we did not found any GFP$^+$-BMCs or GFP$^+$-Sirt3KO-BMCs in hearts of post-MI mice after 14 and 28 days of BMC treatment, suggesting that injected BMCs had not differentiated into Sca1$^+$/c-kit$^+$ cells in ischemic hearts and Sca1$^+$/c-kit$^+$ cells was not coming from injected BMCs. CXCR-4 has been identified as a key mediator that regulates vascular progenitor cell homing into the ischemic area and contributes to improvement of cardiac function after MI [36]. CXCR-4 has been shown to protect the heart after myocardial infarction via promoting stem cell recruitment [36–38]. We therefore speculated that these increased Sca1$^+$/c-kit$^+$ cells may be recruited to ischemic area due to releasing CXCR-4 after BMC treatment. This notion was confirmed by our data that the basic levels of CXCR-4 were significantly reduced in Sirt3KO-EPCs. Our data suggest that impairment of CXCR-4 expression in EPCs and reduction of number of Sca1$^+$/c-kit$^+$ stem cell in infarcted area may be responsible for impairment of cardiac repair in Sirt3KO-BMC treatment in post-MI mice.

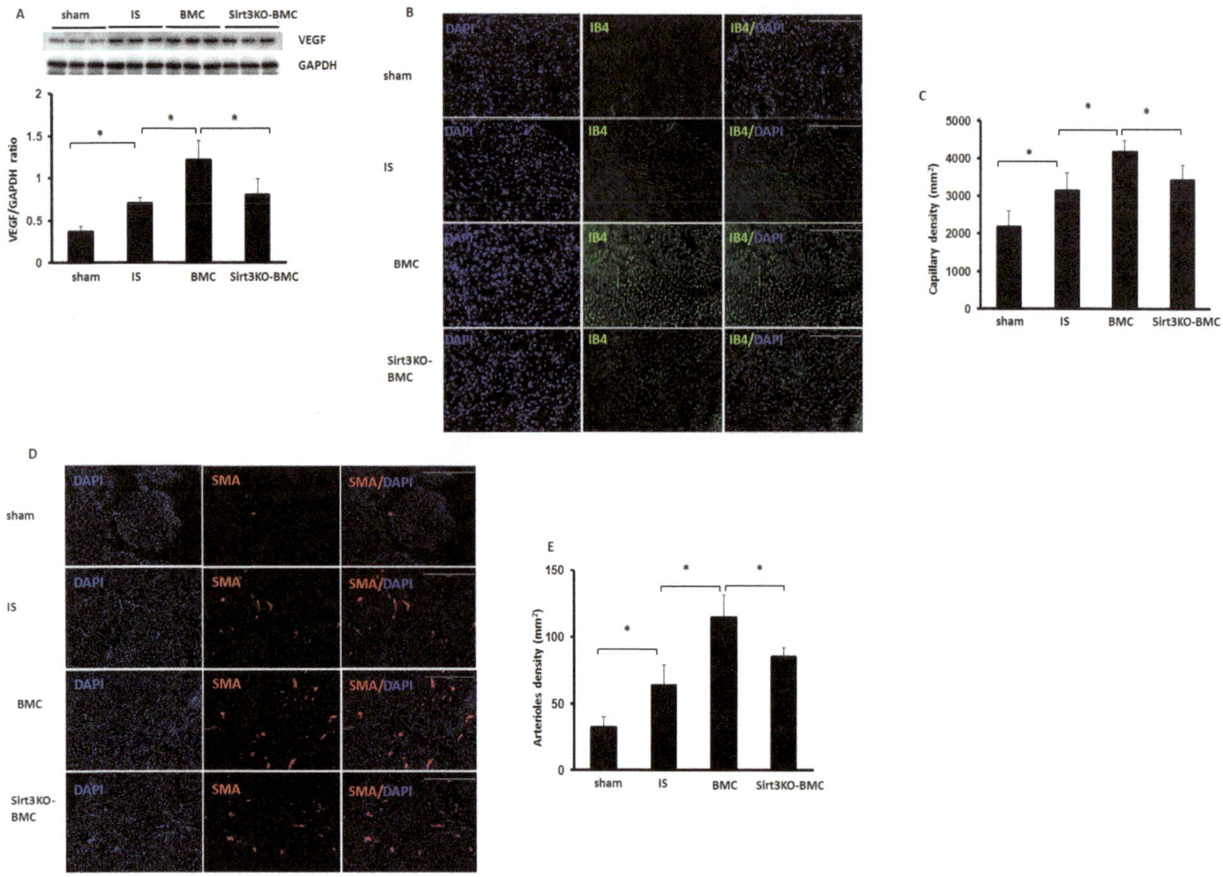

Figure 5. Loss of Sirt3 blunts BMC-induced VEGF expression and angiogenesis. A. Western blot analysis showing that VEGF expression was significantly reduced in Sirt3KO-BMC treated mice compared to BMC treated mice. n = 6 mice; $^*p<0.05$. **B and C**. Myocardial ischemia significantly increased myocardial capillary density by IB4 staining (green, 10x). BMC treatment significantly increased capillary formation compared to post-MI mice. Myocardial capillary density was significant reduced in Sirt3KO-BMC treated mice compared to BMC treated mice. n = 5 mice; $^*p<0.05$. **D and E**. BMC treatment significantly increased myocardial arteriole density by SMA staining (Red, 10x) in post-MI mice. Myocardial arteriole density was significantly decreased in post-MI mice treatment with Sirt3KO-BMCs compared to mice treated with BMCs. n = 5 mice; $^*p<0.05$.

The present study provides evidence that basal Sirt3 activity in EPC is required for the protective effects afforded by BMC therapy in post-MI. Myocardial ischemia has been shown to induce rapid mobilization of bone marrow derived vascular progenitor cells from the bone marrow niches [39,40]. BMCs are recruited into the sites of infarcted area and promote cardiac repair in the infarcted hearts [16,17]. BMCs have been shown to promote cardiac repair and improve functional recovery of post-MI; however, the molecular mechanisms by which BMCs promotes cardiac repair are incompletely understood. Our data, for the first time, showed that Sirt3 levels were reduced in ischemic hearts. Recently studies have directly linked loss of Sirt3 contributing to the ROS formation. Sirt3-deficient cells subjected to metabolic stress lead to a significant increase in ROS formation [41]. Furthermore, cardiomyocytes cultured from Sirt3KO mice show increased ROS production. Sirt3 protects cardiomyocytes from oxidative stress-mediated cell death [15]. Therefore, reduction of Sirt3 levels in the ischemic hearts may contribute to increased ROS formation and apoptosis. In line with these studies, our present data also showed that loss of Sirt3 in EPCs enhanced ROS formation and increased stress-induced cell apoptosis. Furthermore, treatment with NADPH oxidase inhibitor reduced Sirt3KO-EPC apoptosis *in vitro*. Overexpression of Sirt3 protected EPCs against stress-induced cell apoptosis. Our data further showed that intramyocardial delivery of BMCs in infarcted area led to a significant increase in Sirt3 expression. Moreover, treatment with BMCs resulted in a significant reduction of NADPH oxidase $p47^{phox}$ and $gp91^{phox}$ expression and ROS formation. In contrast, loss of Sirt3 in BMCs significantly blunted BMC therapy mediated upregulation of Sirt3 and suppression of ROS formation. This was accompanied by a significant increase in myocardial apoptosis in post-MI mice. Taken together, our data suggest that a critical role of Sirt3 in the regulation of ROS formation and apoptosis in the ischemic heart and that increased ROS formation and apoptosis in Sirt3KO-EPCs maybe also contribute, at least in part, to the failure of Sirt3KO-BMC treatment in post-MI.

Aging bone marrow cells have been shown to fail to promote cardiac angiogenesis and improve cardiac function [42]. Both experimental and clinical studies demonstrated that aging interferes with bone marrow derived progenitor cell functions [43–45]. ROS formation is increased in HSCs with age. This is accompanied by an increased HSCs apoptosis and impairment of self-renewal capacity. Sirt3 expression is significantly reduced with the aging in skeletal muscle [9]. A recent study also shows a significant reduction of Sirt3 levels with age or stress in HSCs. Overexpression of Sirt3 in HSCs from old mice improves regenerative capacity of aged HSCs [21]. In this study, we

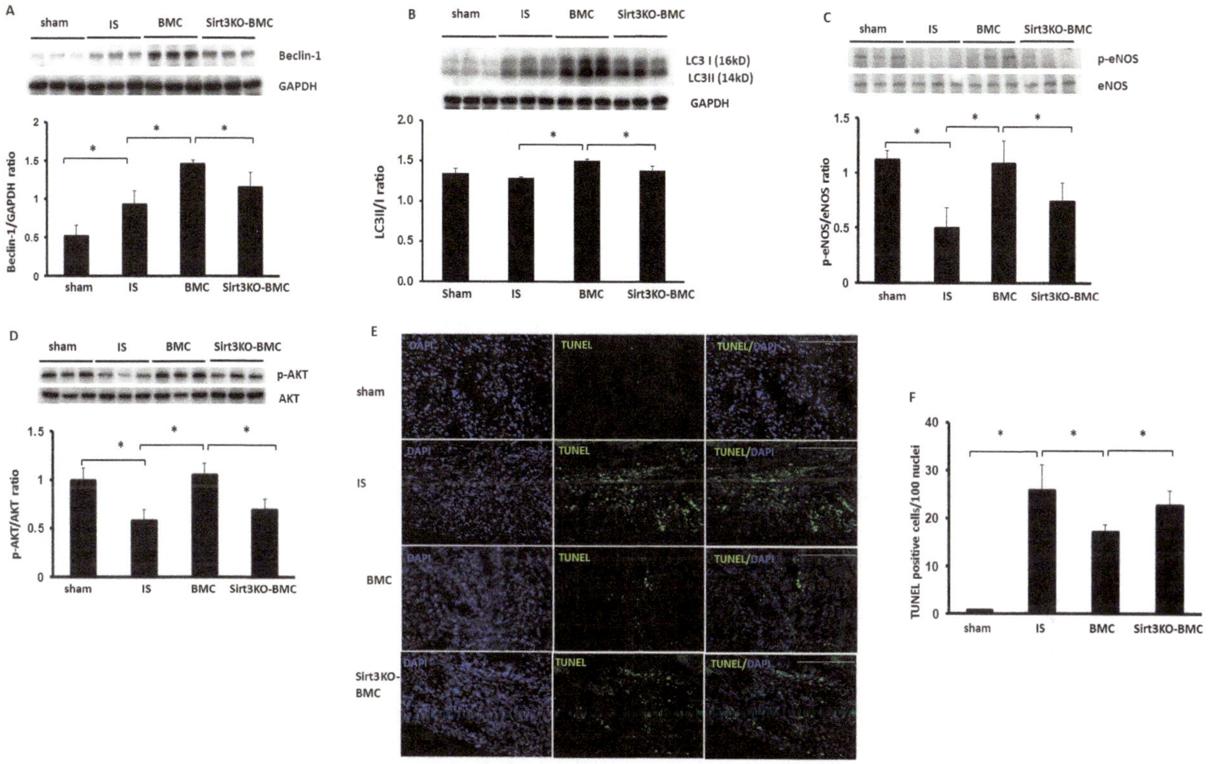

Figure 6. Loss of Sirt3 in BMCs increases apoptosis in post-MI mice. A and B. Western blot analysis demonstrating that treatment of post-MI mice with BMCs resulted in a significant increase in autophagy gene beclin-1 expression (A) and LC3-II/I ratio (B). Beclin-1 expression and LC3-II/I ratio were significantly decreased in Sirt3KO-BMC treated mice compared to BMC treated post-MI mice. n = 6 mice; $*p<0.05$. **C and D**. Western blot analysis demonstrating that BMC treatment significantly increased phosphorylation levels of eNOS (C) and Akt (D) in post-MI mice. The phosphorylation levels of Akt and eNOS were significantly decreased in Sirt3KO-BMC treated mice compared to BMC treated mice. n = 6 mice; $*p<0.05$. **E and F**. Apoptotic cells in the infarcted area of the left ventricle were identified by TUNEL staining (green, 10x). Treatment of post-MI mice with BMCs significantly decreased TUNEL$^+$ cells in ischemic area. TUNEL$^+$ cells were significantly increased in Sirt3KO-BMC treated mice when compared with BMC treated mice. n = 5 mice; $*p<0.05$.

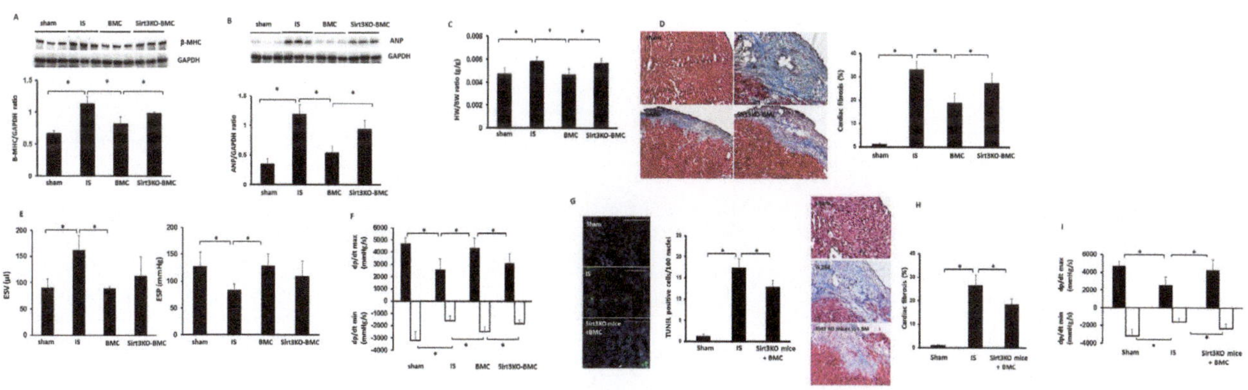

Figure 7. Loss of Sirt3 abolished BMC-mediated cardiac repair in post-MI mice. A and B. Western blot analysis showing that treatment of post-MI mice with BMCs significantly reduced hypertrophic marker β-MHC (A) and ANP (B) expression. Treatment of post-MI mice with Sirt3KO-BMC failed to suppression of β-MHC and ANP expression compared to BMC treated mice. n = 6 mice, $*p<0.05$. **C**. BMC treatment significantly reduced HW/BW ratio in post-MI mice. Treatment with Sirt3KO-BMCs failed to significant reduction of HW/BW ratio compared to BMC treated mice. n = 6 mice, $*p<0.05$. **D**. Representative images of cardiac fibrosis in the infarction zone and quantitative analysis of fibrotic area in mice (Masson's trichrome). BMC treatment significantly reduced the area of cardiac fibrosis. Sirt3KO-BMC treatment significantly increased cardiac fibrosis area compared to BMC treated mice. n = 5 mice; $*p<0.05$. **E**. The end-systolic volume (ESV) was significantly increased in post-MI mice. BMC treatment significantly decreased ESV. The end-systolic pressure (ESP) was decreased in post-MI mice. Treatment with BMCs significantly increased ESP whereas treatment of post-MI mice with Sirt3KO-BMC had little effects on ESV and ESP. n = 5–7 mice, $*p<0.05$. **F**. BMC therapy led to a significant improvement of maximum +dP/dt and minimum -dP/dt pressures compared to control post-MI mice. Sirt3KO-BMC treatment failed to improve maximum +dP/dt and minimum -dP/dt pressures in post-MI mice. n = 5–7 mice, $*p<0.05$. **G**. Treatment of Sirt3KO post-MI mice with WT-BMCs significantly reduced TUNEL$^+$ cells in ischemic area. Apoptotic cells in the infarcted area of the left ventricle were identified by TUNEL staining (green, 10x). n = 6 mice; $*p<0.05$. **H**. Treatment of Sirt3KO post-MI mice with WT-BMCs significantly reduced the area of cardiac fibrosis (Masson's trichrome). n = 5 mice; $*p<0.05$. **I**. Treatment of Sirt3KO post-MI mice with WT-BMCs significantly improved maximum +dP/dt and minimum -dP/dt pressures in post-MI mice. n = 5–6 mice, $*p<0.05$.

hypothesized that loss of Sirt3 in BM stem cells; similar as aged BM derived HSCs, fails to improve angiogenesis and cardiac repair in post-MI. To substantiate this notion, we first compared the expression of angiogenic growth factor and angiogenesis between WT-EPCs and Sirt3KO-EPCs *in vitro*. Our data showed that loss of Sirt3 in EPCs reduced VEGF and VEGFR2 expression. Moreover, treatment with NADPH oxidase inhibitor or overexpression of Sirt3 rescued impaired VEGF and VEGFR2 expression. In addition, the basal proliferation and angiogenic capacities were significantly reduced in Sirt3KO-EPCs. Our study *in vivo* further confirmed that BMC treatment increased VEGF expression and elevated phosphorylation levels of eNOS and Akt. This was accompanied by increased myocardial vascular densities and improved cardiac function in post-MI mice. In contrast, knockout of Sirt3 in BMCs reduced BMC-mediated VEGF expression and neovascularization. Furthermore, loss of Sirt3 in BMCs abolished BMC-mediated cardiac repair and improvement of cardiac function in post-MI mice. These findings indicate that lack of Sirt3 and increased ROS formation in aged EPCs maybe contribute to the failure of aged BMC treatment in post-MI.

Autophagy is a dynamic process of intracellular bulk degradation in which cytosolic proteins and organelles are fused with lysosomes for degradation. Under stressed conditions, autophagy selectively removes damaged mitochondria which prevent activation of apoptotic machinery [46,47]. Overexpression of autophagy gene beclin-1 has been shown to protect cardiac myocyte against ischemia/reperfusion injury [48]. Autophagy has been shown to have a protective role in the heart following myocardial ischemia/reperfusion *in vivo* [49–52]. Our previous study indicates that Sirt3 is necessary for apelin-BMC therapy-mediated upregulation of autophagy. Loss of Sirt3 attenuates apelin-induced autophagy gene marker beclin-1 and LC3-I/II expression [20]. In the present study, we demonstrate that treatment with BMCs led to a significant increase in autophagy gene beclin-1 expression and elevation of LC3-II/I ratio. These are accompanied by a dramatic reduction of cardiac apoptosis. However, treatment with Sirt3KO-BMC had no effect on LC3-II/I ratio and Beclin-1 expression in post-MI mice. This was associated with a significant higher number of apoptotic cells in ischemic hearts. Our studies revealed a novel molecular mechanism of BMC stem cell therapy which stem cells may attenuate apoptosis via regulation of autophagy in post-MI. Autophagy also has been shown to promote stem cell generation and differentiation. Inhibition of autophagy reduces stem cell self-renewal and differentiation [53–57]. A recent study further underscores the important role of autophagy in EPC survival, proliferation and differentiation. Inhibition of autophagy reduces proliferation and differentiation of EPCs. In contrast, increasing autophagy enhances EPC survival under hypoxic conditions [53]. Our data showed that loss of Sirt3 reduced autophagic gene LC3-II levels whereas overexpression of Sirt3 or treated with NADPH oxidase inhibitor increased LC3-II levels in EPCs. Overexpression of Sirt3 further attenuated EPCs apoptosis. These data suggest that reduction of autophagy may be contributed to the higher apoptosis of Sirt3KO-EPCs. Although Sirt3 has been shown to rejuvenate HSCs [21], so far, it remains unanswered what is necessary for Sirt3 to complete its rejuvenation. In addition, if Sirt3-induced HSCs rejuvenation requires to removing additional damaged organelles such as mitochondria via regulation of autophagy remains unknown. Further studies are warranted to elucidate the molecular mechanisms by which Sirt3 regulates autophagy in stem cell rejuvenation.

In summary, the current study provides evidence that basal levels of Sirt3 in stem cells contribute the therapeutic effects of BMCs in post-MI mice. Since the levels of Sirt3 were reduced in aging and aged stem cells, our findings implicate that reduced levels of Sirt3 may contribute to the failure of BMC therapy in aging patients. Our findings further suggest that augmentation of Sirt3 activity in stem cells may represent a novel therapeutic approach for the improvement of stem cell therapy for the ischemic heart diseases.

Author Contributions

Conceived and designed the experiments: JXC. Performed the experiments: HZ LL. Analyzed the data: HZ LL. Contributed reagents/materials/analysis tools: JXC. Contributed to the writing of the manuscript: JXC.

References

1. Tanno M, Kuno A, Horio Y, Miura T (2012) Emerging beneficial roles of sirtuins in heart failure. Basic Res Cardiol 107: 273. 10.1007/s00395-012-0273-5 [doi].
2. Pillai VB, Sundaresan NR, Jeevanandam V, Gupta MP (2010) Mitochondrial SIRT3 and heart disease. Cardiovasc Res 88: 250–256. cvq250 [pii];10.1093/cvr/cvq250 [doi].
3. Tanno M, Sakamoto J, Miura T, Shimamoto K, Horio Y (2007) Nucleocytoplasmic shuttling of the NAD+-dependent histone deacetylase SIRT1. J Biol Chem 282: 6823–6832. M609554200 [pii];10.1074/jbc.M609554200 [doi].
4. Becatti M, Taddei N, Cecchi C, Nassi N, Nassi PA, et al (2012) SIRT1 modulates MAPK pathways in ischemic-reperfused cardiomyocytes. Cell Mol Life Sci 69: 2245–2260. 10.1007/s00018-012-0925-5 [doi].
5. Donato AJ, Magerko KA, Lawson BR, Durrant JR, Lesniewski LA, et al (2011) SIRT-1 and vascular endothelial dysfunction with ageing in mice and humans. J Physiol 589: 4545–4554. jphysiol.2011.211219 [pii];10.1113/jphysiol.2011.211219 [doi].
6. Sundaresan NR, Pillai VB, Gupta MP (2011) Emerging roles of SIRT1 deacetylase in regulating cardiomyocyte survival and hypertrophy. J Mol Cell Cardiol 51: 614–618. S0022-2828(11)00029-0 [pii];10.1016/j.yjmcc.2011.01.008 [doi].
7. Zu Y, Liu L, Lee MY, Xu C, Liang Y, et al (2010) SIRT1 promotes proliferation and prevents senescence through targeting LKB1 in primary porcine aortic endothelial cells. Circ Res 106: 1384–1393. CIRCRESAHA.109.215483 [pii];10.1161/CIRCRESAHA.109.215483 [doi].
8. Vinciguerra M, Santini MP, Martinez C, Pazienza V, Claycomb WC, et al (2012) mIGF-1/JNK1/SirT1 signaling confers protection against oxidative stress in the heart. Aging Cell 11: 139–149. 10.1111/j.1474-9726.2011.00766.x [doi].
9. Lanza IR, Short DK, Short KR, Raghavakaimal S, Basu R, et al (2008) Endurance exercise as a countermeasure for aging. Diabetes 57: 2933–2942. db08-0349 [pii];10.2337/db08-0349 [doi].
10. Bellizzi D, Rose G, Cavalcante P, Covello G, Dato S, et al (2005) A novel VNTR enhancer within the SIRT3 gene, a human homologue of SIR2, is associated with survival at oldest ages. Genomics 85: 258–263. S0888-7543(04)00308-8 [pii];10.1016/j.ygeno.2004.11.003 [doi].
11. Rose G, Dato S, Altomare K, Bellizzi D, Garasto S, et al (2003) Variability of the SIRT3 gene, human silent information regulator Sir2 homologue, and survivorship in the elderly. Exp Gerontol 38: 1065–1070. S0531556503002092 [pii].
12. Palacios OM, Carmona JJ, Michan S, Chen KY, Manabe Y, et al (2009) Diet and exercise signals regulate SIRT3 and activate AMPK and PGC-1alpha in skeletal muscle. Aging (Albany NY) 1: 771–783.
13. Someya S, Yu W, Hallows WC, Xu J, Vann JM, et al (2010) Sirt3 mediates reduction of oxidative damage and prevention of age-related hearing loss under caloric restriction. Cell 143: 802–812. S0092-8674(10)01138-4 [pii];10.1016/j.cell.2010.10.002 [doi].
14. Pillai VB, Sundaresan NR, Kim G, Gupta M, Rajamohan SB, et al (2010) Exogenous NAD blocks cardiac hypertrophic response via activation of the SIRT3-LKB1-AMP-activated kinase pathway. J Biol Chem 285: 3133–3144. M109.077271 [pii];10.1074/jbc.M109.077271 [doi].
15. Sundaresan NR, Gupta M, Kim G, Rajamohan SB, Isbatan A, et al (2009) Sirt3 blocks the cardiac hypertrophic response by augmenting Foxo3a-dependent antioxidant defense mechanisms in mice. J Clin Invest 119: 2758–2771. 39162 [pii];10.1172/JCI39162 [doi].
16. Isner JM, Asahara T (1999) Angiogenesis and vasculogenesis as therapeutic strategies for postnatal neovascularization. J Clin Invest 103: 1231–1236.

17. Yoon YS, Johnson IA, Park JS, Diaz L, Losordo DW (2004) Therapeutic myocardial angiogenesis with vascular endothelial growth factors. Mol Cell Biochem 264: 63–74.
18. Schachinger V, Assmus B, Britten MB, Honold J, Lehmann R, et al (2004) Transplantation of progenitor cells and regeneration enhancement in acute myocardial infarction: final one-year results of the TOPCARE-AMI Trial. J Am Coll Cardiol 44: 1690–1699.
19. Sack MN (2012) The role of SIRT3 in mitochondrial homeostasis and cardiac adaptation to hypertrophy and aging. J Mol Cell Cardiol 52: 520–525. S0022-2828(11)00470-6 [pii];10.1016/j.yjmcc.2011.11.004 [doi].
20. Li L, Zeng H, Hou X, He X, Chen JX (2013) Myocardial Injection of Apelin-Overexpressing Bone Marrow Cells Improves Cardiac Repair via Upregulation of Sirt3 after Myocardial Infarction. PLoS One 8: e71041. 10.1371/journal.pone.0071041 [doi]; PONE-D-13-05351 [pii].
21. Brown K, Xie S, Qiu X, Mohrin M, Shin J, et al (2013) SIRT3 reverses aging-associated degeneration. Cell Rep 3: 319–327. S2211-1247(13)00012-0 [pii];10.1016/j.celrep.2013.01.005 [doi].
22. Li L, Zeng H, Chen JX (2012) Apelin-13 increases myocardial progenitor cells and improves repair postmyocardial infarction. Am J Physiol Heart Circ Physiol 303: H605–H618. ajpheart.00366.2012 [pii];10.1152/ajpheart.00366.2012 [doi].
23. Tuo QH, Zeng H, Stinnett A, Yu H, Aschner JL, et al (2008) Critical role of angiopoietins/Tie-2 in hyperglycemic exacerbation of myocardial infarction and impaired angiogenesis. Am J Physiol Heart Circ Physiol 294: H2547–H2557.
24. Chen JX, Stinnett A (2008) Ang-1 gene therapy inhibits hypoxia-inducible factor-1alpha (HIF-1alpha)-prolyl-4-hydroxylase-2, stabilizes HIF-1alpha expression, and normalizes immature vasculature in db/db mice. Diabetes 57: 3335–3343.
25. Zeng H, Li L, Chen JX (2012) Overexpression of Angiopoietin-1 Increases CD133+/c-kit+ Cells and Reduces Myocardial Apoptosis in db/db Mouse Infarcted Hearts. PLoS One 7: e35905. 10.1371/journal.pone.0035905 [doi]; PONE-D-11-21968 [pii].
26. Li L, Crockett E, Wang DH, Galligan JJ, Fink GD, et al (2002) Gene transfer of endothelial NO synthase and manganese superoxide dismutase on arterial vascular cell adhesion molecule-1 expression and superoxide production in deoxycorticosterone acetate-salt hypertension. Arterioscler Thromb Vasc Biol 22: 249–255.
27. Chen JX, Stinnett A (2008) Critical role of the NADPH oxidase subunit p47(phox) on vascular TLR expression and neointimal lesion formation in high-fat diet-induced obesity. Lab Invest 88: 1316–28.
28. Chen JX, Tuo Q, Liao DF, Zeng H (2012) Inhibition of protein tyrosine phosphatase improves angiogenesis via enhancing Ang-1/Tie-2 signaling in diabetes. Exp Diabetes Res 2012: 836759. 10.1155/2012/836759 [doi].
29. Chen JX, Zeng H, Tuo QH, Yu H, Meyrick B, et al (2007) NADPH oxidase modulates myocardial Akt, ERK1/2 activation and angiogenesis after hypoxia/reoxygenation. Am J Physiol Heart Circ Physiol 292: H1664–H1674.
30. Chen JX, Zeng H, Lawrence ML, Blackwell TS, Meyrick B (2006) Angiopoietin-1-induced Angiogenesis is Modulated by Endothelial NADPH Oxidase. Am J Physiol Heart Circ Physiol 291: H1563–72.
31. Assmus B, Iwasaki M, Schachinger V, Roexe T, Koyanagi M, et al (2012) Acute myocardial infarction activates progenitor cells and increases Wnt signalling in the bone marrow. Eur Heart J 33: 1911–1919. ehr388 [pii];10.1093/eurheartj/ehr388 [doi].
32. Giralt A, Villarroya F (2012) SIRT3, a pivotal actor in mitochondrial functions: metabolism, cell death and aging. Biochem J 444: 1–10. BJ20120030 [pii];10.1042/BJ20120030 [doi].
33. Liu Y, Zhang D, Chen D (2011) SIRT3: Striking at the heart of aging. Aging (Albany NY) 3: 1–2. 100256 [pii].
34. Sack MN (2011) Emerging characterization of the role of SIRT3-mediated mitochondrial protein deacetylation in the heart. Am J Physiol Heart Circ Physiol 301: H2191–H2197. ajpheart.00199.2011 [pii];10.1152/ajpheart.00199.2011 [doi].
35. Sundaresan NR, Samant SA, Pillai VB, Rajamohan SB, Gupta MP (2008) SIRT3 is a stress-responsive deacetylase in cardiomyocytes that protects cells from stress-mediated cell death by deacetylation of Ku70. Mol Cell Biol 28: 6384–6401. MCB.00426-08 [pii];10.1128/MCB.00426-08 [doi].
36. Askari AT, Unzek S, Popovic ZB, Goldman CK, Forudi F, et al (2003) Effect of stromal-cell-derived factor 1 on stem-cell homing and tissue regeneration in ischaemic cardiomyopathy. Lancet 362: 697–703. S0140-6736(03)14232-8 [pii];10.1016/S0140-6736(03)14232-8 [doi].
37. Hu X, Dai S, Wu WJ, Tan W, Zhu X, et al (2007) Stromal cell derived factor-1 alpha confers protection against myocardial ischemia/reperfusion injury: role of the cardiac stromal cell derived factor-1 alpha CXCR4 axis. Circulation 116: 654–663. CIRCULATIONAHA.106.672451 [pii];10.1161/CIRCULATIONAHA.106.672451 [doi].
38. Frederick JR, Fitzpatrick JR, III, McCormick RC, Harris DA, Kim AY, et al (2010) Stromal cell-derived factor-1alpha activation of tissue-engineered endothelial progenitor cell matrix enhances ventricular function after myocardial infarction by inducing neovasculogenesis. Circulation 122: S107–S117. 122/11_suppl_1/S107 [pii];10.1161/CIRCULATIONAHA.109.930404 [doi].
39. Massa M, Rosti V, Ferrario M, Campanelli R, Ramajoli I, et al (2005) Increased circulating hematopoietic and endothelial progenitor cells in the early phase of acute myocardial infarction. Blood 105: 199–206. 10.1182/blood-2004-05-1831 [doi];2004-05-1831 [pii].
40. Wojakowski W, Tendera M (2005) Mobilization of bone marrow-derived progenitor cells in acute coronary syndromes. Folia Histochem Cytobiol 43: 229–232.
41. Kim HS, Patel K, Muldoon-Jacobs K, Bisht KS, Aykin-Burns N, et al (2010) SIRT3 is a mitochondria-localized tumor suppressor required for maintenance of mitochondrial integrity and metabolism during stress. Cancer Cell 17: 41–52. S1535-6108(09)00428-0 [pii];10.1016/j.ccr.2009.11.023 [doi].
42. Edelberg JM, Tang L, Hattori K, Lyden D, Rafii S (2002) Young adult bone marrow-derived endothelial precursor cells restore aging-impaired cardiac angiogenic function. Circ Res 90: E89–E93.
43. Fan M, Chen W, Liu W, Du GQ, Jiang SL, et al (2010) The effect of age on the efficacy of human mesenchymal stem cell transplantation after a myocardial infarction. Rejuvenation Res 13: 429–438. 10.1089/rej.2009.0986 [doi].
44. Khan M, Mohsin S, Khan SN, Riazuddin S (2011) Repair of senescent myocardium by mesenchymal stem cells is dependent on the age of donor mice. J Cell Mol Med 15: 1515–1527. JCMM998 [pii];10.1111/j.1582-4934.2009.00998.x [doi].
45. Zhang H, Fazel S, Tian H, Mickle DA, Weisel RD, et al (2005) Increasing donor age adversely impacts beneficial effects of bone marrow but not smooth muscle myocardial cell therapy. Am J Physiol Heart Circ Physiol 289: H2089–H2096. 289/5/H2089 [pii];10.1152/ajpheart.00019.2005 [doi].
46. Gottlieb RA, Finley KD, Mentzer RM Jr (2009) Cardioprotection requires taking out the trash. Basic Res Cardiol 104: 169–180. 10.1007/s00395-009-0011-9 [doi].
47. Kim I, Rodriguez-Enriquez S, Lemasters JJ (2007) Selective degradation of mitochondria by mitophagy. Arch Biochem Biophys 462: 245–253. S0003-9861(07)00162-2 [pii];10.1016/j.abb.2007.03.034 [doi].
48. Hamacher-Brady A, Brady NR, Gottlieb RA (2006) Enhancing macroautophagy protects against ischemia/reperfusion injury in cardiac myocytes. J Biol Chem 281: 29776–29787. M603783200 [pii];10.1074/jbc.M603783200 [doi].
49. Hamacher-Brady A, Brady NR, Gottlieb RA, Gustafsson AB (2006) Autophagy as a protective response to Bnip3-mediated apoptotic signaling in the heart. Autophagy 2: 307–309. 2947 [pii].
50. Hamacher-Brady A, Brady NR, Logue SE, Sayen MR, Jinno M, et al (2007) Response to myocardial ischemia/reperfusion injury involves Bnip3 and autophagy. Cell Death Differ 14: 146–157. 4401936 [pii];10.1038/sj.cdd.4401936 [doi].
51. Kanamori H, Takemura G, Goto K, Maruyama R, Ono K, et al (2011) Autophagy limits acute myocardial infarction induced by permanent coronary artery occlusion. Am J Physiol Heart Circ Physiol 300: H2261–H2271. ajpheart.01056.2010 [pii];10.1152/ajpheart.01056.2010 [doi].
52. Przyklenk K, Undyala VV, Wider J, Sala-Mercado JA, Gottlieb RA, et al (2011) Acute induction of autophagy as a novel strategy for cardioprotection: getting to the heart of the matter. Autophagy 7: 432–433. 14395 [pii].
53. Wang HJ, Zhang D, Tan YZ, Li T (2013) Autophagy in endothelial progenitor cells is cytoprotective in hypoxic conditions. Am J Physiol Cell Physiol 304: C617–C626. ajpcell.00296.2012 [pii];10.1152/ajpcell.00296.2012 [doi].
54. Chen J, Xavier S, Moskowitz-Kassai E, Chen R, Lu CY, et al (2012) Cathepsin cleavage of sirtuin 1 in endothelial progenitor cells mediates stress-induced premature senescence. Am J Pathol 180: 973–983. S0002-9440(11)01102-3 [pii];10.1016/j.ajpath.2011.11.033 [doi].
55. Lee Y, Jung J, Cho KJ, Lee SK, Park JW, et al (2013) Increased SCF/c-kit by hypoxia promotes autophagy of human placental chorionic plate-derived mesenchymal stem cells via regulating the phosphorylation of mTOR. J Cell Biochem 114: 79–88. 10.1002/jcb.24303 [doi].
56. Phadwal K, Watson AS, Simon AK (2013) Tightrope act: autophagy in stem cell renewal, differentiation, proliferation, and aging. Cell Mol Life Sci 70: 89–103. 10.1007/s00018-012-1032-3 [doi].
57. Zhuang W, Li B, Long L, Chen L, Huang Q, et al (2011) Induction of autophagy promotes differentiation of glioma-initiating cells and their radiosensitivity. Int J Cancer 129: 2720–2731. 10.1002/ijc.25975 [doi].

13

In Vivo Tracking and Comparison of the Therapeutic Effects of MSCs and HSCs for Liver Injury

Qiang Li❡, Xinmin Zhou❡, Yongquan Shi❡, Jinge Li, Linhua Zheng, Lina Cui, Jun Zhang, Lu Wang, Zheyi Han, Ying Han*, Daiming Fan

State Key Laboratory of Cancer Biology, Xijing Hospital of Digestive Diseases, The Fourth Military Medical University, Xi'an, Shaanxi Province, China

Abstract

Background: Mesenchymal stem cells (MSCs) and hematopoietic stem cells (HSCs) have been studied for damaged liver repair; however, the conclusions drawn regarding their homing capacity to the injured liver are conflicting. Besides, the relative utility and synergistic effects of these two cell types on the injured liver remain unclear.

Methodology/Principal Findings: MSCs, HSCs and the combination of both cells were obtained from the bone marrow of male mice expressing enhanced green uorescent protein(EGFP)and injected into the female mice with or without liver fibrosis. The distribution of the stem cells, survival rates, liver function, hepatocyte regeneration, growth factors and cytokines of the recipient mice were analyzed. We found that the liver content of the EGFP-donor cells was significantly higher in the MSCs group than in the HSCs or MSCs+HSCs group. The survival rate for the MSCs group was significantly higher than that of the HSCs or MSCs+HSCs group; all surpassed the control group. After MSC-transplantation, the injured livers were maximally restored, with less collagen than the controls. The fibrotic areas had decreased to a lesser extent in the mice transplanted with HSCs or MSCs+HSCs. Compared with mice in the HSCs group, the mice that received MSCs had better improved liver function. MSCs exhibited more remarkable paracrine effects and immunomodulatory properties on hepatic stellate cells and native hepatocytes in the treatment of the liver pathology. Synergistic actions of MSCs and HSCs were most likely not observed because the stem cells in liver were detected mostly as single cells, and single MSCs are insufficient to provide a beneficial niche for HSCs.

Conclusions/Significance: MSCs exhibited a greater homing capability for the injured liver and modulated fibrosis and inflammation more effectively than did HSCs. Synergistic effects of MSCs and HSCs were not observed in liver injury.

Editor: Aditya Bhushan Pant, Indian Institute of Toxicology Reserach, India

Funding: This work was supported by National Natural Science Foundation of China (No. 81070326) and "863" Program of China (2011AA020111). The funders had no role in study design, data collection and analysis, decision to publish, or preparation of the manuscript.

Competing Interests: The authors have declared that no competing interests exist.

* E-mail: hanying1@fmmu.edu.cn

❡ These authors contributed equally to this work.

Introduction

Liver transplantation remains the definitive treatment option for end-stage liver disease. But the mismatch between the number of patients requiring transplantation and the amount of available organs is set to grow, highlighting the need to develop new strategies to reduce liver scarring and stimulate liver regeneration. Stem cell replacement strategies are therefore being investigated as an attractive alternative approach to liver repair.

To date, there are several published human clinical studies investigating the effects of stem cell therapy in patients with liver disease and most of the studies yielded positive results. The cells mostly used to transplant were derived from bone marrow including MSCs [1], HSCs [2,3] and unsorted mononuclear cells [4].Among them, MSCs and HSCs can be obtained in a great quantity. MSCs represent excellent candidates for the following reasons: MSCs (i) can contribute to the direct production of new hepatocytes [4,5,6], (ii) can promote tissue repair by secreting trophic molecules [6,7,8], (iii) are immunomodulatory and hypoimmunogenic (i.e., in addition to not being recognized by the recipient's immune system unlike allogeneic transplants, MSCs could manage exacerbated inflammatory process) [9,10,11] and (iv) have anti-fibrotic properties and inhibit the activation of hepatic stellate cells [11,12]. HSCs represent another significant stem cell type in bone marrow. The liver-specific phenotypic and functional changes of a highly enriched population of HSCs in response to injury in vitro or *in vivo* have been examined. HSCs also have many of the following advantages: HSCs (i) can also be converted into functional hepatocytes without fusion and therefore contribute to the regeneration of injured liver [13], (ii) can reduce liver injury through paracrine effects [4] and (iii) can modulate the immune system of the recipient [14]. However, which cell type is more effective in treating the injured liver remains to be determined. Moreover, several recent studies have suggested that MSCs and HSCs function synergistically for the therapy of diabetes and heart failure and for vascularizing bioengineered tissues [15,16,17]. Whether these cells can work synergistically in the injured liver is unclear.

Our aim was to evaluate the biodistribution of the stem cells after the peripheral infusion of MSCs or HSCs into liver injured mice. The anti-inflammatory and anti-fibrotic activities of these two stem cells were also evaluated. In addition, whether MSCs and HSCs exhibit synergistic effects in treating liver injury was studied. We hope that these findings contribute to better understanding of the interactions between stem cells and the environment that leads to homing and integration into livers.

Materials and Methods

Mice

Breeding pairs of C57BL/6-Ly5.1 mice and EGFP transgenic mice (C57BL/6-Ly5.2 background) were purchased from the animal center of the Fourth Military Medical University. All aspects of the animal research were approved by the Animal Care and Use Committee of the Fourth Military Medical University (Approval ID 12008) and were in compliance with Guidelines for the Care and Use of Laboratory Animals, as published by the National Academy Press.

Isolation and Characterization of HSCs

Male EGFP transgenic mice 9 to 12 weeks old (19 to 24 grams) were used as bone marrow donors. The mice were humanely sacrificed and the bone marrow cells were flushed from the tibiae and femurs, pooled, and washed twice with Ca^{2+}/Mg^{2+}-free phosphate-buffered saline (PBS; Beyotime, China) containing 0.1% bovine serum albumin (BSA) (Beyotime, China). Single-cell suspensions were produced after repeated pipetting and filtering through a 50-μm nylon mesh. Bone marrow cells with densities ranging from 1.063 to 1.077 g/ml were collected by gradient separation using Nycodenz (Sigma, USA). Lineage negative (Lin^-) bone marrow cells of the EGFP-expressing mice were prepared by incubating the cells with anti- macrophage-1(Mac-1), anti-granulocyte receptor-1 (Gr-1), anti-TER-119, anti-B220, anti-CD4 and anti-CD8 and then by removing the positive cells with immunomagnetic beads (Dynabeads M-450 coupled to sheep anti-rat IgG) (Dynal, Great Neck). The resulting Lin^- cells were stained with phycoerythrin (PE)-conjugated anti-Sca-1, biotin-conjugated anti-CD34, and APC-conjugated anti-c-kit and with the mouse lineage panel of antibodies, followed by streptavidin-conjugated PharRed. After the addition of propidium iodide at a concentration of 1 μg/ml, the cells were washed twice, resuspended in PBS containing 0.1% BSA, and maintained on ice for cell sorting. Five-color and cell sorting analyses were performed using a FACS Vantage (BD Biosciences, San Jose) with the appropriate isotype-matched controls. The Lin^- cells were enriched for the c-kit$^+$/Sca-1$^+$ population. Single Lin^-, c-kit$^+$, Sca-1$^+$, CD34$^-$ cells were deposited into round-bottomed 96-well plates using the CloneCyt system (Becton Dickinson, USA). The plates were incubated at 37°C in a humidified 5% CO_2 atmosphere. At 20 hours after cell deposition, the wells containing single cells were marked and incubated for 7 days. We selected clones consisting of no more than 20 cells for clonal cell transplantation [18].

Isolation, Expansion, and Phenotypic Characterization of MSCs

Bone marrow cells were collected after flushing the tibiae and femurs of 6-week-old male EGFP-expressing mice (14 to 17 grams) using sterile PBS. The cells were aspirated with a 29-gauge needle to disrupt aggregates, and then the entire aspirate was centrifuged at 2,000 rpm for 10 min. The pellet was seeded into a 25-cm^2 culture plate with alpha-MEM (Gibco, USA) medium supplemented with 10% selected fetal bovine serum (FBS; Gibco, USA), an anti-mycotic solution (Sigma, USA) and 1% antibiotics. The non-adherent cells were removed after 72 h by changing the medium, and the medium was entirely replaced every 5 days [19]. When the foci reached conuence, the adherent cells were detached with 0.25% trypsin in 2.65 mM EDTA (Sigma, USA), centrifuged, and subcultured at 7,000 cells/cm^2. After two subcultures, the adherent cells were detached, and the cells were stained with PE - conjugated antibodies against CD29, CD90, CD34, CD105, CD80 and CD45 (Santa Cruz, USA) antigens for 30 min and then washed and analyzed using a flow cytometer (BD Biosciences, USA).

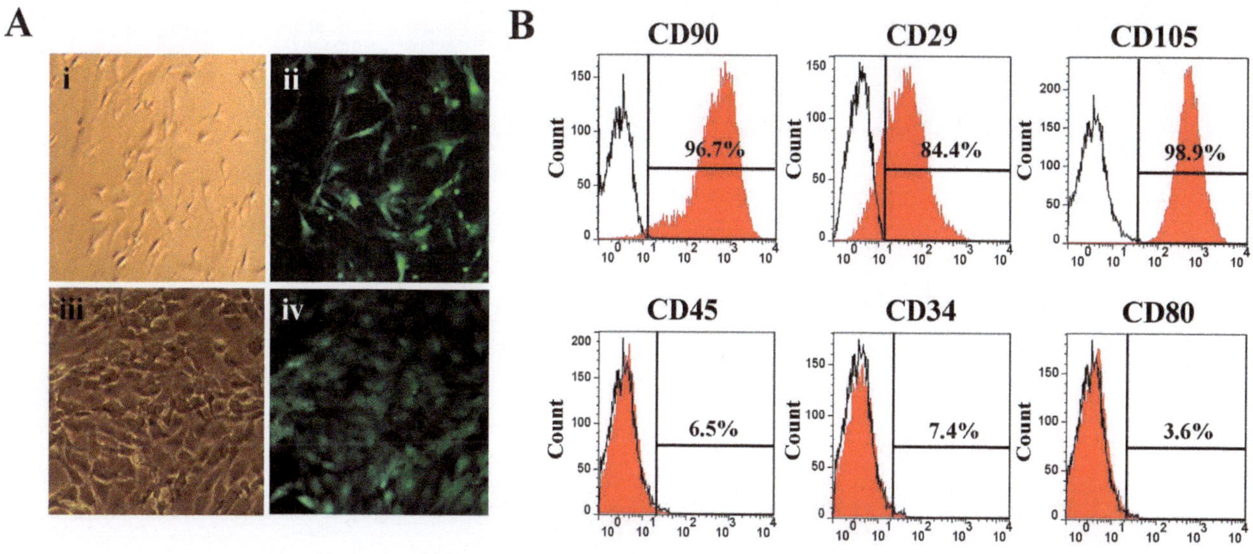

Figure 1. Characterization of MSCs derived from EGFP-transgenic mice. (A) The morphologies of the MSCs in the first (i and ii) and third (iii and iv) passages observed under bright field and fluorescence microscopy, respectively. (B) Surface molecule characterization of the MSCs performed by FACS analyses after incubation with PE-conjugated antibodies (CD90, CD29, CD105, CD45, CD34 and CD80).

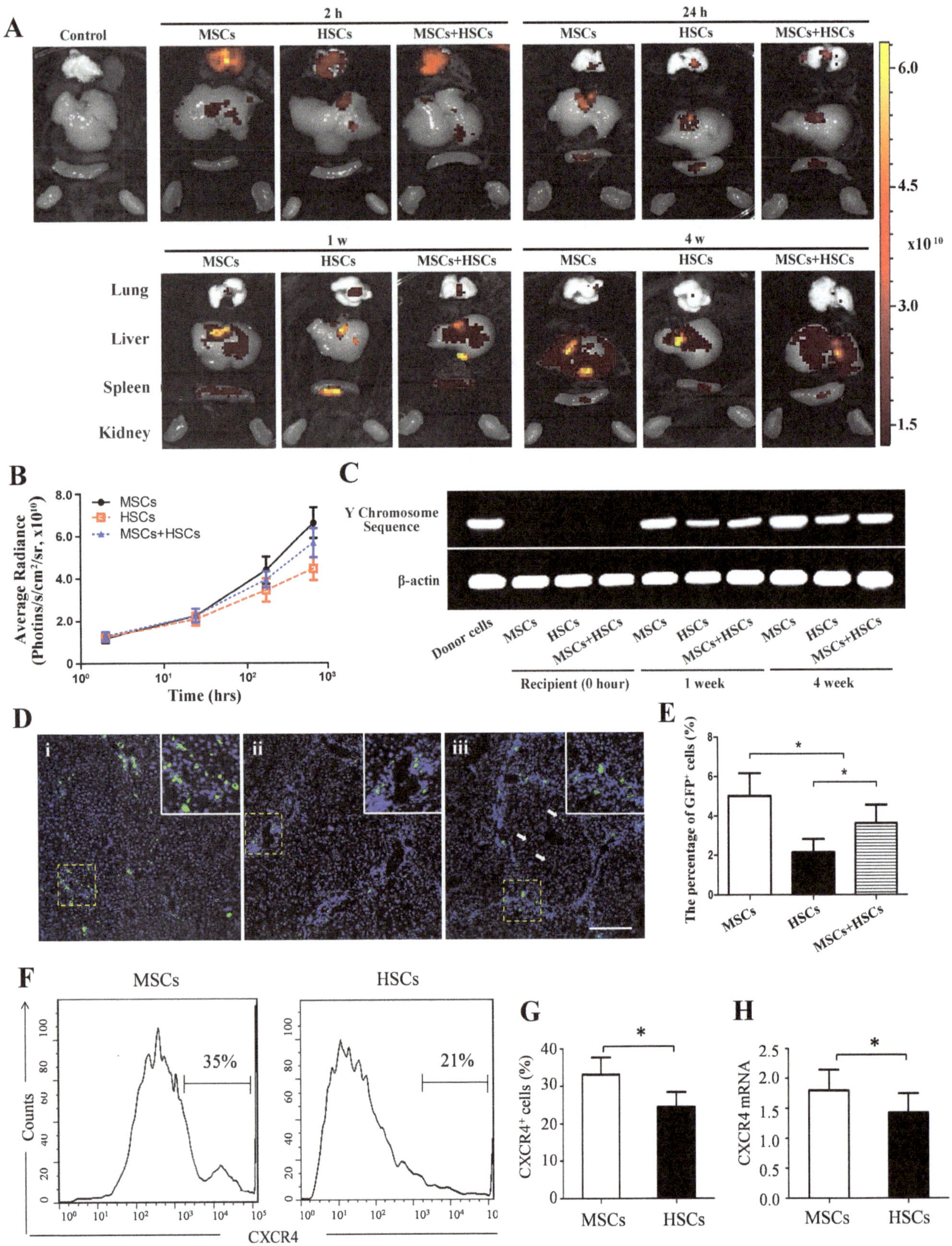

Figure 2. Detection of EGFP$^+$ cells in recipient livers after transplantation. EGFP$^+$ cells from male donors were injected into liver-injured female mice. (A) After transplantation of stem cells, the EGFP fluorescence in the lung, liver, spleen and kidney was examined using bio- imaging system. A luminescent image from red (least intense) to yellow (most intense) represents the spatial distribution of the detected photons emitted

from EGFP+ cells within the organs. The EGFP signal was not detected in the control mice. (B) Average radiance was quantified in the liver after stem cells transplantation. (C) PCR-based detection of donor-derived cells in the livers of different recipients. (D and E) Liver sections were stained with DAPI, and the distribution of EGFP+ cells in the portal lobe was quantified in the different groups at 4 weeks. White arrows show the stem cells in the sinusoids (i, MSCs group; ii, HSCs group; and iii, MSCs+HSCs group). (F and G) FACS analyses of CXCR4 expression on MSCs and HSCs. (H) mRNA levels of CXCR4 on MSCs and HSCs. $P<0.05$, n = 3.

Induction of Liver-injured Mice

Wild-type mice (9 to 12 weeks old, 19 to 24 grams) were housed in a sterile animal facility with a 12-hour dark/light cycle and free access to food and water. Advanced liver fibrosis was induced in adult female mice with intraperitoneal injections of 7 ml/kg body weight of a 1:4 solution of CCl_4 (Sigma, USA) and olive oil (Sigma, USA) twice a week for 3 months [18]. After final CCl_4 injection at 3 months, the mice except the control group were randomly divided into different groups to ensure that the relatively level of liver disfunction was constant. Mice from the same cohort were randomly allocated to receive different cell stem cells via injections of the tail vein. The candidate stem cells from age- and strain-matched mice were suspended in 0.1 ml of PBS. CCl_4 administration continued for an additional four weeks.

Cell Transplantation

A total of 1×10^6 cells were resuspended in 100 μL PBS and slowly infused via the tail vein (Normal, MSCs, HSCs and MSCs+HSCs treatment groups). Following transplantation, CCl_4 was re-administered for four additional weeks (7 ml/kg body weight of a 1:4 solution of CCl_4 and olive oil twice a week) to enable the transplanted cells to engraft and differentiate. CCl_4 was not administered to the normal group (n = 20) during this period. In the CCl_4 group (n = 22), the mice were injected with CCl_4 alone, without cell transplantation. In the MSCs group (n = 22), the mice were administered MSCs. In the HSCs group (n = 22), the mice were administered HSCs. In the MSCs+HSCs group (n = 22), the mice were administered 5×10^5 MSCs and 5×10^5 HSCs [17].

Bio-imaging

The mice were sacrificed under deep anesthesia at 2 h, 24 h, 7 d and 28 d after transplantation. The lungs, livers, spleens and kidneys were isolated and were directly imaged by CCD with its excitation wavelength at 465/430 nm and emission filter at 560 nm with the following parameters: binning: 4, F/Stop: 1, exposure time: 1 min. The bio-imaging was conducted with a NightOWL imaging system using WinLight software (Berthold, Germany).

Tissue Preparation and Confocal Microscopy

To observe the engrafted cells in the liver, the tissues were prepared as frozen sections. The livers were dissected into pieces, immediately incubated with an optimal cutting temperature compound (OCT compound, Sakura Fine technical, Japan) and held in liquid nitrogen for 20 seconds. The frozen liver sections were directly transferred to a −70°C freezer. For observing the EGFP-expressing cells in the tissue, the frozen sections were allowed to melt in distilled water for 3 min and placed on slides.

Polymerase Chain Reaction

Genomic DNA was isolated from each liver using a lysis buffer containing 1% of 0.2 mg/ml proteinase K (Sigma, USA) and SDS followed by phenol-chloroform extraction. The genomic DNA was used to detect a Y chromosome sequence with the forward primer CTGGAGCTCTACAGTGATGA and the reverse primer CAGTTACCAATCAACACATCAC. The PCR products were

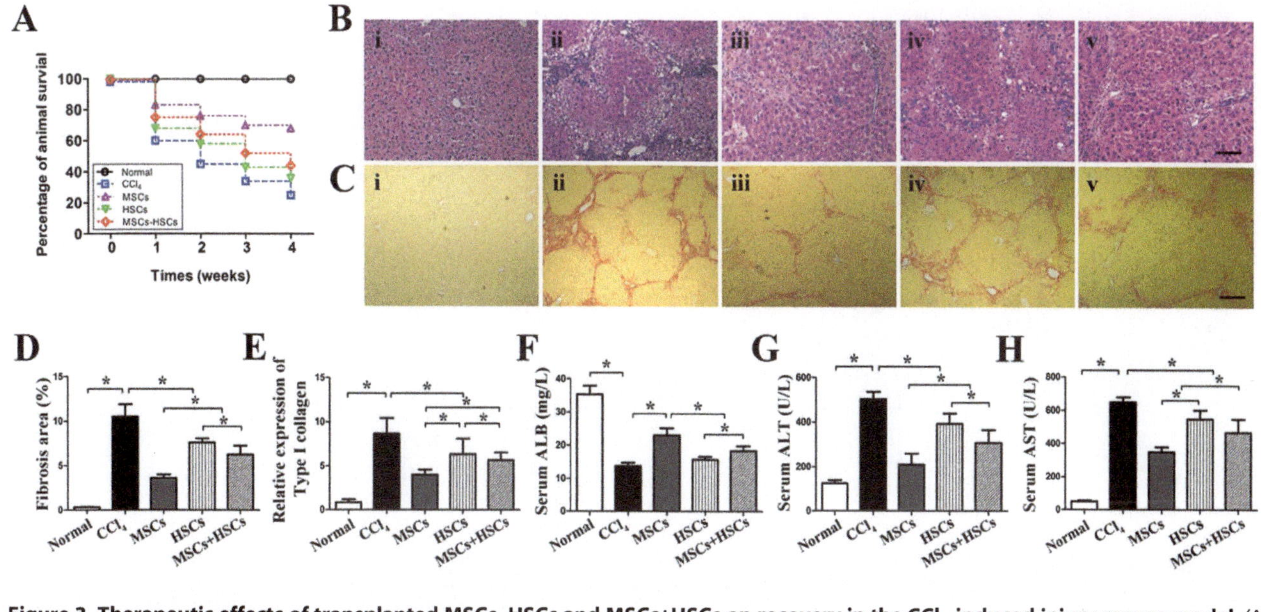

Figure 3. Therapeutic effects of transplanted MSCs, HSCs and MSCs+HSCs on recovery in the CCl_4-induced injury mouse model. (A) A survival curve for the injured mice that underwent intravenous cell transplantation. Representative photomicrographs of H&E-stained (B) and Sirius red-stained (C) mouse livers from the different groups (i, normal mice; ii, CCl_4 group; iii, MSCs group; iv, HSCs group; and v, MSCs+HSCs group).(D) Analyses of the fibrosis percentage using Image J software. (E) mRNA levels of *type I collagen*. (F, G and H) Serum ALB, ALT and AST levels in the experimental groups. $P<0.05$, n = 5, Scale bars: B = 150 μm, C = 200 μm.

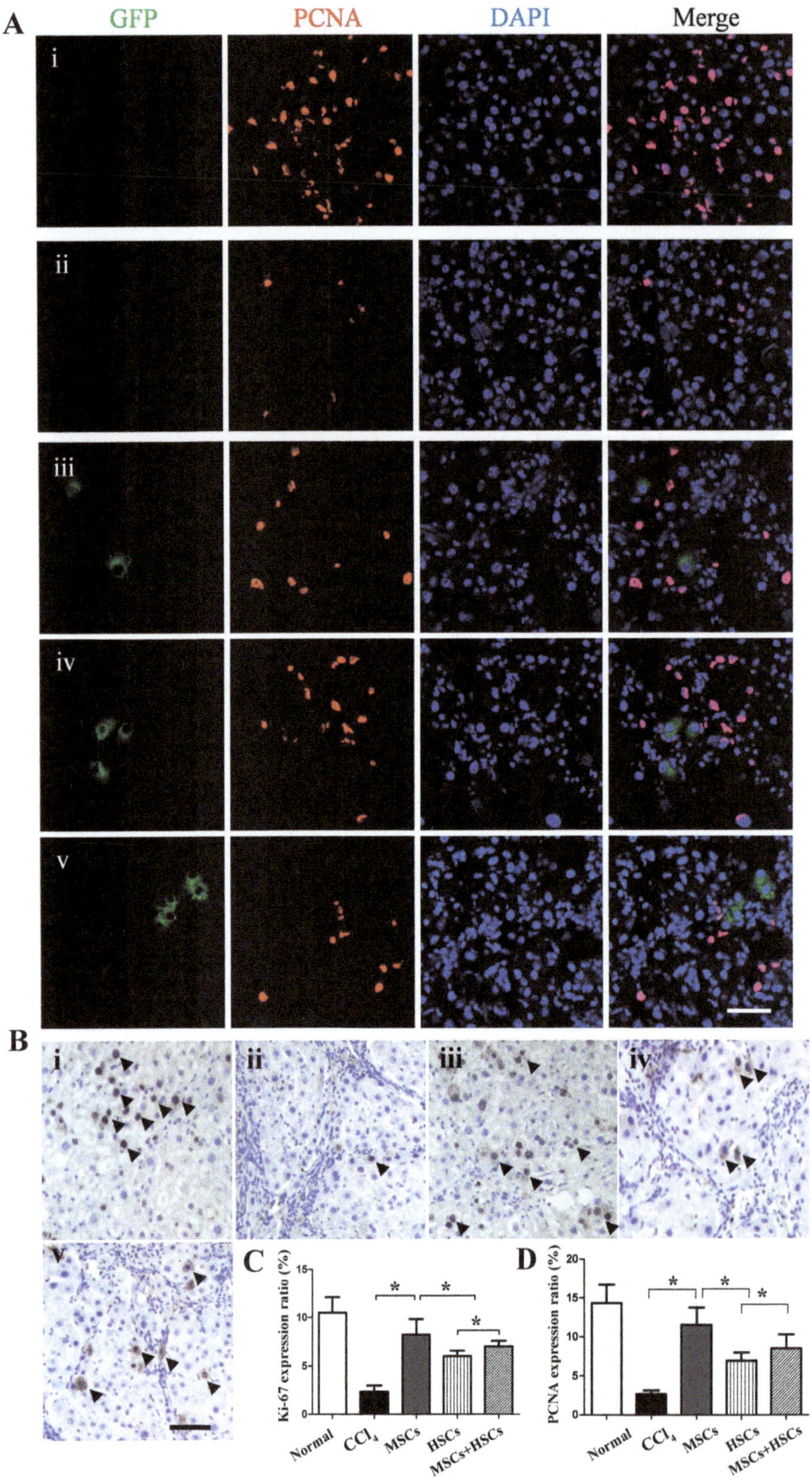

Figure 4. Effect of MSC, HSC and MSC+HSC transplantation on hepatocyte regeneration. (A) Immunohistochemistry analyses of PCNA expression in liver tissues (i, normal mice; ii, CCl_4 group; iii, MSCs group; iv, HSCs group; and v, MSCs+HSCs group). (B) Immunohistochemistry analyses of Ki-67 expression in liver tissues (i, normal mice; ii, CCl_4 group; iii, MSCs group; iv, HSCs group; and v, MSCs+HSCs group). (C) Quantitative image analyses of the percentage of Ki-67$^+$ cells. (D) Quantitative image analyses of the percentage of PCNA$^+$ cells. $P<0.05$, n = 5, Scale bar = 100 μm.

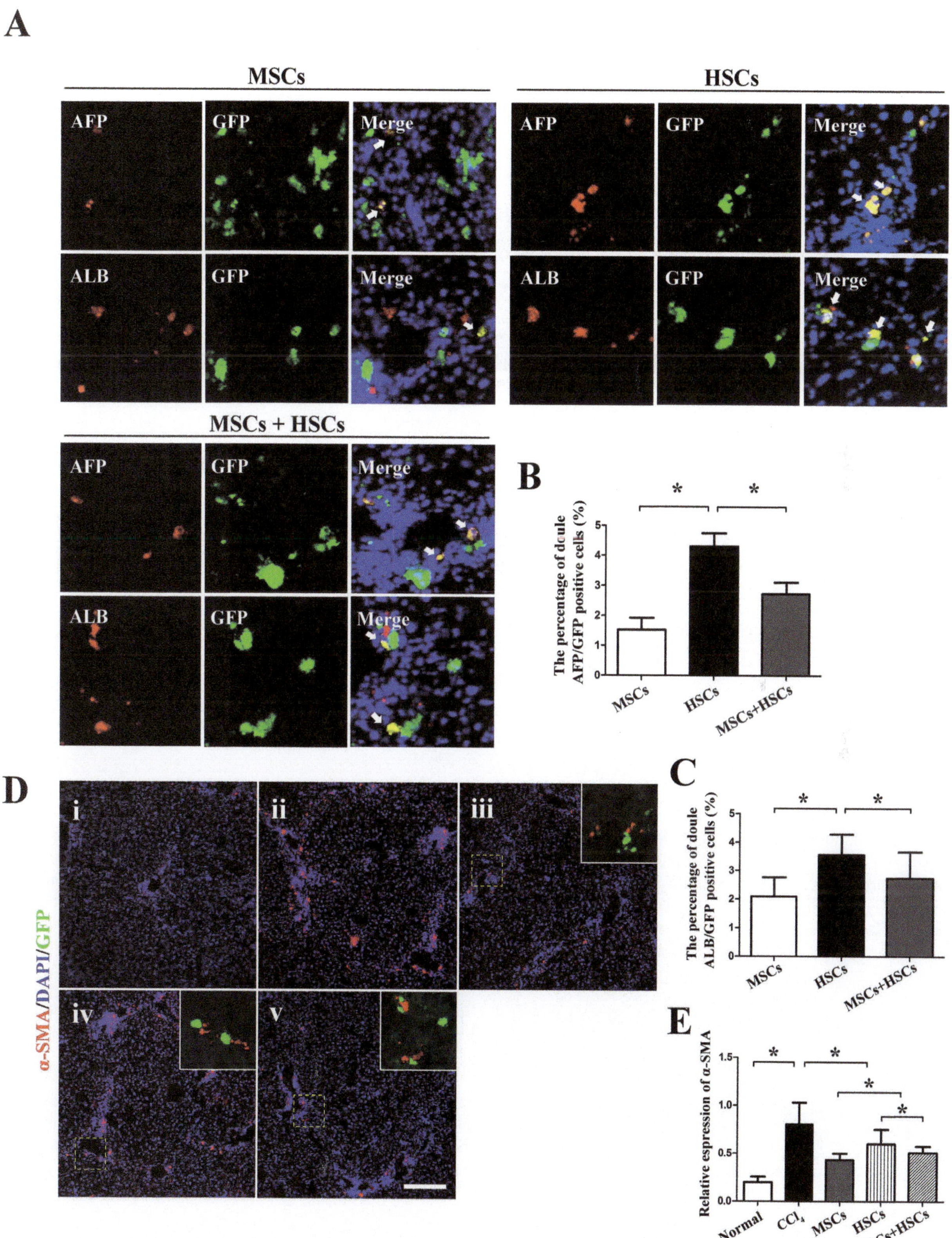

Figure 5. Differentiation of transplanted MSCs, HSCs and MSCs+HSCs in CCl_4-induced injured livers at 4 weeks. The liver sections were observed under fluorescence microscopy. (A, B and C) The liver sections were co-stained with either AFP or ALB to detect the differentiation of transplanted cells (white arrowhead) in the different groups. (D) The expression of α-SMA in the livers of CCl_4-induced injured mice in the different groups (i, normal mice; ii, CCl_4 group; iii, MSCs group; iv, HSCs group; and v, MSCs+HSCs group). (E) mRNA levels of α-SMA. $P<0.05$, n = 5, Scale bars: A = 50 μm, D = 150 μm.

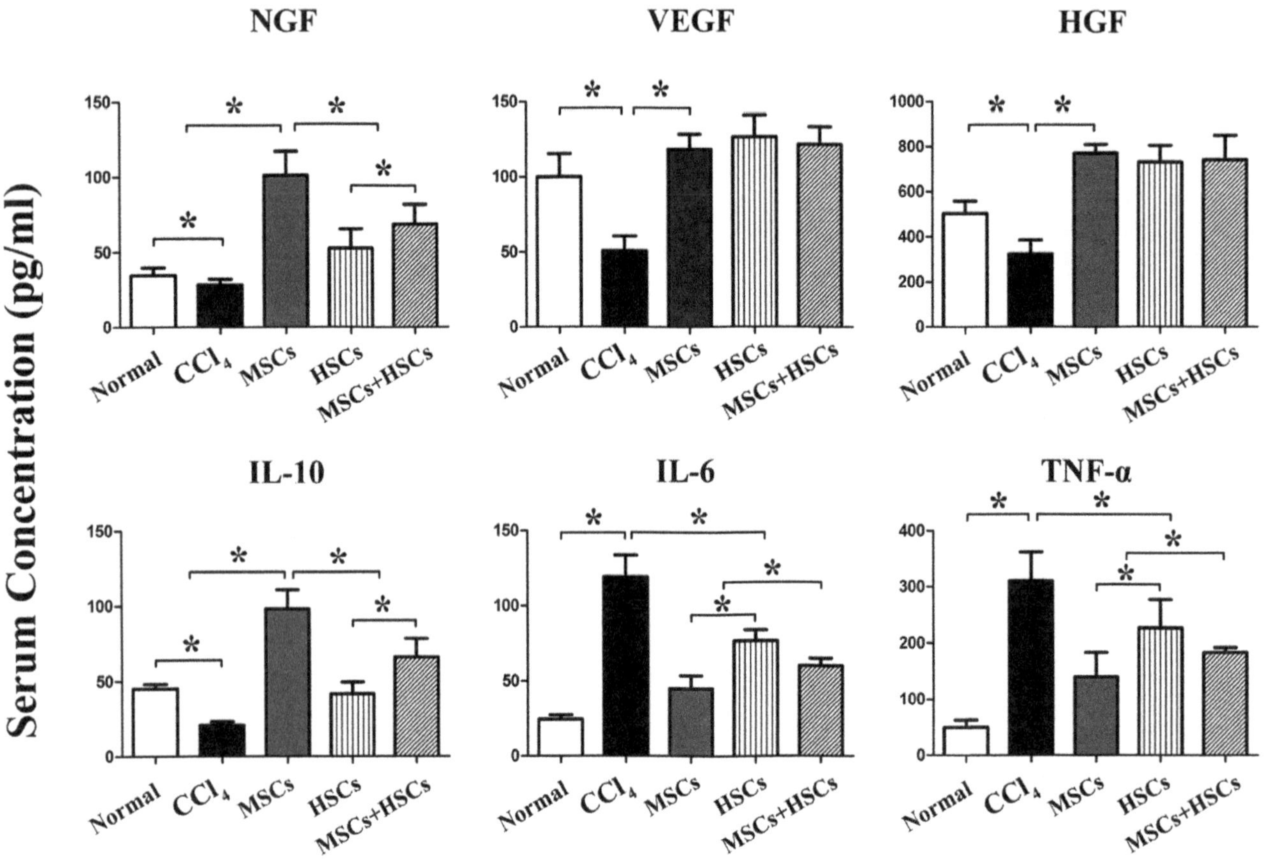

Figure 6. The concentration of growth factors and cytokines in each group 4 w after transplantation of MSCs, HSCs or MSCs+HSCs. $P<0.05$, $n=5$. NGF, nerve growth factor; VEGF, vascular endothelial growth factor; HGF, hepatocyte growth factor; IL, interleukin; TNF-α, tumor necrosis factor-alpha.

fractionated by 1.2% agarose gel electrophoresis and visualized under UV illumination after staining with ethidium bromide.

Biochemical and Liver Enzyme Assays

Serum was collected to analyze aspartate aminotransferase (AST), alanine aminotransferase (ALT), and albumin (ALB) using a chemistry analyzer (Abbott Architect c8000, USA).

Histopathological Staining

For the histological examinations, formalin-fixed livers were dehydrated and embedded in paraffin. The liver sections were deparaffinized and stained using hematoxylin and the eosin stain or Sirius red. For immunohistochemical studies, after microwave-based antigen retrieval, the sections were treated with 0.3% H_2O_2 in PBS to quench the endogenous peroxidase and then incubated with 5% goat serum (Beyotime, China) to block the non-specific sites. Monoclonal primary antibodies against mouse Ki-67 (1:100, Abcam, USA) were applied when incubating at 4°C for 12 h, followed by incubation with the corresponding secondary antibodies (Beyotime, China) at 37°C for 30 min. The specimens were then incubated with a diaminobenzidine peroxidase substrate and then subsequently counterstained with hematoxylin. For the immunofluorescence analyses, the tissue samples were fixed using 4% paraformaldehyde (Beyotime, China) and then permeabilized using 100% acetone. The samples were blocked using 5% BSA and then incubated overnight at 4°C with alpha-fetoprotein(AFP) (1:200, Abcam, USA), ALB (1:200, Abcam, USA), proliferating cell nuclear antigen(PCNA)(1:500, Abcam, USA) and alpha-smooth muscle actin (α-SMA) (1:300, Abcam, USA) antibodies diluted in antibody dilution solutions (Beyotime, China). The excess primary antibody was removed by washing five times in PBS, and the samples were incubated with a PE-conjugated secondary antibody (1:500) at room temperature for 2 hours. The sections were incubated with 4′, 6-diamidino-2-phenylindole (DAPI; Beyotime, China) to label the nuclei. The slides were mounted in propidium iodide-containing mounting medium (Beyotime, China) for visualization using a confocal microscope (FV-1000, Olympus, Japan).

Fluorescence-Activated Cell Sorting (FACS)

The MSCs and HSCs suspension was incubated with PE-labeled anti-mouse CXCR4 antibody (1:500,BD Pharmingen, USA) on ice for 1 h, washed with staining buffer, and fixed with 2% paraformaldehyde. FACS data were acquired using a flow cytometer (BD Biosciences, USA) [20]. Data were analyzed using CellQuest software.

Real-time PCR

Total RNA was isolated from fresh liver tissue and stem cells using Trizol reagent (Life Technologies, USA), and the ratio between the absorbance values at 260 and 280 nm provided an estimate of the RNA purity. Real-time PCR was performed using a one-step kit (Takara, Japan) with the following primers: α-SMA: forward: GGGAGTAATGGTTGGAATGG, reverse: GGCAG-

TAGTCACGAAGGAATAG; *type I collagen*: forward: AACTTTGCTTCCCAGATGTCCT, reverse: TCGGTGTCCCTTCATTCCAG; CXCR4: forward :CTCATCCTGGCCTTCATCAGC, reverse :TCAGCCAG-CAGTTTCCTTGG; and β-actin: forward: TTCCTTCTTGGGTATGGAAT, reverse: GAGCAAT-GATCTTGATCTTC following the manufacturer's instructions, using a Eppendorf Thermal Cycler (Takara, Japan) and using the appropriate cycle profiles.

Enzyme-linked Immunosorbant Assays

Quantification of the mouse serum levels of nerve growth factor (NGF), hepatocyte growth factor (HGF), vascular endothelial growth factor (VEGF), IL-10, IL-6 and tumor necrosis factor-alpha (TNF-α) was determined using enzyme-linked immunosorbent assays kits per the manufacturer's instructions (R&D Systems, USA), and the wells were read at 450 nm on an optical plate reader. Standard curves were prepared using purified cytokine standards. Each experimental sample was run in duplicate.

Data Analyses

The data are expressed as the mean with the standard error of the mean. The statistical significances were determined using SPSS 12.0 software (SPSS, USA). The statistical significances between the control and test groups were determined using Student's t-test. For the analyses of multiple groups, the P-values were adjusted using the Bonferroni method and a $P<0.05$ was used for statistical significance. All of the procedures were performed by blinded investigators.

Results

Characterization of EGFP-positive MSCs

The MSCs were isolated from the bone marrow of EGFP-transgenic mice. In the first passage, the cells derived from the donors emitted heterogeneous levels of green fluorescence when observed under the fluorescence microscope and were of various sizes, as observed in bright field. In the third passage, the EGFP signal intensity was uniform among the cells, and the cells exhibited a homogeneous morphology (Fig. 1A). Flow cytometry analyses were used to characterize the surface markers of the cultured cells. Most of the cells expressed the standard MSC surface markers, CD90 (95.2±6.1%), CD29 (85.6±3.5%) and CD105 (96.5±4.3%), whereas they were negative for CD45 (6.3±2.2%), CD34 (7.2±1.8%) and CD80 (3.8±1.2%) (Fig. 1B). MSCs were also successfully isolated for use in the following experiments.

Detection of Donor-derived Cells

After intravenous infusion, the EGFP signals first accumulated in the lung, but by 2 h, those signals began to decrease, whereas they started to accumulate in the liver and spleen at 2 h after infusion. During the following hours to days, the EGFP signal intensity gradually increased in the liver. From 24 h to 7 d, the EGFP signal intensity gradually increased in the spleen and then decreased. The EGFP signal was barely detectable in the kidney. These trends were similar in the MSCs, HSCs and MSCs+HSCs groups (Fig. 2A). The EGFP intensity of the livers in the MSCs group was significantly higher than in the HSCs or MSCs+HSCs group (Fig. 2B). The number of homing stem cells to the CCl_4-induced cirrhotic liver was significantly higher than that in the normal group (Fig. S1). The donor-derived signals were stronger in the livers 4 w after the cell injection than at 2 w (Fig. 2C). These results were confirmed by the percentage of EGFP-positive cells detected after nuclear staining with DAPI (Fig. 2D and 2E). From 2 w on, most of the transplanted cells were located infiltrating the areas around the liver's portal tracts and interlobular connective tissue. Only a few of the cells migrated toward the central region of the hepatic lobes and can be detected in the sinusoids (Fig. 2D). FACS analyses of CXCR4 expression on MSCs and HSCs shows that CXCR4 expressed on MSCs (in the third passage)(33.2±8.1%) was significantly higher than HSCs(24.5±6.8%, $P<0.05$). Real-time PCR revealed that CXCR4 mRNA expression was higher in MSCs (in the third passage) than in HSCs (Fig. 2F, 2G and 2H).

Morphology and Functional Evaluation of Injured Liver

pone.0062426The mice in the normal group survived the observation period. The survival of the mice in the three groups that underwent stem cell transplantation was significantly higher than in the group treated with CCl_4. The survival percentage in the MSCs group (68.2%) was significantly higher than in the HSCs (36.4%) or MSCs+HSCs (45.5%) group, while the survival percentage in the MSCs+HSCs group was significantly higher than in the HSCs group (Fig. 3A). The histological sections of the CCl_4-injured mice demonstrated that when compared with normal mice, a large number of inflammatory cells had infiltrated the sinusoids and centrilobular regions, and the coagulation necrosis of hepatocytes was observed (H&E staining in Fig. 3B). In addition, liver fibrosis had increased significantly, characterized by fibrotic septum formation starting in the portal areas (Sirius red staining in Fig. 3C). After transplantation of MSCs, the injured livers showed maximal restoration with thinner fibrotic areas and decreased collagen depositions (3.6±0.7%). However, the fibrotic areas had decreased to a lesser extent in the mice transplanted with HSCs (7.6±0.8%) or MSCs+HSCs (6.2±1.7%, $P<0.05$) (Fig. 3D). These results were confirmed by the expression of *type I collagen* (4.0±1.1%, 6.3±3.1%, 5.7±1.5%, respectively, $P<0.05$) (Fig. 3E). Moreover, when compared with the mice in the HSCs group and the MSCs+HSCs group, the mice receiving MSCs showed the best improvement of liver function, however, liver function in MSCs group was still inferior than in the normal cohort, as demonstrated by the ALB, ALT and AST levels of the peripheral blood (Figs. 3F, 3G and 3H). To evaluate whether stem cell transplantation enhances the proliferation of hepatocytes in cirrhotic livers, the PCNA and Ki-67 expression levels were assessed by immunofluorescence and immunohistochemistry. In the MSCs-transplanted livers, the percentage of $PCNA^+$ (11.5±3.4%) and $Ki\text{-}67^+$ (8.2±2.7%) cells was increased significantly when compared with those of the HSCs (6.9±1.8%, 6.0±1.1%, respectively) and MSCs+HSCs (8.5±3.1%,7.2±1.9%, respectively, $P<0.05$) groups (Figs. 4A,4B,4C and 4D).

Derivation of Transplanted Cells

The immunofluorescence staining in the HSCs group revealed a higher percentage of double-labeled GFP^+/AFP^+ and GFP^+/ALB^+ cells in the host livers (4.3±0.6% and 3.5±0.7%, respectively), compared with the MSCs (1.4±0.5% and 2.1±0.3%, respectively) or the MSCs+HSCs (2.8±0.4% and 2.4±0.6%, respectively,$P<0.05$) groups (Figs. 5A, 5B and 5C). The expression of α-SMA was significantly decreased in the livers of mice transplanted with any cell type. The α-SMA expression in the MSCs group was significantly lower than in the other two groups (Figs. 5D and 5E). Double-labeled $GFP^+/\alpha\text{-}SMA^+$ cells were not found in any of the groups.

Growth Factor and Cytokine Levels in Each Group

Four weeks after cell transplantation, the serum showed a significant increase in NGF in the MSCs (101±12 pg/ml) group when compared with the HSCs and MSCs+HSCs groups (53±5 pg/ml and 69±7 pg/ml, respectively, $P<0.05$). The levels of HGF and VEGF were not significantly different among the cell transplantation groups. The expression of IL-10 in the MSCs group (98±22 pg/ml) was higher than in the other cell transplantation groups (HSCs and MSCs+HSCs groups: 42±14 pg/ml, 66±21 pg/ml, respectively) or the CCl_4 group (21±4 pg/ml, $P<0.05$), whereas the concentration of IL-6 (45±15 pg/ml) and TNF-α (140±75 pg/ml) in the MSCs group was lower than that of HSCs (IL-6:77±13 pg/ml, TNF-α: 227±87 pg/ml) and MSCs+HSCs (IL-6:60±9 pg/ml, TNF-α: 183±15 pg/ml, $P<0.05$) groups (Fig. 6).

Discussion

In this study, we conducted a comparison of MSCs and HSCs *in vivo* engraftment capacities, as well as the administration of MSCs, HSCs and MSCs+HSCs for functional liver repair using the same liver fibrosis mouse model. The major findings of our study were the following: (i) when compared with HSCs alone and a combination of HSCs and MSCs, MSCs had the greatest homing capability to the injured liver; (ii) MSCs showed the greatest capability to promote hepatocyte proliferation, inhibit hepatic stellate cell activation *in vivo* and facilitate mice survival when compared with other two groups; and (iii) synergistic effects of MSCs and HSCs were not observed in CCl_4-induced liver injury.

Cell homing and engraftment into the host liver are integral to cell-based therapies. The mechanism that governs the recruitment of bone marrow stem cells is complicated. Several signaling pathways [21,22,23] and growth factors [24] have been shown to contribute to the recruitment of bone marrow cells to the injured organs. Stress in the liver results in the release of various chemokines/cytokines including stromal-derived factor-1(SDF-1) and HGF. These increase stem cells mobility through the cell surface receptors [e.g. chemokine receptor-4 (CXCR4) and hepatocyte growth factor receptor (c-met)], facilitating stem cells homing to sites of wound healing [25,26,27]. SDF-1-CXCR4 and HGF-c-met axes may be involved in recruitment of expanded MSCs to damaged tissues while only SDF-1-CXCR4 axes is involved in stress-induced HSCs recruitment to the injured liver [26,27].

Previous studies regarding the homing capacity of stem cells to the injured liver have reported conflicting conclusions. Some reported that the stem cells gradually accumulated in the liver [25,28]; however, one study demonstrated that the transplanted MSCs reached a peak amount in the liver 1 day after perfusion, and then the MSCs content gradually declined [29]. A few of the studies provided information concerning the biodistribution of the cells in major organs or compared the homing capacities of MSCs and HSCs to the injured liver.

To overcome these limitations, for the present investigation, highly purified and functionally active $EGFP^+$ MSCs and HSCs isolated from bone marrow were transplanted intravenously into mice with CCl_4-induced chronic liver injury. Our data showed that the EGFP signals were detected not only in livers but also in other tissues, such as the lungs and spleens. The distributions of these two stem cells were similar and both migrated effectively to the liver rather than to the lung, spleen and kidney in mice with CCl_4-induced liver injury. The MSCs and HSCs gradually accumulated in the liver after they were first observed there. Furthermore, we found that MSCs exhibited a more superior homing ability to the injured liver in comparison with HSCs, which would help them exert their effects there. Meanwhile, fluorescence-activated cell sorting and Real-time PCR analyses of CXCR4 expression on both stem cells revealed that CXCR4 expression in MSCs(in the third passage)was higher than in HSCs.

As for the localization of stem cells in the liver, Sakaida et al. transplanted bone marrow cells into mice with liver fibrosis and found that the cells predominantly engrafted to the periportal area [30]. Additionally, di Bonzo et al. determined the location of human MSCs that were transplanted into mice and reported that the human cells were detected mostly around the portal tracts in injured livers [31]. Our results showed that in both injured and non-injured livers, the stem cells were detected mainly as single cells around the portal tracts and in the interlobular connective tissue. The stem cells were also detected in the sinusoids. This finding is most likely the result of the space of Disse being composed of sinusoidal endothelial cells and these cells being able release SDF-1 [32] and SDF-1can attract stem cells.

The transdifferentiation ability of bone marrow-derived stem cells into hepatocytes may play a significant role in the repair of the injured liver [13,33,34]. AFP and ALB typically expressed in stem cells-derived hepatocytes [28,35,36].However, our data showed that at 4 weeks after transplantation, the standard time of most research studies [31], the number of $EGFP^+$ cells with a hepatocyte phenotype was relatively low, and the apparent number of HSCs in this model was slightly higher than that of MSCs. These cells were limited to a small portion of the total liver mass and were not sufficient to reverse the injuries (a percentage of 2.5–5% is necessary for this process [37]). Therefore, the differentiation of these two stem cells into hepatocytes cannot explain the improvement of liver function in this study.

The concept of stem cell transplantation exerting a paracrine proliferative effect on endogenous hepatocytes is gaining support. Hepatocytes in fibrotic livers have reached replicative senescence after many rounds of injury and repair, and they have reduced proliferative capacity [38,39,40,41]. Stem cell infusion may increase the intrinsic ability of hepatocytes to proliferate by facilitating the breakdown of scar tissue or by the release of proliferative cytokines, thereby removing a block to proliferation [20]. Diarmaid et al. concluded that HSCs might reduce the liver injury through paracrine effects. The expression of growth factors, including HGF and VEGF, promote liver regeneration and hepatocyte proliferation [4,42]. In addition to their differentiation to hepatocytes, MSCs infusions can also exert a paracrine proliferative effect on endogenous hepatocytes. The primary cytokines include HGF, VEGF, and NGF, which are reported to increase the intrinsic ability of hepatocytes to proliferate or to facilitate the breakdown of scar tissue [20,43,44,45]. In this study, the expression levels of NGF in the MSCs group were higher than in the HSCs group, while the levels of HGF and VEGF were not significantly different in each group. We also discovered more significant hepatocyte proliferation in the MSCs group relative to the other groups. Therefore, the improvement of liver functions by MSC transplantation was partially contributed to enhancing hepatocyte proliferation.

Furthermore, the immunomodulatory properties of MSCs and HSCs can also play a significant role in the extenuation of liver injury. The local down regulation of pro-inflammatory cytokines and up regulation of anti-inflammatory cytokines, such as IL-10, after MSC transplantation has been described in kidney, lung injury and fulminant hepatic failure models [46,47]. Pulavendran et al. found that IL-6 and TNF-α, which can be down regulated by IL-10 and are promoters of liver fibrosis, were lower in the

MSCs group when compared with the control group in acute CCl_4-induced liver injury. These authors also reported that HSCs did not seem to have any role in immunosuppression, and hence, its therapeutic application in the case of liver fibrogenic diseases is doubtful [48]. While Li et al. reported that autologous HSCs transplantation was positively correlated with the concentrations of serum IL-10 but negatively correlated with the levels of serum TNF-α in patients with type 1 diabetes [14]. Whether HSCs exhibit immunomodulatory properties in chronic CCl_4-induced liver cirrhosis remains unclear. Our data show that the concentration of IL-10 was higher in MSCs group than in the HSCs group, although the levels in both groups were higher than that in the control group. However, the concentrations of IL-6 and TNF-α in the MSCs group were lower than those in the HSCs group. Because the proinammatory phase followed by the injury leads to the induction of fibrosis, oppressing the proinammatory cytokines results in the reversal of hepatic fibrosis. Thus, anti-inflammatory cytokines can reduce liver injury and then reduce hepatic fibrosis.

Several studies have demonstrated the synergistic actions of HSCs and MSCs in tissue regeneration and engineering. Moioli et al. found that the co-transplantation of HSCs and MSCs facilitated the neovascularization process in the bioengineered bone [15]. Consistently, These phenomena were also found in heart failure patients treated with a combination of HSCs and MSCs [17]. The rationale for combining these two stem cell populations came from several perspectives that include: (i) both MSCs and HSCs are mobilized after tissue injury, and thus, potentially administering both cell types may produce therapeutic synergistic activity in the injured tissue [49,50,51];(ii) in vitro, MSCs promote the expansion of HSCs; *in vivo*, MSCs provide a microenvironment for HSCs in both the embryonic and postnatal stage [52]. In this study, however, the synergistic actions of MSCs and HSCs had not been observed because the stem cells in liver were detected mostly as single cell which are insufficient to provide a beneficial niche for regulating the self-renewing and differentiation of HSCs.

In conclusion, these results suggest that MSCs have a greater ability to modulate chemically induced inflammation in the fibrotic liver than do HSCs. However, liver fibrosis has different causes, including alcoholic hepatitis, allograft rejection, autoimmune hepatitis and metabolic diseases. All of these disease mechanisms are quite different. In addition, the different stem cells have a variety of putative functional roles; thus, careful thought is required as to what biological action is intended from their infusion. Accordingly, further studies are needed concerning the choice of a cell therapy for specific types of liver injury.

Supporting Information

Figure S1 Comparison of recruitment of stem cells to the CCl_4 -induced cirrhotic liver and the normal liver. Average radiance was quantified in the liver after stem cells transplantation. (A,MSCs group; B, HSCs group; and C, MSCs+HSCs group). $P<0.05$, n = 3.

Author Contributions

Conceived and designed the experiments: QL XMZ YQS YH DMF. Performed the experiments: QL JGL LNC LHZ LW. Analyzed the data: ZYH YQS JZ YH. Contributed reagents/materials/analysis tools: XMZ JGL YH. Wrote the paper: QL YQS YH.

References

1. Peng L, Xie DY, Lin BL, Liu J, Zhu HP, et al. (2011) Autologous bone marrow mesenchymal stem cell transplantation in liver failure patients caused by hepatitis B: short-term and long-term outcomes. Hepatology 54: 820–828.
2. Yannaki E, Anagnostopoulos A, Kapetanos D, Xagorari A, Iordanidis F, et al. (2006) Lasting amelioration in the clinical course of decompensated alcoholic cirrhosis with boost infusions of mobilized peripheral blood stem cells. Exp Hematol 34: 1583–1587.
3. am Esch JS 2nd, Knoefel WT, Klein M, Ghodsizad A, Fuerst G, et al. (2005) Portal application of autologous CD133+ bone marrow cells to the liver: a novel concept to support hepatic regeneration. Stem Cells 23: 463–470.
4. Houlihan DD, Newsome PN (2008) Critical review of clinical trials of bone marrow stem cells in liver disease. Gastroenterology 135: 438–450.
5. Zhao Q, Ren H, Zhu D, Han Z (2009) Stem/progenitor cells in liver injury repair and regeneration. Biol Cell 101: 557–571.
6. Fang B, Shi M, Liao L, Yang S, Liu Y, et al. (2004) Systemic infusion of FLK1(+) mesenchymal stem cells ameliorate carbon tetrachloride-induced liver fibrosis in mice. Transplantation 78: 83–88.
7. Zhao W, Li JJ, Cao DY, Li X, Zhang LY, et al. (2012) Intravenous injection of mesenchymal stem cells is effective in treating liver fibrosis. World J Gastroenterol 18: 1048–1058.
8. van Poll D, Parekkadan B, Cho CH, Berthiaume F, Nahmias Y, et al. (2008) Mesenchymal stem cell-derived molecules directly modulate hepatocellular death and regeneration in vitro and in vivo. Hepatology 47: 1634–1643.
9. Shi Y, Hu G, Su J, Li W, Chen Q, et al. (2010) Mesenchymal stem cells: a new strategy for immunosuppression and tissue repair. Cell Res 20: 510–518.
10. Rasmusson I (2006) Immune modulation by mesenchymal stem cells. Exp Cell Res 312: 2169–2179.
11. Parekkadan B, van Poll D, Megeed Z, Kobayashi N, Tilles AW, et al. (2007) Immunomodulation of activated hepatic stellate cells by mesenchymal stem cells. Biochem Biophys Res Commun 363: 247–252.
12. Zhao DC, Lei JX, Chen R, Yu WH, Zhang XM, et al. (2005) Bone marrow-derived mesenchymal stem cells protect against experimental liver fibrosis in rats. World J Gastroenterol 11: 3431–3440.
13. Jang YY, Collector MI, Baylin SB, Diehl AM, Sharkis SJ (2004) Hematopoietic stem cells convert into liver cells within days without fusion. Nat Cell Biol 6: 532–539.
14. Li L, Shen S, Ouyang J, Hu Y, Hu L, et al. (2012) Autologous hematopoietic stem cell transplantation modulates immunocompetent cells and improves beta-cell function in Chinese patients with new onset of type 1 diabetes. J Clin Endocrinol Metab 97: 1729–1736.
15. Moioli EK, Clark PA, Chen M, Dennis JE, Erickson HP, et al. (2008) Synergistic actions of hematopoietic and mesenchymal stem/progenitor cells in vascularizing bioengineered tissues. PLoS One 3: e3922.
16. Urban VS, Kiss J, Kovacs J, Gocza E, Vas V, et al. (2008) Mesenchymal stem cells cooperate with bone marrow cells in therapy of diabetes. Stem Cells 26: 244–253.
17. Ichim TE, Solano F, Lara F, Rodriguez JP, Cristea O, et al. (2010) Combination stem cell therapy for heart failure. Int Arch Med 3: 5.
18. Miyata E, Masuya M, Yoshida S, Nakamura S, Kato K, et al. (2008) Hematopoietic origin of hepatic stellate cells in the adult liver. Blood 111: 2427–2435.
19. Ezquer M, Ezquer F, Ricca M, Allers C, Conget P (2011) Intravenous administration of multipotent stromal cells prevents the onset of non-alcoholic steatohepatitis in obese mice with metabolic syndrome. J Hepatol 55: 1112–1120.
20. Wang J, Zhou X, Cui L, Yan L, Liang J, et al. (2010) The significance of CD14+ monocytes in peripheral blood stem cells for the treatment of rat liver cirrhosis. Cytotherapy 12: 1022–1034.
21. Belema-Bedada F, Uchida S, Martire A, Kostin S, Braun T (2008) Efficient homing of multipotent adult mesenchymal stem cells depends on FROUNT-mediated clustering of CCR2. Cell Stem Cell 2: 566–575.
22. Li C, Kong Y, Wang H, Wang S, Yu H, et al. (2009) Homing of bone marrow mesenchymal stem cells mediated by sphingosine 1-phosphate contributes to liver fibrosis. J Hepatol 50: 1174–1183.
23. Mendt M, Cardier JE (2012) Stromal-derived factor-1 and its receptor, CXCR4, are constitutively expressed by mouse liver sinusoidal endothelial cells: implications for the regulation of hematopoietic cell migration to the liver during extramedullary hematopoiesis. Stem Cells Dev 21: 2142–2151.
24. Kawai K, Xue F, Takahara T, Kudo H, Yata Y, et al. (2012) Matrix metalloproteinase-9 contributes to the mobilization of bone marrow cells in the injured liver. Cell Transplant 21: 453–464.
25. Chen Y, Xiang LX, Shao JZ, Pan RL, Wang YX, et al. (2010) Recruitment of endogenous bone marrow mesenchymal stem cells towards injured liver. J Cell Mol Med 14: 1494–1508.
26. Kollet O, Shivtiel S, Chen YQ, Suriawinata J, Thung SN, et al. (2003) HGF, SDF-1, and MMP-9 are involved in stress-induced human CD34+ stem cell recruitment to the liver. J Clin Invest 112: 160–169.
27. Son BR, Marquez-Curtis LA, Kucia M, Wysoczynski M, Turner AR, et al. (2006) Migration of bone marrow and cord blood mesenchymal stem cells in vitro is regulated by stromal-derived factor-1-CXCR4 and hepatocyte growth

factor-c-met axes and involves matrix metalloproteinases. Stem Cells 24: 1254–1264.
28. Cho KA, Ju SY, Cho SJ, Jung YJ, Woo SY, et al. (2009) Mesenchymal stem cells showed the highest potential for the regeneration of injured liver tissue compared with other subpopulations of the bone marrow. Cell Biol Int 33: 772–777.
29. Kanazawa H, Fujimoto Y, Teratani T, Iwasaki J, Kasahara N, et al. (2011) Bone marrow-derived mesenchymal stem cells ameliorate hepatic ischemia reperfusion injury in a rat model. PLoS One 6: e19195.
30. Sakaida I, Terai S, Yamamoto N, Aoyama K, Ishikawa T, et al. (2004) Transplantation of bone marrow cells reduces CCl4-induced liver fibrosis in mice. Hepatology 40: 1304–1311.
31. di Bonzo LV, Ferrero I, Cravanzola C, Mareschi K, Rustichell D, et al. (2008) Human mesenchymal stem cells as a two-edged sword in hepatic regenerative medicine: engraftment and hepatocyte differentiation versus profibrogenic potential. Gut 57: 223–231.
32. Sawitza I, Kordes C, Reister S, Haussinger D (2009) The niche of stellate cells within rat liver. Hepatology 50: 1617–1624.
33. Krause DS, Theise ND, Collector MI, Henegariu O, Hwang S, et al. (2001) Multi-organ, multi-lineage engraftment by a single bone marrow-derived stem cell. Cell 105: 369–377.
34. Newsome PN, Johannessen I, Boyle S, Dalakas E, McAulay KA, et al. (2003) Human cord blood-derived cells can differentiate into hepatocytes in the mouse liver with no evidence of cellular fusion. Gastroenterology 124: 1891–1900.
35. Aurich H, Sgodda M, Kaltwasser P, Vetter M, Weise A, et al. (2009) Hepatocyte differentiation of mesenchymal stem cells from human adipose tissue in vitro promotes hepatic integration in vivo. Gut 58: 570–581.
36. Sato Y, Araki H, Kato J, Nakamura K, Kawano Y, et al. (2005) Human mesenchymal stem cells xenografted directly to rat liver are differentiated into human hepatocytes without fusion. Blood 106: 756–763.
37. Aurich I, Mueller LP, Aurich H, Luetzkendorf J, Tisljar K, et al. (2007) Functional integration of hepatocytes derived from human mesenchymal stem cells into mouse livers. Gut 56: 405–415.
38. Korbling M, Estrov Z (2003) Adult stem cells for tissue repair - a new therapeutic concept N Engl J Med 349: 570–582.
39. Orlic D (2003) Adult bone marrow stem cells regenerate myocardium in ischemic heart disease. Ann N Y Acad Sci 996: 152–157.
40. Fassett J, Tobolt D, Hansen LK (2006) Type I collagen structure regulates cell morphology and EGF signaling in primary rat hepatocytes through cAMP-dependent protein kinase A. Mol Biol Cell 17: 345–356.
41. Trak-Smayra V, Contreras J, Dondero F, Durand F, Dubois S, et al. (2004) Role of replicative senescence in the progression of fibrosis in hepatitis C virus (HCV) recurrence after liver transplantation. Transplantation 77: 1755–1760.
42. Majka M, Janowska-Wieczorek A, Ratajczak J, Ehrenman K, Pietrzkowski Z, et al. (2001) Numerous growth factors, cytokines, and chemokines are secreted by human CD34(+) cells, myeloblasts, erythroblasts, and megakaryoblasts and regulate normal hematopoiesis in an autocrine/paracrine manner. Blood 97: 3075–3085.
43. Hsiao ST, Asgari A, Lokmic Z, Sinclair R, Dusting GJ, et al. (2012) Comparative analysis of paracrine factor expression in human adult mesenchymal stem cells derived from bone marrow, adipose, and dermal tissue. Stem Cells Dev 21: 2189–2203.
44. Sassoli C, Pini A, Chellini F, Mazzanti B, Nistri S, et al. (2012) Bone marrow mesenchymal stromal cells stimulate skeletal myoblast proliferation through the paracrine release of VEGF. PLoS One 7: e37512.
45. Weil BR, Markel TA, Herrmann JL, Abarbanell AM, Meldrum DR (2009) Mesenchymal stem cells enhance the viability and proliferation of human fetal intestinal epithelial cells following hypoxic injury via paracrine mechanisms. Surgery 146: 190–197.
46. Togel F, Hu Z, Weiss K, Isaac J, Lange C, et al. (2005) Administered mesenchymal stem cells protect against ischemic acute renal failure through differentiation-independent mechanisms. Am J Physiol Renal Physiol 289: F31–42.
47. Ortiz LA, Dutreil M, Fattman C, Pandey AC, Torres G, et al. (2007) Interleukin 1 receptor antagonist mediates the antiinflammatory and antifibrotic effect of mesenchymal stem cells during lung injury. Proc Natl Acad Sci U S A 104: 11002–11007.
48. Pulavendran S, Vignesh J, Rose C (2010) Differential anti-inflammatory and anti-fibrotic activity of transplanted mesenchymal vs. hematopoietic stem cells in carbon tetrachloride-induced liver injury in mice. Int Immunopharmacol 10: 513–519.
49. Herrera MB, Bussolati B, Bruno S, Fonsato V, Romanazzi GM, et al. (2004) Mesenchymal stem cells contribute to the renal repair of acute tubular epithelial injury. Int J Mol Med 14: 1035–1041.
50. Seebach C, Henrich D, Tewksbury R, Wilhelm K, Marzi I (2007) Number and proliferative capacity of human mesenchymal stem cells are modulated positively in multiple trauma patients and negatively in atrophic nonunions. Calcif Tissue Int 80: 294–300.
51. Ciulla MM, Ferrero S, Gianelli U, Paliotti R, Magrini F, et al. (2007) Direct visualization of neo-vessel formation following peripheral injection of bone marrow derived CD34+ cells in experimental myocardial damage. Micron 38: 321–322.
52. Wang XY, Lan Y, He WY, Zhang L, Yao HY, et al. (2008) Identification of mesenchymal stem cells in aorta-gonad-mesonephros and yolk sac of human embryos. Blood 111: 2436–2443.

n-Butylidenephthalide (BP) Maintains Stem Cell Pluripotency by Activating Jak2/Stat3 Pathway and Increases the Efficiency of iPS Cells Generation

Shih-Ping Liu[1,2*], Horng-Jyh Harn[3], Ying-Jiun Chien[1], Cheng-Hsuan Chang[1], Chien-Yu Hsu[1], Ru-Huei Fu[2,4], Yu-Chuen Huang[5,6], Shih-Yin Chen[5,6], Woei-Cherng Shyu[2], Shinn-Zong Lin[2]

1 Graduate Institute of Basic Medical Science, China Medical University, Taichung, Taiwan, **2** Center for Neuropsychiatry, China Medical University Hospital, Taichung, Taiwan, **3** Department of Pathology, China Medical University Hospital, Taichung, Taiwan, **4** Graduate Institute of Immunology, China Medical University, Taichung, Taiwan, **5** Genetics Center, Department of Medical Research, China Medical University Hospital, Taichung, Taiwan, **6** Graduate Institute of Chinese Medical Science, China Medical University, Taichung, Taiwan

Abstract

In 2006, induced pluripotent stem (iPS) cells were generated from somatic cells by introducing Oct4, Sox2, c-Myc and Klf4. The original process was inefficient; maintaining the pluripotency of embryonic stem (ES) and iPS cell cultures required an expensive reagent–leukemia induced factor (LIF). Our goal is to find a pure compound that not only maintains ES and iPS cell pluripotency, but also increases iPS cell generation efficiency. From 15 candidate compounds we determined that 10 μg/ml n-Butylidenephthalide (BP), an *Angelica sinensis* extract, triggers the up-regulation of Oct4 and Sox2 gene expression levels in MEF cells. We used ES and iPS cells treated with different concentrations of BP to test its usefulness for maintaining stem cell pluripotency. Results indicate higher expression levels of several stem cell markers in BP-treated ES and iPS cells compared to controls that did not contain LIF, including alkaline phosphatase, SSEA1, and Nanog. Embryoid body formation and differentiation results confirm that BP containing medium culture was capable of maintaining ES cell pluripotency after six time passage. Microarray analysis data identified PPAR, ECM, and Jak-Stat signaling as the top three deregulated pathways. We subsequently determined that phosphorylated Jak2 and phosphorylated Stat3 protein levels increased following BP treatment and suppressed with the Jak2 inhibitor, AG490. The gene expression levels of cytokines associated with the Jak2-Stat3 pathway were also up-regulated. Last, we used pou5f1-GFP MEF cells to test iPS generation efficiency following BP treatment. Our data demonstrate the ability of BP to maintain stem cell pluripotency via the Jak2-Stat3 pathway by inducing cytokine expression levels, at the same time improving iPS generation efficiency.

Editor: Pranela Rameshwar, University of Medicine and Dentistry of New Jersey, United States of America

Funding: The authors wish to thank the Taiwan Department of Health Clinical Trial and Research Center of Excellence (DOH101-TD-B-111-004), National Science Council plan (NSC 101-2314-B-039-009-MY2) and China Medical University (CMU99-COL-45) for their financial support. The funders had no role in study design, data collection and analysis, decision to publish, or preparation of the manuscript.

Competing Interests: The authors have declared that no competing interests exist.

* E-mail: spliu@mail.cmu.edu.tw

These authors contributed equally to this work.

Introduction

Stem cells are currently being used for many clinical therapeutic purposes. For example, the combination of hematopoietic stem cells (HSCs) and transplanted bone marrow is applied to treat leukemia, hemophilia, and anemia. Cells that respond to ischemia or injuries and are part of revascularization processes are known as mesenchymal stem cells (MSCs) [1]. Embryonic stem cells (ESCs), which are pluripotent cells derived from the inner cell masses of mammalian blastocysts, are capable of differentiating into the endodermal, mesodermal, and ectodermal cells of embryos [2]. ESCs are viewed as having significant potential for clinical cell therapies due to their ability to self-renew and differentiate into a wide range of specialized cell types [3]. However, they have two major drawbacks for therapeutic use: immune rejection and challenges based on ethical concerns.

Induced pluripotent stem (iPS) cells can be generated from human and mice fibroblasts to which four genes have been introduced: Oct4, Sox2, c-Myc and Klf4 [4,5]. They are similar to ESCs in terms of proliferation, morphology, gene expression, surface antigens, the epigenetic status of pluripotent cell-specific genes, and telomerase activity. ES and iPS cell pluripotency gives them exciting potential for use in tissue repair and replacement therapies [6], but the inefficiency of reprogramming primary human cells makes it difficult to generate patient-specific iPS cells from a small starting population [7]. In addition, maintaining ES and/or iPS cell pluripotency requires treatment with leukemia inhibitory factor (LIF), an expensive reagent. LIF signaling (via Jaks) involves the activation of Stat3 (a signal transducer and activator of transcription 3) [8], which is essential for LIF-dependent ES cell self-renewal [9]. LIF transmits signals via LIF receptors and gp130, a co-receptor of the IL-6 cytokine family that also includes IL-11, CNTF, and OSM. [10,11,12,13,14] The gp130 and LIF receptors lack kinase catalytic domains, but they are capable of binding to and activating one or more members of the Jak-Stat tyrosine kinase family [15,16,17]. The cytokine superfamily uses the Jak-Stat pathway as a major signaling pathway into cell nuclei [18].

Angelica sinensis (called *danggui* in Chinese), one of the most commonly used traditional Chinese medicines, is variously prescribed as a tonic, hemopoetic, spasmolytic, and analgesic [19]. n-Butylidenephthalide (BP), a compound derived from Angelica sinensis chloroform extract, has been identified as having a strong antitumoral effect, arresting the growth and apoptosis of malignant brain tumors in vitro and in vivo [20,21]. However, while these findings indicate that BP holds potential as an anti-cancer compound for clinical applications, little is known about BP function in terms of stem cell activity.

Our goal in this study was to determine whether a pure compound extracted from a traditional Chinese medicine is capable of maintaining ES and iPS cell pluripotency while increasing iPS cell generation efficiency. Our main finding is that BP is capable of maintaining ES and iPS cell pluripotency via Jak2 and Stat3 activation. We also determined that BP treatment increased iPS cell generation efficiency.

Results

MTT Assays for BP Treatment

MTT assays were used to identify MEF cell viability following BP treatment. According to data from tests using eight concentrations of BP (5, 10, 20, 40, 80, 160, 320 and 640 μg/ml), cell survival rates significantly decreased in the 80, 160, 320 and 640 μg/ml treatment groups 24 h and 72 h post-treatment (Fig. 1A). We therefore selected the 5, 10, 20 and 40 μg/ml BP concentrations for our experiments.

BP Treatment Resulted in Oct4 and Sox2 Gene Expression Up-Regulation

Our data indicated significant up-regulation of Oct4 and Sox2 gene expression levels in MEF cells following treatment with BP, especially in the 10 μg/ml treatment group (Fig. 1B). The effect decreased as concentration increased to 20 and 40 μg/ml. Oct4 and Sox 2 expression levels were higher than in the solvent control group; in contrast, c-Myc and Klf4 gene expression levels were not.

BP Treatment Maintained ES and iPS Cell Self-Renewal

After detecting Oct4 and Sox2 gene expression up-regulation following BP treatment, we performed tests to determine whether BP treatment also maintained ES and iPS cell self-renewal and pluripotency as an alternative to LIF. Alkaline phosphatase (AP), Nanog, and SSEA1 were used to identify ES and iPS cell pluripotency. According to results from our ES cell analyses, quantities of AP-positive clones in ES cells treated with 10, 20 or 40 μg/ml of BP were all significantly higher than in the controls that did not contain LIF (LIF OUT control) (Fig. 2A). However, quantities of AP-positive clones in ES cells treated with 5 μg/ml BP were lower compared to the LIF OUT cells (Fig. 2A). In addition, data for Nanog and SSEA-1 levels in ES cells indicated that those treated with 10, 20 or 40 μg/ml BP had higher quantities compared to the LIF OUT control cells (Figs 2B and C).

Mouse iPS cells (from Nanog-GFP transgenic mice) were obtained from the laboratory of Dr. Shinya Yamanaka. [5] AP staining analysis data were similar to ES cell results, with quantities of AP-positive clones in 10, 20 and 40 μg/ml BP-treated iPS cells being significantly greater compared to the LIF OUT control cells (Fig. 3A). As shown in Fig. 3B, GFP signaling (Nanog expression) in the 10, 20 and 40 μg/ml BP-treated cells was higher than in the LIF OUT controls. The SSEA-1 stem cell marker staining showed that 40 μg/ml BP-treated cells was higher than in the LIF OUT controls (Fig. 3C). Combined, the data indicate the ability of BP to maintain ES and iPS cell self-renewal and pluripotency.

Embryoid Body Formation and Differentiation

In another test designed to determine the ability of BP to serve as a LIF substitute in ES cell cultures, embryoid body formation was used to determine ES cell pluripotency by passaging cells six times in culture containing BP (LIF replacement) medium. We observed differentiation initiated by ES cells after adding the embryoid bodies to cell culture plates. Immunofluorescent staining was used to detect cells that were positive for Tuj1 (ectoderm marker), α-smooth muscle actin (α-SMA, mesoderm marker), and Gata4 (endoderm marker). As shown in Fig. 4A, we observed ES cells differentiation in all three germ layer cell types, including neuronal cells (Tuj1), hepatocyte (Gata4) and α-smooth muscle actin (muscle cells). The LIF control differentiation in all three germ layer cell types was showed in Fig. 4B. These data also suggest that BP can serve as a substitute for LIF in ES cell cultures to maintain cell pluripotency after passaging.

Microarray Analysis of Genes Involved in Various Pathways in BP-Treated MEF Cells

Microarray analyses were performed to evaluate gene expression profiles in various pathways in MEF cells treated with 10 μg/ml BP, 40 μg/ml BP, and DMSO (control). All data were submitted to GEO system and the accession numbers are MEF (GSM886449), MEF cells treated with 10 μg/ml BP (GSM886450), and MEF cells treated with 40 μg/ml BP (GSM886451), respectively. A Babelomics web tool and KEGG pathways database were used as part of this procedure. Numbers and percentages of significantly deregulated genes with known biological functions are shown in Table 1. Higher percentages of deregulated genes post-BP treatment were observed in PPAR, ECM-receptor interaction, and Jak-Stat signaling pathways. Since the Jak-Stat signaling pathway is the most important in terms of stem cell maintenance and renewal, we selected it to study stem cell self-renewal and pluripotency maintenance mechanisms.

Phosphorylated Jak2 and Stat3 Levels in BP-Treated ES Cells

Jak2 and Stat3 are known as key Jak-Stat pathway proteins for maintaining stem cell self-renewal. We used real-time PCR and western blot assays to determine Jak2 and Stat3 gene mRNA and protein levels in BP-treated ES cells. As shown in Fig. 5A, both Jak2 and Stat3 mRNA levels significantly increased after treatment with 5 or 10 μg/ml BP. Compared to the LIF OUT controls, Jak2 and Stat3 protein levels increased in ES cells treated with 5 and 10 μg/ml BP (Figs. 5B and C). In addition, the phosphorylated Jak2 and Stat3 forms (active forms) also increased following treatment with 5 and 10 μg/ml BP. According to these data, BP maintains ES cell self-renewal by activating the Jak2-Stat3 signaling pathway. It was also observed that the Jak2 inhibitor, AG490 suppressed the phosphorylated Jak2 and phosphorylated Jak2 expression in ES cells treated with BP (Fig. 4D).

Cytokine Regulation in BP-Treated ES Cells

To understand why BP treatment resulted in Jak2-Stat3 signaling up-regulation, we used real-time PCR analysis to determine the mRNA levels of cytokine genes involved in the Jak2-Stat3 pathway: LIF, EGF, EPO, IL-5, IL-11 and OSM. As

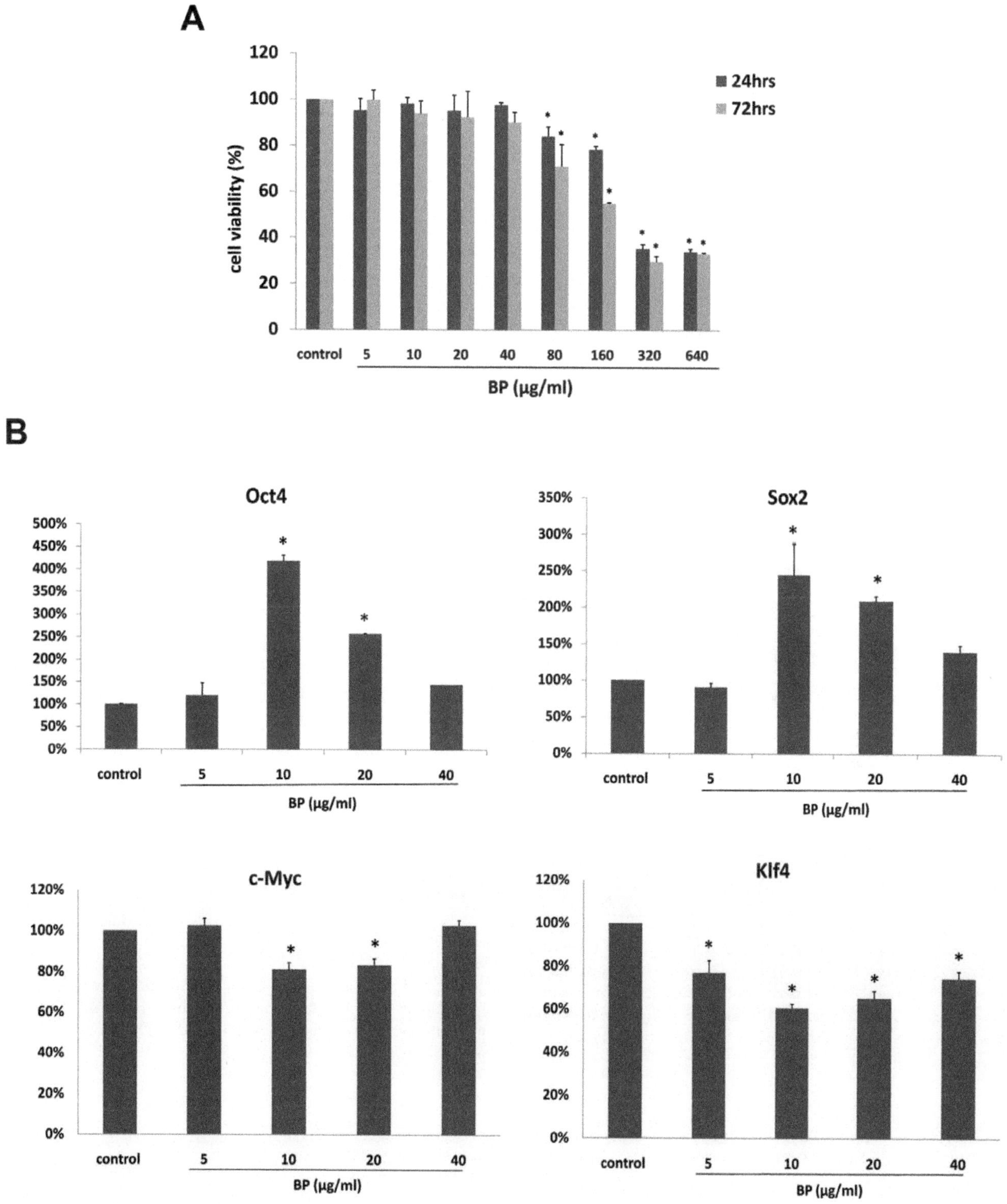

Figure 1. Cell viability and iPS cell related gene expression levels in MEF cells treated with BP. (A) MTT assay of MEF cells treated with various concentrations of BP. (B) Oct4, Sox2, c-Myc and Klf4 gene expression levels in MEF cells treated with various concentrations of BP (real-time PCR).

shown in Fig. 6A, significant up-regulation following treatment with 10 μg/ml BP was observed for LIF, EGF, EPO, IL-5, IL-11 and OSM. These data match those showing the highest levels of phosphorylated Jak2 and Stat3 following treatment with 10 μg/ml BP. Combined, the results suggest that BP is capable of increasing Jak2-Stat3-related cytokine levels, thus activating Jak2 and Stat3 proteins to maintain stem cell pluripotency.

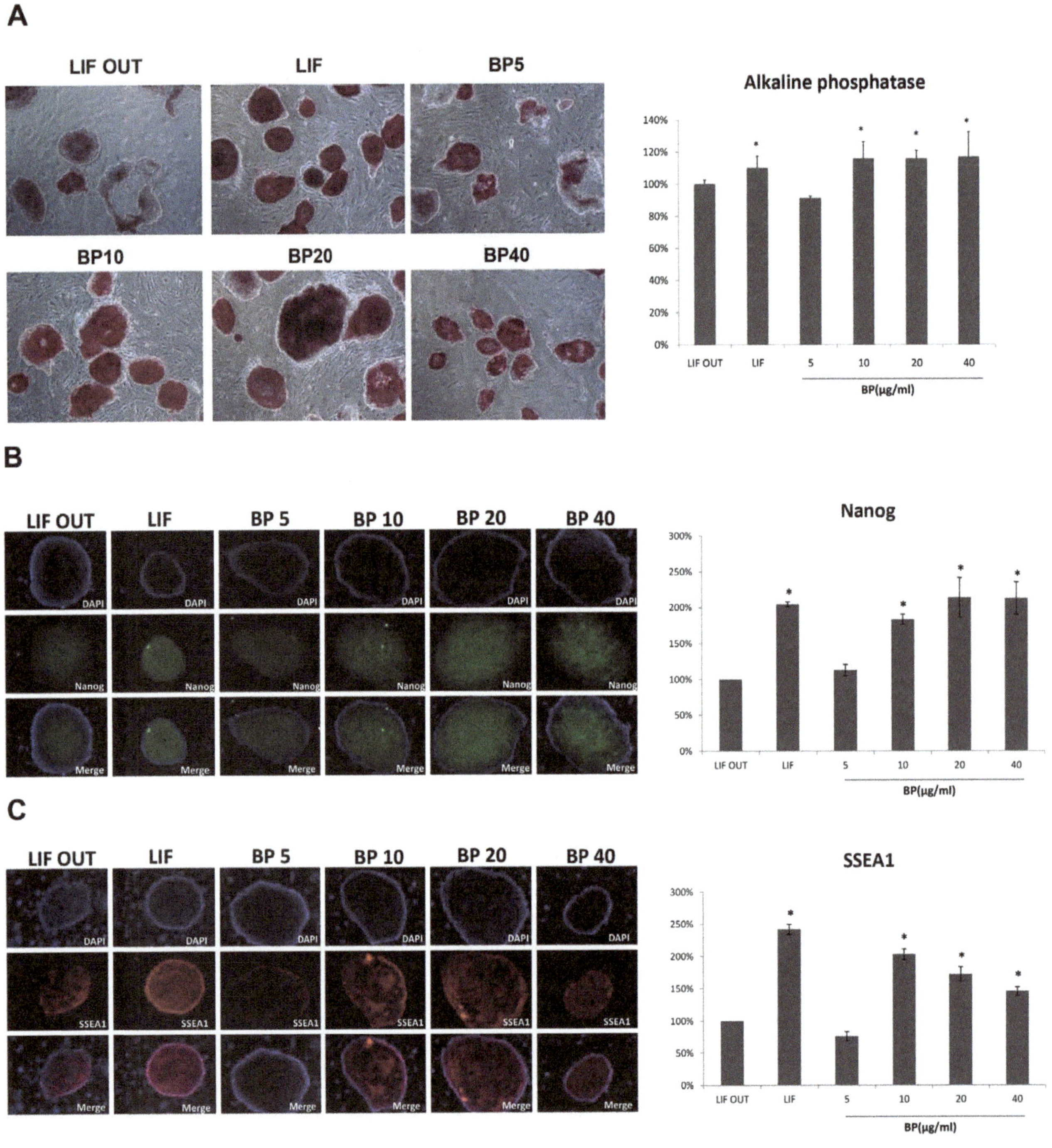

Figure 2. Stem cell markers staining in ES cells treated with BP. (A) Alkaline phosphatase staining ES cells treated with various concentrations of BP and the quantities of AP-positive clones in ES cells. (B) SSEA1 immunocytochemistry in ES cells treated with various concentrations of BP and the quantities of SSEA1 expression levels in ES cells. (C) Nanog immunocytochemistry in ES cells treated with various concentrations of BP and the quantities of SSEA1 expression levels in ES cells. $^*p < 0.05$ versus control.

Improved iPS Cell Generation Efficiency

One of our primary goals was to determine if BP can be used to enhance the efficiency of iPS cell generation. We therefore used MEF obtained from Pou5f1-GFP mice to introduce plasmids containing Oct4, Sox2, c-Myc and KlF4 genes once every two days (4 times total) to generate iPS cells. As part of this process, 10 μg/ml of BP-containing medium was used to culture MEF cells 6 h post-transfection. We established two groups after seeding feeder cells on day 9: BP (T), meaning that treatment was limited to BP-only following transfection, and BP (T+P), meaning that BP treatment occurred both after transfection and after passaging feeder cells. The data show that regardless of treatment/no treatment with 10 μg/ml BP after seeding, the efficiency of iPS cell generation (GFP positive clones) significantly increased in the BP

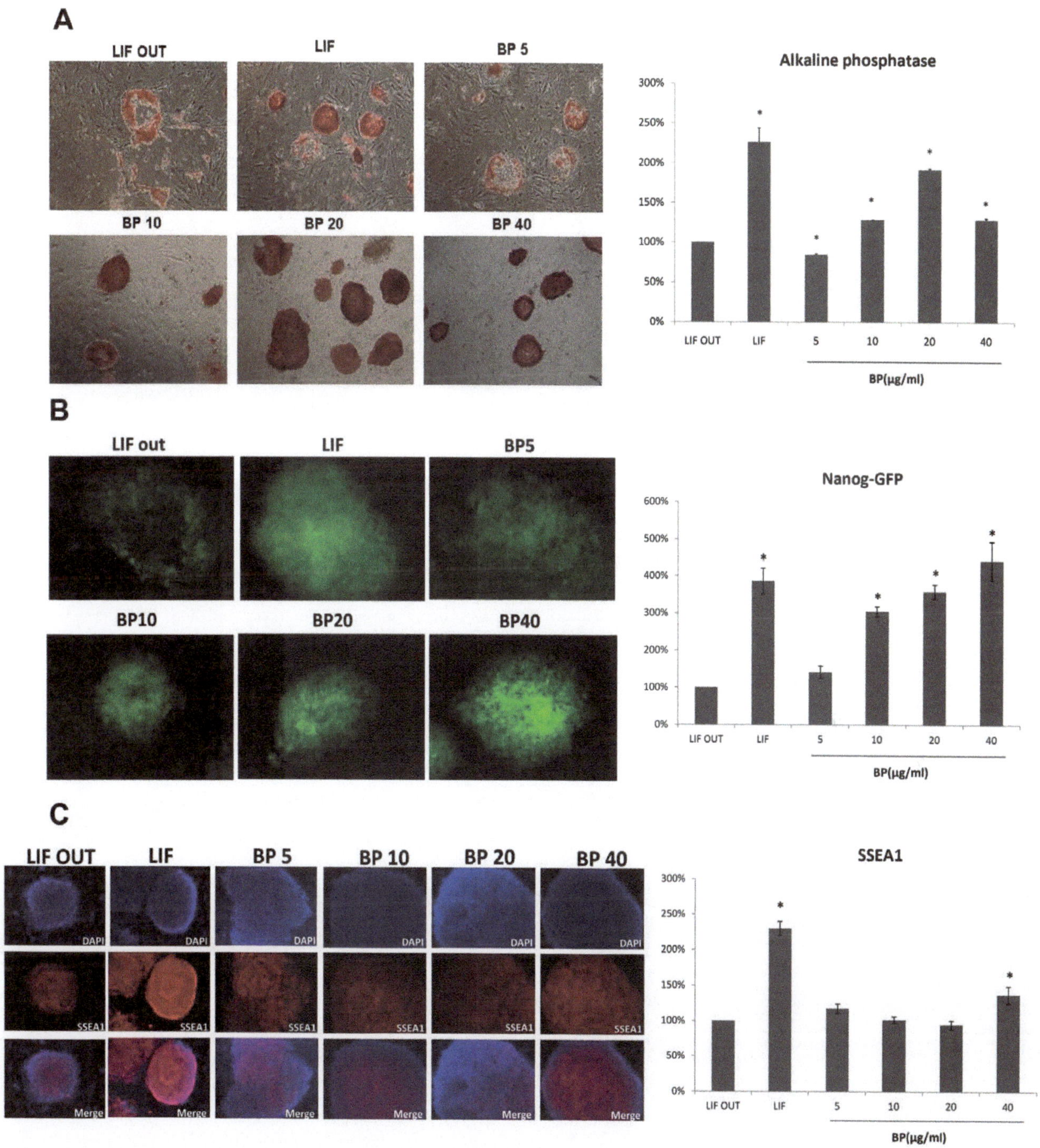

Figure 3. Stem cell markers staining in iPS cells treated with BP. (A) Alkaline phosphatase staining iPS cells treated with various concentrations of BP and the quantities of AP-positive clones in iPS cells. (B) SSEA1 immunocytochemistry in iPS cells treated with various concentrations of BP and the quantities of SSEA1 expression levels in iPS cells. (C) Nanog immunocytochemistry in iPS cells treated with various concentrations of BP and the quantities of SSEA1 expression levels in iPS cells. $*p < 0.05$ versus control.

treatment group (Fig. 6B). In summary, we found evidence that BP treatment enhances iPS cell generation efficiency.

Characterizations of the Novel iPS Cells Generated by Adding BP (iPS-BP)

iPS-BP cells generated by the method mentioned above (BP[T+P]) were determined the pluripotentcy of these novel cells. We used immunofluorescent staining to determine stem cell marker expression in iPS-BP cells. iPS-BP cells were stained with mouse antibodies against Nanog, SSEA1 and alkaline phosphotase. As shown in Fig. 6C, iPS-BP cells tested positive for all four stem cell markers. In addition, we used embryoid body formation to determine iPS-BP cell pluripotency. The results indicated that iPS-OSH cells formed embryoid bodies in

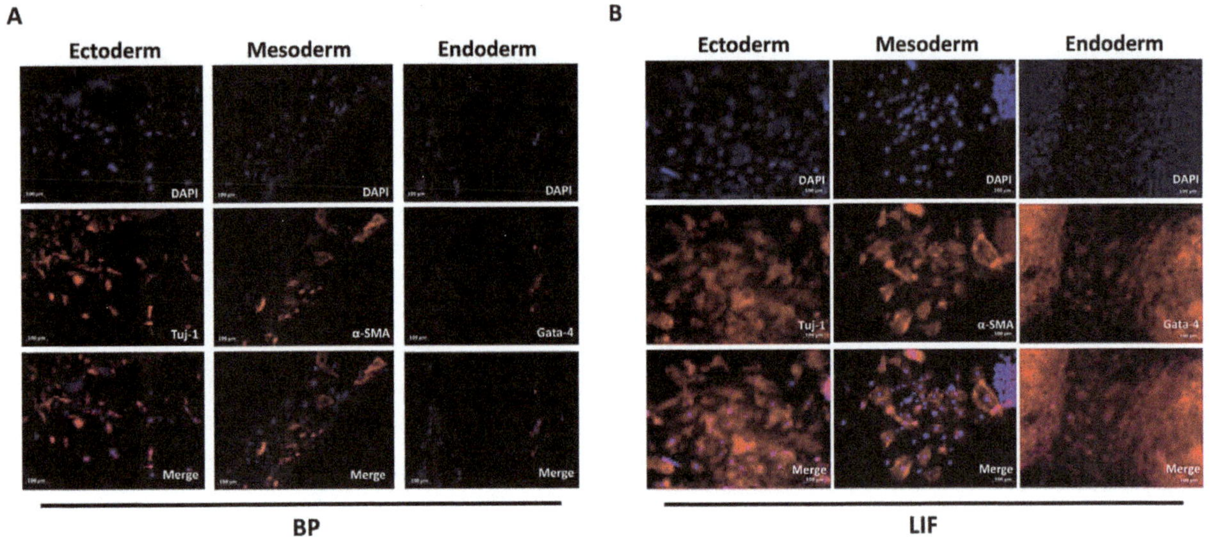

Figure 4. Embryoid body-mediated differentiation of ES cells passaged three times in (A) BP containing medium (replaced LIF) and (B) LIF containing medium as control. Immunofluorescent staining of Tuj1 (ectoderm marker), α-smooth muscle actin (α-SMA, mesoderm marker) and Gata4 (endoderm marker). Nuclei were stained with DAPI (blue).

low-attached plates. We observed differentiation initiated by the iPS-BP cells after placing the embryoid bodies in cell culture plates. Next, we used immunofluorescent staining to detect cells that were positive for Gata4 (endoderm marker), α-smooth muscle actin (mesoderm marker), and Tuj1 (ectoderm marker). As shown in Figure 6D, we observed iPS-BP cells differentiation

Table 1. Numbers of significantly deregulated genes with known biological functions classified according to KEGG and Babelomics databases.

	BP10				BP40			
Function/Pathway	**I**	**D**	**No.**	**%**	**I**	**D**	**No.**	**%**
Signal transduction								
PPAR signaling pathway	14	14	28/74	37.8	12	14	27/74	36.5
ECM-receptor interaction	3	18	21/74	28.4	3	16	19/74	25.7
JAK-STAT signaling pathway	9	30	39/140	27.9	8	27	35/140	25
Calcium signaling pathway	4	28	32/146	21.9	2	30	32/146	21.9
TGF-beta signaling pathway	3	2	5/45	11.1	5	1	6/45	13.3
MAPK signaling	7	17	24/227	10.6	5	19	24/227	10.6
Wnt signaling pathway	2	5	7/66	10.6	0	6	6/66	9.1
Insulin signaling pathway	3	7	10/102	9.8	2	3	5/102	4.9
VEGF signaling pathway	2	0	2/31	6.5	0	0	0/31	0
Cell proliferation								
Cell communication	3	17	20/92	21.7	2	18	20/92	21.7
Cell cycle	0	4	4/163	2.5	0	6	6/163	3.7
Metabolism								
Lipid metabolism	3	15	18/109	16.5	2	12	14/109	12.8
Amino acid metabolism	0	1	1/23	4.3	0	2	2/23	8.7
Cell adhesion								
Cell adhesion molecules	13	20	33/150	22	13	21	34/150	22.7
Tight junction	2	7	9/66	13.6	2	9	11/66	16.7
Focal adhesion	3	16	19/173	11	2	20	22/173	12.7
Apoptosis	3	8	11/94	11.7	1	6	7/94	7.4

I: number of up-regulated genes; D: number of down-regulated genes.

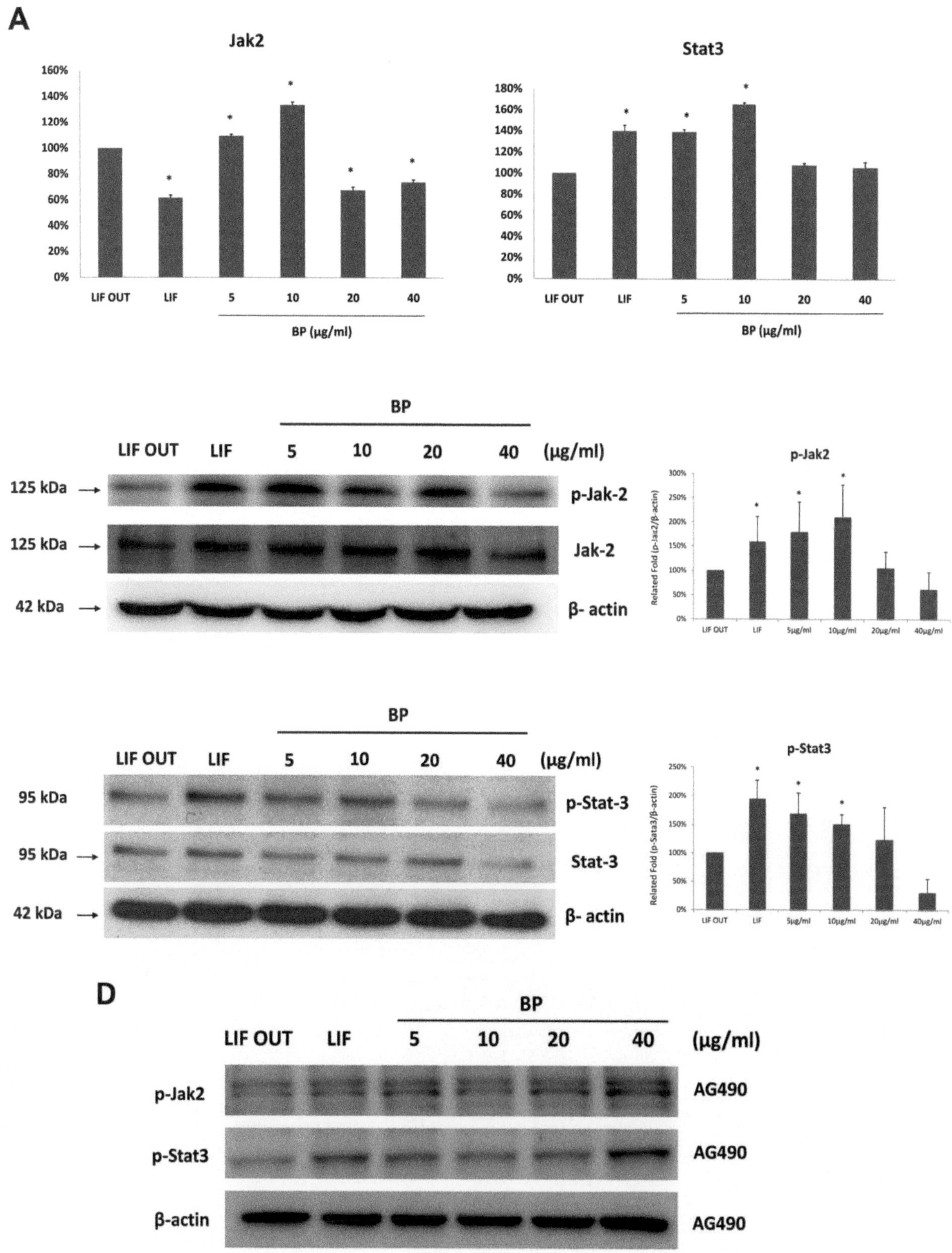

Figure 5. Jak2 and Stat3 expression levels in BP-treated ES Cells. (A) Jak2 and Stat3 gene expression levels in ES cells treated with various concentrations of BP (real-time PCR). (B) Jak2 and phosphorylated-Jak2 protein levels in ES cells treated with BP (western blot). (C) Stat3 and phosphorylated-Stat3 protein levels in ES cells treated with BP (western blot). (D) phosphorylated-Jak2 and phosphorylated-Stat3 protein levels in ES cells treated with BP and Jak2 inhibitor, AG490 (western blot).

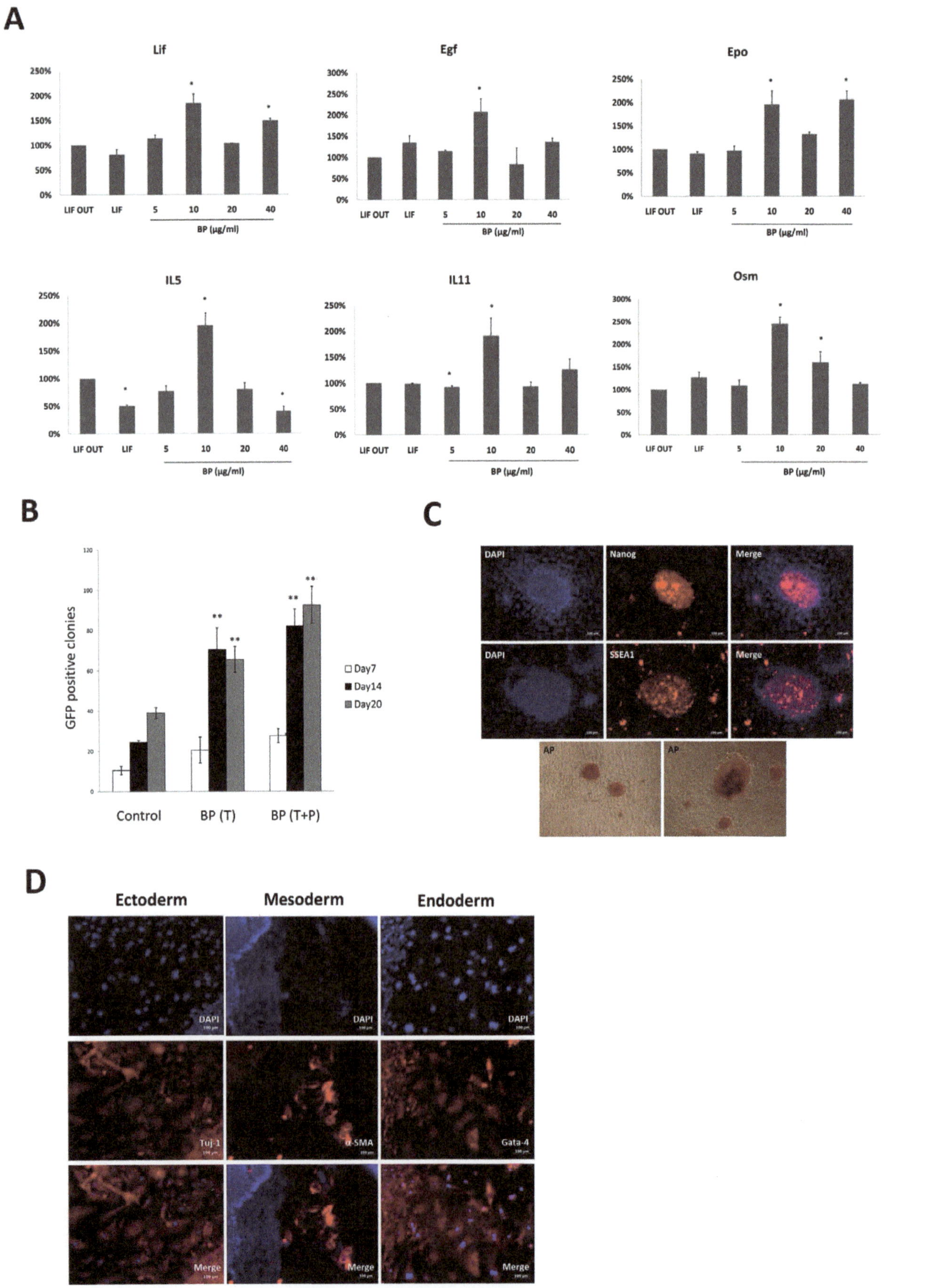
A
Lif
Egf
Epo
IL5
IL11
Osm
LIF OUT
LIF
5
10
20
40
BP (µg/ml)
B
GFP positive clonies
Day7
Day14
Day20
Control
BP (T)
BP (T+P)
C
DAPI
Nanog
Merge
SSEA1
AP
D
Ectoderm
Mesoderm
Endoderm
Tuj-1
α-SMA
Gata-4

Figure 6. Cytokine regulation in BP-treated ES cells and characterizations of the novel iPS cells generated by adding BP (iPS-BP). (A) Real-time PCR results for cytokine gene expression levels associated with the Jak2-Stat3 pathway. Significant up-regulation of six cytokine genes was noted following BP treatment (10 μg/ml). (B) Pou5f1-GFP MEF cells were introduced with plasmids containing Oct4, Sox2, c-Myc and Klf4 to test iPS cell-generation efficiency. BP(T): treatment with BP post-transfection. BP(T+P): treated with BP post-transfection and following passage to feeder cells. Control group was introduced with plasmids containing Oct4, Sox2, c-Myc and Klf4 to generate non-BP treated iPS cells. (C) The stem cell markers (Nanog, SSEA1 and alkaline phosphotase) staining of iPS-BP cells. Nuclei were stained with DAPI (blue). (D) Immunofluorescent staining in iPS-BP cells: Gata4 (endoderm marker), α-smooth muscle actin (mesoderm marker), and Tuj1 (ectoderm marker). Nuclei were stained with DAPI (blue).

in all three germ layer cells, including endoderm (Gata4), mesoderm (α-smooth muscle actin), and ectoderm (Tuj1).

Discussion

Stem cells are capable of self-renewal and differentiation into a wide range of cell types with multiple clinical and therapeutic applications [22]. Of the stem cells we cultured, ES cells have the characteristics of pluripotency and the ability to differentiate into three primary germ layer derivatives. These are important attributes for producing healthy cells for therapeutic purposes. Induced pluripotent (iPS) cells generated from somatic cells via the introduction of four transcriptional factors (Oct4, Sox2, c-Myc and Klf4)6 positively address two important issues associated with ES cells: immune rejection and medical ethics. However, embryonic stem cells (ESCs) and iPS cells require treatment with LIF (an expensive reagent) via the Jak-Stat pathway in order to maintain pluripotency. [8] In addition, low iPS cell generation efficiency is considered an important limitation of iPS cell technology. The BP compound that we tested in this study, derived from Angelica sinensis, has potential to not only serve as a substitute for LIF, but to also improve that efficiency.

BP significantly arrests malignant brain tumor growth and apoptosis in vitro and in vivo. [20,21] However, this anti-tumoral effect requires high concentrations of BP–between 50 and 100 μg/ml. [23,24] Little effort has been made to determine the effects of low BP concentrations such as those used in the present study (40 μg/ml or less), or the mechanisms involved. Results from our MTT assays indicate that treatment at less than 40 μg/ml does not lead to cell death (Fig. 1A), which supports our decision to use lower concentrations in our experiments.

According to Nichols et al., pluripotent stem cell formation in mammalian embryos depends on Oct4;28 later studies revealed that members of the Sox family are also key genes in mammalian development [25]. Specifically, Oct4 and Sox2 have been identified as key proteins in maintaining stem cell pluripotency. We observed up-regulation of Oct4 and Sox2 expression levels in MEF cells following BP treatment at concentrations as low as 10 μg/ml, and were surprised to find decreases in this effect at 20 μg/ml and 40 μg/ml. We therefore used microarray analyses to determine differences in gene expression profiles at 10 μg/ml and 40 μg/ml concentrations, and found no differences between the two in terms of KEGG pathway (Table 1). Our conclusion is that BP treatment at 10 μg/ml is optimal for Oct4 and Sox2 up-regulation.

As Ezashi et al. report, human ES cells grow equally well under hypoxic and normoxic conditions [6]. Tejedo et al. reported that the exposure of LIF-deprived mESCs to low levels of NO prevent the loss of self-renewal gene expression (i.e., Oct4, Nanog and Sox2) [26]. Sasaki et al. recently described the contribution of LacdiNAc (GalNAcβ1-4GlcNAc) to mouse embryonic stem cell self-renewal via the regulation of leukemia inhibitory factor/Stat3 signaling [27]. However, none of these researchers considered the potential of natural compounds to maintain stem cell self-renewal. We used BP, an extract of Angelica sinensis, as a substitute for LIF when culturing ES and iPS cells, and then used several stem cell markers containing AP, Nanog, and SSEA-1 to determine the effects of that substitution. Third, we used embryoid body formation, a powerful method used in iPS studies for confirming stem cell pluripotency [4,5,28,29], to determine ES cell pluripotency by passaging cells three times in culture containing 20 μg/ml BP medium. Our results indicate that the cells maintained pluripotency and the ability to differentiate into three germ layers.

To identify related mechanisms, we used microarray analyses to identify three major pathways affected by BP: PPAR signaling, ECM-receptor interaction, and Jak-Stat signaling. Mo et al. have described how peroxisome proliferator-activated receptors (PPARγ, a nuclear receptor transcription factor) regulates the LIF-induced growth and self-renewal of mouse ES cells via the Tyk2-Stat3 pathway [30]. We measured PPARγ levels in BP-treated and control ES cells, and failed to find any significant differences (data not shown). Focusing on the Jak-Stat signaling pathway, we observed increases in phosphorylated-Jak2 and phosphorylated-Stat3 protein levels following BP treatment (Figs. 5B and C). According to Niwa et al., [8] LIF signaling via Jak2 involves Stat3 activation, which is essential for LIF-dependent ES cell self-renewal [9]. Our data indicate that BP treatment activates Jak2 and Stat3 proteins to maintain ES cell self-renewal; furthermore, the Jak-Stat signaling pathway is activated by cytokines such as LIF, EGF, EPO, IL5, IL11 and OSM. [10,11,12,13,14] After detecting the gene expression levels of these cytokines, we observed the up-regulation of these genes following treatment with 10 μg/ml BP (Fig. 6A). This finding fits well with the observation that the highest levels of phosphorylated Jak2 and phosphorylated Stat3 resulted from BP treatment at 10 μg/ml. We determined that BP treatment increased Jak2-Stat3 related cytokine levels to activate Jak2 and Stat3 proteins, which in turn maintained stem cell pluripotency.

Finally, we introduced four factors via plasmid transfection to test our hypothesis that the BP-associated up-regulation of Oct4 and Sox2 levels also enhances iPS cell generation efficiency. Valproic acid (VPA) is one of several factors that have been tested in an effort to increase the efficiency of iPS cell generation [7]. Our results indicate that VPA treatment did not increase Oct4 and Sox2 expression, but did increase the virus insertion rate. According to Esteban et al., treatment with vitamin C increases the generation of both mouse and human iPS cells [31]. In the present study we used pou5f1-GFP MEF cells to introduce with Oct4, Sox2, c-Myc and Klf4 genes to compare iPS cell generation efficiency with and without BP treatment. Our data indicate that BP treatment produced a 5-fold to 8-fold increase, especially when treatment continued following seeding to feeder cells (Fig. 6B). Our current plans include identifying other factors combined with BP treatment that enhance safety and efficiency when generating iPS cells. In summary, our data demonstrated the ability of BP to maintain stem cell pluripotency via the Jak2-Stat3 pathway by inducing cytokine expression levels, at the same time improving iPS generation efficiency.

Materials and Methods

Mouse Embryonic Fibroblast Cell Cultures

Primary mouse embryonic fibroblast (MEF) cells were isolated from the 13.5 d-old embryos of C57BL/6 mice. Embryos were retrieved by Cesarean section, and internal organs, legs, and heads were removed. The remaining embryo parts were minced with fine scissors and placed in tubes containing Trypsin for cell digestion. MEF cells were cultured in DMEM (GIBCO BRL) with 10% heat-inactivated FBS (GIBCO BRL), penicillin (100 U/ml), streptomycin (100 μg/ml), non-essential amino acids (0.1 mM), and L-glutamine (2 mM) in a humidified incubator (37°C, 5% CO_2). Experimental protocols were approved by the Institutional Animal Care and Use Committee of China Medical University (100-57-N).

Mouse iPS and Embryonic Stem Cell Cultures

Mouse iPS cells were a gift from Dr. Shinya Yamanaka of the Riken Bioresource Center (Ibaraki, Japan) [5]. These and ES cells were cultured in DMEM (GIBCO BRL) with 15% heat-inactivated FBS (Hyclone), non-essential amino acids (0.1 mM), L-glutamine (2 mM), β-mercaptoethanol (0.2 mM) and LIF (4 ng/ml) in a humidified 37°C incubator with 5% CO_2.

BP Treatment

BP (Angelica sinensis extract; mol weight 188.23, 95%) was purchased from Lancaster Synthesis Ltd. (Newgate Morecambe, UK). BP was dissolved in DMSO to a concentration of 100 mg/ml and stored at −20°C until used as a stock solution. MTT assays were performed using 5, 10, 20, 40, 80, 160, 320 and 640 μg/ml concentrations. Concentrations at 5, 10, 20, and 40 μg/ml were used to determine gene expression profiles and for maintaining stem cell self-renewal.

MTT Assays

MEF cells were re-seeded in 96-well plates and treated with different concentrations of BP. Prior to each assay, culture medium in each well was replaced by 100 μL fresh medium containing 10 μL of 5 mg/mL MTT (Sigma-Aldrich) stock solution. After 4 h of labeling cells with MTT, medium was removed and replaced with 100 μL DMSO in each well for 10 min at 37°C. Samples were mixed and absorbances read at 540 nm [32].

Real-Time PCR and RT-PCR

TRIzol (Invitrogen, Carlsbad, CA) was used to extract total RNA from MEF and ES cells; concentrations were determined by spectrophotometry. Complementary DNA was produced from mRNA (5 μg) using a SuperScript III Reverse Transcriptase Kit (Invitrogen). Real-time PCR was performed as previously described [33,34] to determine the gene expression levels of Oct4, Sox2, c-Myc, Klf4, Jak2, Stat3, LIF, EGF, EPO, IL5, IL-11 and OSM. Primer sequences are listed in Table 2.

Alkaline Phosphatase Staining and Indirect Immunofluorescent Antibody Assays (IFA)

Alkaline phosphatase staining was performed using a Vector Leukocyte Alkaline Phosphatase kit. For IFA, cells cultured in slides were treated with fixing solution I (4% paraformaldehyde, 400 mM sucrose in PBS) and held for 30 min at 37°C prior to treatment with fixing solution II (fixing solution I plus 0.5% Triton X-100) at room temp for 15 min. After washing with PBS, slides were treated with blocking buffer (0.5% BSA in PBS) and held at room temp for 1 hr, then washed 3 times with PBS prior to reacting with anti-Nanog (Novus) or anti-SSEA1 (Millipore) antibodies at a dilution of 1:100 (4°C overnight or 37°C for 1 hr). Slides were then washed with cold PBS 5 times prior to reacting with FITC-conjugated anti-mouse IgG or TRITC-conjugated anti-rabbit IgG (Sigma-Aldrich) at a dilution of 1:500. After 5 additional washings with cold PBS, slides were mounted and observed using a fluorescence microscope. DAPI (Invitrogen) was used to stain the DNA and localize the nucleus.

Embryoid Body Formation and Differentiation

Embryoid body formation and differentiation experiments were used to determine whether BP can serve as a LIF replacement for ES cell cultures. ES cells were passaged six times onto 0.1% gelatin-coated tissue culture dishes with feeder layer and cultured with BP-containing medium. After final passages, ES cells were harvested by trypsinization and transferred to ultra-low attached culture dishes in ES medium without LIF. After 3 days, aggregated cells were placed into gelatin-coated tissue culture dishes and incubated for another 3 days before staining with anti-alpha smooth muscle actin monoclonal (Chemicon), anti-Gata4 polyclonal (Abcam), or anti-Tuj1 monoclonal antibodies (Chemicon) plus DAPI.

DNA Microarrays

Total RNA from MEF or MEF treated with either 10 μg/ml or 40 μg/ml BP were labeled with Cy3. Samples were hybridized to Agilent Mouse G3 Whole Genome Oligo 8×60K microarrays according to the manufacturer's instructions. Arrays were scanned with a Microarray Scanner System and data analyzed using GeneSpring GX software (both from Agilent, Santa Clara, CA).

AG490 Treatment

AG490 was purchased from Sigma-Aldrich. AG490 was dissolved in methanol to a concentration of 5 mM and stored at −20°C until used as a stock solution. AG490 were used to inhibit the Jak2 that performed using 5 μM concentrations in the medium.

Western Blot (WB) Assays

Western blot procedures have been described previously [34]. Rabbit anti-Jak2 (cell signaling), mouse anti-phospho-Jak2 (cell signaling), rabbit anti-Stat3 (BD), and rabbit anti-phospho-Stat3 (cell signaling) were used in this research.

Determining iPS Cell Generation Efficiency

MEF cells isolated from Pou5f1-GFP transgenic mice (Jackson Lab) were used to determine iPS cell generation efficiency. pCX-OKS-2A and pCX-cMyc plasmid DNA (supported by Addgene) [28] were transfected into Pou5f1-GFP MEF cells once every two days (4 times total). Medium was changed to 10 μg/ml of BP-containing medium 6 h post-transfection, and held until the next transfection. On day 9, the MEF cells were passaged onto feeder cells and observed to detect ES cell-like clones. After seeding, medium was changed to iPS cell culture medium with or without 10 μg/ml BP, and the two types were separated. One type continued to be treated with 10 μg/ml BP (BP[T+P]), while the other was used with iPS cell culture medium only (BP[T]).

Table 2. Primers used for real-time PCR.

Primer	Forward Sequence	Reverse Sequence	Taqman Probe
Sox2	AGGGCTGGACTGCGAACTG	TTTGCACCCCTCCCAATTC	
Myc	CATTCAAGCAGACGAGCA	CGAGTTAGGTCAGTTTATGCAC	
Klf4	CCTTTCAGTGCCAGAAGT	ACTACGTGGGATTTAAAAGTGC	
Jak2	CAATGATAAACAAGGGCAAATGAT	CTTGGCAATCTTCCGTTGCT	
Stat3	CCCCGTACCTGAAGACCAAGT	CCGTTATTTCCAAACTGCATCA	
EGF	GAGTCTGCCTGCGGATGGT	GCTGCAGGGAGGGAGACA	
EPO	CCCCCACGCCTCATCTG	TGCCTCCTTGGCCTCTAAGA	
IL5	TCCCTGCTACTCTCCCCAAA	CAACCTTCTCTCTCCCCAAGAA	
IL11	CATGCCACACCCCAAACAA	CCCCTCACCCAGGTCTACTG	
LIF	CCTACCTGCGTCTTACTCCATCA	TGTTTTCCCCAAAGGCTCAA	
OSM	CGGTCCACTACAACACCAGATG	GCGATGGTATCCCCAGAGAA	
β-actin	GTGCGTGACATCAAAGAGAAGC	TGGATGCCACAGGATTCCATAC	
Taqman-Oct4	GAGGCTACAGGGACACCTTTC	GTGCCAAAGTGGGGACCT	Roche Universal Probes 6
Taqman-β-actin	CTAAGGCCAACCGTGAAAAG	ACCAGAGGCATACAGGGACA	Roche Universal Probes 64

Statistical Analysis

Results are expressed as mean ± SD. Student's t-tests were used to evaluate mean differences between control and treatment groups. Data lacking normal distribution were analyzed by one-way ANOVA; statistical significance was established as $p<0.05$.

Author Contributions

Conceived and designed the experiments: S-PL S-ZL. Performed the experiments: S-PL Y-JC C-HC C-YH. Analyzed the data: S-PL H-JH Y-JC C-HC C-YH R-HF Y-CH S-YC W-CS S-ZL. Contributed reagents/materials/analysis tools: H-JH W-CS. Wrote the paper: S-PL.

References

1. Liu SP, Fu RH, Yu HH, Li KW, Tsai CH, et al. (2009) MicroRNAs regulation modulated self-renewal and lineage differentiation of stem cells. Cell Transplant 18: 1039–1045.
2. Blakaj A, Lin H (2008) Piecing together the mosaic of early mammalian development through microRNAs. J Biol Chem 283: 9505–9508.
3. Marson A, Levine SS, Cole MF, Frampton GM, Brambrink T, et al. (2008) Connecting microRNA genes to the core transcriptional regulatory circuitry of embryonic stem cells. Cell 134: 521–533.
4. Takahashi K, Tanabe K, Ohnuki M, Narita M, Ichisaka T, et al. (2007) Induction of pluripotent stem cells from adult human fibroblasts by defined factors. Cell 131: 861–872.
5. Takahashi K, Yamanaka S (2006) Induction of pluripotent stem cells from mouse embryonic and adult fibroblast cultures by defined factors. Cell 126: 663–676.
6. Ezashi T, Das P, Roberts RM (2005) Low O2 tensions and the prevention of differentiation of hES cells. Proc Natl Acad Sci U S A 102: 4783–4788.
7. Huangfu D, Osafune K, Maehr R, Guo W, Eijkelenboom A, et al. (2008) Induction of pluripotent stem cells from primary human fibroblasts with only Oct4 and Sox2. Nat Biotechnol 26: 1269–1275.
8. Niwa H, Burdon T, Chambers I, Smith A (1998) Self-renewal of pluripotent embryonic stem cells is mediated via activation of STAT3. Genes Dev 12: 2048–2060.
9. Raz R, Lee CK, Cannizzaro LA, d'Eustachio P, Levy DE (1999) Essential role of STAT3 for embryonic stem cell pluripotency. Proc Natl Acad Sci U S A 96: 2846–2851.
10. Metcalf D (1991) The leukemia inhibitory factor (LIF). Int J Cell Cloning 9: 95–108.
11. Ip NY, Nye SH, Boulton TG, Davis S, Taga T, et al. (1992) CNTF and LIF act on neuronal cells via shared signaling pathways that involve the IL-6 signal transducing receptor component gp130. Cell 69: 1121–1132.
12. Wijdenes J, Heinrich PC, Muller-Newen G, Roche C, Gu ZJ, et al. (1995) Interleukin-6 signal transducer gp130 has specific binding sites for different cytokines as determined by antagonistic and agonistic anti-gp130 monoclonal antibodies. Eur J Immunol 25: 3474–3481.
13. Gearing DP, Thut CJ, VandeBos T, Gimpel SD, Delaney PB, et al. (1991) Leukemia inhibitory factor receptor is structurally related to the IL-6 signal transducer, gp130. EMBO J 10: 2839–2848.
14. Guschin D, Rogers N, Briscoe J, Witthuhn B, Watling D, et al. (1995) A major role for the protein tyrosine kinase JAK1 in the JAK/STAT signal transduction pathway in response to interleukin-6. EMBO J 14: 1421–1429.
15. Ihle JN (1995) Cytokine receptor signalling. Nature 377: 591–594.
16. Narazaki M, Witthuhn BA, Yoshida K, Silvennoinen O, Yasukawa K, et al. (1994) Activation of JAK2 kinase mediated by the interleukin 6 signal transducer gp130. Proc Natl Acad Sci U S A 91: 2285–2289.
17. Boulton TG, Zhong Z, Wen Z, Darnell JE, Jr., Stahl N, et al. (1995) STAT3 activation by cytokines utilizing gp130 and related transducers involves a secondary modification requiring an H7-sensitive kinase. Proc Natl Acad Sci U S A 92: 6915–6919.
18. Johnston JA, Bacon CM, Finbloom DS, Rees RC, Kaplan D, et al. (1995) Tyrosine phosphorylation and activation of STAT5, STAT3, and Janus kinases by interleukins 2 and 15. Proc Natl Acad Sci U S A 92: 8705–8709.
19. Wang H, Chen R, Xu H (1998) [Chemical constituents of radix Angelicae Sinensis]. Zhongguo Zhong Yao Za Zhi 23: 167–168, inside backcover.
20. Tsai NM, Lin SZ, Lee CC, Chen SP, Su HC, et al. (2005) The antitumor effects of Angelica sinensis on malignant brain tumors in vitro and in vivo. Clin Cancer Res 11: 3475–3484.
21. Tsai NM, Chen YL, Lee CC, Lin PC, Cheng YL, et al. (2006) The natural compound n-butylidenephthalide derived from Angelica sinensis inhibits malignant brain tumor growth in vitro and in vivo. J Neurochem 99: 1251–1262.
22. Liu SP, Fu RH, Huang YC, Chen SY, Chien YJ, et al. (2011) Induced pluripotent stem (iPS) cell research overview. Cell Transplant 20: 15–19.
23. Lin PC, Chen YL, Chiu SC, Yu YL, Chen SP, et al. (2008) Orphan nuclear receptor, Nurr-77 was a possible target gene of butylidenephthalide chemotherapy on glioblastoma multiform brain tumor. J Neurochem 106: 1017–1026.
24. Chen YL, Jian MH, Lin CC, Kang JC, Chen SP, et al. (2008) The induction of orphan nuclear receptor Nur77 expression by n-butylenephthalide as pharmaceuticals on hepatocellular carcinoma cell therapy. Mol Pharmacol 74: 1046–1058.
25. Kiefer JC (2007) Back to basics: Sox genes. Dev Dyn 236: 2356–2366.
26. Tejedo JR, Tapia-Limonchi R, Mora-Castilla S, Cahuana GM, Hmadcha A, et al. (2010) Low concentrations of nitric oxide delay the differentiation of embryonic stem cells and promote their survival. Cell Death Dis 1: e80.
27. Sasaki N, Shinomi M, Hirano K, Ui-Tei K, Nishihara S (2011) LacdiNAc (GalNAcbeta1-4GlcNAc) contributes to self-renewal of mouse embryonic stem cells by regulating leukemia inhibitory factor/STAT3 signaling. Stem Cells 29: 641–650.
28. Okita K, Nakagawa M, Hyenjong H, Ichisaka T, Yamanaka S (2008) Generation of mouse induced pluripotent stem cells without viral vectors. Science 322: 949–953.
29. Stadtfeld M, Nagaya M, Utikal J, Weir G, Hochedlinger K (2008) Induced pluripotent stem cells generated without viral integration. Science 322: 945–949.

30. Mo C, Chearwae W, Bright JJ (2010) PPARgamma regulates LIF-induced growth and self-renewal of mouse ES cells through Tyk2-Stat3 pathway. Cell Signal 22: 495–500.
31. Esteban MA, Wang T, Qin B, Yang J, Qin D, et al. (2010) Vitamin C enhances the generation of mouse and human induced pluripotent stem cells. Cell Stem Cell 6: 71–79.
32. Carmichael J, DeGraff WG, Gazdar AF, Minna JD, Mitchell JB (1987) Evaluation of a tetrazolium-based semiautomated colorimetric assay: assessment of radiosensitivity. Cancer Res 47: 943–946.
33. Lee CM, Chen SY, Lee YC, Huang CY, Chen YM (2006) Benzo[a]pyrene and glycine N-methyltransferse interactions: gene expression profiles of the liver detoxification pathway. Toxicol Appl Pharmacol 214: 126–135.
34. Liu SP, Li YS, Chen YJ, Chiang EP, Li AF, et al. (2007) Glycine N-methyltransferase−/− mice develop chronic hepatitis and glycogen storage disease in the liver. Hepatology 46: 1413–1425.

15

Reversal of Ischemic Cardiomyopathy with Sca-1$^+$ Stem Cells Modified with Multiple Growth Factors

Ning Li[1], **Zeeshan Pasha**[2], **Muhammad Ashraf**[2]*

1 Department of Physiology and Cell Biology, The Ohio State University, Columbus, Ohio, United States of America, **2** Department of Pharmacology, University of Illinois at Chicago, Chicago, Illinois, United States of America

Abstract

Background: We hypothesized that bone marrow derived Sca-1$^+$ stem cells (BM Sca-1$^+$) transduced with multiple therapeutic cytokines with diverse effects will induce faster angiomyogenic differentiation in the infarcted myocardium.

Methods and Results: BM Sca-1$^+$ were purified from transgenic male mice expressing GFP. Plasmids encoding for select quartet of growth factors, i.e., human IGF-1, VEGF, SDF-1α and HGF were prepared and used for genetic modification of Sca-1$^+$ cells (^{GF}Sca-1$^+$). Scramble transfected cells (^{Sc}Sca-1$^+$) were used as a control. RT-PCR and western blotting showed significantly higher expression of the growth factors in ^{GF}Sca-1$^+$. Besides the quartet of the therapeutic growth factors, PCR based growth factor array showed upregulation of multiple angiogenic and prosurvival factors such as Ang-1, Ang-2, MMP9, Cx43, BMP2, BMP5, FGF2, and NGF in ^{GF}Sca-1$^+$ ($p<0.01$ *vs* ^{Sc}Sca-1$^+$). LDH and TUNEL assays showed enhanced survival of ^{GF}Sca-1$^+$ under lethal anoxia ($p<0.01$ *vs* ^{Sc}Sca-1$^+$). MTS assay showed significant increased cell proliferation in ^{GF}Sca-1$^+$ ($p<0.05$ *vs* ^{Sc}Sca-1$^+$). For in vivo study, female mice were grouped to receive the intramyocardial injection of 15 μl DMEM without cells (group-1) or containing 2.5×10^5 ^{Sc}Sca-1$^+$ (group-2) or ^{GF}Sca-1$^+$ (group-3) immediately after coronary artery ligation. As indicated by *Sry* gene, a higher survival of ^{GF}Sca-1$^+$ in group-3 on day4 (2.3 fold higher *vs* group-2) was observed with massive mobilization of stem and progenitor cells (cKit$^+$, Mdr1$^+$, Cxcr4$^+$ cells). Heart tissue sections immunostained for actinin and Cx43 at 4 weeks post engraftment showed extensive myofiber formation and expression of gap junctions. Immunostaining for vWF showed increased blood vessel density in both peri-infarct and infarct regions in group-3. Infarct size was attenuated and the global heart function was improved in group-3 as compared to group-2.

Conclusions: Administration of BM Sca-1$^+$ transduced with multiple genes is a novel approach to treat infarcted heart for its regeneration.

Editor: Toru Hosoda, Tokai University, Japan

Funding: This work was supported by NIH grants HL-087246(MA). The funders had no role in study design, data collection and analysis, decision to publish, or preparation of the manuscript.

Competing Interests: The authors have declared that no competing interests exist.

* E-mail: ashrafm2@uic.edu

Introduction

Stem cell based cell therapy offers a potentially therapeutic option for ischemic heart disease [1]. Bone marrow-derived stem cells (BMSCs) have been widely studied for use in cardiac repair due to their favorable properties including multipotency, transdifferentiation, immunomodulation and free from the risks of teratoma formation. Promising results have been reported in preclinical and clinical studies [2–5]. The results show that BMSCs not only differentiate into cardiomyocytes and vascular cells, but also secrete multiple growth factors and cytokines which may mediate endogenous regeneration via activation of resident cardiac stem cells and neovascularization, and reduce apoptosis [6]. Nevertheless, current evidence supports that efficacy of BMSC was limited due to the poor viability and massive death of the engrafted cells in the infarcted myocardium. The heart cell therapy with BMSC to compensate for loss of functional cardiomyocytes during the ischemic episode may be less meaningful without restoration of the regional blood flow in the ischemic myocardium. Hence, it would be practical to combine cell transplantation with therapeutic gene delivery to the heart to achieve maximum benefits of stem cell therapy.

In this study, we hypothesized that a combined approach involving BM Sca-1$^+$ cells genetically modified to express multiple specific therapeutic genes including vascular endothelial growth factor (VEGF), insulin like growth factor-1 (IGF-1), hepatocyte growth factor (HGF) and stromal cell derived factor-1α (SDF-1α) would be more effective in promoting new growth and preservation of the global heart functions. The BM derived Sca-1$^+$ cells would serve as reservoirs of multiple growth factors to support angiomyogenic repair of the infarcted heart. Moreover, expression of growth factors in the heart would create a gradient to favor mobilization of resident stem/progenitor cells from the BM, peripheral circulation and the heart via specific ligand/receptor interaction for participation in the angiomyogenic repair of the infarcted heart.

Materials and Methods

Ethics Statement

All animal experimental procedures conform to the Guide for the Care and Use of Laboratory Animals published by the US National Institutes of Health (NIH Publication #85-23, revised 1996) and were conducted according to a protocol approved by the Institutional Animal Care and Use Committee, University of Cincinnati.

In vitro Studies

BM Sca-1$^+$ selection. BM was harvested from 6–8 weeks old transgenic male mice expressing GFP. Sca-1$^+$ cells were purified by EasyStep (Stem cell Technology Inc.) isolation kit according to the manufacturer's instruction. Sca-1 surface marker was confirmed by flow cytometry and fluorescent immunostaining as described earlier [7] and detailed in Text S1.

Preparation of plasmids and nano-particle based cell transfection. Plasmids encoding for select quartet of growth factors, i.e., human IGF-1(pCMV-IGF), VEGF (pCMV-VEGF), SDF-1α (pORF-hSDF-1α) and HGF (pBLAST49-hHGF) were prepared and used for genetic modification of Sca-1$^+$ cells (^{GF}Sca-1$^+$) as in Figure S1. The list of primers used are described in Table S1. Cells were separately transfected with one of the 4 plasmids using Polyethyleneimine (PEI, Polysciences Inc.) based on our optimized protocol as described in Text S1. After 48 hours in culture, the cells transfected with respective growth factor were pooled together and cultured for further 24 hours before use for *in vitro* as well as *in vivo* studies. Scramble transfected Sca-1$^+$ cells (^{Sc}Sca-1$^+$) were used as control.

***In vitro* characterization of transfected Sca-1$^+$ cells.** Transfection and expression efficiencies were determined by RT-PCR, western blotting and fluorescent immunostaining 48 hours after their transfection with each plasmid.

Estimation of cytoprotection and cell proliferation. The cytoprotective action of growth factors transfection was assessed by treating the ^{GF}Sca-1$^+$ and ^{Sc}Sca-1$^+$ or co-cultured cardiomyocytes (CM) under anoxia with glucose and serum free DMEM. At different time points, the supernatant of different group were collected. Lactate dehydrogenase (LDH) release was measured using CytoTox-ONE^{TM} homogeneous membrane integrity assay (Promega) as an indicator of cell membrane integrity and cell viability. The experimental protocol is detailed in Text S1.

In addition to the LDH assay, terminal dUTP Nick-End Labeling Assay (TUNEL) was performed on paraformaldehyde fixed cells with in-situ cell death detection kit (TMR red; Roche Inc.) per manufacturer's instructions. The degree of apoptotic cell death was determined by counting total number of TUNEL positive nuclei per microscopic field (400x).

The cell proliferation was evaluated in vitro with the use of the MTS [3-(4,5-dimethylthiazol-2-yl)-5-(3-carboxymethoxyphenyl)-2-(4-sulfophenyl)-2H-tetrazolium] assay according to the manufacturer's recommendations (Promega) as previously described [8].

Tube formation and cell migration assays. Conditioned medium from transfected pooled Sca-1$^+$ cells were used for matrigel angiogenesis assay and transwell cell migration assay as detailed in Text S1.

In vivo Studies

Experimental animal model of myocardial infarction. Myocardial infarction model was developed in 8–12 weeks old female C57BL/6J mice as described earlier [9]. Briefly, the animals were anesthetized (Ketamine/Xylazine 0.05 ml intraperitoneally), intubated, and mechanically ventilated. Minimally invasive thoracotomy was performed for permanent ligation of left anterior descending coronary artery. Myocardial ischemia was confirmed by color change of left ventricular wall. The animals were grouped to receive 15 μl DMEM without cells (group-1) or containing 2.5×10^5 ^{Sc}Sca-1$^+$ cells (group-2) or growth factor transfected pooled cells (group-3) by direct intramyocardial injections in and around the infarction area. The chests were closed and the animals were allowed to recover. Buprenex (0.1 mg/kg per 12 hours) was administered for 24 hours to alleviate pain. The animals were euthanized on 4 days and 4 weeks after their respective treatment. The heart tissue samples were used for molecular, histological and immunohistological studies.

Estimation of cells survival post transplantation. PCR for *sry*-gene expression [7] was performed on myocardial tissue samples from various treatment groups of the animals 4 days after their respective treatment ($n = 4$ per group).

Histological studies. Histological and immunohistochemical studies were carried out as detailed in Text S1. The antibodies and their concentrations used are given in Table S2.

The heart function studies and infarction size measurement. The heart function evaluation was performed in all the animals by transthoracic echocardiography 4 weeks after their respective treatment ($n = 7$/group) as described in Text S1. Histological tissue sections were stained with H&E and Masson's Trichrome for infarction size measurement.

Statistical Analysis

All the data were described as mean±SEM. To analyze the data statistically, we performed Student's *t*-test and one-way ANOVA with post-hoc analysis and considered a value of $p<0.05$ as statistically significant.

Results

In vitro Studies

Efficiency of growth factor transfection in purified BMSca-1$^+$. BM Sca-1$^+$ were successfully isolated from male donor mice. Flow cytometry (Figure-1 Panel A; A1–A4) and fluorescent immunostaining (Figure-1B) demonstrated more than 96% pure Sca-1$^+$ population. Plasmid encoding for human IGF-1(pCMV-IGF), VEGF (pCMV-VEGF), SDF-1α (pORF-hSDF-1α) and HGF (pBLAST49-hHGF) were successfully constructed and used for transfection of Sca-1$^+$. Expression of the respective transgene was increased in Sca-1$^+$ after transfection with the individual growth factor plasmid which was evident from RT-PCR (Figure 1C), western blotting (Figure 1D) and fluorescence immunostaining (Figure 1E, E1–E4).

Gene expression by ^{GF}Sca-1$^+$. Besides multi-fold increase by overexpression of the four growth factors, real-time PCR using growth factor array showed up-regulation of multiple angiogenic and pro-survival factors including angiopoietin-1 & 2 (Ang-1 & Ang-2), matrix metalloproteinase-9 (MMP9), connexin-43 (Cx43), bone morphogenetic protein-2 & -5 (BMP-2 & 5), fibroblast growth factor-2 (FGF-2), and nerve growth factor (NGF) in pooled ^{GF}Sca-1$^+$ cells compared to native Sca-1$^+$ cells (Figure 2A).

Effect of ^{GF}Sca-1+ on cytoprotection, proliferation and angiogenesis. Transfection of Sca-1$^+$ with growth factors was protective against anoxia under co-culture conditions and prevented cell death. LDH release, a marker of cellular injury was significantly reduced in ^{GF} Sca-1$^+$ and co-cultured cardiomyocytes (CM) (Figure 2B). Also TUNEL positive cells were reduced in ^{GF}Sca-1$^+$ (13.52% ±3.08%) as compared to ^{Sc}Sca-1$^+$ cells (24.98% ±2.60%, $p<0.05$, Figure 2D) under 8 hours lethal anoxia. To examine whether growth factor transfection had any

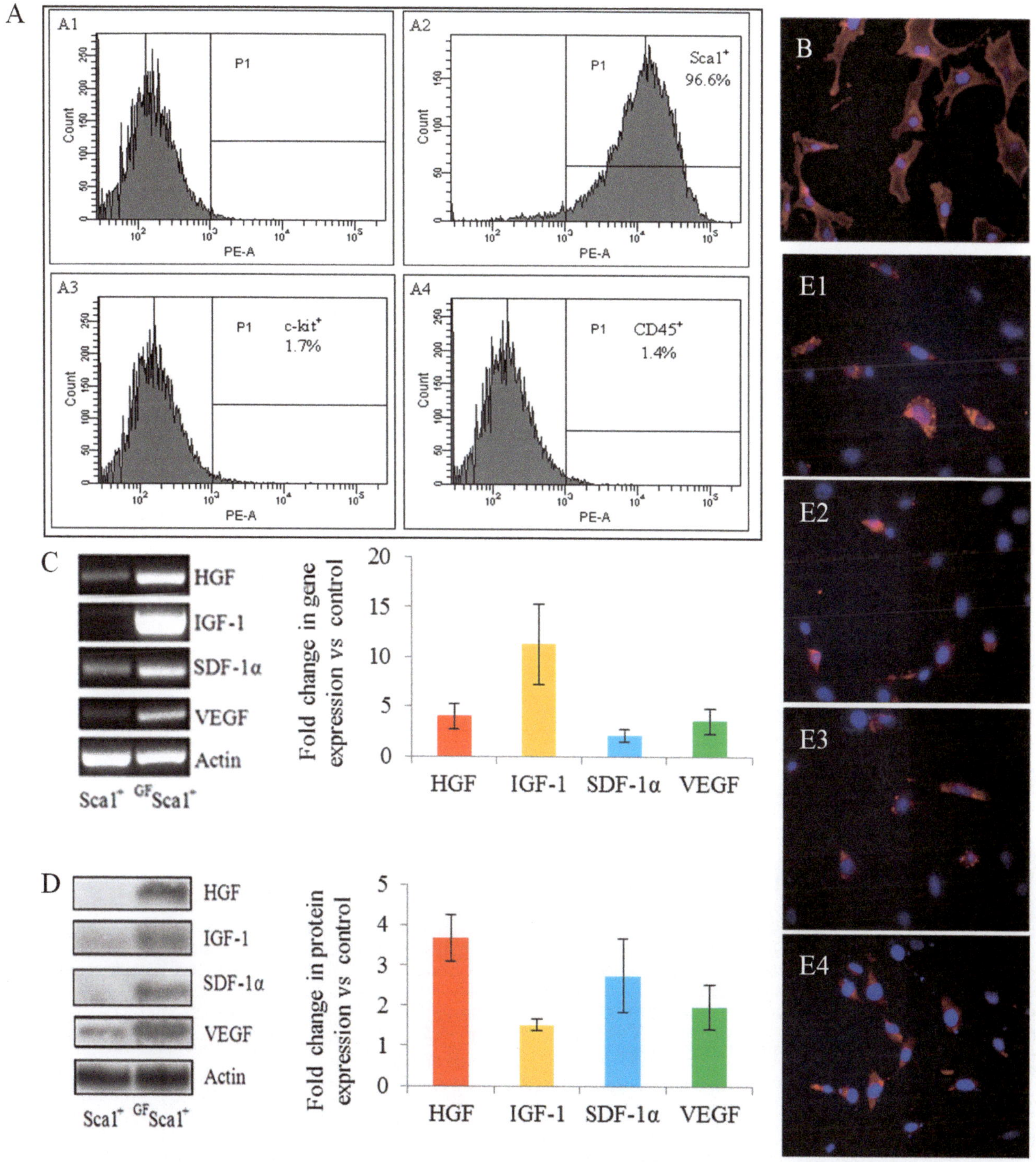

Figure 1. Flow cytometry of the purified mouse BM Sca-1^{+} cells for surface markers analysis. (A) unlabeled Sca-1^{+} cells (control; A1), labeled cells showing 96.6% Sca-1^{+} (A2),1.7% c-kit^{+}(A3), and 1.4% CD45^{+}(A4) cell populations. (B) Immunostaining of cells for Sca-1 antigen (red = Sca-1; blue = DAPI; magnification = 400x). (C–E) Representative figures of *in vitro* characterization of BM Sca-1^{+} cells transfected for expression of the respective transgene. RT-PCR (C) and western blot (D) for HGF, IGF-1, SDF-1α, and VEGF transgene expression *in vitro*. In each case, non-transfected native Sca-1^{+} cells were used as a control. Densitometry showed significantly up-regulated expression of the respective growth factor gene and protein in the transfected Sca-1^{+} cells 48 hours after transfection. (E) Fluorescence images of the Sca-1^{+} cells immunostained for respective growth factor HGF (E1, red), IGF-1(E2, red), SDF-1α (E3, red), VEGF (E4, red). DAPI was used to visualize the nuclei (blue, magnification = 400x).

effect on proliferation rate, MTS assay was used to determine proliferative activity *in vitro*. The proliferation rate of ^{GF} Sca-1^{+} group was significantly higher compared with ^{Sc}Sca-1^{+} (0.59±0.06 *vs*. 0.42±0.04 absorbance at 490 nm, Figure 2C).

The conditioned medium from ^{GF} Sca-1^{+} ($^{GF\text{-}Sca1}$CMD) also promoted human umbilical vein endothelial cells (HUVEC) migration *in vitro* in the trans-well migration system as compared with medium from ^{SC}Sca-1^{+}($^{Sc\text{-}Sca1}$CMD) (15±2.38 *vs* 8±2.35, $p<$

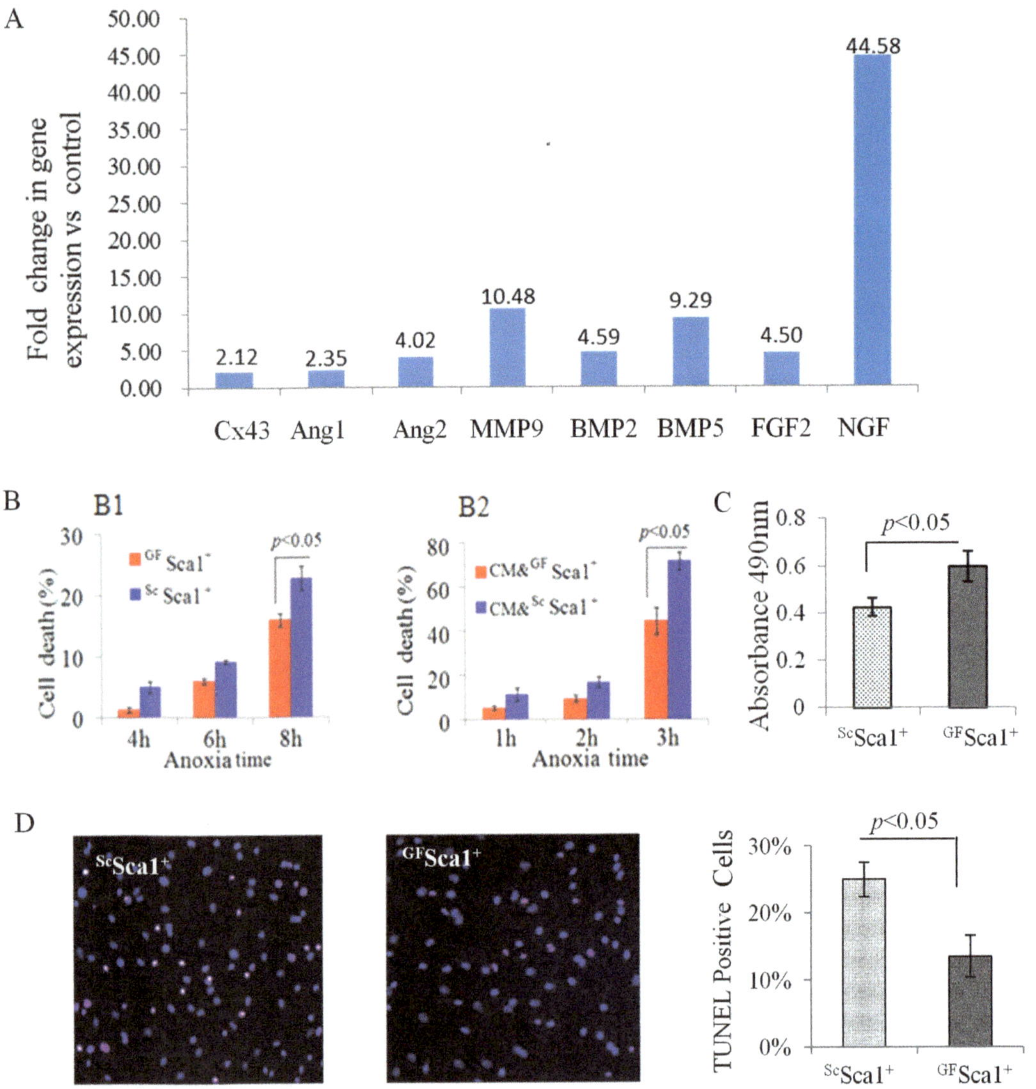

Figure 2. (A) Real-time PCR based growth factor array showing up-regulation of multiple angiogenic and pro-survival factors including connexin-43 (Cx43), angiopoietin-1 & 2 (Ang-1 & Ang-2), matrix metalloproteinase-9 (MMP9), Bone morphogenetic protein 2 & 5 (BMP2 & 5), fibroblast growth factor-2 (FGF-2), and Nerve growth factor (NGF) in pooled ^{GF}Sca-1^{+} cells compared to control Sca-1^{+} cells. (B; B1–B2) LDH release assay showed significant reduction of cell death in ^{GF}Sca-1^{+} as compared to ^{Sc}Sca-1^{+} after 8 hours cultured under lethal anoxia (B1) and co-cultured cardiomyocytes after 3 hours under lethal anoxia (B2). (C) MTS assay was used to determine proliferative activity in vitro. The proliferation rate of ^{GF}Sca-1^{+} group was significantly higher as compared to ^{Sc}Sca-1^{+} (0.59±0.06 vs. 0.42±0.04 absorbance at 490 nm) (D) TUNEL staining confirmed that 8 hours lethal anoxia caused higher TUNEL positivity in ^{Sc}Sca-1^{+} as compared to ^{GF}Sca-1^{+} cells.

0.01, Figure 3A). Similarly, *in vitro* tube formation assay on matrigel showed higher branch points (58±5.8) per microscope field (100x) at 16 hours after incubation with $^{GF\text{-}Sca1}$CMD as compared with $^{Sc\text{-}Sca1}$CMD (39±4.6, $p<0.05$, Figure 3B).

In vivo Studies

Sca-1^{+} transplantation and differentiation in the heart. RT-PCR of heart tissue samples showed higher expression of transfected growth factors including HGF, IGF-1, SDF-1α and VEGF 4 days after engraftment (Figure 4A). The transplanted male ^{GF}Sca-1^{+} showed better survival in the infarcted heart. PCR for *sry*-gene in female mice heart on day 4 revealed 2.3 times increased survival of the donor male cells in ^{GF} Sca-1^{+} group as compared to ^{Sc}Sca-1^{+} group ($P<0.01$, Figure 4B). We did not observe *sry*-gene expression in DMEM injected female animal hearts which served as a negative control. These results were supported by immunohistological studies which confirmed extensive presence of GFP^{+} positive cells at the site of cell graft in ^{GF} Sca-1^{+} group at day 4 after transplantation (Figure 4C).

Stem/progenitor cells homing into myocardium. The histological sections of heart tissues from different treatment groups were immunostained for the presence of stem and progenitor cells of various lineages which were mobilized into the infarcted heart. Figure 4D shows numerous c-Kit^{+}, CXCR4^{+}, MDR1^{+} (Figure 4D; D1–D3) cells respectively which mobilized into the infarcted heart in ^{GF} Sca-1^{+} treatment group. The number of mobilized ckit^{+} cells was significantly higher in group-3 as

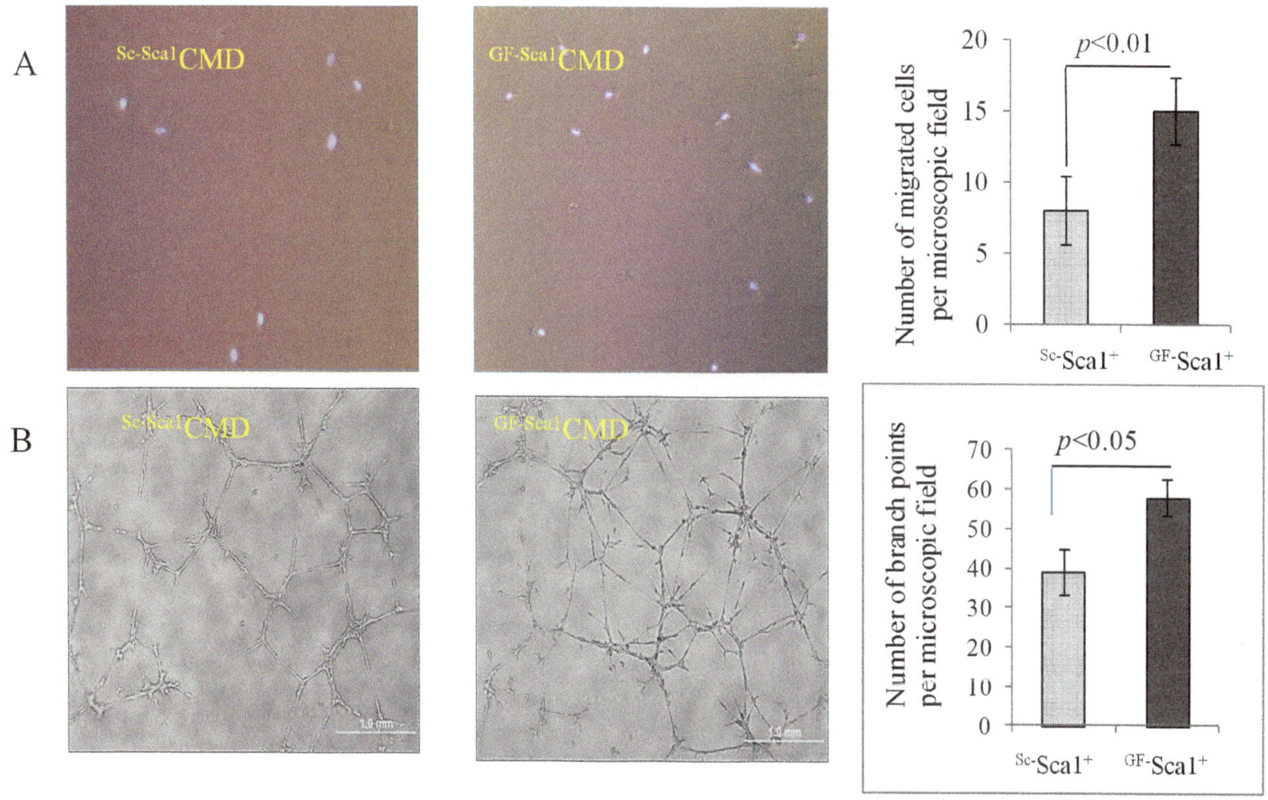

Figure 3. Cell migration and angiogenic effects of CMD *in vitro*. (A)Trans-well cell migration assay using HUVECs showed that the number of migrating cells per microscopic field (200x) in response to $^{GF\text{-}Sca1}$CMD was significantly higher compared to $^{Sc\text{-}Sca1}$CMD as a control (15±2.38 *vs* 8±2.35, $p<0.01$). Nuclei were stained with DAPI (blue). (B) Tube formation assay on matrigel using HUVECs treated with conditioned medium (CMD) from ^{GF}Sca-1^{+} cells ($^{GF\text{-}Sca1}$CMD) or ^{Sc}Sca-1^{+} cells ($^{Sc\text{-}Sca1}$CMD). $^{GF\text{-}Sca1}$CMD showed significantly higher number of branch points per microscopic field (100x) as compared to $^{Sc\text{-}Sca1}$CMD.

compared to group-2 (55.50±9.87 *vs.* 20.88±8.53/microscope field, $p<0.05$).

Histological evaluation of transplanted cells in the infarcted heart. Immunostaining of the heart tissue for subcellular structures revealed increased angiomyogenic differentiation of the transplanted Sca-1^{+} in the infarct and peri-infarct regions 4 weeks after transplantation (GFP, green; Figure 5A1, B1), α-actinin (red; Figure 5A2) and vWF (red; Figure 5B2). Fluorescence immunostaining also showed Cx43 positivity for gap junctions formed between ^{GF}Sca-1^{+} and the host cardiomyocytes in the infarcted area (Figure 5C). These observations were consistent with our *in vitro* data that ^{GF}Sca-1^{+} showed two fold increase in Cx43 (Figure 2A) and Cx43 formation between co-cultured ^{GF} Sca-1^{+} and cardiomyocytes (Figure 5D).

^{GF}Sca-1^{+} improve angiogenesis in the infarcted myocardium. Blood vessel density was measured in the infarct and peri-infarct regions after immunostainning of histological sections with vWF (red) to detect vascular structures. A large number of blood vessels were observed in the cell transplanted areas in both ^{GF} Sca-1^{+} and ^{Sc}Sca-1^{+} treatment groups. Blood vessel count per microscope field (200x; Figure 6) was the highest in both peri-infarct (34.9±4.7) and infarct (24.6±3.2) regions in ^{GF} Sca-1^{+} group ($P<0.01$) as compared to the peri-infarct and infarct regions in ^{Sc}Sca-1^{+} group (24.4±1.7 and 16±2.7 respectively) and in DMEM group (15.6±2.9 and 9.3±1.5 respectively, Figure 6).

Infarct size and heart function. Histological sections at mid-papillary muscle level followed by Masson's trichrome staining showed transmural infarction in all the animals at 4 weeks after coronay artery ligation (Figure 7A). Noticeable left ventricle wall thinning was observed at 4 weeks in DMEM treated group whereas left ventricle wall thickness was considerably retained in ^{GF}Sca-1^{+} group. Infarct size was significantly lower at 4 weeks in ^{GF}Sca-1^{+} group (21.2% ±3.6%; $p<0.01$) as compared to DMEM group (48.9%±4.3%) and ^{Sc}Sca-1^{+} group (35.6% ±1.3%, Figure 7B). Transthoracic echocardiography of animals at 4 weeks after their respective treatment showed significant differences in the indices of left ventricle function amongst various treatment groups. Left ventricle ejection fraction (55.9±6.2%) and left ventricle fractional shortening (24.0±3.6%) were significantly improved in ^{GF}Sca-1^{+} group as compared to ^{Sc}Sca-1^{+} group (42.6±3.6% and 16.9±1.7% respectively) and DMEM injected group (31.4±6.2% and 11.9±2.7% respectively, $p<0.01$, Figure 7C–D).

Discussion

For last several years, BMSCs have been extensively used to regenerate infarcted myocardium [10,11] with heterogeneous results. More recently attempts have been made to render these cells smarter by modifying them exogenous growth factors or gene transfer strategies. Previously we have shown the effectiveness of several individual growth factors for treatment of experimental myocardial infarction [12–15] and in this study we transfected the BM Sca-1^{+} cells with individual transgenic plasmids and then pooled the transfected cells before transplantation. This multi-modal non-viral gene therapy approach resulted in activation and

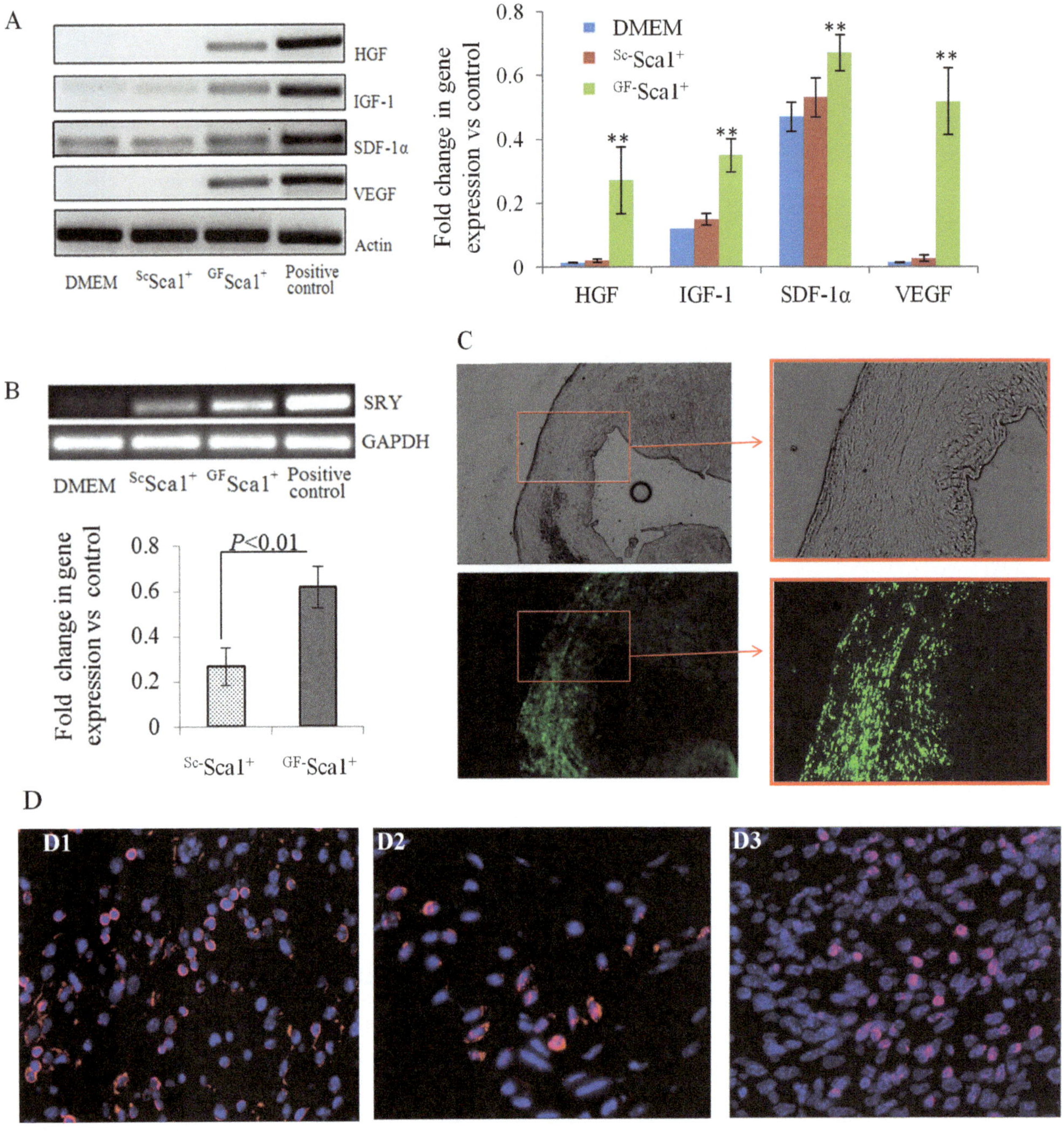

Figure 4. Cell survival and release of growth factors at 4 day post transplantation. (A) RT-PCR using the heart tissue samples on day 4 from different treatment groups. The results showed continued expression of each growth factor until day 4 post transplantation. (**$P<0.01$, ^{GF}Sca-1^{+} group *vs* control groups). (B) *Sry*-gene expression in female mouse heart revealed extensive survival of the male donor cells in ^{GF}Sca-1^{+} group as compared to ^{Sc}Sca-1^{+} group. (C) Fluorescent immunostaining for GFP expression showed extensive presence of ^{GF}Sca-1^{+} cells in the infarct and peri-infarct regions at 4 days after transplantation (magnification 40x, 100x). (D) Immunostaining for c-kit (D1, red), CXCR4 (D2, red) and MDR1(D3, red) showed extensive presence of mobilized cells in the infarct and peri-infarct regions at 4 days after ^{GF}Sca-1^{+} transplantation (magnification 200x). The nuclei are stained with DAPI (blue).

participation of the intrinsic ligand/receptor systems: VEGF/VEGFR, HGF/cMet, IGF-1/IGF-1R and SDF/CXCR4, in the repair processes of infarcted myocardium. Our multipronged strategy activated the multiple signaling mechanisms resulting in reduced cell apoptosis, increased cell proliferation, massive stem/progenitor cell homing to the site of ischemic injury. The transplanted and mobilized cells appeared to differentiate into cardiomyocytes in the infarcted area with significant improvement in the left ventricular ejection fraction.

The rationale of selecting various growth factors was based on their distinct functional role in the myocardial biology. These growth factors have pleiotropic functions. IGF-1 interacts with its

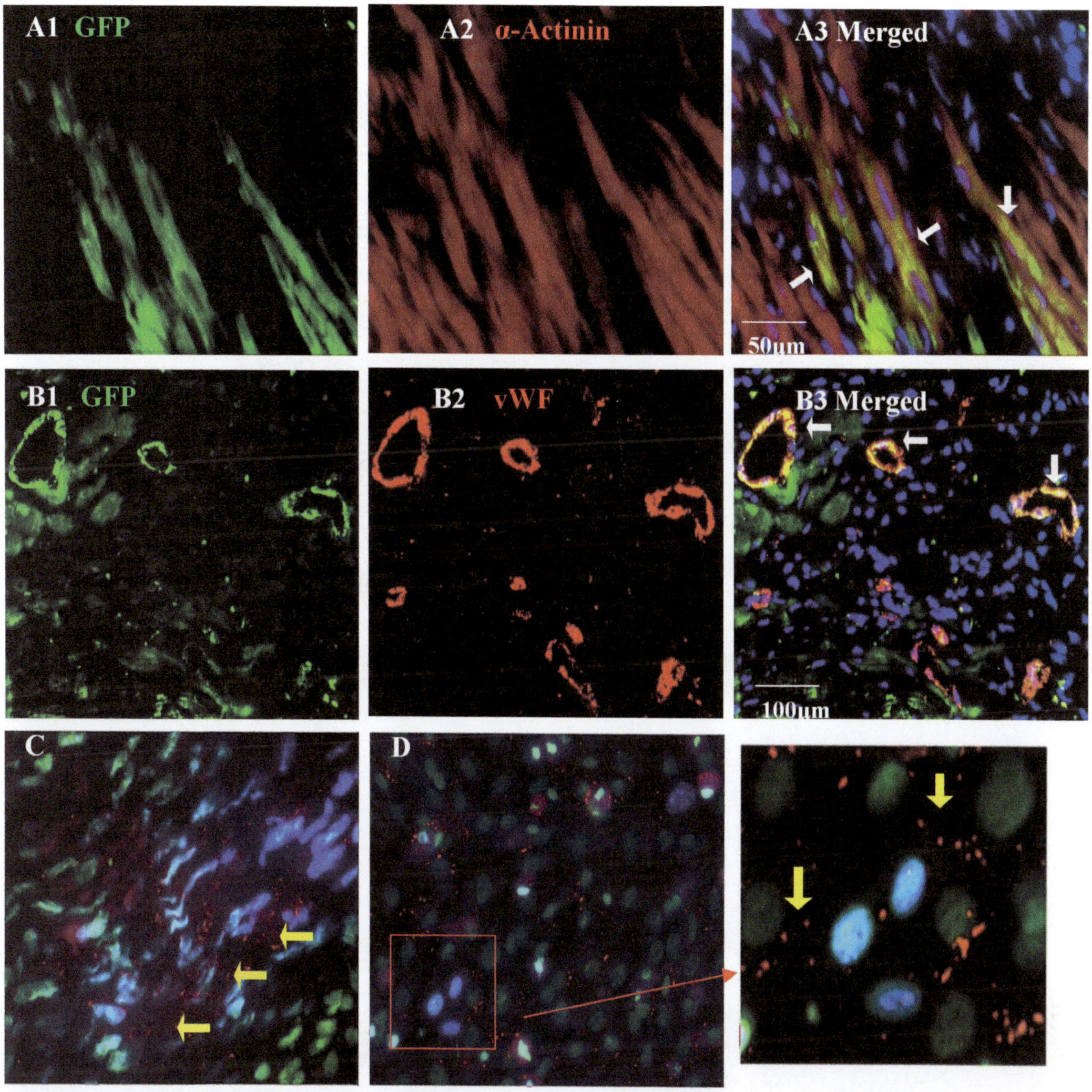

Figure 5. Microscopic images from recipient mouse hearts 4 weeks post-transplantation. Fluorescence immunostaining of the heart tissue revealed increased angiomyogenic differentiation of the transplanted Sca-1$^+$ in the infarct and peri-infarct regions 4 weeks after transplantation (GFP (green 5A1,5B1), α-actinin (red; Figure 5A2) and vWF (red; Figure 5B2)). Arrows show the merged color of red and green indicating the colocalization of α-actinin and GFP (5A3) or vWF and GFP (5B3). The nuclei were stained with DAPI (blue) (C) Immunostaining of the heart tissue revealed Cx43 (red) expression between ^{GF}Sca-1$^+$ and the host cardiomyocytes (indicated by yellow arrows). (D) Cx43 (red) expression between ^{GF}Sca-1$^+$ and the co-cultured cardiomyocytes at day 4 after co-culture. (C–D) ^{GF}Sca-1$^+$ stem cells were prelabeled with DAPI (blue), and the nuclei were double stained with SYTOX (green).

IGF-1R receptors to impart its cytoprotective, pro-proliferative and pro-differentiation effects [16,17]. We have reported that IGF-1 preconditioning of Sca-1$^+$ cells promotes Cx43 induction and their translocation into mitochondria for its dual role of stem cell integration as well as protection post-engraftment [7]. HGF is a mesenchyme derived pleiotropic factor that is thought to mediate the interaction between epithelial and endothelial cells via autocrine and paracrine functions, promoting angiogenesis. In addition, HGF exerts its anti-cell death effects by blocking apoptosis [18]. The HGF receptor is a membrane spanning tyrosine kinase encoded by the c-Met proto-oncogene [19]. Myocardial HGF/c-Met is upregulated in the heart following myocardial infarction [20,21]. HGF does not influence the mitosis of cardiac cells, but prevents post-infarction remodeling of heart. Administration of IGF/HGF improved cardiomyocyte survival, and reduced fibrosis and cardiomyocytes reactive hypertrophy. It significantly activated resident endogenous cardiac stem/progenitor cells and fostered the generation of new myocytes and microvasculature in infarcted and peri-infarct/border areas [22]. VEGF is well known for its ability to stimulate proliferation of endothelial cells in vitro and neovasculogenesis in the ischemic heart in vivo [23,24]. In addition VEGF, an angiogenic growth factor can also promote differentiation of stem cells into cardiomyocytes and endothelial cells [25,26]. SDF-1α is a ligand for CXCR4 receptors expressed on BMSC. SDF was transiently upregulated in infarct and peri-infarct regions in experimental

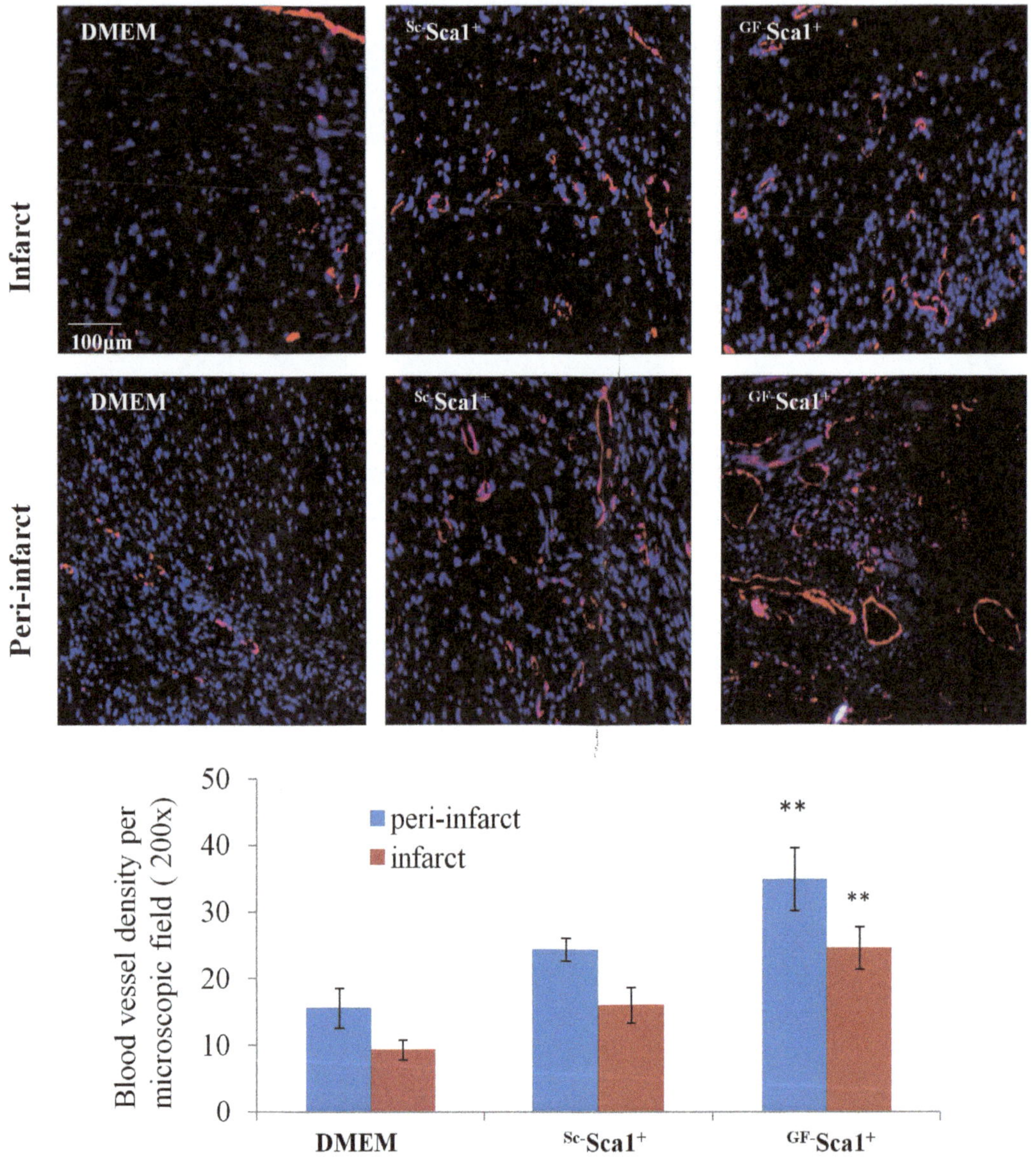

Figure 6. Blood vessel density analyzed by fluorescence immunostaining for vWFactor-VIII (red). Blood vessel density was significantly higher in ^{GF}Sca-1^{+} group both in the infarct and peri-infarct regions as compared to control groups (**$p<0.01$, magnification 200x).

animal models of myocardial infarction and SDF-1/CXCR4 interactions played a crucial role in the recruitment of BMSCs to the infarcted myocardium [8]. The SDF-1/CXCR4 axis seems to be a novel therapeutic approach to improve post-infarction therapy by attracting circulating stem cells to the site of injury where they differentiated into cardiac cells in the infarcted heart [27]. We have previously shown that transplantation of MSCs overexpressing IGF-1 caused massive stem cell mobilization into infarcted hearts through paracrine signaling of SDF-1α resulting in extensive angiogenesis with better cardiac function [12]. A recent study has suggested that SDF-1α and VEGF form a synergistic angiogenic pathway that is critical for endothelial progenitor cell induced neovascularization [28].

Given the distinct functional role of these cytokines, our novel strategy of multiple growth factor transgene delivery is aimed to harness their combined beneficial effects for treatment of infarcted heart. This study provides evidence that the selected quartet of growth factors created a growth factor gradient for mobilization of recipient's own pool of BMSC, EPC and resident cardiac stem cells for participation in the repair process. In addition MSCs can secrete large amounts of paracrine factors which are angiogenic and anti-apoptotic, alter the restoration of extracellular matrix and also recruit endogenous stem cells [29,30,31]. In this study, we also observed significant upregulation of matrix metalloproteinases (MMP), bone morphogenetic protein (BMP), nerve growth factor (NGF), and fibroblast growth factor (FGF).

MMPs are important for cell migration, invasion, proliferation, and apoptosis prevention. They regulate many developmental processes, including branching morphogenesis, angiogenesis, wound healing, and extracellular matrix degradation [32]. MMP-9 induced in BM cells, releases soluble kit-ligand (sKitL), permitting the transfer of stem cells from the quiescent state to proliferative niche favoring differentiation and reconstitution of the stem/progenitor cell pool. SDF-1 and VEGF both can up-

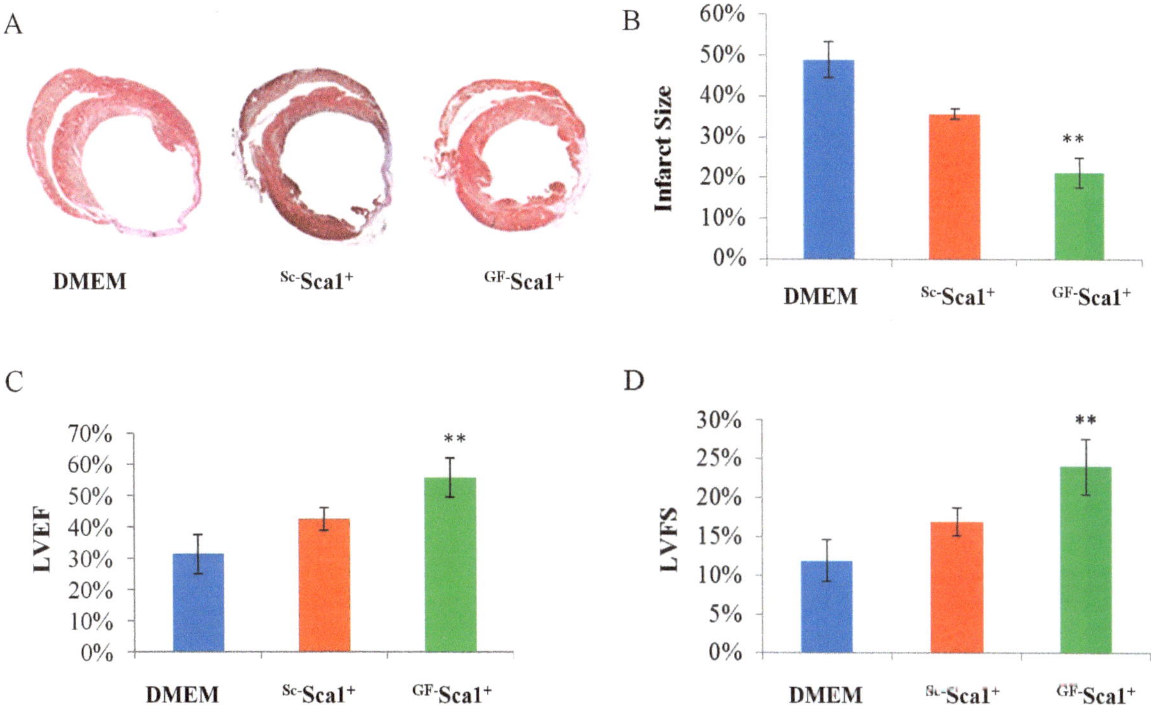

Figure 7. Masson trichrome staining was performed to evaluate the infarction size. Significant attenuation of infarct size was observed in ^{GF}Sca-1^{+} group as compared to other groups. Indices of the left ventricular contractile function including LVEF and LVFS were significantly preserved in ^{GF}Sca-1^{+} group at 4 weeks after cell engraftment as compared to other groups(***P*<0.01).

regulate MMP-9 expression, and cause shedding of sKitL and recruitment of c-Kit^{+} stem/progenitors [33]. BM-derived MMP-9 plays an important role in BM cell mobilization and focal angiogenesis in response to VEGF stimulation [34]. BMPs are part of the transforming growth factor β (TGF-b) superfamily [35] and comprise of a large, evolutionarily conserved family of secreted signaling molecules that are required for numerous developmental processes. BMP-2 is known to play an important role in the process of heart development [36]. BMP signaling also cooperates with VEGF signaling to affect the process of angiogenesis [37]. In vitro, BMP-2 stimulates proliferation of human aortic endothelial cells (HAEC) [38] and pulmonary aortic endothelial cells (PAEC) [39]. BMP-2 also increases the migration and tube formation of human microvascular endothelial cells (HMEC) [40] and human umbilical vein endothelial cells (HUVEC) [38]. NGF is a secreted glycoprotein of the neurotrophin family. It promotes angiogenesis and cardiomyocyte survival, which are both desirable for postinfarction myocardial healing. Recent findings demonstrate that the co-transfection of hNGF+VEGF genes in BMSCs can enhance the angiogenic effect *in vivo* [41]. NGF elicits its biological effects mainly by binding the high affinity TrkA receptor (tropomyosin-related receptor A). The prosurvival/proangiogenic Akt/Foxo pathway mediated the therapeutic benefits of NGF. NGF gene transfer ameliorated endothelial cells and cardiomyocyte survival, promoted neovascularization and improved myocardial blood flow and cardiac function in the infarcted myocardium [42]. FGFs are members of a family of polypeptides and FGF receptor (FGFR) are a family of transmembrane tyrosine kinase. FGF and FGFR system plays a significant role in mitogenic and angiogenic activity. FGF-2 can be induced by VEGF while FGF2 in turn can also induce VEGF expression. Both VEGF and FGF2 activate the Erk-1/2 pathway and share some similarities [43]. Thus FGF-2 is an important regulator of cell proliferation, angiogenesis, collagen synthesis, myocyte hypertrophy, scar contraction, and ultimately improvement of left ventricular contractile function [44]. FGF-2 also plays a pivotal role in the self-regeneration of the heart by accelerating the mobilization and differentiation of resident stem cells for cardiac repair [45]. More importantly, we observed higher Cx43 expression in ^{GF}Sca-1^{+} transplanted animals by real time PCR and immunostaining. This is corroborated with in vitro data showing extensive Cx43 formation at the cell-cell junctions of ^{GF}Sca-1^{+} co-cultured with neonatal cardiomyocytes. There are few reports on Cx43 expression by stem cells post-engraftment and their electromechanical coupling with host cardiomyocytes. However, in the present study, ^{GF}Sca-1^{+} expressed significant amounts of Cx43 until 4 weeks post engraftment, which most likely improved the cells survival and differentiation.

Despite promising results, there are several limitations to this study. We used human growth factors (GFs) for pre-treatment of the BMSCs isolated from mice, although structural homology of these GFs is very high between humans and mice. Sca-1 is a cosignaling molecule that can modify the signaling capacity of receptor complex [46]. To evaluate a single factor in comparison with multiple growth factor requires thorough analysis. Because we treated BMSCs with a combination of four GFs and used non-viral method, it may limit the efficiency of individual GF effect and the understanding of underlying mechanism of individual GF. We did not look into the effect of single growth factor transduction on the gene expression of other growth factors. Although a significant increase in blood vessel density and decrease in infarction area was observed however, the role and fate of the mobilized cells could not be determined due to lack of specific markers on the cells. The cells which co-expressed GFP and cardiac muscle marker were not numerous and these cells may be the consequence of transdifferentiation of BMSCs into cardiomyocyte-like cells, or the result of cell fusion of BMSCs with host cardiomyocytes, or both [47,48]. It is less convincing that low frequent transdifferentiation or cell fusion may contribute to significant cardiac function improvement in this study. The cytokines/growth factors exhibit a wide range of other functions (eg, stem/progenitor cell mobilization, recruitment, antiapoptotic activity, paracrine signaling and proangiogenic effects [49] thus, it is unclear to what extent the induction of cardiac differentiation by stem cells contributed to the therapeutic benefits. In a recent study on cardiac stem cells [50], a detailed mechanistic insight for the understanding of myocardial homeostasis and tissue repair has been discussed.

The present data provide supportive evidence that the gene delivery and paracrine effects mediated by ^{GF}Sca-1$^+$ are significant in attenuation of ventricular remodeling and the improvement of cardiac function during cell transplantation. Given the fact that tissue damage and regeneration are complex in nature, it is more likely that multiple growth factors and trophic factors acting in synergy are involved in tissue healing. Simultaneous use of BMSCs transfected with multiple cytokines with diverse effects is a novel approach for cardioprotection and regeneration of the infarcted myocardium.

Supporting Information

Figure S1 Plasmid sequence of human SDF-1α, HGF, IGF-1, and VEGF used for transgenic overexpression of the respective growth factor ligand in Sca-1$^+$.

Table S1 Primers for conventional RT-PCR.

Table S2 Primary antibodies used for western blotting and immunohistochemistry.

Text S1 Supporting methods.

Author Contributions

Conceived and designed the experiments: NL ZP MA. Performed the experiments: NL ZP. Analyzed the data: NL ZP MA. Contributed reagents/materials/analysis tools: MA. Wrote the paper: NL ZP MA.

References

1. Mozid AM, Arnous S, Sammut EC, Mathur A (2011) Stem cell therapy for heart diseases. Br Med Bull 98: 143–59.
2. Orlic D, Kajstura J, Chimenti S Jakoniuk I, Anderson SM et al. (2001) Bone marrow cells regenerate infarcted myocardium. Nature 410: 701–5.
3. Shake JG, Gruber PJ, Baumgartner WA, Senechal G, Meyers J, et al. (2002) Mesenchymal stem cell implantation in a swine myocardial infarct model: engraftment and functional effects Ann Thorac Surg 73: 1919–1925.
4. Martin-Rendon E, Brunskill SJ, Hyde CJ, Stanworth SJ, Mathur A, Watt SM (2008) Autologous bone marrow stem cells to treat acute myocardial infarction: a systematic review. Eur Heart J 29: 1807–1818.
5. Piepoli MF, Vallisa D, Arbasi M, Cavanna L, Cerri L et al. (2010) Bone marrow cell transplantation improves cardiac, autonomic, and functional indexes in acute anterior myocardial infarction patients. Eur J Heart Fail 12: 172–80.
6. Wen Z, Zheng S, Zhou C, Wang J, Wang T (2011) Repair mechanisms of bone marrow mesenchymal stem cells in myocardial infarction J. Cell. Mol. Med. 15: 1032–1043.
7. Lu G, HK Haider, S Jiang, M Ashraf (2009) Sca-1+ stem cell survival and engraftment in the infarcted heart: dual role for preconditioning-induced connexin-43. Circulation 119: 2587–2596.
8. Pasha Z, Wang Y, Sheikh R, Zhang D, Zhao T, Ashraf M. (2008) Preconditioning enhances cell survival and differentiation of stem cells during transplantation in infarcted myocardium. Cardiovasc Res. 77: 134–42.
9. Pasha Z, Haider K H, M Ashraf (2011) Efficient non viral reprogramming of myoblasts to stemness with a single small molecule for generating cardiac progenitors cells. PLoS One 6: e23667.
10. Nesselmann C, Ma N, Bieback K, Wagner W, Ho A, Konttinen YT et al. (2008) Mesenchymal stem cells and cardiac repair. J Cell Mol Med. 12(5B): 1795–810.
11. Song H, Song BW, Cha MJ, Choi IG, Hwang KC. (2010) Modification of mesenchymal stem cells for cardiac regeneration. Expert Opin Biol Ther. 10: 309–19.
12. Haider H, Jiang S, Idris NM, Ashraf M. (2008) IGF-1-overexpressing mesenchymal stem cells accelerate bone marrow stem cell mobilization via paracrine activation of SDF-1 alpha/CXCR4 signaling to promote myocardial repair. Circ Res 103: 1300–1308.
13. Elmadbouh I, Haider H, Jiang S, Idris NM, Lu G, Ashraf M. (2007) Ex vivo delivered stromal cell-derived factor-1alpha promotes stem cell homing and induces angiomyogenesis in the infarcted myocardium. J Mol Cell Cardiol 42: 792–803.
14. Wang Y, Ahmad N, Wani MA, Ashraf M. (2004) Hepatocyte growth factor prevents ventricular remodeling and dysfunction in mice via Akt pathway and angiogenesis. J Mol Cell Cardiol 37: 1041–1052.
15. Wang Y, Haider HK, Ahmad N, Xu M, Ge R, Ashraf M. (2006) Combining pharmacological mobilization with intramyocardial delivery of bone marrow cells over-expressing VEGF is more effective for cardiac repair. J Mol Cell Cardiol 40: 736–45.
16. Kofidis T, de Bruin JL, Yamane T, Balsam LB, Lebl DR et al. (2004) IGF promotes engraftment, differentiation, and functional improvement after transfer of ESCs for myocardial restoration. Stem Cells. 22: 1239–1245.
17. Kanemitsu N, Tambara K, Premaratne GU, Kimura Y, Tomita S, et al. (2006) IGF-1 enhances the efficacy of myoblast transplantation with its multiple functions in the chronic myocardial infarction rat model. J Heart Lung Transplant. 25: 1253–1262.
18. Jayasankar V, Woo YJ, Pirolli TJ, Bish LT, Berry MF et al. (2005) Induction of angiogenesis and inhibition of apoptosis by hepatocyte growth factor effectively treats postischemic heart failure. J Card Surg 20: 93–101.
19. Gherardi E, Stoker M (1991) Hepatocyte growth factor–scatter factor: Mitogen, motogen, and met. Cancer Cells 3: 227–232.
20. Ono K, Matsumori A, Shioi T, et al. (1997) Enhanced expression of hepatocyte growth factor/c-Met by myocardial ischemia and reperfusion in a rat model. Circulation; 95: 2552–2558.
21. Yasuda S, Goto Y, Baba T, et al. (2005) Enhanced secretion of cardiac hepatocyte growth factor from an infarct region is associated with less severe ventricular enlargement and improved cardiac function. J AmColl Cardiol 36: 115–121.
22. Ellison GM, Torella D, Dellegrottaglie S, Perez-Martinez C, Perez de Prado A et al. (2011) Endogenous Cardiac Stem Cell Activation by Insulin-Like Growth Factor-1/Hepatocyte Growth Factor Intracoronary Injection Fosters Survival and Regeneration of the Infarcted Pig Heart. J Am Coll Cardiol. 58: 977–86.
23. Zisa D, Shabbir A, Suzuki G, Lee T. (2009) Vascular endothelial growth factor (VEGF) as a key therapeutic trophic factor in bone marrow mesenchymal stem cell-mediated cardiac repair. Biochem Biophys Res Commun. 390: 834–8.
24. Gao F, He T, Wang H, Yu S, Yi D, Liu W, Cai Z. (2007) A promising strategy for the treatment of ischemic heart disease: Mesenchymal stem cell-mediated vascular endothelial growth factor gene transfer in rats. Can J Cardiol. 23: 891–8.
25. Song YH, Gehmert S, Sadat S, Pinkernell K, Bai X, Matthias N, Alt E. (2007) VEGF is critical for spontaneous differentiation of stem cells into cardiomyocytes. Biochem Biophys Res Commun. 354(4): 999–1003.
26. Oswald J., Boxberger S., Jorgensen B., Feldmann S., Ehninger G et al. (2004) Mesenchymal stem cells can be differentiated into endothelial cells in vitro Stem Cells, 22: 377–384.
27. Ghadge SK, Mühlstedt S, Ozcelik C, Bader M. (2011) SDF-1α as a therapeutic stem cell homing factor in myocardial infarction. Pharmacol Ther. 129: 97–108.
28. Yu JX, Huang XF, Lv WM, Ye CS, Peng XZ et al. (2009) Combination of stromal-derived factor-1alpha and vascular endothelial growth factor gene-modified endothelial progenitor cells is more effective for ischemic neovascularization. J Vasc Surg. 50: 608–16.
29. Gnecchi M, Zhang Z, Ni A, et al. (2008) Paracrine mechanisms in adult stem cell signaling and therapy. Circ Res 103: 1204–1219.
30. Fedak PW. (2008) Paracrine effects of cell transplantation: modifying ventricular remodeling in the failing heart. Semin Thorac Cardiovasc Surg 20: 87–93.
31. Uemura R, Xu M, Ahmad N, Ashraf M.(2006) Bone marrow stem cells prevent left ventricular remodeling of ischemic heart through paracrine signaling. Circ Res 98: 1414–1421.
32. Vu TH, Werb Z. (2000) Matrix metalloproteinases: effectors of development and normal physiology. Genes Dev 14: 2123–2133.
33. Heissig B, K Hattori, S Dias, M Friedrich, B Ferris et al. (2002) Recruitment of stem and progenitor cells from the bone marrow niche requires MMP-9 mediated release of kit-ligand. Cell 109: 625–637.

34. Hao Q, Su H, Palmer D, Sun B, Gao P et al. (2011) Bone marrow-derived cells contribute to vascular endothelial growth factor-induced angiogenesis in the adult mouse brain by supplying matrix metalloproteinase-9. Stroke. 42: 453–8.
35. Shi Y., Massague J. (2003) Mechanisms of TGF-beta signaling from cell membrane to the nucleus Cell 113: 685–700.
36. Shahab M Danesh, Alethia Villasenor, Diana Chong, Carrie Soukup, Ondine Cleaver (2009) BMP and BMP receptor expression during murine organogenesis Gene Expression Patterns 9: 255–265.
37. David M Wiley, Suk-Won Jin. (2011) Bone Morphogenetic Protein functions as a context-dependent angiogenic cue in vertebrates. Seminars in Cell & Developmental Biology 22 : 1012–1018.
38. Langenfeld EM, Langenfeld J (2004) Bone morphogenetic protein-2 stimulates angiogenesis in developing tumors. Mol Cancer Res 2: 141–9.
39. de Jesus Perez VA, Alastalo TP, Wu JC, Axelrod JD, Cooke JP et al. (2009) Bone morphogenetic protein 2 induces pulmonary angiogenesis via Wnt-beta-catenin and Wnt-RhoA-Rac1 pathways. J Cell Biol 184: 83–99.
40. Rothhammer T, Bataille F, Spruss T, Eissner G, Bosserhoff AK. (2007) Functional implication of BMP4 expression on angiogenesis in malignant melanoma. Oncogene 26: 4158–70.
41. Fan BS, Lou JY (2011) Enhancement of angiogenic effect of co-transfection human NGF and VEGF genes in rat bone marrow mesenchymal stem cells. Gene. 485: 167–71.
42. Meloni M, Caporali A, Graiani G, Lagrasta C, Katare R et al. (2010) Nerve growth factor promotes cardiac repair following myocardial infarction. Circ Res. 106: 1275–84.
43. Maulik N, Thirunavukkarasu M. (2008) Growth factor/s and cell therapy in myocardial regeneration. J Mol Cell Cardiol. 44: 219–27.
44. Virag JA, Rolle ML, Reece J, Hardouin S, Feigl EO, Murry CE (2007) Fibroblast growth factor-2 regulates myocardial infarct repair: effects on cell proliferation, scar contraction, and ventricular function. Am J Pathol. 171: 1431–40.
45. Rosenblatt-Velin N, Lepore MG, Cartoni C, Beermann F, Pedrazzini T. (2005) FGF-2 controls the differentiation of resident cardiac precursors into functional cardiomyocytes. J Clin Invest 115: 1724–33.
46. Holmes C, Stanford WL.Stem cell antigen -1: Expression, Function, and Engima. stem cells 2007. 25: 1339–1347.
47. J Kajstura, M Rota, B Whang, S Cascapera, T Hosoda, C etal. (2005) Bone marrow cells differentiate in cardiac cell lineages after infarction independently of cell fusion Circ Res 96: 127–137.
48. M Alvarez-Dolado, R Pardal, J.M Garcia-Verdugo, J.R Fike, H.O Lee, K etal. (2003) Fusion of bone-marrow-derived cells with Purkinje neurons, cardiomyocytes and hepatocytes Nature 425: 968–97.
49. Chavakis E, Koyanagi M, Dimmeler S. (2010) Enhancing the Outcome of Cell Therapy for Cardiac Repair Progress From Bench to Bedside and Back. Circulation. 121(2): 325–35.
50. Anversa P, Kajstura J, Leri A, Bolli R. (2006) Life and death of cardiac stem cells: a paradigm shift in cardiac biology. Circulation.113(11): 1451–63.

16

Intramyocardial Transplantation and Tracking of Human Mesenchymal Stem Cells in a Novel Intra-Uterine Pre-Immune Fetal Sheep Myocardial Infarction Model: A Proof of Concept Study

Maximilian Y. Emmert[1,2,3]⁹, Benedikt Weber[1,2]⁹, Petra Wolint[1,2], Thomas Frauenfelder[4], Steffen M. Zeisberger[1], Luc Behr[5], Sebastien Sammut[5], Jacques Scherman[3], Chad E. Brokopp[1,2], Ruth Schwartländer[6], Viola Vogel[6], Peter Vogt[7], Jürg Grünenfelder[2,3], Hatem Alkadhi[4], Volkmar Falk[2,3], Andreas Boss[4], Simon P. Hoerstrup[1,2,3]*

1 Swiss Centre for Regenerative Medicine, University of Zurich, Zurich, Switzerland, **2** Department of Surgical Research, University Hospital of Zurich, Zurich, Switzerland, **3** Clinic for Cardiovascular Surgery, University Hospital of Zurich, Zurich, Switzerland, **4** Institute of Diagnostic Radiology, University Hospital Zurich, Zurich, Switzerland, **5** IMM RECHERCHE, Institute Mutualiste Montsouris, Paris, France, **6** Department of Health Science and Technology, Laboratory for Biologically Oriented Materials, ETH Zurich, Zurich, Switzerland, **7** Department of Pathology, University Hospital of Zurich, Zurich, Switzerland

Abstract

Although stem-cell therapies have been suggested for cardiac-regeneration after myocardial-infarction (MI), key-questions regarding the in-vivo cell-fate remain unknown. While most available animal-models require immunosuppressive-therapy when applying human cells, the fetal-sheep being pre-immune until day 75 of gestation has been proposed for the in-vivo tracking of human cells after intra-peritoneal transplantation. We introduce a novel intra-uterine myocardial-infarction model to track human mesenchymal stem cells after direct intra-myocardial transplantation into the pre-immune fetal-sheep. Thirteen fetal-sheep (gestation age: 70–75 days) were included. Ten animals either received an intra-uterine induction of MI only (n = 4) or MI+intra-myocardial injection (IMI;n = 6) using micron-sized, iron-oxide (MPIO) labeled human mesenchymal stem cells either derived from the adipose-tissue (ATMSCs;n = 3) or the bone-marrow (BMMSCs;n = 3). Three animals received an intra-peritoneal injection (IPI;n = 3; ATMSCs;n = 2/BMMSCs;n = 1). All procedures were performed successfully and follow-up was 7–9 days. To assess human cell-fate, multimodal cell-tracking was performed via MRI and/or Micro-CT, Flow-Cytometry, PCR and immunohistochemistry. After IMI, MRI displayed an estimated amount of 1×10^5–5×10^5 human cells within ventricular-wall corresponding to the injection-sites which was further confirmed on Micro-CT. PCR and IHC verified intra-myocardial presence via detection of human-specific β-2-microglobulin, MHC-1, ALU-Sequence and anti-FITC targeting the fluorochrome-labeled part of the MPIOs. The cells appeared viable, integrated and were found in clusters or in the interstitial-spaces. Flow-Cytometry confirmed intra-myocardial presence, and showed further distribution within the spleen, lungs, kidneys and brain. Following IPI, MRI indicated the cells within the intra-peritoneal-cavity involving the liver and kidneys. Flow-Cytometry detected the cells within spleen, lungs, kidneys, thymus, bone-marrow and intra-peritoneal lavage, but not within the heart. For the first time we demonstrate the feasibility of intra-uterine, intra-myocardial stem-cell transplantation into the pre-immune fetal-sheep after MI. Utilizing cell-tracking strategies comprising advanced imaging-technologies and in-vitro tracking-tools, this novel model may serve as a unique platform to assess human cell-fate after intra-myocardial transplantation without the necessity of immunosuppressive-therapy.

Editor: Li Chen, University of Houston, United States of America

Funding: MYE and PW were supported by the SPUM (Special Program University Medicine; Swiss National Science Foundation) and BW was supported by the 7th Framework Programme, Life Valve, European Commission. RS was supported by the European Union 7th Framework Programme as part of the project NanoCARD, grant agreement no. 229294 and the MRC Sandpit Project "Nanomaterials for Stem-cell based Myocardial Regeneration". The funders had no role in study design, data collection and analysis, decision to publish, or preparation of the manuscript.

* E-mail: simon_philipp.hoerstrup@usz.ch

⁹ These authors contributed equally to this work.

Introduction

Stem cells have been repeatedly suggested as a next generation therapeutic approach for the treatment of heart failure due to myocardial infarction or cardiomyopathy [1]. Based on various animal trials, there are increasing numbers of early phase patient trials that aim to demonstrate the feasibility and potential efficacy of stem cell-based therapies in the clinical setting [2–6]. However, despite the plethora of generated data in the field [7], the in-vivo cell fate with specific regards to cell retention and engraftment, survival, and importantly contribution to cardiac regeneration after stem-cell transplantation into the heart remains to be elucidated. One major reason is certainly the too rapid translation from small animal studies or non-comparable large animal studies (mainly pigs and sheep) to clinical human studies, while only a systematic evaluation of the early and late stem cell fate will allow defining the optimal stem cell therapy concept for sustained cardiac regeneration.

To assess the cell fate including cellular in-vivo bio-distribution, engraftment and survival after transplantation, a surrogate animal model is mandatory allowing for sufficient cell tracking in absence of any immunologic rejection [8–11]. However, with the exception of gene-modified murine models, the availability of suitable animal models to assess human stem cell fate and bio-distribution is very limited. As most available animal models are associated with the necessity for immunosuppressive therapy when applying human cells, the clinical relevance of findings obtained from such animal models is compromised.

The fetal sheep has been suggested to be an optimal animal model for the assessment of human cell-fate [8–15]. Although the fetal sheep has a normal functioning immune-system, it is still able to support human cell engraftment and differentiation if the cells are transplanted before day 75 of gestation [8–11,16]. Following ultrasound-guided, intra-peritoneal stem cell transplantation, previous reports have shown that the fetal sheep is immunologically tolerant to human skin grafts and to allogenic or xenogenic stem cells during this "pre-immune" period of development allowing for a significant engraftment of human cells without the necessity of immunosuppressive therapy [8,9,16–23]. Taking this unique advantage of this pre-immune status as well as the large size and the long life-span into account, the fetal sheep represents an highly interesting animal model to study human cell-fate offering experimental opportunities that are not available in murine models [10,11,16].

In this study and for the first time, we investigated the feasibility to use the pre-immune fetal sheep model for the assessment of human stem cell fate after direct intra-myocardial mesenchymal stem cell transplantation following acute myocardial infarction with specific attention to cell retention and early bio-distribution using advanced, imaging-guided cell tracking protocols.

Materials and Methods

Study Animals & Experimental Design

We studied 8 pregnant Pre-Alp ewes between 70 to 75 days' gestation (term 145 days) with a total of 13 fetal-sheep. Ten animals either received an intra-uterine induction of MI only (n = 4) or MI + intra-myocardial injection (IMI; n = 6) using micron-sized, iron-oxide (MPIO) labeled human mesenchymal stem cells either derived from the adipose tissue (ATMSCs; n = 3) or the bone-marrow (BMMSCs; n = 3). Three animals received an intra-peritoneal injection (IPI; n = 3; ATMSCs; n = 2/BMMSCs; n = 1) (figure 1). All animals received humane care in compliance with the "Principles of Laboratory Animal Care" as well as the US National Institutes of Health guidelines for the care and use of animals. All procedures were approved by the Institutional Ethics Committee of the IMM RECHERCHE, Institut Mutualiste Montsouris, Paris, France [Approval-No 10-18/2010]. Figure 1 summarizes the study design and the animal distribution.

Cell Isolation, Characterization and Labeling

Isolation of human bone marrow derived mesenchymal stem cells (BMMSCs). After written informed consent and approval of the institutional review board of the University Hospital of Zurich, 80 mL of bone marrow was aspirated from the sternum of a 52 years old male patient using a bone marrow aspiration syringe with a needle prior to cardiac surgery at the University Hospital Zurich. BMMSCs were isolated as previously described [24]. *Please see supporting file S1 for further details.*

Isolation of human adipose tissue derived mesenchymal stem cells (ATMSCs). ATMSCs (isolated from a 45 years old female patient after written informed consent and approval of the institutional review board of the University Hospital of Zurich) were generously provided by Dr. Maurizio Calcagni from the Clinic for Plastic and Reconstructive Surgery, University Hospital Zurich. The cells were isolated from tissue according to a standard protocol as described elsewhere [25].

Labeling of human BMMSCs and ATMSCs with CM-DiI and MPIOs. The cells were first labeled with a fluorescent dye (CellTracker CM-DiI, Invitrogen, Switzerland) according to the manufacturer's instructions. CM-DiI is a lipophilic carbocyanine tracer and incorporates into membranes. Afterwards the cells were incubated with culture medium supplemented with 1.63 um-large encapsulated super-paramagnetic microspheres (MPIOs; Bang Laboratories, USA) at a final concentration of 2.75 ug/ml iron for 24 to 36 hours as described previously [26]. The particles were co-labeled with Dragon-green fluorochromes for additional analysis using fluorescence-based techniques. *Please see supporting file S1 for further details.*

The Fetal Sheep Cell Transplantation Model

The fetal sheep model has proven to be an appropriate model for the assessment of in-vivo cell fate and bio-distribution of human cells with specific regards to survival and engraftment [10–12,14,16]. In contrast to other animal models either being genetically modified (i.e. nude mice) or requiring immunosuppressive therapy (xenogenic models such as adult porcine or ovine models) for the assessment of human stem cells, the fetal sheep has a normal functioning immune system, but is still able to support cell engraftment and survival if the cells are transplanted before day 75 of gestation (term 145 days). Taking advantage of this "window of opportunity" and this "pre-immune" period of development, this model allows for a significant engraftment of human cells without the necessity of immunosuppressive therapy [8–12,14,16].

Operative Procedures

Following overnight fasting, the animals were sedated by intravenous injection of pentothal (10 mg/kg body weight). The animals were placed in supine position, intubated, ventilated with 100% oxygen and 1–2% isoflurane and monitored continuously throughout the entire procedure. No anti-arrhythmic agents were used during or after the procedure. After the ewes were placed in dorsal recumbency on the operating table, the uterus was exteriorized through a maternal midline laparotomy. Following digital palpation of the fetus, the uterus was opened through a 10 cm incision. The amniotic fluid was collected, stored in a sterile reservoir and out back into the ewe's belly to maintain the temperature until the end of the procedure. While the upper part of the fetal body remained within the maternal uterine cavity, the fetal chest was opened via a left-sided mini-thoracotomy (4th intercostals space). After sharp dissection of the pericardium, the heart was positioned for an optimal access of the anterior wall and the apex.

Induction of Myocardial Infarction. After evaluation of the myocardial vasculature, the left anterior descending coronary artery (LAD) and the diagonal branches (DB) were identified. To

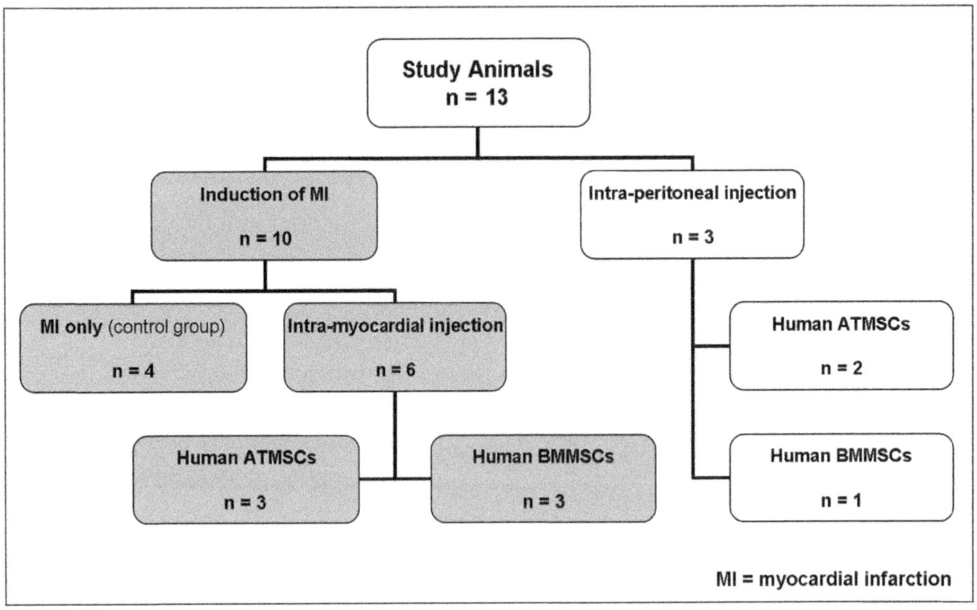

Figure 1. Study Design and Animal Distribution.

achieve a sufficient myocardial infarction involving the anterior wall and the apex, the LAD (± appropriate diagonal branches) were suture ligated using a 7/0 suture. Sufficient ligation was confirmed by instant changes of the regional wall movement and the colour in the anterior-apical area.

In utero Stem Cell Transplantation

Intra-myocardial Injection. After ligation of the coronary vessels and an ischemic period of 15–20 min to achieve sufficient anterior-apical myocardial damage, 6 target zones for cell transplantation were defined in the left-ventricular anterior area above the apex. Following careful exposure of the fetal heart, a mean of $1.25 \pm 0.30 \times 10^6$ cells was slowly injected (60 sec) into the fetal myocardium using a 29 gauge needle (table 1). The exact injection sites were carefully documented. After stem cell delivery, the absence of myocardial bleeding from the injection sites and cardiac arrhythmia was confirmed before layered closing of the fetal chest using Vycril sutures. Thereafter, the fetus was carefully re-positioned with particular regards to the umbilical cord and the amniotic fluid was re-injected before closing the uterus.

Intra-peritoneal Injection. The procedure was performed as previously described [20,21,23]. *Please see supporting file S1 for further details.*

Post-interventional Care, Follow Up and Animal Harvest

After cell transplantation, the fetal heartbeat was monitored to confirm post-procedural wellbeing. After confirming stable hemodynamic conditions the anaesthetics were stopped. The ewes were left to recover and checked daily for their health status and signs of spontaneous abortion. After 7–9 days post transplantation the ovine fetuses were delivered abdominally. In brief, the harvest procedure was as follows: a general anesthesia of the ewe was initiated with a dose of Sodium Thiopental 10 mg/kg (Nesdonal®, Merial, Lyon, France) which was then maintained with Isoflurane 2% in 100% O_2 (Attane™ Isoflurane, JD Medical, Phoenix, United States of America) during the surgery. Thereafter, euthanization was of the ewe was performed by an intravenous injection of Pentobarbital (Vétoquinol, Lure, France) and by intracardiac injection of Dolethal (Vétoquinol, Lure, France) in the ovine fetuses.

Next the fetal organs were harvested and processed for further analysis in compliance with the "Principles of Laboratory Animal Care" as well as the US National Institutes of Health guidelines for the care and use of animals. All sacrifice and harvest procedures were approved by the Institutional Ethics Committee of the IMM RECHERCHE, Institut Mutualiste Montsouris, Paris, France [Approval-No 10-18/2010].

Histology & Immunohistochemistry of myocardial infarction

To assess myocardial infarction, necrosis and cell death, standardized H&E staining, Masson's Trichrome and van Giesson staining was performed. In addition, cleaved Caspase-3 staining (1:300; Cell Signaling Technology) and TUNEL staining was performed to further access the presence of apoptosis. *Please see supporting file S1 for further details.*

Detection and tracking of CM-DiI$^+$/MPIO$^+$ human mesenchymal stem cells in the fetal sheep myocardium

To assess the presence and survival of CM-DiI$^+$/MPIO$^+$ human mesenchymal stem cells a multimodal evaluation strategy was applied comprising Flow Cytometry, PCR and immunohistochemistry. In brief, the heart was processed as follows: After cutting away the apical and basal region, the left ventricular, anterior region (area of injection) was divided into three parts (each approximately 5 mm×5 mm) which were then either processed freshly for Flow-Cytometry (part 1), snap frozen for PCR (part 2) or in formaldehyde for immunohistochemistry (part 3). *Please see supporting file S1 for further details.*

Assessment of intrinsic immune response versus CM-DiI$^+$/MPIO$^+$ human mesenchymal stem cells in the fetal sheep myocardium

To assess a potential intrinsic immune response against the injected CM-DiI$^+$/MPIO$^+$ human mesenchymal stem cells a

Table 1. Preoperative & Procedural Data.

Study Animals	Myocardial Infarction n = 4	MI+Intra-myocardial tem Cell Transplantation = 6	Intra-peritoneal tem Cell Transplantation = 3
Cell Type (ATMSCs/BMMSCs) (n)	n/a	3/3	2/1
Number of Cells (Mio/10^6)	n/a	1.25±0.30	11.75±2.81
Injection Volume (ul)	n/a	100±30	960±60
Number of Injections (n)	n/a	6	1
Injection Needle (gauge)	n/a	29	23
Duration of Procedure (min)	32±4	51±9	33±11
Waiting Time between MI Induction and Cell Injection (min)	n/a	19±1	n/a
Procedural Mortality (%)	0 (0%)	0 (0%)	0 (0%)
Major procedural complications	none	none	none
Mortality during Follow Up	1 (25%)	2 (33%)*	0 (0%)

**the two animals that died had received human BMMSCs.*
Animal Weight (gr) / Size (cm): 331 ± 81 / 20 ± 2. Overall Survival (%): 10 (77%).

detailed immunohistochemical analysis for ovine inflammatory cells was performed using anti-human antibodies reactive to CD3 (Thermo Scientific, Switzerland), CD20 (Cell marque Lifescreen Ltd., Switzerland) and CD68 (Biosystems Switzerland AG, Switzerland) to detect $CD3^+$ T-cells, $CD20^+$ B-lymphocytes and $CD68^+$ macrophages in the fetal myocardial tissue respectively. The detection of ovine inflammatory cells using cross-reactive anti-human antibodies was histologically confirmed on ovine spleen tissue sections.

Magnetic Resonance Imaging (MRI)

All measurements were performed in a Bruker 4.7T Biospec 47/40 with a gradient strength of 60 mT/m and a slew rate of 1100 T/m/s equipped with a circular polarized 1H mouse whole body RF coil. The imaging protocol consisted after gradient-echo (GRE) localizers in 3 spatial directions of 2D T2w fast spin-echo (FSE) sequences (TR/TE 2500 ms/11 ms; effective TE 33 ms; echo train length 8; matrix 256×256; FoV 40×40 mm; slice thickness 1 mm; averages 3) in transversal, sagittal and coronal orientation.

To estimate the amount of MPIO-labelled cells in the myocardium an in vitro dilution series was performed. Therefore labelled human MSCs of various counts (1×10^6, 5×10^5, 1×10^5, 5×10^4, 2×10^4) were mixed in each case with 1×10^6 ovine cardiomyocytes. The cell pellets were embedded in a 2% (w/v) agarose gel in PBS. The measurement was performed as described above.

Micro Computed Tomography (Micro CT)

The Micro-CT analysis was performed using a Mirco50 tomographer (Scanco Medical; 90 kVp) with a voxelsize of 1.2 um.

Disclosures and Freedom of Investigation

The equipment and technology used in the study were purchased using academic funds. The authors had full control of the design of the study, methods used, analysis of data, and production of the written report.

Statictical Analysis

Quantitative data are presented as mean ± standard deviation (SPSS 17.0, IBM, Somers, NY, USA).

Results

Characterization and MPIO Labeling of human BMMSCs and ATMSCs

Cell surface proteins of human BMMSCs and human ATMSCs were evaluated by Flow cytometry. Positive expression of CD44 (mean ± SD, 93.0%±4.3%), CD73 (97.8%±0.3%), CD90 (97.5%±0.7%), CD105 (84.2%±12.5%), and CD166 (94.7%±3.3%) was observed for both MSC cell types, while there was none or only weak expression of CD146 (29.4%±14.7%). Negative levels were detected for CD31 (2.5%±3.5%), CD34 (2.8%±3.9%) and CD45 (2.5%±1.9%) (figure 2A and B). Using the assays described, human ATMSCs and human BMMSCs demonstrated their differentiation potential to the adipogenic, osteogenic and chondrogenic lineages (figure 2C–H).

Successful labeling of human BMMSCs and human ATMSCs was demonstrated on Prussian Blue staining as well as on immunofluorescence clearly showing the presence of MPIOs (figure 3A–G). This was further confirmed on FACS analysis which also demonstrated an excellent cell labeling efficiency in excess of 95% (figure 3H).

Surgical Procedures and Induction of myocardial infarction

The surgical procedure and the induction of myocardial infarction (MI) could be successfully performed in all animals (figure 4A–G) scheduled for intra-myocardial stem-cell transplantation or MI only (n = 10, table 1) and was intra-operatively confirmed by instant changes of the regional wall movement and the colour in the anterior-apical area (figure 5A). The respective animals remained hemodynamically stable during the entire procedure and did not display any major complications (table 1). In three animals, a short bradycardic episode was present immediately after ligation of the coronary vessel, but normalized quickly after a few minutes. In addition, two animals encountered mild bleeding from the stitching channels of the ligation suture which self-terminated after a few minutes or was managed with a hemostatic sealant.

The successful induction of myocardial infarction was also confirmed on histology and immunohistochemistry. On H&E staining, the area of myocardial infarction (MI) could be easily identified showing the typical areas of necrosis when compared to

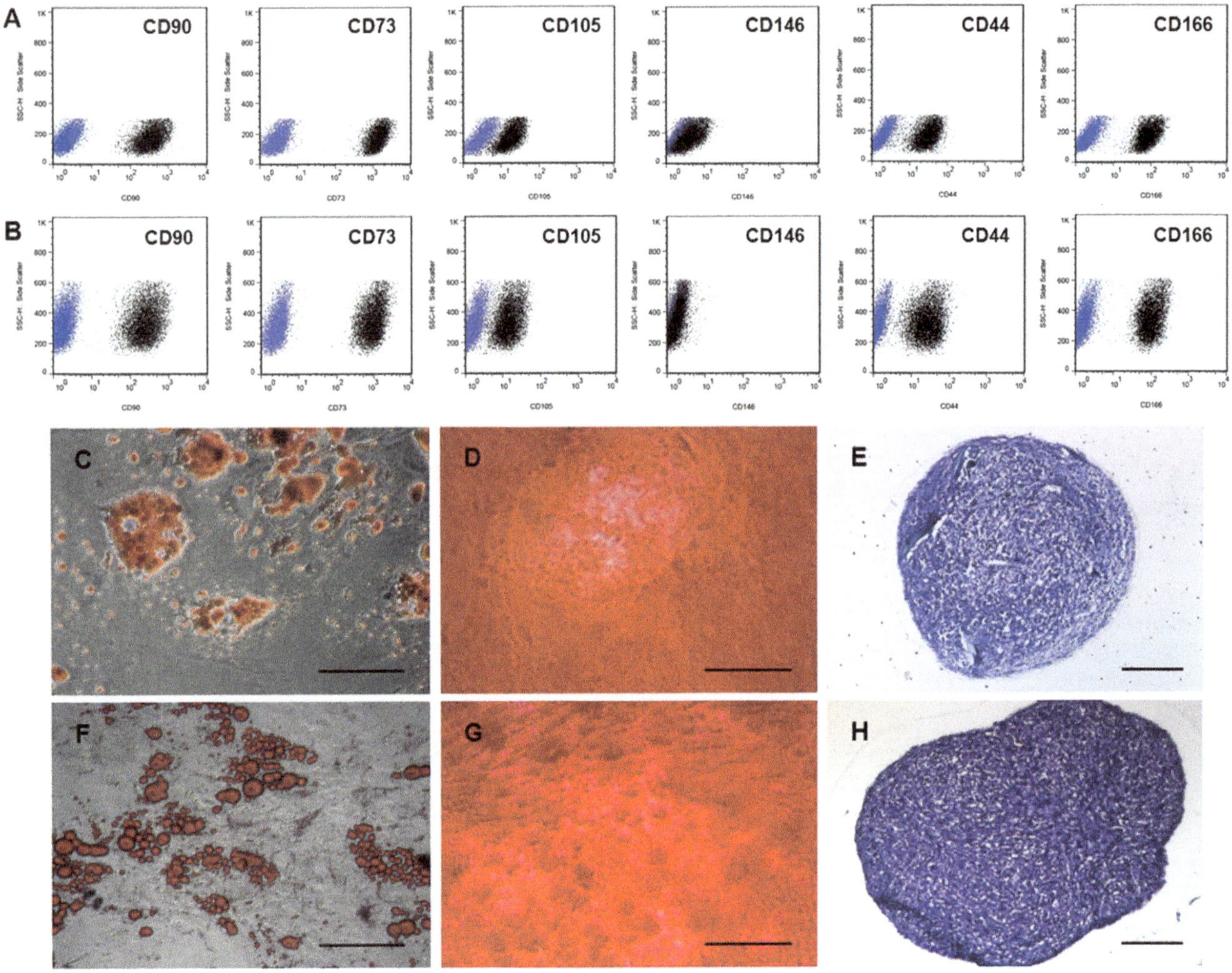

Figure 2. Characterization of human BMMSCs and ATMSCs. Cell surface proteins of human BMMSCs and ATMSCs were evaluated by flow cytometric analysis (**panel A and B**; *blue population represents isotype control*). Positive expression of CD44, CD73, CD90, CD105, CD166 and none/weak expression of CD146 was observed for both MSC cell types. Human BMMSCs (**C–E**) and ATMSCs (**F–H**) demonstrated their differentiation potential to the adipogenic *(Oil Red O Staining, (C, F))*, osteogenic *(Alizarin Red S Staining, (D, G))* and chondrogenic lineages *(Toluidine Blue Staining, (E, H)). Scale Bar: 100 um.*

the surrounding border zone and the healthy myocardium (figure 5B). In detail, loss of the classical myocardial morphology, necrotic cell death with the loss of the nuclei and cardiac proteins as well as the complete loss of entire muscle fibres could be observed (figure 5B). These findings were further confirmed on Masson's Trichrome and van Giesson staining (data not shown). In addition, also an apoptotic component could be confirmed by positive staining for activated caspase-3 and TUNEL suggesting a programmed cell-death in the area of infarction (figure 5C–E).

Intra-myocardial and intra-peritoneal stem cell transplantation. After induction of MI and the definition of 5–6 target zones in the antero-apical area, intra-myocardial transplantation either using human BMMSCs (n = 3) or ATMSCs (n = 3) was successfully performed. All animals tolerated the procedure very well without any major complications. Two animals displayed transient arrhythmia during intra-myocardial stem cell delivery and one animal had minor bleeding from the injection sites.

The intra-peritoneal transplantation was also carried out successfully in all animals (n = 3; ATMSCs; n = 2/BMMSCs; n = 1) and no complications such as bleeding or organ damage occurred (table 1).

Postoperative Complications and Follow Up

Postoperatively, no major complications occurred in the ewes and in the fetuses. However, as known from previous reports [20,21] and despite postoperative monitoring, during follow up, three (23%) of the fetuses of which one animal underwent induction of MI only and two animals had received intramyocardial transplantation of BMMSCs died due to spontaneous abortion and were expelled overnight which was primarily related to the fragile in-utero approach itself. Table 1 summarizes the preoperative and procedural data.

MRI and Micro CT for Stem Cell Tracking

CM-DiI$^+$/MPIO$^+$ human BMMSCs and ATMSCs could be successfully detected after stem cell transplantation on high resolution MRI. In selected animals that had received an intra-myocardial transplantation after MI, the labeled cell clusters were clearly visible as areas of strong focal signal loss in the septal and

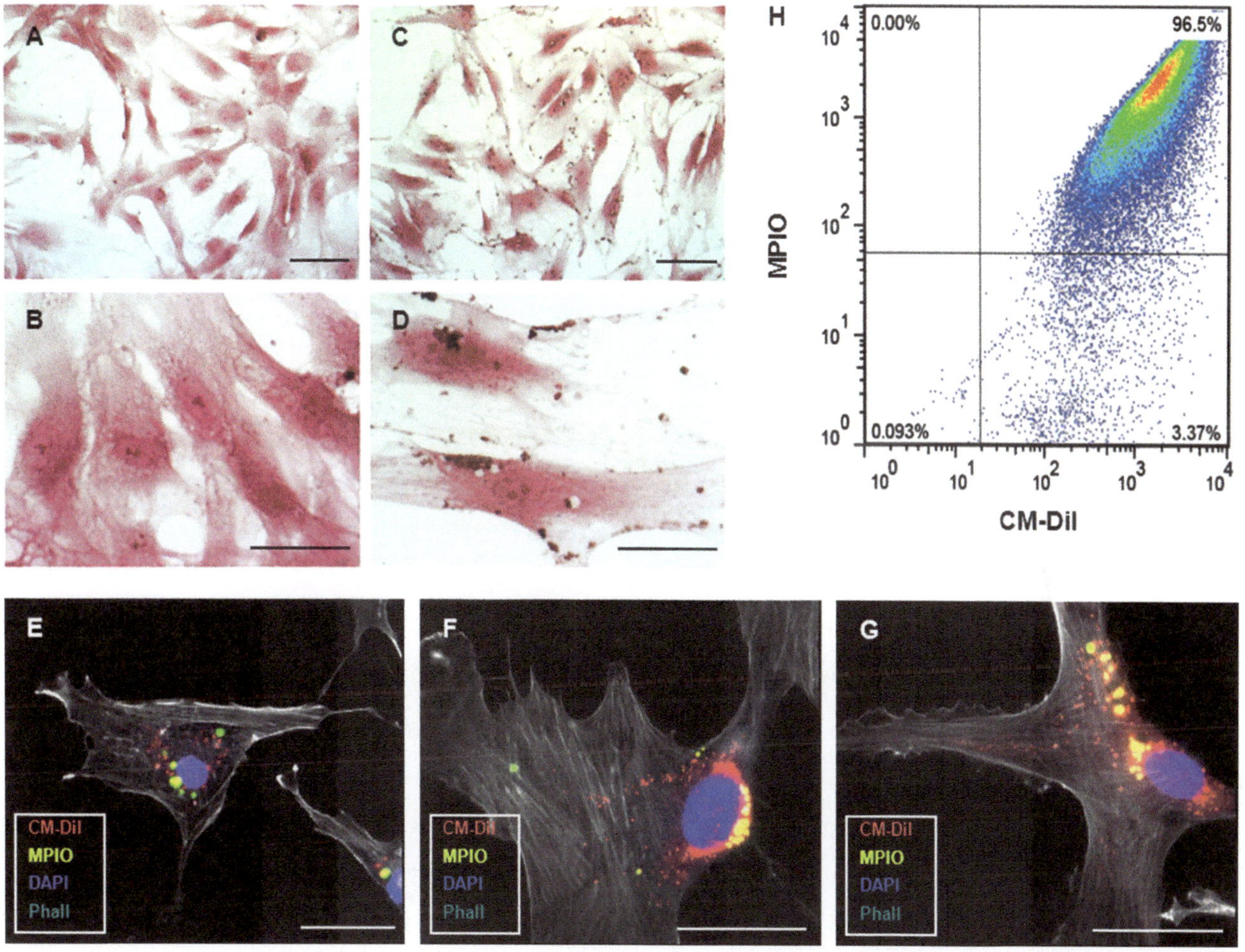

Figure 3. Labeling of human BMMSCs and ATMSCs with superparamagnetic microspheres (MPIOs; co-labeled with Dragon-green fluorochromes) and CM-DiI. Successful labeling of human BMMSCs and ATMSCs was demonstrated on Prussian Blue staining before (**A and B**) and after MPIO labeling (**C and D**) as well as on immunofluorescence clearly showing the presence of MPIOs (**E–G**). After labeling, CM-DiI$^+$/MPIO$^+$ cells displayed excellent cell labeling efficiency in excess of 95% on FACS analysis (**H**). *Scale Bar: 100 um (A and C), 50 um (B, D–G).*

anterior-lateral ventricular wall corresponding to the injection sites (figure 6A and B). In addition, cells were also detectable in the costal area, para-vertebral and in the Sinus Phrenico Costalis indicative that the cells had been distributed in the thorax over time.

Animals that underwent an intra-peritoneal transplantation displayed a distribution within the entire intra-peritoneal cavity involving the liver, the kidneys, the Gerota's Fascia, intestines as well as diffuse cell clusters in the surrounding areas (figure 6C–E).

Comparing the dilution series (figure 6F) of MPIO-labeled cells to the morphological images, the amount of cells in the myocardium could be estimated to approximately 1×10^5–5×10^5 cells and the cells found in the liver to 5×10^5 cells and around the kidneys to 5×10^5–1×10^6 cells.

Three dimensional MRI reconstruction confirmed the previous results on 2D MRI (figure 7A–D) and in selected tissue samples, the intra-myocardial presence of the iron-oxide labeled cells was further confirmed high-resolution Micro CT showing several labeled cell clusters in the myocardium (figure 7E).

Explant Macroscopy

Ten fetuses completed follow-up and could be uneventfully harvested 7–9 days post transplantation prior to ewes euthanization. The fetal heart was carefully excised and the ligation site as well as the area of myocardial infarction could be easily identified. In selected animals, the heart appeared to be slightly adhered to the pericardium and required careful dissection. The other fetal organs including lungs, liver, kidneys, spleen and others also appeared to be in good shape and were harvested for further analysis. Interestingly, most of the animals that had received left-sided thoracotomy displayed almost complete healing of the incision already after 7–9 days indicating an accelerated healing and regeneration potential in the fetal stage.

Assessment of cell fate and early bio-distribution of injected human BMMSCs and ATMSCs

Detection of CM-DiI$^+$/MPIO$^+$ human ATMSCs and BMMSCs in fetal tissue by Flow Cytometry. After intra-myocardial injection, presence of CM-DiI$^+$/MPIO$^+$ human ATMSCs and BMMSCs could be primarily confirmed within the heart and within the spleen (table 2, figure 8 A and B). Positive

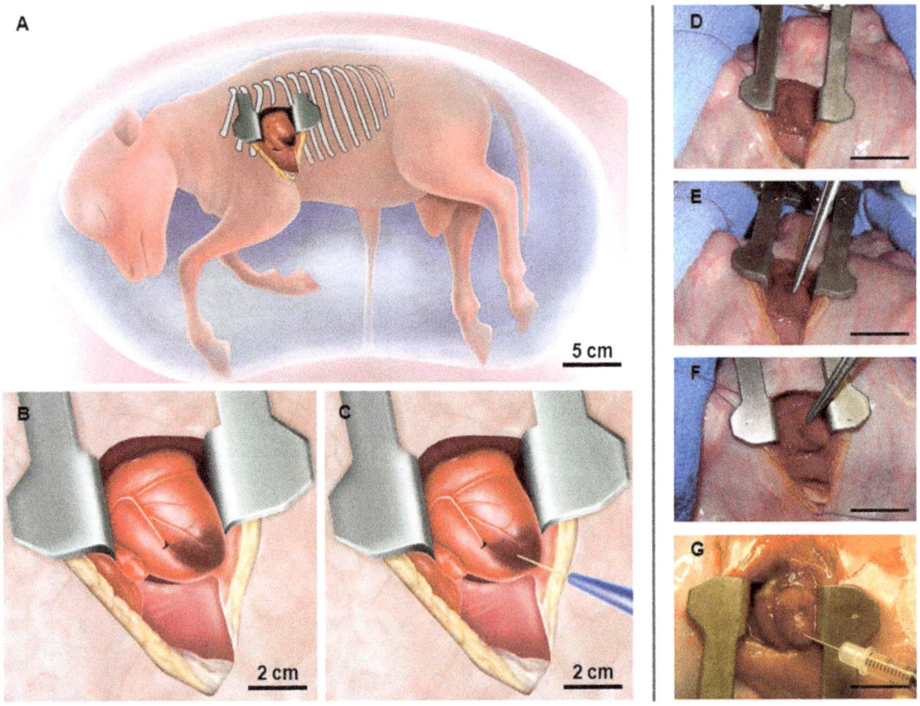

Figure 4. Concept of intra-uterine induction of myocardial infarction and intra-myocardial stem cell delivery into the pre-immune fetal sheep. The uterus was exteriorized through a maternal midline laparotomy. Following digital palpation of the fetus, the uterus was opened through a 10 cm incision. The fetal chest was opened via a left-sided mini-thoracotomy *(4th intercostal space)* (**A**). After sharp dissection of the pericardium, the heart was positioned for an optimal access of the anterior wall and the apex (**A and D**). After evaluation of the myocardial vasculature, the left anterior descending coronary artery (LAD) and the diagonal branches (DB) were identified (**E**). To achieve a sufficient myocardial infarction involving the anterior wall and the apex, the LAD ($\pm$ appropriate diagonal branches) were suture ligated using a 7/0 suture (**B and F**). Sufficient ligation was confirmed by instant changes of the regional wall movement and the colour in the anterior-apical area (**B and F**). After ligation of the coronary vessels, the 5–6 target zones for stem cell delivery were defined. Following careful exposure of the fetal heart, the cells were slowly injected into the fetal myocardium (**C and G**). *Scale Bar: 5 cm (D–G).*

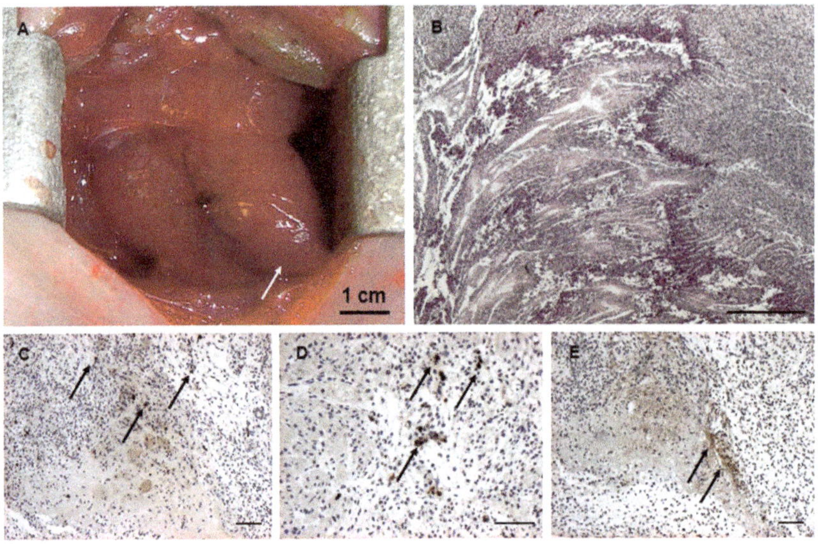

Figure 5. Assessment of intra-uterine induction of myocardial infarction in the pre-immune fetal sheep. Induction of acute myocardial infarction (MI) was confirmed by instant changes of the regional wall movement and the colour in the anterior-apical area (**A**; *purple discolouration; white arrow*). The area of myocardial infarction (MI) could be easily identified showing the typical areas of necrosis when compared to the surrounding border zone and the healthy myocardium. In detail, loss of the classical myocardial morphology, necrotic cell death with the loss of the nuclei and cardiac proteins as well as the complete loss of entire muscle fibres could be observed (**B**; *magnification x2.5*). MI was further confirmed on histology via positive staining for activated caspase-3 (**C and D**; *black arrows*) and TUNEL (**E**; *black arrows*) in fetal infarcts at 7 days suggesting programmed cardiomyocyte death within the infracted region. *Scale Bar: 500 um (B), 100 um (C–E).*

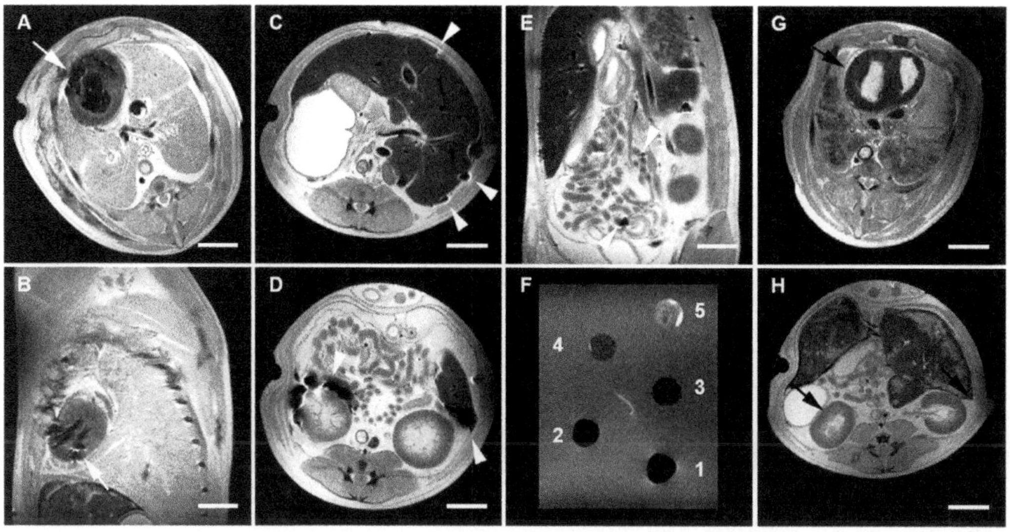

Figure 6. Tracking of CM-DiI^{+}/MPIO^{+} human ATMSCs and BMMSCs on two-dimensional magnetic resonance imaging (MRI). CM-DiI^{+}/MPIO^{+} human ATMSCs and BMMSCs could be detected after stem cell transplantation on high resolution 4.7 Tesla MRI. Exemplary image series of an animal that had received an intra-myocardial injection of CM-DiI^{+}/MPIO^{+} human ATMSCs the labeled cell clusters were clearly visible as areas of strong focal signal loss in the anterior-lateral and septal ventricular wall corresponding to the injection sites (**A and B**; *white arrows; axial and sagittal view*; **and G**; *black arrow; showing a non-injected myocardium as respective negative control*). Next, an exemplary image series of an animal that underwent intra-peritoneal injection of CM-DiI^{+}/MPIO^{+} human BMMSCs displayed the distribution within the entire intra-peritoneal cavity (**C–E**; *white arrow heads*; **and H**; *black arrows; showing a non-injected intra-peritoneal cavity as respective negative control*) involving the liver (**C**; *white arrow heads; axial and view*), the kidneys, the Gerota's fascia (**D**; *white arrow heads; axial view*), diffuse cell clusters in the surrounding intra-peritoneal areas (**D**; *white arrow heads; axial view*) as well as between the intestines (**E**; *white arrow heads; sagittal view*). Comparing the dilution series (**F**) of MPIO labeled mesenchymal stem cells to the morphological images, the amount of cells in the myocardium could be estimated to approximately 1×10^{5}–5×10^{5} cells and the cells found in the liver to 5×10^{5} cells and around the kidneys to 5×10^{5}–1×10^{6} cells (**A–D and F**). *(Cell Dilution: 1×10^{6} [1], 5×10^{5} [2], 1×10^{5} [3], 5×10^{4} [4], 2×10^{4} [5]. Scale Bar: 2.5 cm.*

cells were also found within the bone marrow, the kidneys, lungs and in the brain (table 2).

In animals that had received intra-peritoneal injection, CM-DiI^{+}/MPIO^{+} human ATMSCs and BMMSCs were primarily identified within the lymphatic organs, including spleen, thymus and the bone-marrow, while in none of the animals CM-DiI^{+}/MPIO^{+} cells were found within the heart (table 2, figure 8C and D). Positive cells were also detected in the intra-peritoneal lavage supporting the findings on the MRI analysis. In addition, low levels of positive cells were distributed in different organs, including the liver, lungs and kidneys (table 2).

Detection and tracking of CM-DiI^{+}/MPIO^{+} human ATMSCs and BMMSCs in the pre-immune fetal sheep myocardium via PCR Analysis and IHC. Following detection of CM-DiI^{+}/MPIO^{+} human ATMSCs and BMMSCs in the fetal myocardium in the imaging analysis (MRI and Micro CT scanning) and the FACS assessment, intra-myocardial presence was further confirmed via PCR analysis using human β-2 microglobulin, a component of the class I antigen complex (figure 8E). Negative controls from non-injected sheep hearts as well as human mesenchymal stem cells used as a positive control clearly confirmed the human specificity of the staining for human β-2 microglobulin within the fetal myocardium (figure 8E).

Morphologically the cells could be easily identified within the fetal myocardium, appeared to in physiological shape and viable (figure 9A–I). The cells had integrated within the fetal heart and could be found in clusters as well as in the interstitial and intravascular spaces (figure 9A–I). CM-DiI^{+}/MPIO^{+} human ATMSCs and BMMSCs were positive for both human-cell specific MHC-1 staining (figure 9A and B) as well as anti-FITC staining detecting the MPIOs within the cytoplasma of the injected human cells (figure 9C and D). In addition, double positive staining for the human-specific ALU Sequence in combination with the anti-FITC staining clearly confirmed the presence of the injected CM-DiI^{+}/MPIO^{+} human ATMSCs and BMMSCs within the fetal myocardium (figure 9E–I).

Assessment of intrinsic immune response versus CM-DiI^{+}/MPIO^{+} human ATMSCs and BMMSCs in the fetal sheep myocardium

Detailed assessment for ovine CD3^{+} T-cells, CD20^{+} B-lymphocytes as well as CD68^{+} macrophages in the fetal myocardial tissue did not show a potential intrinsic immune response against the injected CM-DiI^{+}/MPIO^{+} human ATMSCs and BMMSCs (figure 10A–I). The ovine immune cells could be detected in normal, tissue-specific frequencies and appeared to be disseminated within the entire fetal myocardium without any sign of activation due to an immune response against the human cell graft (figure 10D–F). In particular, neither a T- or B-cell infiltration into the area of the human cell graft, nor increased numbers of macrophages in this area could be observed (figure 10D–F). In contrast, CM-DiI^{+}/MPIO^{+} human ATMSCs and BMMSCs could be easily identified morphologically, and by positive staining for human-specific ALU-sequence as well as the specific intracellular, brown colour dots given by the iron-oxide particles clearly indicating that the human cells were in physiological shape, have kept the MPIOs intra-cellular and have not been phagocytized by ovine macrophages (figure 10G–I).

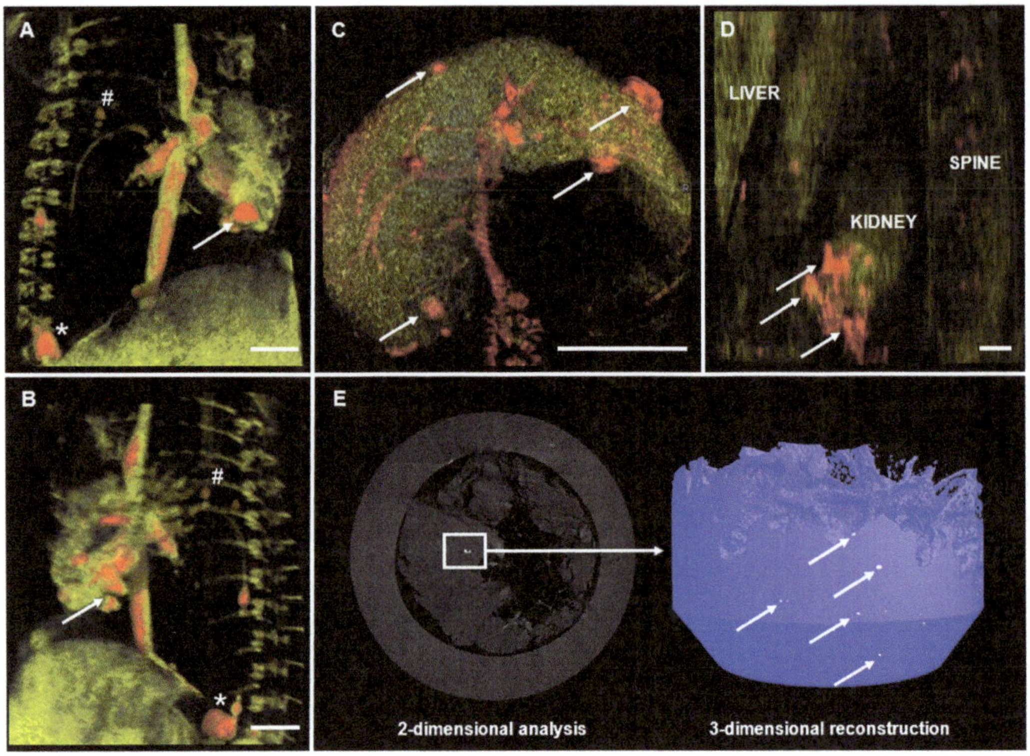

Figure 7. Tracking of CM-Dil$^+$/MPIO$^+$ human ATMSCs and BMMSCs on three-dimensional reconstruction MRI and Micro CT Scanning. Exemplary three-dimensional reconstruction analysis confirmed labeled cell clusters of human CM-Dil$^+$/MPIO$^+$ ATMSCs after intramyocardial transplantation in the anterior-lateral and septal ventricular wall (**A and B**; *white arrows; anterior-lateral and septal view*). In addition, labeled cells were also detectable in the costal/para-vertebral area (**A and B**; *white **) and in the Sinus phrenico-costalis (**A and B**; *white #*) indicative that the cells had been distributed in the fetal thorax. Abdominal 3D reconstruction analysis showed labeled cell clusters of CM-Dil$^+$/MPIO$^+$ human BMMSCs following intra-peritoneal injection within and around the liver (**C**; *white arrows; anterior-inferior view*) as well as in the anterior the Gerota's fascia (**D**; *white arrows; lateral view*). The intra-myocardial presence of CM-Dil$^+$/MPIO$^+$ human ATMSCs could be further confirmed showing several labeled cell clusters in the fetal myocardium on 2-dimensional as well as 3-dimensional high-resolution Micro CT analysis (**E**). *Scale Bar: 2 cm.*

Discussion

Stem cells have shown great promise as a therapeutic strategy for the failing heart after myocardial infarction and based on encouraging preclinical studies [7], there are growing numbers of clinical pilot trials showing the principal feasibility of stem cell-based therapies [2–6]. However, the rapid and often too premature translation into the clinical arena has left many key questions unanswered with special regards to the in-vivo cell fate which is fundamental to fully understand the possible beneficial effect on the diseased heart. In addition, the availability of suitable animal-models to assess human stem-cell fate and bio-distribution in-vivo is limited. As most available large animal models require immunosuppressive therapy when applying human cells to the myocardium [7] the clinical relevance is compromised and a surrogate animal model is mandatory to evaluate human stem-cell fate including bio-distribution, engraftment and survival in the absence of any xenogenic immune response.

To overcome these limitations, for the first time we demonstrate the principal feasibility to use a pre-immune fetal sheep intra-uterine myocardial infarction model for the assessment of human stem-cell fate after direct intra-myocardial stem-cell transplantation. Using advanced, imaging-guided cell-tracking protocols comprising magnetic-resonance imaging (MRI) and micro computed-tomography (Micro CT) as well as in-vitro cell-tracking tools, this novel model offers an excellent platform to evaluate human cell-fate in the absence of immunosuppressive therapy.

The pre-immune fetal sheep model has been previously suggested to represent an appropriate animal model for the assessment of human cell-fate. Before day 75 of gestation the immune-system of the fetal sheep is normal functioning, but is still largely immuno-naïve supporting the engraftment and differentiation of human stem cells without the necessity of immunosuppressive therapy that would also compromise the transplanted cells [8–12,16,17]. In line with that, in our study we did not detect a potential intrinsic xenogenic immune response after direct intra-myocardial injection of human mesenchymal stem cells.

While the pre-immune fetal sheep has been primarily used in the field of experimental haematology to assess in-vivo cell-fate and bio-distribution after ultrasound-guided, intra-peritoneal stem cell delivery [9,20–23,27], this is the first report demonstrating the feasibility of direct intra-myocardial stem-cell transplantation after myocardial infarction. While previous studies reported a limited homing to the myocardium after intra-peritoneal stem-cell transplantation [9,20–23,27] indicating the clear limitations of this model with regards to myocardial regeneration, our novel concept of intra-uterine, direct intra-myocardial stem-cell transplantation appears to be more appropriate in the setting of cardiovascular stem cell therapy concepts and offers a significant improvement with regards to myocardial cell retention.

In this study, we used an advanced concept of imaging-guided tracking methodology including MRI/Micro-CT Scanning as well as multiple analysis tools comprising Flow Cytometry, PCR and IHC to track human cells in the fetal sheep myocardium. Using

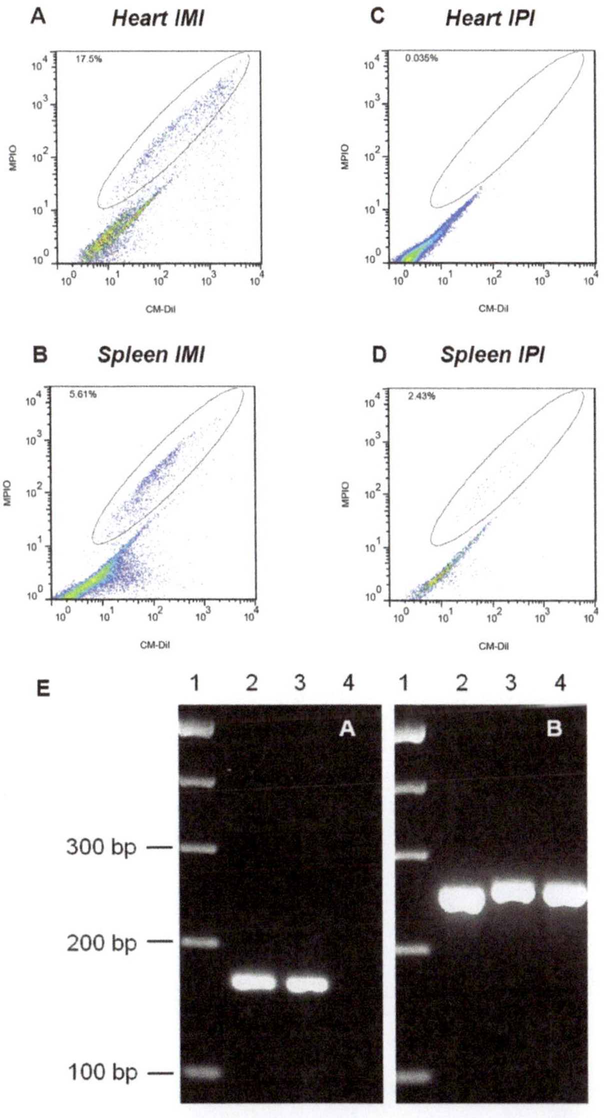

Figure 8. Assessment of cell fate and early bio-distribution of injected of human ATMSCs and BMMSCs via Flow Cytometry and PCR analysis. Flow cytometric analysis in an exemplary recipient after direct intra-myocardial injection (IMI) of human ATMSCs. CM-DiI$^+$/MPIO$^+$ human ATMSCs were primarily detected within the heart and within the spleen (**A and B**). In an animal that had received intra-peritoneal injection (IPI) of human BMMSCs, CM-DiI$^+$/MPIO$^+$ BMMSCs were primarily identified within the lymphatic organs, in particular within spleen, while no CM-DiI$^+$/MPIO$^+$ cells were found within the heart (**C and D**). Intra-myocardial presence of CM-DiI$^+$/MPIO$^+$ human ATMSCs was further confirmed via PCR analysis using human β-2 microglobulin (**E**), a component of the class I antigen complex. Negative controls from non-injected sheep hearts as well as human mesenchymal stem cells used as a positive control clearly confirmed the human specificity of the staining for human β-2 microglobulin within the fetal heart: Agarose gel analysis of human β-2 microglobulin *(left panel, A)* and β -actin *(right panel, B)* PCR products. Lane 1: molecular size marker (100 bp DNA ladder, Genecraft, Germany). Lane 2: human mesenchymal stem cells as a positive control. Lane 3: ovine fetal heart after human ATMSCs injection. Lane 4: tissue from sheep heart as a negative control for the human β-2 microglobulin sequence.

novel micron-sized, Dragon-green fluorochrome labelled iron-oxide particles (MPIO) that had so far been used in the field of liver research [28,29], we were able to evaluate human cell-fate after intra-myocardial stem cell transplantation on MRI which was then followed by detailed Flow Cytometry, PCR and IHC assessment highlighting the advantage of the fluorochrome co-labelled MPIOs. On MRI, the MPIO labeled cell-clusters were clearly visible within the septal and anterior-lateral ventricular-wall corresponding to the injection sites. Thereafter, the intra-myocardial presence was verified on Flow-Cytometry and PCR, before it was further confirmed on IHC via positive staining for human-specific MHC-1, ALU-Sequence as well as anti-FITC detecting the fluorochrome labeled part of the iron-oxide particles. Considering the initially performed dilution series, the amount of cells within the myocardium could be estimated to approximately 1×10^5–5×10^5 providing a brief quantitative estimation of cell-retention after intra-myocardial stem-cell transplantation as a precondition to explain potential beneficial effects of stem-cell therapy concepts.

Advanced imaging technologies for stem-cell tracking with regard to survival, engraftment and differentiation represent a key-requisite to validate functional effects of cell-based therapy concepts [30,31]. Various imaging-concepts are currently under evaluation including MRI imaging with direct cell-labelling of cells using super-paramagnetic agents, PET- or SPECT-imaging using radio-nuclides, as well as reporter-genes [30,31]. While each of these approaches has advantages and disadvantages, most of the recent large animal-studies applied MRI technology to detect super-paramagnetic agents as this technique has become an important endpoint to demonstrate efficacy in clinical pilot studies [30]. It offers detailed morphologic as well as functional cardiac information and therefore appears to be an appropriate imaging tool to comprise both, efficacy evaluation and the capability for cell-tracking. Current studies aim on refining contrast-agents to achieve a maximum signal for minimum labelling [30]. While the lowest cell amount detectable with a conventional MRI scanner was 10^5, a recent study highlighted that this threshold of detection can be further reduced whit the use of high resolution scanners (11.7-T) that may even allow for single cell tracking [32]. In addition, preclinical animal studies applying micron- or nano-sized iron-oxide particles showed the feasibility of non-toxic labelling of mesenchymal stem cells (MSCs) without compromising their trans-differentiation capacity [32–35].

On the other hand, it has been described that the main disadvantage of using super-paramagnetic agents is the fact that the imaging signal is not directly linked to cell-viability and the inability to discriminate between vital, labelled cells and particle-loaded cell-debris or hosting macrophages that may have actively phagocytised the particles after cell-death significantly confounding a sufficient call tracking analysis [31]. However, utilizing a detailed in-vitro tissue evaluation and importantly, advanced super-paramagnetic agents such as micron-sized, Dragon-green fluorochrome co-labeled iron-oxide particles [28,29,36], we were able to proof the MRI findings and to confirm the presence of the transplanted cells within the fetal myocardium. Using specific antibodies to either detect human cells, the fluorochrome co-labeled MPIOs or both, we were able to clearly detect the transplanted cells highlighting the efficacy of this cell-tracking approach.

Finally, while various stem cell sources are currently under evaluation for their ability to promote cardiac repair [37–44], in this study we used human mesenchymal stem cells (MSCs) either derived from the bone-marrow or from the adipose tissue as they are considered clinically safe, easily available and immuno-privileged [45]. In particular, bone-marrow derived MSCs representing a benchmark cell for cardiovascular stem cell therapies have been repeatedly used in preclinical animal models

Table 2. Distribution of CM-DiI$^+$/MPIO$^+$ human mesenchymal stem cells in ovine fetal tissue by Flow Cytometry.

	Intra-myocardial Stem Cell Transplantation		Intra-peritoneal Stem Cell Transplantation	
Tissue	**Analyzed**	**Detected**	**Analyzed**	**Detected**
	n	n (%)	n	n (%)
Heart	4	4 (100%)	3	0 (0%)
Brain	4	3 (75%)	3	0 (0%)
Spleen	4	4 (100%)	3	2 (66%)
Kidneys	4	4 (100%)	3	2 (66%)
Lavage	4	0 (0%)	3	3 (100%)
Bone Marrow	2	2 (100%)	3	2 (66%)
Liver	4	1 (25%)	3	1 (33%)
Lung	4	3 (75%)	3	2 (66%)
Thymus	4	1 (25%)	2	2 (100%)

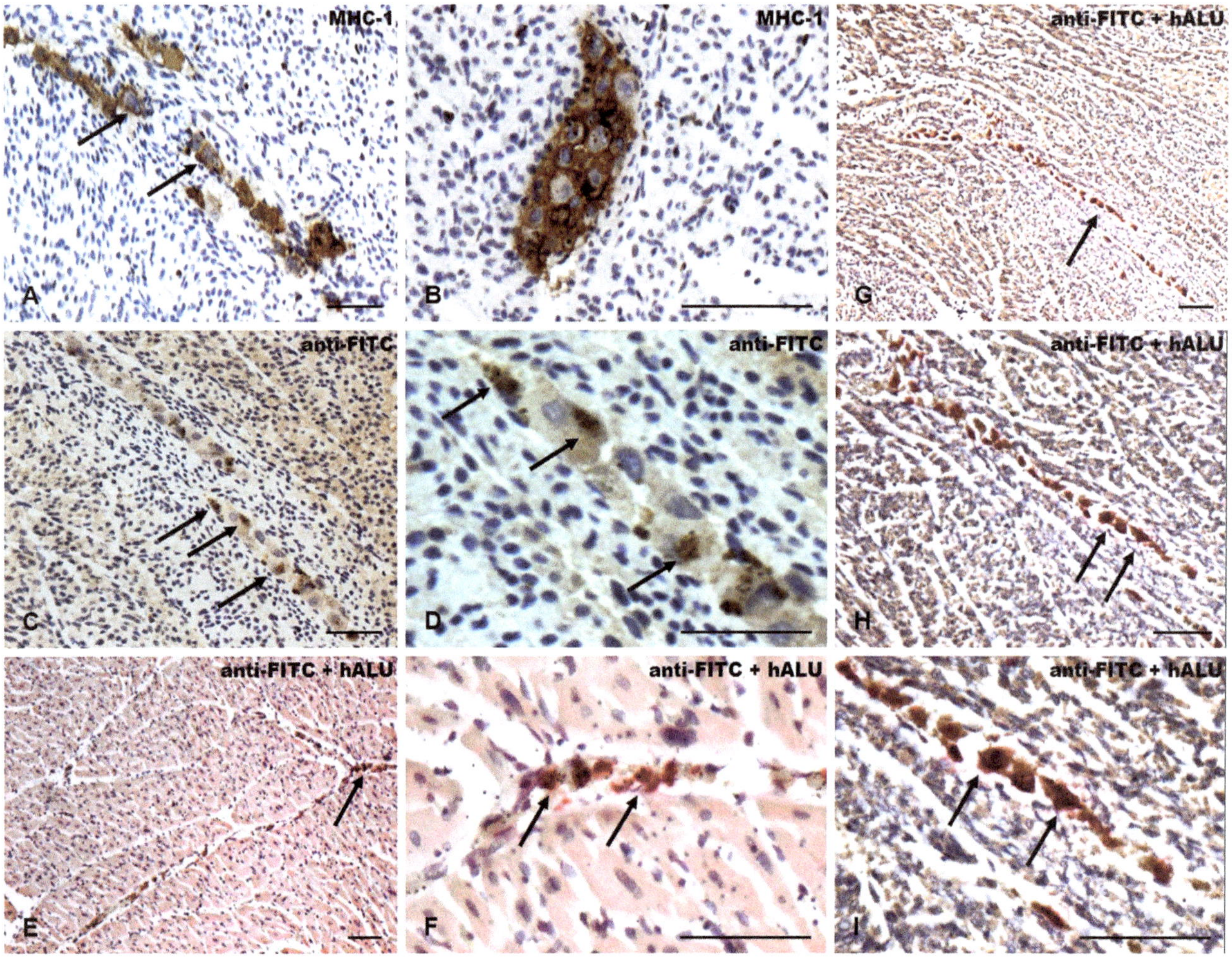

Figure 9. Detection of CM-DiI$^+$/MPIO$^+$ human ATMSCs in the pre-immune fetal sheep myocardium via immunohistochemistry. Exemplary image series post intramyocardial transplantation of human ATMSCs into the fetal myocardium. Morphologically CM-DiI$^+$/MPIO$^+$ human ATMSCs could be easily identified within the fetal heart tissue suggesting to be in good shape and viable. The cells appeared to be integrated within the fetal myocardium and could be found as clusters as well as in the interstitial and intravascular spaces (**A–I**; *black arrows*). The cells stained positive for human specific Major Histo Compatibility Complex 1 (MHC-1) clearly confirming the human origin (**A and B**; *black arrows*) and also stained positive for anti-FITC detecting the Dragon Green fluorochrome labelled MPIOs within the human cells (**C and D**; *black arrows*). In addition, double staining for ALU Sequence and anti-FITC further confirmed the presence of the injected CM-DiI$^+$/MPIO$^+$ human ATMSCs within healthy heart tissue (**E and F**; *black arrows*) as well as in infarcted myocardium (**G–I**; *black arrows) Scale Bar: 100 um (A–C, E, G, H), 50 um (D, F, I).*

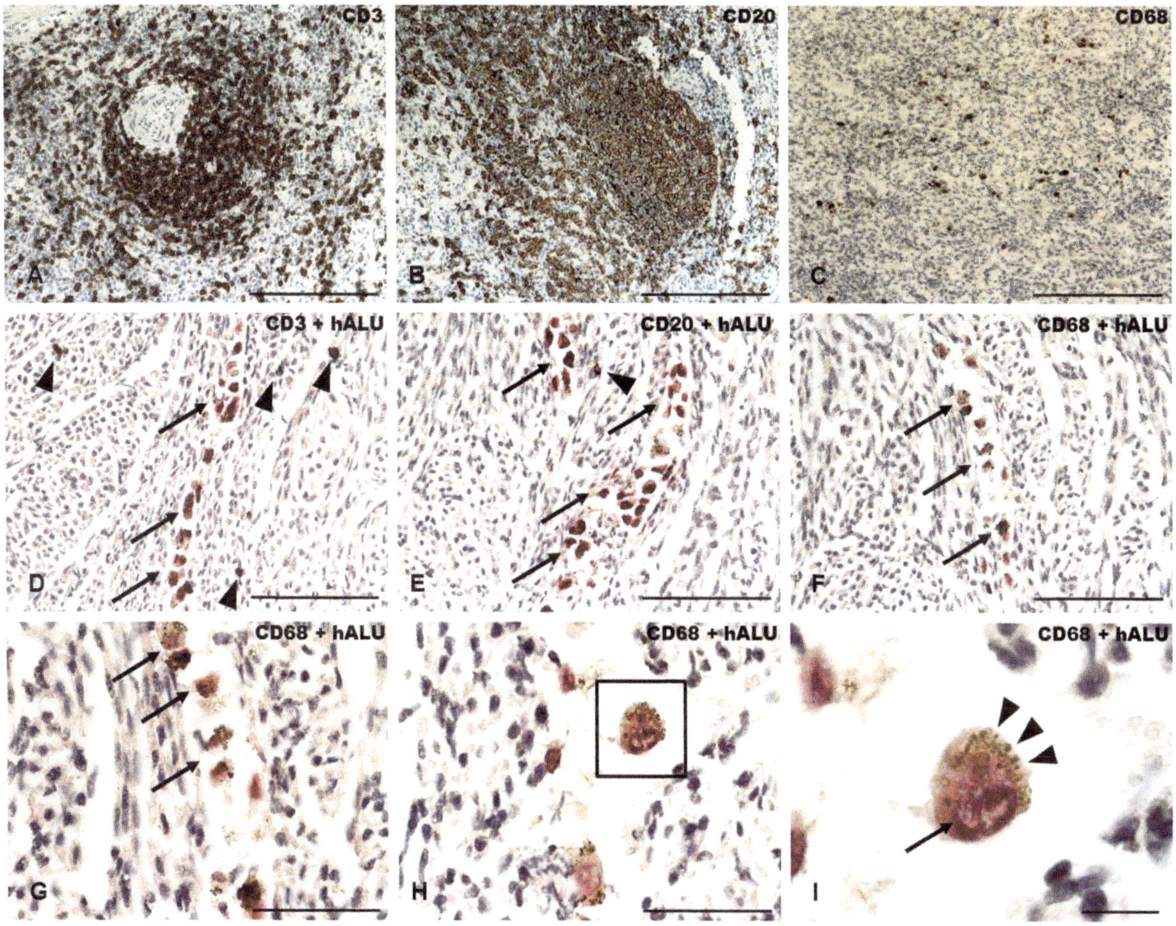

Figure 10. Assessment of intrinsic immune response versus CM-DiI$^+$/MPIO$^+$ human ATMSCs in the pre-immune fetal sheep myocardium. Control staining of CD3$^+$ T-cells, CD20$^+$ B-lymphocytes and CD68$^+$ macrophages in ovine spleen tissue (**A–C**) After direct intra-myocardial injection of CM-DiI$^+$/MPIO$^+$ human ATMSCs (**D–I**; black arrows), CD3$^+$ T-cells (**D**; black arrow heads) CD20$^+$ B-lymphocytes (**E**; black arrow head) could be detected in normal, tissue-specific frequencies and appeared to be disseminated within the entire fetal myocardium without any sign of activation due to an immune response against the human cell graft. In particular, neither a T- or B-cell infiltration into the area of the human cell graft, nor increased numbers of CD68$^+$ macrophages (**F**) in this area could be observed. In contrast, CM-DiI$^+$/MPIO$^+$ ATMSCs could be easily identified morphologically and by positive staining for human-specific ALU-sequence (**D–I**; black arrows). The human cells were in physiological shape (**G and H**; black arrows and black frame) have kept the MPIOs intra-cellular (**I**; black arrow heads) and have not been phagocytized by ovine macrophages. *Scale Bar: 100 um (A–F), 50 um (G and H) and 20 um (I).*

[46–49] as well as in clinical pilot trials [2,5,45]. Importantly, beside their suggested function through multiple paracrine effects [2] recent reports indicate that bone-marrow MSCs can be programmed into a cardiac committed stage increasing their clinical relevance and potential [50]. Similarly, the adipose tissue also represents a rich source of MSCs being well comparable to those derived from the bone marrow [51] with regards to the surface marker profile and the differentiation capacity. Furthermore, due to their abundant availability and the even lesser invasive access, adipose tissue derived MSCs should be considered as an attractive and clinically highly relevant stem cell source for future therapy concepts [52].

There are several limitations in our report that need to be addressed in further studies: First, since it was the overall aim of this proof-of-concept study to establish a novel model of intra-uterine induction of MI and direct intra-myocardial stem cell transplantation in a technically challenging and very delicate fetal environment, the number of animals was low and the follow-up time was relatively short. Therefore, further studies with an increased number of animals and a longer follow-up are mandatory in order to address the important aspect of long-term engraftment and bio-distribution. Secondly, in the context of the presented multimodal approach for an advanced cell tracking comprising Flow-Cytometry, PCR and IHC, it is to mention that the Flow Cytometry results may have been influenced by the location and size of the harvested cardiac sample as well as the gating strategy. To address the key aspect of cellular retention more accurately including absolute number of cells, the analysis of the entire heart with only a single assessment strategy either using Flow-Cytometry or qPCR would be necessary. Third, in this study two different types of human mesenchymal stem cells were used potentially influencing the study results to some extent. In addition, and although the cardiovascular differentiation potential of human mesenchymal stem cells is of particular interest, this was beyond of this proof-of-concept study. Moreover, besides the infarction histology, a functional assessment (i.e. with echo) of the fetal heart after myocardial infarction was not performed in this first feasibility study. Finally, the MRI analysis was based on T2w sequences. To further enhance cell number estimation accuracy, T2w and T2*w relaxation curves of the tissue will be performed.

Conclusions

For the first time we demonstrate the principal feasibility of direct intra-myocardial stem-cell transplantation following intra-uterine induction of myocardial infarction in the pre-immune fetal sheep. Using an advanced cell-tracking strategy comprising *state-of-the-art* imaging techniques including magnetic-resonance imaging (MRI) and micro computed-tomography (Micro CT) as well as multimodal in-vitro tracking tools, this model offers a unique platform to evaluate human cell-fate and to track human stem cells in a relevant large animal-model without the necessity of immunosuppressive therapy.

Supporting Information

File S1 provides further details on materials and methods.

Acknowledgments

We thank Scanco Medical for performing the Micro-CT analyses and Silvia Behnke for her excellent support on immunohistochemical analysis.

Author Contributions

Conceived and designed the experiments: MYE PW SPH. Performed the experiments: MYE BW PW LB SS AB. Analyzed the data: MYE PW TF SZ JS CEB PV JG HA VF AB SPH. Contributed reagents/materials/analysis tools: RS VV. Wrote the paper: MYE SPH.

References

1. Segers VF, Lee RT (2008) Stem-cell therapy for cardiac disease. Nature 451: 937–942.
2. Williams AR, Trachtenberg B, Velazquez DL, McNiece I, Altman P, et al. (2011) Intramyocardial stem cell injection in patients with ischemic cardiomyopathy: functional recovery and reverse remodeling. Circ Res 108: 792–796.
3. Assmus B, Honold J, Schachinger V, Britten MB, Fischer-Rasokat U, et al. (2006) Transcoronary transplantation of progenitor cells after myocardial infarction. N Engl J Med 355: 1222–1232.
4. Bolli R, Chugh AR, D'Amario D, Loughran JH, Stoddard MF, et al. (2011) Cardiac stem cells in patients with ischaemic cardiomyopathy (SCIPIO): initial results of a randomised phase 1 trial. Lancet 378: 1847–1857.
5. Hare JM, Traverse JH, Henry TD, Dib N, Strumpf RK, et al. (2009) A randomized, double-blind, placebo-controlled, dose-escalation study of intravenous adult human mesenchymal stem cells (prochymal) after acute myocardial infarction. J Am Coll Cardiol 54: 2277–2286.
6. Schachinger V, Erbs S, Elsasser A, Haberbosch W, Hambrecht R, et al. (2006) Intracoronary bone marrow-derived progenitor cells in acute myocardial infarction. N Engl J Med 355: 1210–1221.
7. van der Spoel TI, Jansen of Lorkeers SJ, Agostoni P, van Belle E, Gyongyosi M, et al. (2011) Human relevance of pre-clinical studies in stem cell therapy: systematic review and meta-analysis of large animal models of ischaemic heart disease. Cardiovasc Res 91: 649–658.
8. Flake AW, Harrison MR, Adzick NS, Zanjani ED (1986) Transplantation of fetal hematopoietic stem cells in utero: the creation of hematopoietic chimeras. Science 233: 776–778.
9. Liechty KW, MacKenzie TC, Shaaban AF, Radu A, Moseley AM, et al. (2000) Human mesenchymal stem cells engraft and demonstrate site-specific differentiation after in utero transplantation in sheep. Nat Med 6: 1282–1286.
10. Porada CD, Park P, Almeida-Porada G, Zanjani ED (2004) The sheep model of in utero gene therapy. Fetal Diagn Ther 19: 23–30.
11. Porada GA, Porada C, Zanjani ED (2004) The fetal sheep: a unique model system for assessing the full differentiative potential of human stem cells. Yonsei Med J 45 Suppl: 7–14.
12. Airey JA, Almeida-Porada G, Colletti EJ, Porada CD, Chamberlain J, et al. (2004) Human mesenchymal stem cells form Purkinje fibers in fetal sheep heart. Circulation 109: 1401–1407.
13. Chamberlain J, Yamagami T, Colletti E, Theise ND, Desai J, et al. (2007) Efficient generation of human hepatocytes by the intrahepatic delivery of clonal human mesenchymal stem cells in fetal sheep. Hepatology 46: 1935–1945.
14. Almeida-Porada G, Porada C, Zanjani ED (2004) Plasticity of human stem cells in the fetal sheep model of human stem cell transplantation. Int J Hematol 79: 1–6.
15. Almeida-Porada G, Porada CD, Tran N, Zanjani ED (2000) Cotransplantation of human stromal cell progenitors into preimmune fetal sheep results in early appearance of human donor cells in circulation and boosts cell levels in bone marrow at later time points after transplantation. Blood 95: 3620–3627.
16. Zanjani ED, Almeida-Porada G, Flake AW (1996) The human/sheep xenograft model: a large animal model of human hematopoiesis. Int J Hematol 63: 179–192.
17. Flake AW, Zanjani ED. (1993) In utero transplantation of hematopoietic stem cells. Crit Rev Oncol Hematol 15: 35–48.
18. Zanjani ED, Flake AW, Rice H, Hedrick M, Tavassoli M (1994) Long-term repopulating ability of xenogeneic transplanted human fetal liver hematopoietic stem cells in sheep. J Clin Invest 93: 1051–1055.
19. Zanjani ED, Silva MR, Flake AW (1994) Retention and multilineage expression of human hematopoietic stem cells in human-sheep chimeras. Blood Cells 20: 331–338; discussion 338–340.
20. Schoeberlein A, Holzgreve W, Dudler L, Hahn S, Surbek DV (2004) In utero transplantation of autologous and allogeneic fetal liver stem cells in ovine fetuses. Am J Obstet Gynecol 191: 1030–1036.
21. Schoeberlein A, Holzgreve W, Dudler L, Hahn S, Surbek DV (2005) Tissue-specific engraftment after in utero transplantation of allogeneic mesenchymal stem cells into sheep fetuses. Am J Obstet Gynecol 192: 1044–1052.
22. Surbek DV, Young A, Danzer E, Schoeberlein A, Dudler L, et al. (2002) Ultrasound-guided stem cell sampling from the early ovine fetus for prenatal ex vivo gene therapy. Am J Obstet Gynecol 187: 960–963.
23. Shaw SW, Bollini S, Nader KA, Gastadello A, Mehta V, et al. (2011) Autologous transplantation of amniotic fluid-derived mesenchymal stem cells into sheep fetuses. Cell Transplant 20: 1015–1031.
24. Pittenger MF, Mackay AM, Beck SC, Jaiswal RK, Douglas R, et al. (1999) Multilineage potential of adult human mesenchymal stem cells. Science 284: 143–147.
25. De Ugarte DA, Alfonso Z, Zuk PA, Elbarbary A, Zhu M, et al. (2003) Differential expression of stem cell mobilization-associated molecules on multi-lineage cells from adipose tissue and bone marrow. Immunol Lett 89: 267–270.
26. Shapiro EM, Skrtic S, Koretsky AP (2005) Sizing it up: cellular MRI using micron-sized iron oxide particles. Magn Reson Med 53: 329–338.
27. Young AJ, Holzgreve W, Dudler L, Schoeberlein A, Surbek DV (2003) Engraftment of human cord blood-derived stem cells in preimmune ovine fetuses after ultrasound-guided in utero transplantation. Am J Obstet Gynecol 189: 698–701.
28. Raschzok N, Billecke N, Kammer NN, Morgul MH, Adonopoulou MK, et al. (2009) Quantification of cell labeling with micron-sized iron oxide particles using continuum source atomic absorption spectrometry. Tissue Eng Part C Methods 15: 681–686.
29. Raschzok N, Muecke DA, Adonopoulou MK, Billecke N, Werner W, et al. (2011) In vitro evaluation of magnetic resonance imaging contrast agents for labeling human liver cells: implications for clinical translation. Mol Imaging Biol 13: 613–622.
30. Beeres SL, Bengel FM, Bartunek J, Atsma DE, Hill JM, et al. (2007) Role of imaging in cardiac stem cell therapy. J Am Coll Cardiol 49: 1137–1148.
31. Ruggiero A, Thorek DL, Guenoun J, Krestin GP, Bernsen MR (2012) Cell tracking in cardiac repair: what to image and how to image. Eur Radiol 22: 189–204.
32. Shapiro EM, Skrtic S, Sharer K, Hill JM, Dunbar CE, et al. (2004) MRI detection of single particles for cellular imaging. Proc Natl Acad Sci U S A 101: 10901–10906.
33. Hill JM, Dick AJ, Raman VK, Thompson RB, Yu ZX, et al. (2003) Serial cardiac magnetic resonance imaging of injected mesenchymal stem cells. Circulation 108: 1009–1014.
34. Kraitchman DL, Heldman AW, Atalar E, Amado LC, Martin BJ, et al. (2003) In vivo magnetic resonance imaging of mesenchymal stem cells in myocardial infarction. Circulation 107: 2290–2293.
35. Kraitchman DL, Tatsumi M, Gilson WD, Ishimori T, Kedziorek D, et al. (2005) Dynamic imaging of allogeneic mesenchymal stem cells trafficking to myocardial infarction. Circulation 112: 1451–1461.
36. Raschzok N, Morgul MH, Pinkernelle J, Vondran FW, Billecke N, et al. (2008) Imaging of primary human hepatocytes performed with micron-sized iron oxide particles and clinical magnetic resonance tomography. J Cell Mol Med 12: 1384–1394.
37. Asahara T, Kawamoto A (2004) Endothelial progenitor cells for postnatal vasculogenesis. Am J Physiol Cell Physiol 287: C572–579.
38. Balsam LB, Wagers AJ, Christensen JL, Kofidis T, Weissman IL, et al. (2004) Haematopoietic stem cells adopt mature haematopoietic fates in ischaemic myocardium. Nature 428: 668–673.
39. Haase A, Olmer R, Schwanke K, Wunderlich S, Merkert S, et al. (2009) Generation of induced pluripotent stem cells from human cord blood. Cell Stem Cell 5: 434–441.
40. Kehat I, Khimovich L, Caspi O, Gepstein A, Shofti R, et al. (2004) Electromechanical integration of cardiomyocytes derived from human embryonic stem cells. Nat Biotechnol 22: 1282–1289.

41. Mauritz C, Schwanke K, Reppel M, Neef S, Katsirntaki K, et al. (2008) Generation of functional murine cardiac myocytes from induced pluripotent stem cells. Circulation 118: 507–517.
42. Menasche P (2002) Myoblast transplantation: feasibility, safety and efficacy. Ann Med 34: 314–315.
43. Messina E, De Angelis L, Frati G, Morrone S, Chimenti S, et al. (2004) Isolation and expansion of adult cardiac stem cells from human and murine heart. Circ Res 95: 911–921.
44. Oyama T, Nagai T, Wada H, Naito AT, Matsuura K, et al. (2007) Cardiac side population cells have a potential to migrate and differentiate into cardiomyocytes in vitro and in vivo. J Cell Biol 176: 329–341.
45. Williams AR, Hare JM (2011) Mesenchymal stem cells: biology, pathophysiology, translational findings, and therapeutic implications for cardiac disease. Circ Res 109: 923–940.
46. Hatzistergos KE, Quevedo H, Oskouei BN, Hu Q, Feigenbaum GS, et al. (2010) Bone marrow mesenchymal stem cells stimulate cardiac stem cell proliferation and differentiation. Circ Res 107: 913–922.
47. Quevedo HC, Hatzistergos KE, Oskouei BN, Feigenbaum GS, Rodriguez JE, et al. (2009) Allogeneic mesenchymal stem cells restore cardiac function in chronic ischemic cardiomyopathy via trilineage differentiating capacity. Proc Natl Acad Sci U S A 106: 14022–14027.
48. Schuleri KH, Amado LC, Boyle AJ, Centola M, Saliaris AP, et al. (2008) Early improvement in cardiac tissue perfusion due to mesenchymal stem cells. Am J Physiol Heart Circ Physiol 294: H2002–2011.
49. Schuleri KH, Feigenbaum GS, Centola M, Weiss ES, Zimmet JM, et al. (2009) Autologous mesenchymal stem cells produce reverse remodelling in chronic ischaemic cardiomyopathy. Eur Heart J 30: 2722–2732.
50. Behfar A, Yamada S, Crespo-Diaz R, Nesbitt JJ, Rowe LA, et al. (2010) Guided cardiopoiesis enhances therapeutic benefit of bone marrow human mesenchymal stem cells in chronic myocardial infarction. J Am Coll Cardiol 56: 721–734.
51. Lee RH, Kim B, Choi I, Kim H, Choi HS, et al. (2004) Characterization and expression analysis of mesenchymal stem cells from human bone marrow and adipose tissue. Cell Physiol Biochem 14: 311–324.
52. Valina C, Pinkernell K, Song YH, Bai X, Sadat S, et al. (2007) Intracoronary administration of autologous adipose tissue-derived stem cells improves left ventricular function, perfusion, and remodelling after acute myocardial infarction. Eur Heart J 28: 2667–2677.

Testing Stem Cell Therapy in a Rat Model of Inflammatory Bowel Disease: Role of Bone Marrow Stem Cells and Stem Cell Factor in Mucosal Regeneration

Bo Qu[1], Guo-Rong Xin[2], Li-Xia Zhao[1], Hui Xing[1], Li-Ying Lian[1], Hai-Yan Jiang[3], Jia-Zhao Tong[1], Bei-Bei Wang[1], Shi-Zhu Jin[1]*

1 Department of Gastroenterology and Hepatology, The Second Affiliated Hospital, Harbin Medical University, Harbin, China, **2** Anorectal Department, The First Affiliated Hospital, JiaMuSi Medical University, JiaMuSi, China, **3** Infections Department, JiaMuSi Central Hospital, JiaMuSi, China

Abstract

Background: The gastrointestinal (GI) mucosal cells turnover regularly under physiological conditions, which may be stimulated in various pathological situations including inflammation. Local epithelial stem cells appear to play a major role in such mucosal renewal or pathological regeneration. Less is clear about the involvement of multipotent stem cells from blood in GI repair. We attempted to explore a role of bone marrow mesenchymal stromal cells (BMMSCs) and soluble stem cell factor (SCF) in GI mucosa regeneration in a rat model of inflammatory bowel diseases (IBD).

Methods: BMMSCs labelled with the fluorescent dye PKH26 from donor rats were transfused into rats suffering indomethacin-induced GI injury. Experimental effects by BMMSCs transplant and SCF were determined by morphometry of intestinal mucosa, double labeling of PKH26 positive BMMSCs with endogenous proliferative and intestinal cell markers, and western blot and PCR analyses of the above molecular markers in the recipient rats relative to controls.

Results: PKH26 positive BMMSCs were found in the recipient mucosa, partially colocalizing with the proliferating cell nuclear antigen (PCNA), Lgr5, Musashi-1 and ephrin-B3. mRNA and protein levels of PCNA, Lgr5, Musashi-1 and ephrin-B3 were elevated in the intestine in BMMSCs-treated rats, most prominent in the BMMSCs-SCF co-treatment group. The mucosal layer and the crypt layer of the small intestine were thicker in BMMSCs-treated rats, more evident in the BMMSCs-SCF co-treatment group.

Conclusion: BMMSCs and SCF participate in but may play a synergistic role in mucosal cell regeneration following experimentally induced intestinal injury. Bone marrow stem cell therapy and SCF administration may be of therapeutic value in IBD.

Editor: Michelina Plateroti, University Claude Bernard Lyon 1, France

Funding: This work was supported by Harbin Municipality Science and Technology Youth Innovation Funding (2018RFQYS103 to Bo Qu). The funders had a role in the preparation of the manuscript, but had no other role in the study design, data collection and analysis, or decision to publish.

Competing Interests: The authors have declared that no competing interests exist.

* Email: shizhujinsci@hotmail.com

Introduction

Inflammatory bowel diseases (IBD) are common gastroenterological (GI) disorders among both the paediatric and adult populations [1]. This group of diseases is featured by a chronic, relapsing or remitting course of GI complaints such as diarrhea, rectal bleeding and abdominal cramping, which can result in malnutrition, weakness, electrolyte imbalances, and delayed growth, especially in children [2]. Histopathlogically, inflammation, cell degeneration, ulceration and fistulation may be found in the small and large intestines. Notably, excessive cellular regeneration appears also a part of the chronic pathogenic process, manifested as activation/proliferation of crypt cells and angiogenesis [3]. The etiology and pathogenic mechanism underlying IBD remain largely elusive. Autoimmunity may play a central role in the pathogenic interplay between genetic predisposition, microbial infection and environmental insults [4,5]. Serological and other non-gastrointestinal signatures characteristic of autoimmune attack are commonly seen in IBD, especially evident in ulcerative colitis and Crohn's disease [6].

The mucosa of the GI tract undergoes physiological turnover occurring in every 2–7 days, which may be enhanced under pathological conditions [7]. Polipotent precursors located at the crypts are considered the major reservoir for the normal cellular renewal or pathological regeneration. These local stem cells may proliferate and differentiate into epithelium and glandulous cells in the mucosa [8]. Notably, recent studies suggest that bone marrow mesenchymal stromal cells (BMMSCs) may participate in GI regeneration especially under pathological conditions, including in IBD [9]. However, less is clear about the differentiation and early integration of BMMSCs in injured intestine. A better understanding of the role of blood-derived stem cells and soluble factors in GI histological and functional repair may shed new light on clinical management of IBD. In the present study we isolated BMMSCs from adult rats, pre-labelled them with a lipophilic red fluorescence dye PKH26. These labelled BMMSCs were then transplanted into recipient rats that suffered GI injury induced by indomethacin, an experimental model of IBD. The involvement of transplanted BMMSCs in cellular repair of the injured GI tract was studied by morphological and biochemical approaches. A potential synergetic therapeutic benefit by BMMSCs transplant and stem cell factor (SCF) administration was explored using our model system.

Materials and Methods

Animals and indomethacin treatment

Male Sprague-Dawley rats aged at 2 months were purchased from the Animal facility of Second Affiliated Hospital of Harbin Medical University. Animals were housed at constant temperature and humidity, with a 12/12 hr light/dark illumination cycle and free access to food chow and water. All experimental procedures were approved by the Harbin Medical University Administrative Panel on Laboratory Animal Care.

The experimental model of IBD was established by indomethacin administration in adult rats according to previous reports [10]. The animals were treated by subcutaneous injections of indomethacin (7.5 mg/kg in 5% sodium bicarbonate) 24 hours apart for 3 consecutive days. Four indomethacin-treated rats were killed under anesthetized condition with pentobarbital (50 mg/kg, i.p.) to evaluate the establishment of IBD model with histology. Additional animals (n = 18) were divided into 4 groups: (1) Receiving BMMSCs transplant and SCF treatment (BMMSCs-SCF group, n = 16); (2) receiving BMMSCs transplant (BMMSCs group, n = 16); (3) receiving SCF treatment (SCF group, n = 16); and (4) vehicle control with saline infusion (saline group, n = 16). The experimental effects were examined at day 7, 14, 21 and 28 days after BMMSCs, SCF or saline treatment.

BMMSCs isolation, cultivation and fluorescent pre-labeling

BMMSCs isolation, cultivation and fluorescent pre-labeling were carried out according to previously described methods [11]. Briefly, 2 month-old adult rats (n = 4, also served as naive controls) were anesthetized by sodium pentobarbital (50 mg/kg, i.p.), and the femoral bones were removed. The medullar cavity was bathed by heparin (50 U/mL) in normal saline, and the marrow was aspirated and suspended in a lymphocyte isolation medium, followed by centrifuge at 2000 rpm for 20 minutes. The cell pellets were subsequently diluted with DMEM/F12 medium (DMEM/F12, 15% FBS, 100,000 U/L penicillin, pH = 7.4) to yield a density of more than 5×10^7 cells/mL. Cells were collected after centrifuging at 1000 rpm for 10 minutes. BMMSCs were expanded *in vitro* after 10–15 days of culture. The cultured cells were then pre-labelled by the red-fluorescent lipophilic dye PKH26 according to the manufacturer's instruction [12]. The density of PKH26-labeled cell suspension was adjusted to approximately 3×10^7 cells/mL. Subsequently, the harvested cells with viability greater than 95% measured by trypan blue exclusion were used for cell transplant studies.

Fluorescence-activated cell sorting analysis

Samples of cultured cells were used for fluorescence-activated cell sorting (FACS) analysis in order to determine the expression of various signature antigen markers of mesenchymal stem cells. Approximately 1×10^6 cells were incubated in 2% fetal bovine serum in 0.01 M phosphate-buffered saline (PBS, pH = 7.2) at 4°C for 30 minutes with 1 µl of monoclonal antibody specific for CD29 (BioLegend, 555005), CD34 (Santa Cruz, sc-7324), CD44 (BioLegend, 550947), CD45 (BioLegend, 554878), CD90 (BioLegend, 561409). Negative control was processed by incubating the cells in buffer without primary antibodies. The immunofluorescent signal was analyzed using the FACS Calibur with CellQuest software (Becton Dickinson, USA).

Tissue preparation

Animals were perfused via the ascending aorta with PBS under overdose aesthesia (sodium pentobarbital 100 mg/kg, i.p.). The entire small intestine, appendix and colon were removed from the abdominal cavity for visual inspection and tissue sampling. Four small intestine segments, each approximately 2 mm in length, were blocked immediately proximal to the appendix, and further proximally to it, at approximately every 20 cm intervals. These intestine segments were rinsed in cold PBS for several times to remove contents, and then immersed in 4% paraformaldehyde for 24 hours. The former intestinal samples were embedded in paraffin, sectioned at 6 µm thickness, and used for histology and immunohistochemistry. For each animal case, at least one set of equally-spaced (approximately 60 µm in part) paraffin sections (10 sections mounted on one microslide) was used for a given histological or immunohistochemical stain. The segment left between the first and second sampled parts of the above was snap-frozen in liquid nitrogen for western blot and polymerase chain reaction (PCR) assays.

Histology and Immunohistochemistry

Consecutive intestine sections were stained with H&E and immunoflourescence. Sections were de-waxed, and rehydrated after descending ethanol incubations prior to staining. PKH26-labeled cells in the intestine were examined with and without immunofluorescent colabeling. For double fluorescent labeling, dewaxed sections were preincubated in PBS containing 5% normal donkey serum for 30 minutes. followed by incubation with mouse anti-proliferating cell nuclear antigen (PCNA) (Chemicon, MAB424R, 1:1000), rabbit anti-Lgr5 (Sigma-Aldrich-China, HPA012530, 1:400), rabbit anti-Msi-1 (Sigma-Aldrich-China, SAB4200581, 1:400) and eprhin-B3 (Abcam, ab101699, 1:200) antibodies, respectively, at 4°C overnight. The specific immunoreactive signal was visualized by further incubation with DyLight 488 goat anti-mouse or rabbit IgG (H+L) (1:200, Earthox, E032220) for 1 hr at room temperature. Sections were then counterstained with DAPI, washed and mounted with anti-fading medium before microscopic examination using a laser confocal scanning microscope (LSM 510 META; Zeiss, Germany).

Western Blot

Frozen intestine samples were homogenized in an extraction buffer (1×4 w/v) containing a cocktail of protease inhibitors (Solarbio, Beijing, China), with the resulting lysates centrifuged at 10,000×g at 4°C for 10 min. The supernatants were collected, with protein concentrations determined by DC protein assay (Bio-Rad Laboratories, Hercules, CA, USA). Extract containing 50 µg protein was loaded on each lane in 5% SDS-PAGE gels. The polypeptides were separated by electrophoresis and further transferred into nitrocellulose membrane (Bio-Rad Laboratories). Nitrocellulose membranes were then incubated with the aforementioned antibodies to β-actin (1:4000), PCNA (1:200), Lgr5 (1:200), Msi-1 (1:250) and ephrin-B3 (1:200). Bound proteins were visualized with HRP-conjugated goat anti-mouse or rabbit IgG (1:20,000, Bio- Rad Laboratories), and the Western-Bright ECL (APG BIO-China, Shanghai, China). Immunoblot images were captured in an Omega-Lum G imaging system (APG BIO-China).

Realtime fluorescence quantitative polymerase chain reaction (qPCR)

RNA extraction was carried out using the Trizol extraction kit according to manufacturer' s instruction (Life Technologies-China, Shanghai, China), with purified RNA samples stored at −70°C until use. cDNAs were produced by retrograde transcription using commercial kits including Accupower, RocketScript and RT PreMix (Bioneer-China, Shanghai, China). Specific DNA fragments were amplified using RealMasterMix (Probe) and miRcute miRNA with a Stratagene cycler (Tiangen Biomart, Beijing, China). The following forward (F) and reverse (R) primers were used for qPCR amplifications (synthesized by Bioneer-China). LGR5-F: 5′-TGCCATTATTCACCCCAAC-3′; LGR5-R: 5′-CACAGCACTGGTAAGCGTATG-3′; ephrin-B3-F: 5′-GCTGCTGTTAGGTTTTGCG-3′; ephrin-B3-R: 5′-CCCCG-ATCTGAGGGTAAAG-3′; Msi-1-F: 5′-AGGGGTTTCGGC-TTCGT-3′; Msi-1-R: 5′-TGACCATCTTAGGCTGTGCTC-3′; PCNA-F:5′-GGTGAAGTTTTCTGCGAGTG-3′; PCNA-F: 5′-GGTGAAGTTTTCTGCGAGTG-3′; rat β-actin-F: 5′-GTCAGGTCATCACTATCGGCAAT-3′; and rat β-actin-R: 5′-AGAGGTCTTTACGGATGTCAACGT-3′.

Imaging, data processing and statistic testing

H.E. stained sections were examined on an Olympus (BX60) fluorescent microscope equipped with a digital camera and image analysis system (CellSens, Olympus, Japan). Images (1600×1200 pixels) covering the entire intestine cross-section were taken using the 4× objective for morphometric analyses. The areas of the mucosal layer as well as the layer occupied by crypts were measured across the entire cross-sectional field of the intestine using the Image-J software, with the outside perimeter of the intestine also measured correspondingly. Fluorescent labeling was examined on a Leica DM500 Microscope using the 10× objective, with 10 non-overlapping microscopic fields taken randomly for each sample. Images were merged, and cells exhibiting PKH26 fluorescence (red) and those colocalizing PKH26 signal with immunoflourescence (green) were recorded for each merged image. These records were summed, and the mean rate of colocalization for a given maker was calculated for each case, and the each experimental animal group. Western blot images were analyzed by measuring optic densities of immunoblotted protein bands, which were normalized to the internal standard (β-actin) for statistical comparison. Similarly, qPCR data were normalized to internal reference (β-actin) before calculation of means. Statistical comparisons were carried out using one or two-way ANOVA with Bonferroni's multiple comparison (Prism GraphPad 4.1, San Diego, CA, USA). The minimal significance level was set at $P<0.05$. Figure panels were assembled using Photoshop 7.1.

Results

Evaluation of indomethacin-induced intestinal injury in rats

We established a rat IBD model by indomethacin treatment in the present study [10]. To confirm the occurrence of GI injury, a group of animals (n = 4) were sacrificed 4 days after indomethacin administration (Fig. 1). By visual inspection, the small and large intestines of the indomethacin-treated rats appeared dark blue with 1% Evans blue stain (Fig. 1A), whereas the intestine of healthy control rats appeared pink (Fig. 1B). The total length of small intestine was shortened in the indomethacin-treated rats (127±9.5 cm) than normal controls (171.4±4.4 cm) (Student-t test, $P<0.05$). There was noticeable extension in the diameter of the small intestine in the indomethacin-treated group (Fig. 1A, B). Haematoxylin and eosin (H&E) stain demonstrated histological disruption, atrophy of mucosa, and thinning of villi, with ulcerative lesions penetrating the proper muscle layer. Edema, epithelial exfoliation and infiltration of leukocytes appeared to be also present in the indomethacin-treated intestine relative to control (Fig. 1D, E).

Homing and differentiation of transplanted BMMSCs in injured rat intestine

Details of *in vitro* preparation, characterization and fluorescent labeling of BMMSCs were described in previous studies [11]. In the present study donor bone Fluorescence-activated cell sorting (FACS) analysis was used to confirm the expression of signature antigen markers of mesenchymal stem cells. Thus, 98.83% of the isolated cells expressed CD29; 98.22% expressed CD34, a myeloid progenitor cell antigen that is also present in endothelial cells and some fibroblasts; 34.88% expressed CD44, stem cell marker; and 85.30% expressed CD45, a hematopoietic and leukocyte marker [13]; 38.79% expressed CD90. Overall, the BMMSCs harvested from SD rat femur appeared to be largely bone marrow mesenchymal stem cells [14].

PKH26 fluorescent labeling of BMMSCs was carried out following 10–15 days of *in vitro* expansion (Fig. 2A). The success of labeling for each batch of cells was confirmed under microscope (Fig. 2B) before *in vivo* application. Red fluorescent cells were could be detected *in situ* in the wall of small and large intestines for about 3 weeks post-transplantation. In normal intestinal tissue from naive control animals, only a few fluorescent cells were encountered (data not shown). In contrast, a considerable amount of red fluorescent cells was present in the injured intestines from indomethacin-treated rats, predominantly in the villi areas (Fig. 2C, D).

We evaluated a potential *in situ* proliferation of the transplanted BMMSCs in the GI tract at 2 and 3 weeks post cell transplantation by double labeling of PKH26-labeled cells with an endogenous proliferative marker, PCNA. PKH26-labeled cells were found to colocalize with PCNA at both surviving points (Fig. 3A–D). Cell counts were carried out in 4 animals surviving 2 weeks using 10 double-labelled small intestine sections per animal. PKH26-labeled cells were found to colocalize with PCNA at an estimated mean rate of 50.1±12.2% (137/235; 204/327; 131/354 and 114/267, respectively, for the 4 individual animals).

To explore the possibility of transdifferentiation of transplanted bone marrow cells into gut mucosal cells, we examined the colocalization of the PKH26-labeled cells with markers of immature crypt cells of the GI, including Msi-1, leucine-rich-repeat-containing-protein-coupled receptors (Lgr5) and ephring-B3 [15–17]. Subpopulations of PKH26-labeled cells were found to co-express Msi-1 (Fig. 3E–H), Lgr5 (Fig. 3I–L) and ephrin-B3 (Fig. 3M–P) in indomethacin-treated rats surviving 2 and 3 weeks after infusion of the PKH26-labeled bone marrow cells. Cell counts were performed in double-labelled small intestine sections from 4 indomethacin-treated animals surviving 3 weeks after cell transplant. PKH26-labeled cells were colocalized with Msi-1 at a measured mean rate of 53.3±14.5% (101/231; 119/304; 221/314 and 166/277, respectively, for the 4 individual animals, same format below); with Lgr5 at a measured mean rate of 45.2±7.5% (125/244; 102/253; 101/194 and 87/233); and with ephring-B3

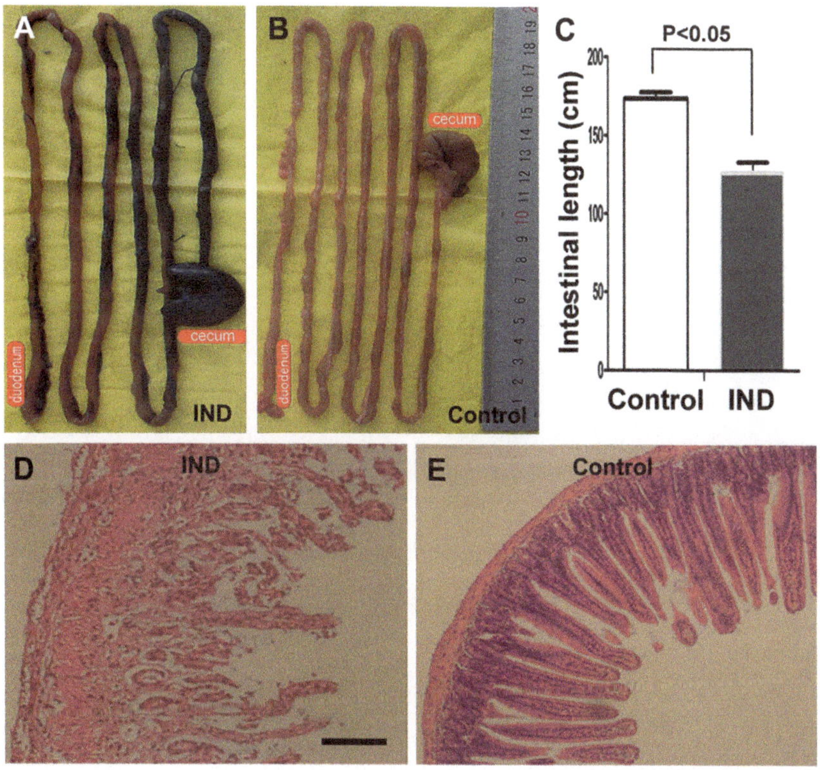

Figure 1. Evaluation of intestinal lesion after indomethacin administration. Panels (A) and (B) show that gross appearance of intestines from rats received indomethacin treatment (A) relative to control (B, saline-treated) surviving 4 days, as stained with Evans blue. The small intestines and appendix from the former appear in dark blue (A), whereas the intestine from healthy controls appears pink (B). The intestinal length is shortened in the indomethacin-treated rats relative to controls (C). In H. & E. preparation, the injured intestine show loss and disruption of the villi, the swelling of the intestinal wall, and the ulcerative lesion with fibrinoid plaques, edema, epithelial exfoliation, and granulocyte infiltration of leukocytes (D), relative to normal control (F). Scale bar =500 µm in (D) applying for (E).

with a mean rate of 58.3±9.2% (153/237; 112/217; 121/246 and 165/244).

Effect of BMMSCs and SCF administration on intestinal mRNA expression

The mRNA levels of proliferative and mucosal molecular markers were determined by qPCR in intestinal tissues at 1, 2, 3 and 4 weeks after experimental treatments (Fig. 4). Levels of ephrin-B3 mRNA expression (expressed by fluorescent absorbent signal) were different among the groups by two-way ANOVA test ($P = 0.002$, $F = 7.53$, $DF = 9, 3, 3, 32$, same statistic test below), exhibiting treatment ($P<0.0001$, $F = 63.4$, $DF = 3, 32$) and time ($P<0.0001$, $F = 22.3$, $DF = 3, 32$) dependent overall differences. Post-hoc analysis indicated that there were differences between the saline group relative to each of the other groups at all time points except for the saline vs SCF groups at the 1 week point. The SCF vs BMMSCs groups showed statistically significant differences at the week 3 but other time point, whereas the SCF vs BMMSCs-SCF groups were different at all time points. The BMMSCs vs BMMSCs-SCF exhibited differences at the week 1, 3 and 4 time points (Fig. 4A, B). Levels of Lgr5 mRNA expression showed treatment ($P<0.0001$; $F = 73.4$; $DF = 3, 32$) and time ($P<0.0001$; $F = 18.6$; $DF = 3, 32$) dependent difference. Significant differences existed for the saline group relative to all other groups at all time points, for the SCF groups relative to the BMMSCs as well as the BMMSCs-SCF groups at all time points. For the BMMSCs vs BMMSCs-SCF groups, difference existed at the week 1 time point (Fig. 4C). Levels of Msi-1 mRNA expression showed an overall treatment ($F = 56.8$, $P<0.0001$) and time ($F = 26.6$, $P<0.0001$) dependent difference ($P<0.0001$, $F = 13.6$, $DF = 9, 3, 3, 32$). Post-hot tests indicated statistically significant differences between the saline vs BMMSCs groups, between the saline and BMMSCs-SCF groups, between the BMMSCs vs SCF groups, and between the SCF and BMMSCs-SCF groups, at all time points ($P<0.0001$ to $P<0.001$ for paired mean comparisons). For the BMMSCs in comparison with BMMSCs-SCF groups, significant differences existed at the week 2 and week 4 surviving points (Fig. 4D).

The levels of PCNA mRNA expression in small intestinal tissue lysate were also affected by BMMSCs and SCF treatments. There was an overall difference among the groups ($P<0.0001$, $F = 7.27$, $DF = 9, 3, 3, 32$), which related to significant treatment ($P<0.0001$; $F = 57.0$; $DF = 3, 32$) and time ($P<0.0001$; $F = 32.5$; $DF = 3, 32$) dependent variations. Post-hoc tests indicated significant differences between the saline group relative to each of other groups at all time points ($P<0.05$ to $P<0.0001$). The means between the SCF and BMMSCs groups were different at all ($P<0.0001$) but the week 3 time points. The means between the BMMSCs and BMMSCs-SCF groups were different at all time points ($P<0.0001$) (Fig. 4E). The mRNA levels of the β-actin housekeeping gene were assayed as an internal control for the qPCR quantifications of the above molecular markers. The means did not show an overall difference among the groups by two-way ANOVA ($P = 0.95$, $F = 2.35$, $DF = 9, 3, 3, 32$). There existed no treatment-dependent significant difference ($P = 0.55$, $F = 1.59$,

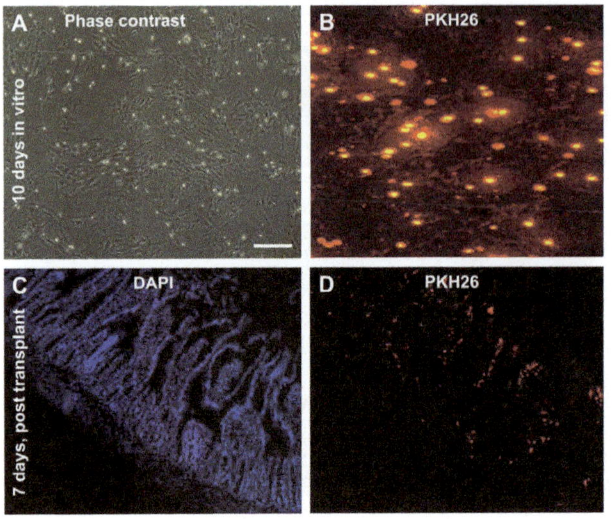

Figure 2. ***In vitro*** **expansion and PKH26 labeling of bone marrow mesenchymal stem cells (BMMSCs) harvested from a donor rat, and homing of these labeled cells to the small intestine in recipient rat with indomethacin.** (A) A large number of cells adhering to a culture dish with spindle-shape morphology at 10 days *in vitro*. (B) Culture cells pre-labeled by PKH26 show bright red fluorescence. Panels (C) and (D) show the distribution of PKH26 positive BMMSCs in the wall of the small intestine 7 days after infusion, with the cells mostly localized to the mucosal region. Scale bar = 500 μm in (A) applying for (C, D), equivalent to 125 μm for (B).

DF = 3, 32), although there was a time-dependent difference (P< 0.0001, F = 72.2, DF = 3, 32). However, Bonferroni post-tests indicated that there were no significant differences between each of the two treatment groups at any time point (Fig. 4F).

Effect of BMMSCs and SCF administration on intestinal protein expression

To further confirm an effect of the experimental treatments on the expression of proteins, we quantitatively assayed the PCNA and intestinal markers by western blot using tissues from 4 week surviving groups (Fig. 5A). Levels of PCNA protein were significantly different among the groups (P<0.001, F = 20.8, DF = 3, 12, one-way ANOVA), with Bonferroni's multiple comparison (post-hoc) indicating differences exciting between the saline group relative the SCF and the BMMSCs groups, and between the BMMSCs group relative to all other groups (Fig. 5B). Levels of Msi-1 protein were significantly different between the 4 groups (P = 0.001, F = 10.5, DF = 3, 12, one-way ANOVA). Post-hoc test indicated statistical difference between the saline group relative to the BMMSCs and BMMSCs-SCF groups (Fig. 5C). Levels of Lgr5 showed an overall difference among the groups (P = 0.0001, F = 16.8, DF = 3, 12, one-way ANOVA). Post-hoc test indicated statistically significant differences between the saline vs other groups, BMMSCs-SCF vs other groups, and the BMMSCs vs the BMMSCs-SCF groups (Fig. 5D). There were differences among the groups for the levels of ephrin-B3 protein (P<0.0001, F = 26.4, DF = 3, 12, one-way ANOVA). Specifically, statistically significant difference occurred between the saline relative to other groups, between the BMMSCs-SCF relative to the SCF and the BMMSCs groups (Fig. 5E).

Improved histological recovery following BMMSCs transplant

Given that transplanted BMMSCs resided mostly in the mucosa (villi and crypts), we determined whether the therapies might improve the histological recovery of the injured intestine. This analysis was carried out in the 4 week surviving groups (Fig. 6). The cross-sectional mucosal areas and outside perimeters were obtained at 4 locations of the small intestine with equal distance (Fig. 6A–H). The ratio of mucosal area divided by perimeter was calculated for each sample, yielding an index of relative mucosa area (Fig. 6I). There was a significant difference in relative mucosal area among the groups (P<0.0001, F = 23.8, df = 3, 12, one-way ANOVA). Post-hoc test indicated a significant reduction of the relative mucosal area in the saline group relative to the BMMSCs and BMMSCs-SCF groups. The relative mucosal area was also significantly increased in the BMMSCs-SCF group relative to the SCF and the BMMSCs groups (Fig. 6I). Considering the crypts are the principal proliferative elements for mucosal regeneration, we further calculated the ratio of the area of occupied by the crypts relative to the total mucosal area in each sample, yielding means for individual animals and groups (Fig. 6E–H). The mean relative crypt to total mucosal areas tended to increase in the SCF and BMMSCs treatment groups relative to the saline group (Fig. 6J), showing significant difference by one-way ANOVA test (P<0.0001, F = 18.3, df = 3, 12). Specifically, Bonferroni's multiple comparison tests indicated statistical difference in the saline group relative to all other groups, and between the SCF and the BMMSCs-SCF groups (Fig. 6J).

Discussion

Bone marrow stem cells may be used as a form of autograft, and are of potential therapeutic utility for tissue regeneration/organ repair. Evidence suggests that BMMSCs therapy facilitates tissue regeneration and improve functional recovery in a number of central and peripheral systems after acute or chronic injury [18,19]. Bone marrow stem cell transplantation has been considered for the treatment of IBD and certain degenerative gastrointestinal disorders [20,21]. For instances, previous studies have shown that transplanted BMMSCs labelled with BrdU, retrovirus or Y chromosome may home into the digestive tracts [22]. A few studies also suggest that transplanted bone marrow cells may colocalize with some gastrointestinal cell markers, such as Lgr5 and Ascl-2 [23].

In the present study, we observe that rats treated with indomethacin show gross anatomical appearance, histological changes and gastrointestinal symptoms of IBD. Thus, there is significant mucosal degeneration in the small and large intestines in the experimental rats as determined by histological evaluation. The height and area of the villi are dramatically reduced within a few days after indomethacin treatment. These histological and functional data are consistent with previous findings regarding the usefulness of indomethacin in inducing an experimental of model of IBD in rodents [10].

To explore the therapeutical potential of bone marrow stem cell transplant in IBD, BMMSCs are isolated from adult SD rats and expanded *in vitro* in the current study. The haemopoietic nature of the cultured cells is confirmed with a panel of specific membrane markers of bone marrow stem cells, including CD29, CD34, CD44, CD45 and CD90 [13]. The *in vitro* expanded BMMSCs are efficiently labelled with PKH26 in order to track these cells *in vivo* for a few weeks following transplantation.

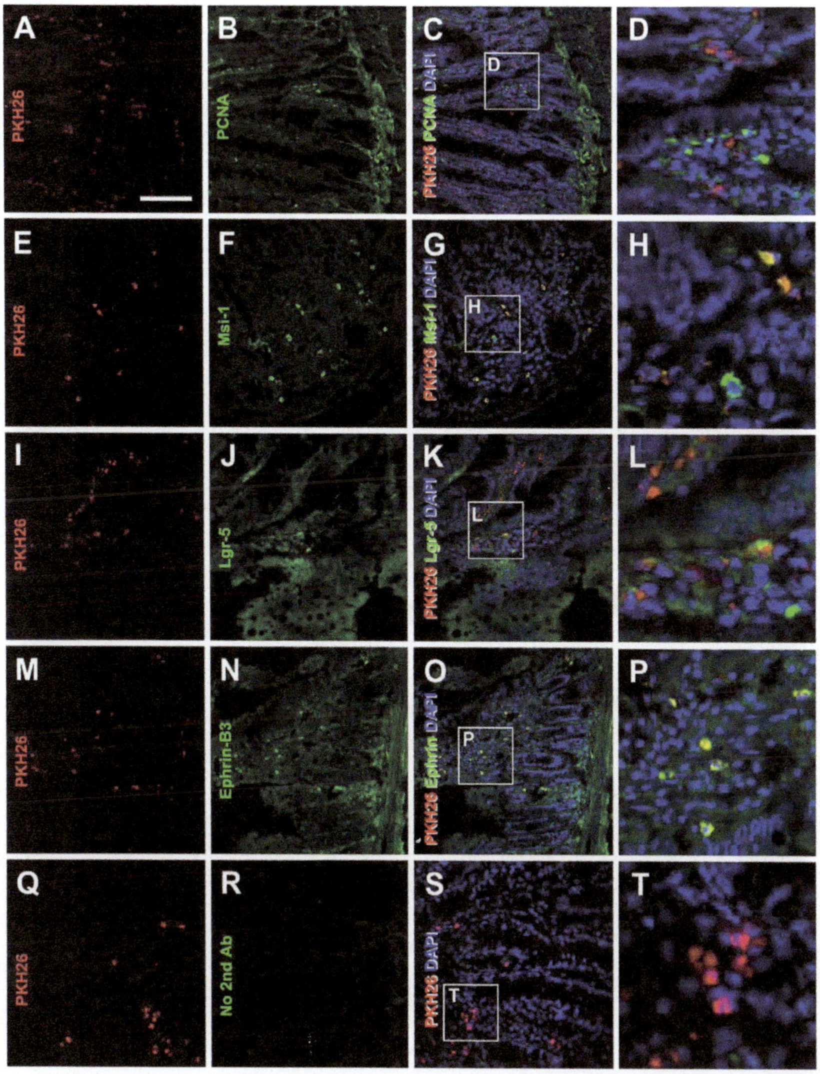

Figure 3. Fluorescent microscopic images showing potential *in situ* proliferation and transdifferentiation of transplanted BMMSCs in recipient rat small intestine with indomethacin injury. The left panels (A, E, I, M, Q) show PKH26 positive BMMSCs in the intestinal mucosa. These cells show partial colocalization with the endogenous proliferative cell nuclear antigen (PCNA) (B–D), and with molecular markers of mucosal stem cells or precursor including Msi-1 (E-H), Lgr5 (I–L) and Ephrin-3 (M–P). Panels (Q–T) show an example of negative assay control, indicating the lack of green fluorescence in the tissue by excluding the DyLight 488-conjugated secondary antibody in immunohistochemical preparation. Scale bar = 250 μm in (A) applying to the 3 left panels, equivalent to 50 μm for the most right panels.

In rats subjected to experimental IBD, transplanted PKH26-labeled cells are found to reside in the injured intestinal areas, mostly evidently in the crypts and villi. These cells show a considerable extent of colocalization with PCNA, implicating that they undergo *in situ* proliferation in the gut. Importantly, the transplanted bone barrow cells may co-express several molecular markers of precursors or immature intestine mucosal cells including Lgr5, Msi-1 and ephrin-B3. Lgr5 is expressed in precursors of intestinal secretory cells [24]. Msi-1 is an RNA-binding protein that has been shown to be a critical regulator of asymmetric division in stem cells. Msi-1 is expressed in both intestinal stem cells and their early progeny. Ephrin B3 is strongly expressed on cells at the bottom of the crypt, peaking at cell positions 4–6 (putative intestinal stem cells location). Thus, ephrin B3 is a useful cell surface marker to study intestinal stem cells [25]. Together, these histological and morphological data suggest that transplanted BMMSCs may home to in injured gut and transdifferentiate into stem-like intestinal stem cells in rats. The homing and early transdifferentiation of the transplanted BMMSCs in the injured gut appear to play a beneficial role in mucosal regeneration. Compared to untreated animals, BMMSCs transplantation leads to increases in the overall thickness of the mucosa as well as the relative areas of the crypts in the intestine. The crypts are considered the proliferative niche for mucosal renewal. Therefore, our findings support the notion that bone marrow stem cell therapy may be useful in the treatment of inflammatory bowl diseases [26,27].

Certain soluble growth factors may play a critical role for mobilization, survival and successful integration of circulating multipotent stem cells in the target organs/tissue, which may be also of value in stem cell therapy for degenerative diseases [28,29]. Adjunctive use of stem cell trophic factors has been shown to improve the outcome of BMMSCs therapy in a number of disease conditions, therefore are recommended in clinical trials of stem

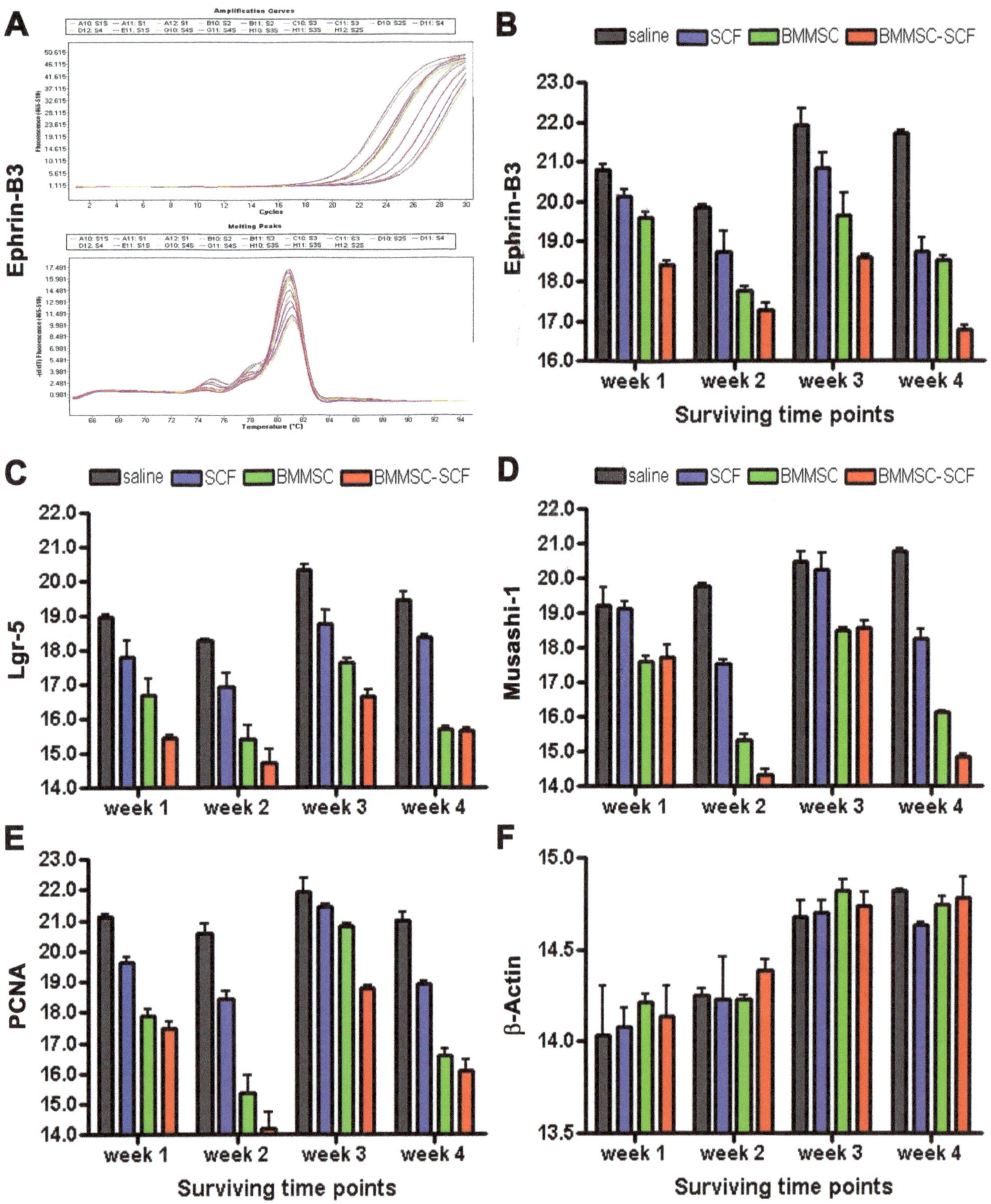

Figure 4. Effect of BMMSCs and stem cell factor (SCF) therapies on intestinal mRNA expression following indomethacin injury in rats. Panel (A) shows an example of the traces of real-time quantitative PCR (qPCR) for, with the top reflecting fluorescent signals vs amplification cycles and the ephrin-B3 bottom vs temperature changes. Panel (B) blots the means of fluorescent absorbent signals in the 4 animal groups at 4 surviving time points. Other panels (C–F) show data for the levels of mRNAs of additional markers as indicated. For details regarding statistical analyses, referring to the result section.

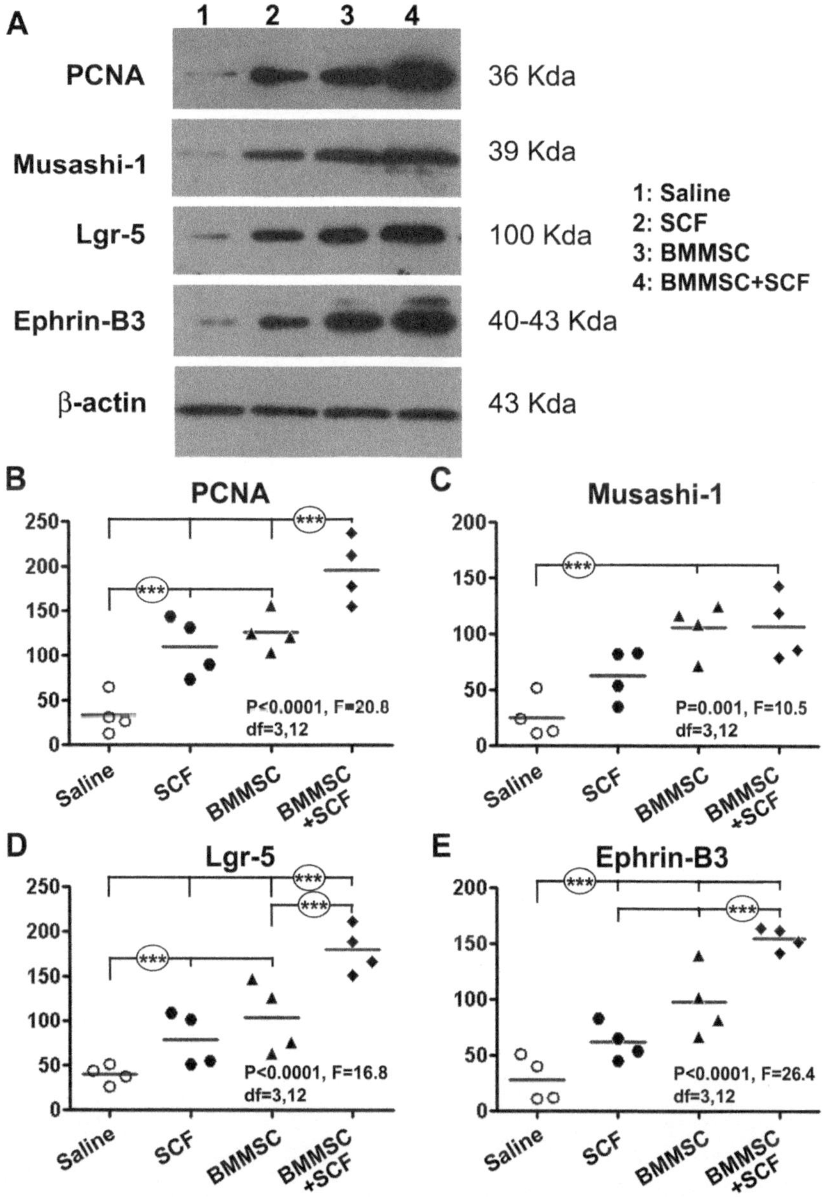

Figure 5. Effect of BMMSCs and stem cell factor (SCF) therapies on intestinal protein expression following indomethacin injury in rats surviving 4 weeks. Panel (A) shows western blot images of protein levels from one set of comparing groups of animals. Note the increased protein levels of PCNA, Msi-1, Lgr5 and ephrin-B3 relative to the internal control β-actin in the BMMSCs and SCF treated groups in comparison with saline group. The quantitative data and statistics for individual proteins are summarized as panels (B–E) as indicated. All data are expressed as the percent levels of β-actin in the same tissue lysate loads.

cell therapy. In the present study we demonstrate that the stem cell factor alone can improve the intestinal mucosal regeneration following indomethacin-induced injury for a certain degree. Co-administration of this factor appears to exhibit a synergistic beneficial effect on BMMSCs transplant to intestinal recovery. This effect includes a stimulation of the expression of cell proliferative antigen (PCNA) as well as particular molecules of mucosal stem cells or precursors including Msi-1, Lgr5 and ephrin-B3. Furthermore, mucosal and crypt areas are increased to a greatest extent in the BMMSCs-SCF treatment group after intestinal injury.

Conclusions

The present study shows *in vitro* expanded BMMSCs relocate to but exhibit local proliferation in the injured intestinal tissues in a rat model of IBD. They also appear to transdifferentiate into intestinal mucosal cells during the first few weeks post-transplantation. The BMMSCs therapy is associated with improved mucosal regeneration in indomethacin injured rat intestines. Importantly, adjunct administration of stem cell factor facilitates the beneficial role of bone marrow stem cell therapy in intestinal mucosal regeneration in this model of IBD.

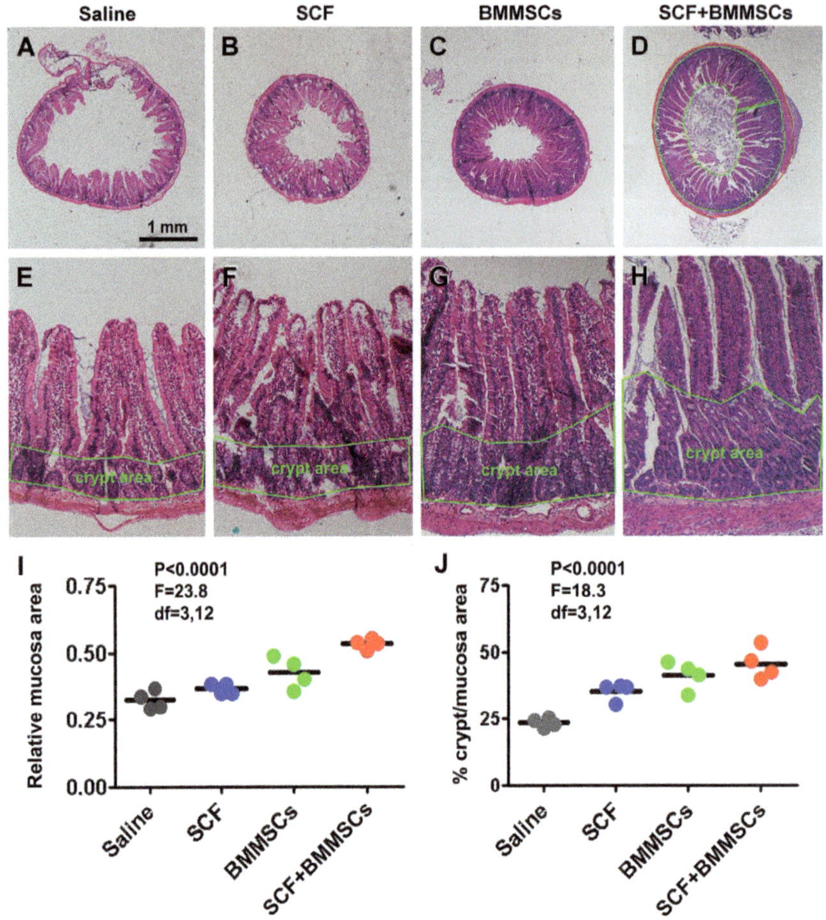

Figure 6. Effect of BMMSCs and stem cell factor (SCF) therapies on intestinal mucosal regeneration following indomethacin injury in rats examined at the 4 week surviving time point. Image panels (A–H) show representative low and high magnification H. & E. stain images of the small intestines from the 4 animal groups, sampled at the mid-point between beginning of the duodenum and appendix. The relative mucosal area is defined by the measured mucosal area divided by the outside perimeter of small intestine (as illustrated in panel D). There exists a trend of increase in the relative mucosal area in the BMMSCs and SCF treated groups, most evident in the BMMSCs and SCF co-treatment group (I). Relative area of the crypt layer to the total mucosal layer is defined by the ratio in the given intestinal cross-section (as illustrated in panel H). The means of the ration tend to increase in BMMSCs and SCF treated groups, especially in the BMMSCs and SCF co-treatment group (J). Scale bar =1 mm in (A) applying to (B–D), equivalent to 250 μm for (E–H).

Acknowledgments

We thank Prof. Ruibo Zhao and Prof. Qi Huang, Teaching Experimental Center of Morphology, Harbin Medical University, for their skilled technical assistance.

Author Contributions

none. Conceived and designed the experiments: BQ GRX. Performed the experiments: LXZ HX LYL. Analyzed the data: HYJ JZT. Contributed reagents/materials/analysis tools: BBW. Wrote the paper: SZJ.

References

1. Benchimol EI, Guttmann A, Griffiths AM, Rabeneck L, Mack DR, et al. (2009) Increasing incidence of paediatric inflammatory bowel disease in Ontario, Canada: evidence from health administrative data. Gut 58: 1490–1497.
2. Heyman MB, Kirschner BS, Gold BD, Ferry G, Baldassano R, et al. (2005) Children with early-onset inflammatory bowel disease (IBD): analysis of a pediatric IBD consortium registry. The Journal of Pediatrics 146: 35–40.
3. Saurer L, Mueller C (2009) T cell-mediated immunoregulation in the gastrointestinal tract. Allergy 64: 505–519.
4. Bernstein CN, Blanchard JF (2000) Viruses and inflammatory bowel disease: Is there evidence for a causal association?. Inflammatory Bowel Diseases 6: 34–39.
5. Danese S, Sans M, Fiocchi C (2004) Inflammatory bowel disease: the role of environmental factors. Autoimmunity Reviews 3: 394–400.
6. Thoree VC, Golby SJC, Boursier L, Hackett M, Dunn-Walters DK, et al. (2002) Related IgA1 and IgG producing cells in blood and diseased mucosa in ulcerative colitis. Gut 51: 44–50.
7. Woltman AM, De Fijter JW, Zuidwijk K, Vlug AG, Bajema IM, et al. (2007) Quantification of dendritic cell subsets in human renal tissue under normal and pathological conditions. Kidney International 71: 1001–1008.
8. Atuma C, Strugala V, Allen A, Holm L (2001) The adherent gastrointestinal mucus gel layer: thickness and physical state in vivo. American Journal of Physiology-Gastrointestinal and Liver Physiology 280: G922–G929.
9. Mimeault M, Batra SK (2012) Great promise of tissue-resident adult stem/progenitor cells in transplantation and cancer therapies. In Stem Cell Transplantation: 171–186.
10. Wang L, Li J, Li Q, Zhang J, Duan XL (2003) Morphological changes of cell proliferation and apoptosis in rat jejunal mucosa at different ages. World Journal of Gastrology 9: 2060–2064.
11. Sauerzweig S, Baldauf K, Braun H, Reymann KG (2009) Time-dependent segmentation of BrdU-signal leads to late detection problems in studies using BrdU as cell label or proliferation marker. Journal of Neuroscience Methods 177: 149–159.
12. Qi S, Wu D (2013) Bone marrow-derived mesenchymal stem cells protect against cisplatin-induced acute kidney injury in rats by inhibiting cell apoptosis. International Journal of Molecular Medicine 32: 1262.
13. Ohishi M, Schipani E (2010) Bone marrow mesenchymal stem cells,Journal of Cellular Biochemistry 109: 277–282.

14. Suh KT, Kim SW, Roh HL, Youn MS, Jung JS (2005) Decreased osteogenic differentiation of mesenchymal stem cells in alcohol-induced osteonecrosis. Clinical Orthopaedics And Related Research 431: 220–225.
15. Leedham SJ, Brittan M, McDonald SAC, Wright NA (2005) Intestinal stem cells. Journal of Cellular And Molecular Medicine 9: 11–24.
16. Shaker A, Rubin DC (2010) Intestinal stem cells and epithelial–mesenchymal interactions in the crypt and stem cell niche. Translational Research 156: 180–187.
17. Markel TA, Crisostomo PR, Lahm T, Novotny NM, Rescorla FJ, et al. (2008) Stem cells as a potential future treatment of pediatric intestinal disorders. Journal of Pediatric Surgery 43: 1953–1963.
18. Kim BC, Yoon JH, Choi B, Lee J (2013) Mandibular Reconstruction With Autologous Human Bone Marrow Stem Cells and Autogenous Bone Graft in a Patient With Plexiform Ameloblastoma. Journal of Craniofacial Surgery 24: e409–e411.
19. Telukuntla KS, Suncion VY, Schulman IH, Hare JM (2013) The Advancing Field of Cell-Based Therapy: Insights and Lessons From Clinical Trials. Journal of the American Heart Association 2: e000338.
20. Brittan M, Alison MR, Schier S, Wright NA (2007) Bone Marrow Stem Cell–Mediated Regeneration in IBD: Where Do We Go From Here?.Gastroenter-Gastroenterology 132: 1171–1173.
21. Tayman C, Uckan D, Kilic E, Ulus AT, Tonbul A, et al. (2011) Mesenchymal stem cell therapy in necrotizing enterocolitis: a rat study. Pediatric Research 70: 489–494.
22. Lin Y-T, Chern Y, Shen C-KJ, Wen H-L, Chang Y-C, et al. (2011) Human mesenchymal stem cells prolong survival and ameliorate motor deficit through trophic support in Huntington's disease mouse models. PLoS One 6: e22924.
23. Van der Flier LG, Van Gijn ME, Hatzis P, Kujala P, Haegebarth A, et al. (2009) Transcription factor achaete scute-like 2 controls intestinal stem cell fate. Cell 136: 903–912.
24. Buczacki SJ, Zecchini HI, Nicholson AM, Russell R, Vermeulen L, et al. (2013) Intestinal label-retaining cells are secretory precursors expressing Lgr5. Nature 495: 65–69.
25. Giannakis M, Stappenbeck TS, Mills JC, Leip DG, Lovett M, et al. (2006) Molecular properties of adult mouse gastric and intestinal epithelial progenitors in their niches. Journal of Biological Chemistry 281: 11292–11300.
26. Ditschkowski M, Einsele H, Schwerdtfeger R, Bunjes D, Trenschel R, et al. (2003) Improvement of inflammatory bowel disease after allogeneic stem-cell transplantation. Transplantation 75: 1745–1747.
27. Duijvestein M, Vos ACW, Roelofs H, Wildenberg ME, Wendrich BB, et al. (2010) Autologous bone marrow-derived mesenchymal stromal cell treatment for refractory luminal Crohn's disease: results of a phase I study. Gut 59: 1662–1669.
28. Cook D, Genever P (2013) Regulation of mesenchymal stem cell differentiation. In Transcriptional and Translational Regulation of Stem Cells: 213–229.
29. Hoggatt J, Pelus LM (2014) New G-CSF agonists for neutropenia therapy. Expert Opinion On Investigational Drugs 23: 21–35.

Interactions of Anaerobic Bacteria with Dental Stem Cells: An *In Vitro* Study

Anne Biedermann[1], Katja Kriebel[1], Bernd Kreikemeyer[2], Hermann Lang[1]*

1 Department of Operative Dentistry and Periodontology, University of Rostock, Rostock, Germany, **2** Institute of Med. Microbiology, Virology and Hygiene, University of Rostock, Rostock, Germany

Abstract

Background: In patients with periodontitis, it is highly likely that local (progenitor) cells encounter pathogenic bacteria. The purpose of this *in vitro* study was to elucidate how human dental follicle stem cells (hDFSC) react towards a direct challenge with anaerobic periodontal pathogens under their natural oxygen-free atmosphere. HDFSC were compared to human bone marrow mesenchymal stem cells (hBMSC) and differentiated primary human gingival fibroblasts (hGiF), as well as permanent gingival carcinoma cells (Ca9-22).

Methodology/Principal Findings: The different cell types were investigated in a co-culture system with *Porphyromonas gingivalis (P. gingivalis)* and *Fusobacterium nucleatum (F. nucleatum)*. The viability of the cells and pathogens under anaerobic conditions, as well as interactions in terms of adherence and internalization, were examined. Additionally, the release of pro-inflammatory interleukin-8 (IL-8) and anti-inflammatory interleukin-10 (IL-10) was quantified via enzyme-linked immunosorbent assay. The bacteria adhered less efficiently to hDFSC compared to Ca9-22 (*P. gingivalis*: 0.18% adherence to hDFSC; 3.1% adherence to Ca9-22). Similar results were observed for host cell internalization (*F. nucleatum*: 0.002% internalization into hDFSC; 0.09% internalization into Ca9-22). Statistically significantly less IL-8 was secreted from hDFSC after stimulation with *F. nucleatum* and *P. gingivalis* in comparison with hGiF (*F. nucleatum*: 2080.0 pg/ml – hGiF; 19.7 pg/ml – hDFSC). The IL-10 response of the differentiated cells was found to be low in relation to their pro-inflammatory IL-8 response.

Conclusions/Significance: The results indicate that dental stem cells are less prone to interactions with pathogenic bacteria than differentiated cells in an anaerobic environment. Moreover, during bacterial challenge, the stem cell immune response seems to be more towards an anti-inflammatory reaction. For a potential future therapeutic use of hDFSC, these findings support the idea of a save application.

Editor: Salomon Amar, Boston University, United States of America

Funding: The authors have no support or funding to report.

Competing Interests: The authors have declared that no competing interests exist.

* Email: hermann.lang@uni-rostock.de

Introduction

To determine the biological role of dental stem cells in their natural environment and to examine the save potential therapeutic application for tissue regeneration, the understanding of cell–bacteria interactions is of great importance [1].

The periodontal tissues are composed of a variety of different cell types. Recently, stem cells have been reported to be part of this cellular environment [2,3]. Generally, stem cells have two major characteristics that distinguish them from other cells: they are capable of self-renewal, and upon division, they give rise to cells that have the potential to differentiate [4].

The dental follicle harbors human dental follicle stem cells (hDFSC), and is present in impacted teeth, which are commonly extracted and disposed of medical waste in dental practice [5]. Therefore, these cells are an easily accessible cell source for experimental and future clinical applications in periodontal tissue or bone regeneration approaches [6,5].

The periodontal pocket is predisposed towards interdependencies of local cells and putative pathogens, which have been one focus of periodontal research [7]. More than 700 different bacterial species can be found in the oral cavity [8]. The majority settles in multiple different forms of biofilm patterns. Once detached from the biofilm, some of the bacteria penetrate into deeper tissues and may disturb host cell homeostasis [7]. Direct contact between bacteria and human cells triggers the expression of a variety of immune response mediators [9,7]. Interleukin (IL)-8 is a pro-inflammatory cytokine and is predominantly secreted by epithelial cells. It has been suggested to be involved in the local host defense mechanisms leading to neutrophil migration [9]. This chemoattractant activates and attracts neutrophils to the site of infection or injury [10,11,9]. As a regulatory mechanism, the expression of the anti-inflammatory IL-10 terminates the secretion of pro-inflammatory cytokines and reduces the expression of major histocompatibility complex class II and co-stimulatory molecules [12]. The critical role of IL-10 in infectious disease appears to be the modulation of inflammatory responses to microbial pathogens [13].

The pathogenic bacteria *F. nucleatum* and *P. gingivalis* can frequently be found in diseased periodontal tissues and they play important roles in the initiation and progression of periodontitis [14,15]. *P. gingivalis* is able to induce strong cytokine and chemokine expression in gingival epithelial and other host cells, which has been shown to positively correlate with the adhesive/invasive potential of the infecting strain [16]. *F. nucleatum* and *P. gingivalis* are obligate anaerobic bacteria that can only exist in areas of the oral cavity with low oxygen partial pressure (*e.g.*, the dental pocket). However, functional interaction studies of hDFSC and oral pathogenic bacteria in their natural anaerobic atmosphere are lacking.

Thus, the focus of our study was to investigate the interactions of hDFSC with periodontal pathogens. The results were compared with cells at various stages of differentiation: human bone marrow mesenchymal stem cells (hBMSC), human gingival fibroblasts (hGiF), and the permanent gingival carcinoma cell line Ca9-22. We hypothesized that hDFSC might show a different immunological response than differentiated cells after co-culture with periodontal pathogens because of their anti-inflammatory and tissue regenerative function in the injured or inflamed tissue [17]. Stem cells have to remain operative even in sites of chronic inflammation [1], and therefore, they might also be more tolerant towards bacterial stimuli.

Material and Methods

Cell Isolation and Culture

The stem cells were obtained from donors undergoing surgical treatment at either the Department of Cardiac Surgery or the Department of Oral Surgery at the University of Rostock. This study conforms to the Declaration of Helsinki, and all cell donors gave their informed written consent. The experimental protocol and further experiments were reviewed and approved by the Ethics Committee of the University of Rostock (No. A 2011 119 and No. A 2011 91).

Ca9-22 was provided by the Japanese Collection of Research Bioresources Cell Bank Osaka, Japan. The cells were cultured in DMEM supplemented with 10% FCS in cell culture flasks (Greiner Bio-one, Frickenhausen, Germany) (5% CO_2, 37°C).

HGiF were cultured in DMEM supplemented with 10% FSC (10% CO_2, 37°C). Isolation and culturing of hBMSC was performed as described by Gaebel *et al.* [18], and hDFSC as described by Haddouti *et al.* [19]. The authenticity of the stem cells was confirmed by plastic adherence, and flow cytometric analysis (monoclonal antibodies against CD29, CD44, CD45, CD73 and CD90 using a FACS scan flow cytometer LSRII with CellQuest-Software, Becton Dickinson). Multipotency of the cells was shown through induction into osteoblasts, chondroblasts, and adipocytes as previously described [20].

Bacteria

P. gingivalis W50 and W83 (DSMZ, Braunschweig, Germany), *F. nucleatum* ATCC 23726, and ATCC 25586 (American Type Culture Collection, Manassas, USA) were purchased from commercial providers. The bacteria were grown in PYG medium supplemented with 5 µg/ml hemin and 1% vitamin K in an anaerobic atmosphere (10% CO_2, 10% H_2, 80% N_2).

Anaerobic Co-Culture

Bacteria were tested for their survival in cell culture medium prior to the co-culture experiments. They were grown in PYG medium supplemented with 5 µg/ml hemin and 1% vitamin K to the stationary phase. Subsequently, bacteria were centrifuged, washed in PBS, and each bacterial suspension was diluted in DMEM 1:10. The optical density (OD) was measured at 600 nm from timepoint 0 every hour to timepoint 12 h with a final measurement after 24 h.

Cells were tested for their tolerance towards oxygen-free conditions. Cell-cultures at the 4th to 6th passage were harvested and seeded at a density of 8×10^3 cells/well in a 24-well culture plate, and maintained in 2 ml of medium. Subsequently, cells were grown in an anaerobic workstation (miniMACS, DWS Meintrupp, Lahden-Holte, Germany) at 37°C for 24, 48, and 76 h, and numbers of viable cells were counted in a Neubauer hemacytometer after trypan blue staining.

For co-culture, hDFSC, hBMSC, hGiF, and Ca9-22 were seeded at a density of 8×10^3–1×10^4 cells/well in a 24-well culture plate, and maintained in 1 ml medium. For each experiment, the final concentration of the bacterial suspension was determined by measurement of the optical density at 600 nm (OD_{600}) to obtain 1×10^8 cells/ml, and dilutions were made to achieve the desired MOI. The bacterial inoculum was confirmed by counting of the colony-forming units (CFU). After reaching confluence, the cells were infected with live bacteria with a multiplicity of infection (MOI) of 1:100 and incubated at 37°C in an anaerobic atmosphere.

For fluorescence microscopy (BX60 microscope, Olympus, Hamburg, Germany) the samples were stained with BacLight Live/Dead (Molecular Probes, Eugene, USA). The staining was documented with an attached digital camera (Leica, Solms, Germany).

Adherence and Internalization

The eukaryotic cells (8×10^3 cells/ml) were cultured in DMEM and grown to a monolayer. The bacteria were grown in PYG medium at 37°C under an anaerobic atmosphere. The bacterial density was adjusted to MOI 1:100 in DMEM and added to the cell monolayer. After 2 h, the cells were washed and subsequently detached by adding 200 µl 0.25% trypsin/0.5 mM EDTA for 5 min. To quantify bound bacteria, the cells were lysed with distilled water and the number of bacteria in the lysate was assessed by viable counts. For quantification of internalized bacteria, the monolayers were washed with PBS. Medium containing gentamicin (300 mg/ml) and metronidazole (200 mg/ml) was added after 2 h and incubated for another 2 h. After 4 h, the same procedure was performed to quantify the internalized bacteria. The number of viable bacteria was confirmed by counting CFUs on agar plates incubated under an anaerobic atmosphere at 37°C.

For scanning electron microscopy (SEM), coverslips with hDFSC co-cultures were fixed in a 2.5% glutardialdehyde solution. They were subsequently washed with 0.1 M sodium acetate buffer (pH 7.3) and dehydrated in a graded series of ethanol. Finally, coverslips were subjected to critical-point drying with CO_2 (Critical Point Dryer, Emitech, Ashford, UK), sputter-coated with gold, and examined with an electron microscope (Zeiss DSM 960A, Jena, Germany).

Cytokine Secretion

IL-8 and IL-10 concentrations were measured in the supernatant of the bacteria–cell co-culture after 1, 2, 4, and 24 h of incubation by enzyme-linked immunosorbent assay (ELISA), using a commercially available kit (BD OptEIA, BD Biosciences, San Diego, USA) according to the manufacturer's instructions.

Statistical Analysis

All results are presented as means ± standard deviation (SD). Statistical analyses were carried out by T-test (Prism 6 for Windows, Version 6.01, GraphPad Software, Inc., San Diego, CA, USA). Differences were considered statistically significant at $^*p<0.05$, $^{**}p<0.01$. All experiments were performed with at least four biological replicates (n = 4).

Results

Anaerobic Co-Culture

The establishment of the anaerobic co-culture system as described by Kriebel *et al.* [21] included the verification of the cell survival under an anaerobic atmosphere prior to a bacterial challenge. Survival of the cells was evaluated after 12, 24 and 72 h (Fig. 1). Compared to aerobic conditions, 81.7% of the hDFSC, 83.2% of the hBMSC, 54.0% of the Ca9-22 cells, and 58.9% of the hGiF remained viable 24 h of an oxygen-free incubation. After 2 days, the hDFSC still showed higher survival rates than the Ca9-22: only 33.8% of Ca9-22 cells survived the 48-h incubation, whereas 50.9% of hDFSC cells were alive. The cell numbers decreased significantly for all cells after 72 h: hDFSC showed a survival of 40.2%, hBMSC 35.0%, hGiF 16.1%, and Ca9-22 13.5% (Fig. 1).

The bacterial growth was determined in cell culture medium under anaerobic conditions. All four tested species showed normal bacterial growth (Fig. 2). From these results, we concluded that co-culture with anaerobic bacteria and hDFSC (and hBMSC, hGiF, or Ca9-22) was possible. Live/dead staining and fluorescence microscopy confirmed the survival of bacteria and hDFSC in the co-culture system (data not shown).

Adherence and Internalization

The human cells were separately challenged with live *P. gingivalis* W50, W83, *F. nucleatum* ATTC 23726, or ATTC 25586, employing a MOI of 1:100. Attachment to and invasion of hDFSC by oral pathogenic bacteria was measured, and was compared with the results obtained with hBMSC, hGiF, and Ca9-22. All bacterial species were able to adhere to and invade the hDFSC. Notably, in all co-cultures, Ca9-22 cells showed the highest susceptibility to bacterial adhesion, which was statistically significant for *F. nucleatum* ATCC 2326 and *P. gingivalis* W50. Roughly the same number of bacteria adhered to the fibroblasts compared to the hDFSC. In contrast, in all co-cultures, the hBMSC were statistically significantly less prone to bacterial adhesion. SEM confirmed bacterial attachment to the hDFSC (Fig. S1 and S2).

The efficiency of the periodontal pathogens to internalize into the hDFSC was very weak. In hGiF and Ca9-22 co-cultures with *F. nucleatum* ATCC 23726 and *F. nucleatum* ATCC 25586, statistically significantly higher bacterial invasion rates were found compared with hDFSC. Both *P. gingivalis* strains appeared to be hardly invasive in all four cell lines (Fig. 3 and Table S1).

IL-8 and IL-10 Secretion

After co-culture with either live *F. nucleatum* or *P. gingivalis* (MOI 1:100) for 1, 2, 4, and 24 h, IL-8 and IL-10 was detected in the supernatant of all four cell lines by ELISA (Fig. 4 and Tables S2 and S3). Unchallenged cells were used as a negative control.

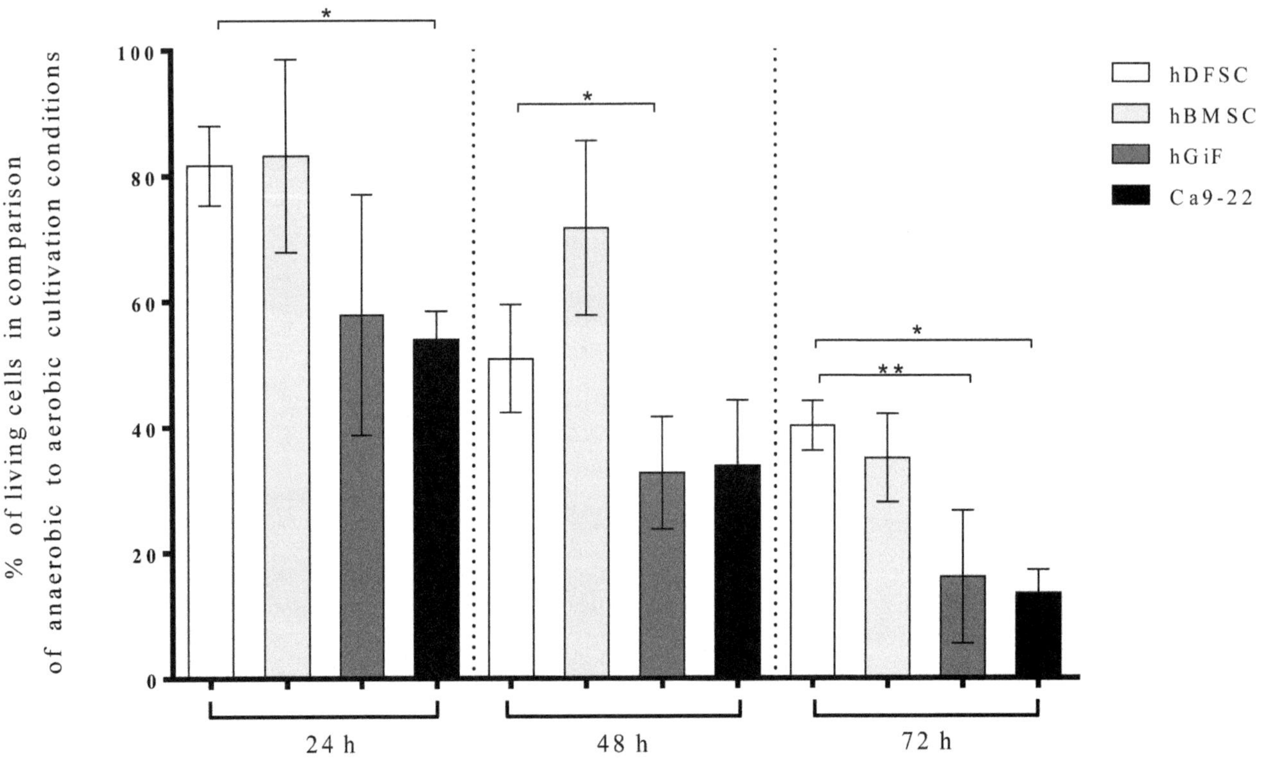

Figure 1. Cell survival under anaerobic conditions. Survival of hDFSC, hBMSC, hGiF and Ca9-22 cells under anaerobic compared with aerobic conditions. Anaerobic atmosphere: (10% CO_2, 10% H_2, 80% N_2) and 37°C. Viable cells were counted at timepoints of 24, 48 and 72 h in a Neubauer hemacytometer by exclusion of trypan blue. Values are expressed as means ± SD (standard deviation), $^*p<0.05$ (T-test), n = 4. Asterisks indicate statistically significant differences between hDFSC and hBMSC, hGiF, and Ca9-22.

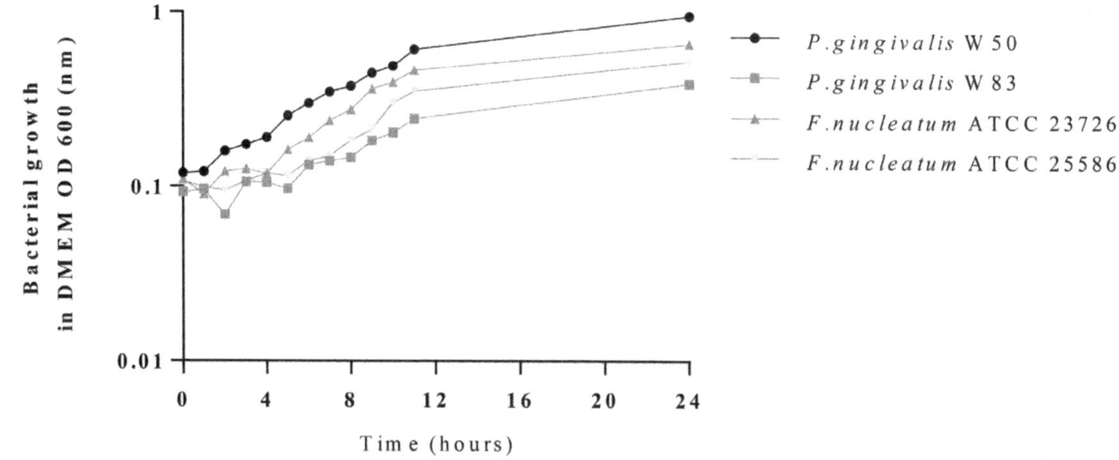

Figure 2. Bacterial growth in cell culture medium. Growth rates of bacteria in DMEM cell culture medium under anaerobic conditions (10% CO_2, 10% H_2, 80% N_2), 37°C. Bacteria were grown in PYG medium supplemented with 5 µg/ml hemin and 1% vitamin K to the stationary phase. Subsequently, they were centrifuged, washed in PBS, and each bacterial suspension was diluted in DMEM 1:10. The OD_{600} was measured at 600 nm from timepoint 0 every hour to timepoint 12 h with a final measurement after 24 h.

Of note, there were statistically significantly lower IL-8 secretion levels in hDFSC and hBMSC supernatants compared with hGiF and Ca9-22. At all timepoints, hGiF consistently showed the highest IL-8 response (with the exception of at 1 h with *F. nucleatum* ATCC 255886 infection). Overall, *F. nucleatum* was the most efficient stimulatory bacterium for the secondary cytokine response. The strongest IL-8 response was noted after 24 h for

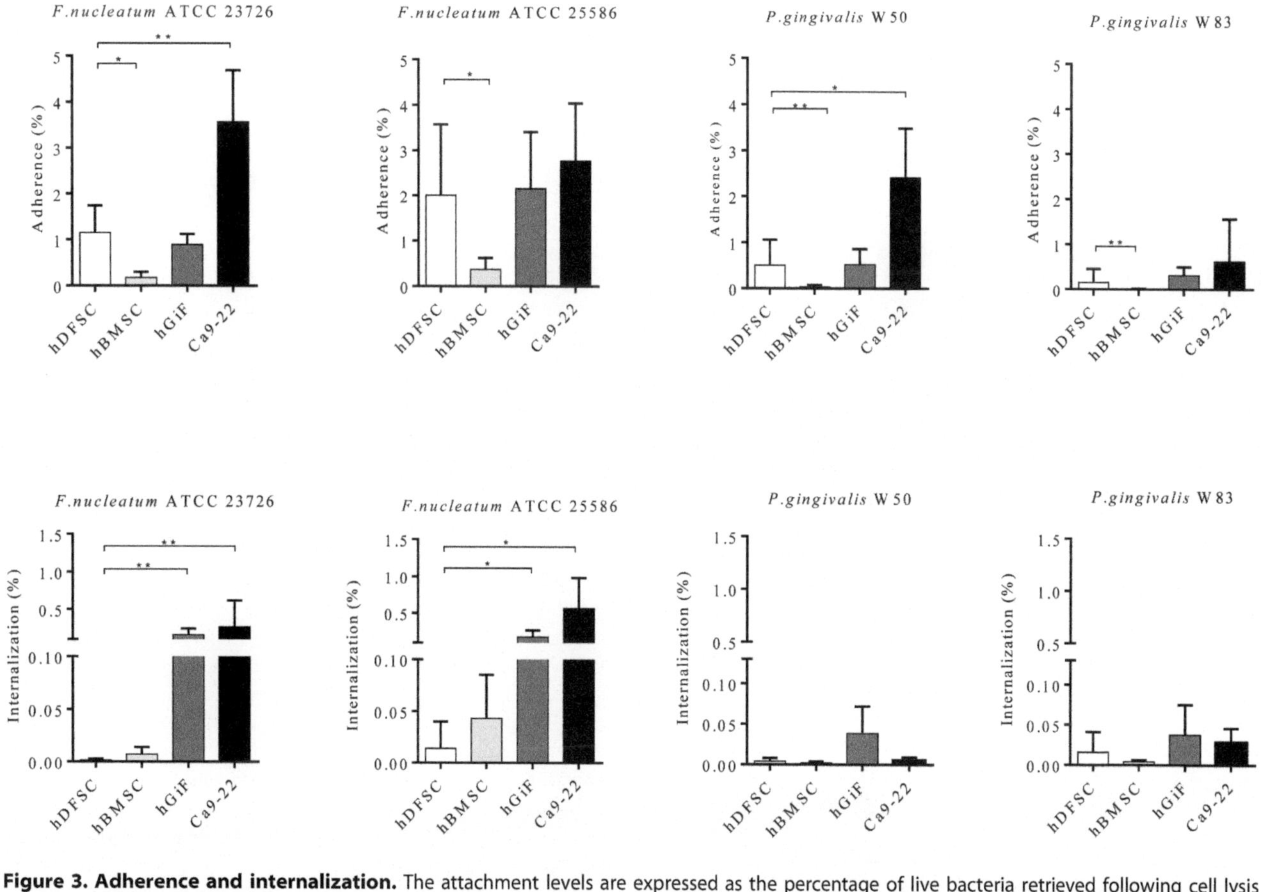

Figure 3. Adherence and internalization. The attachment levels are expressed as the percentage of live bacteria retrieved following cell lysis relative to the total number of bacteria at each timepoint. The assays were carried out as described in the text. Attachment levels of hDFSC were compared to hBMSC, hGiF, and Ca9-22. Values represent the means ± SD, *$p<0.05$, **$p=0.01$ (T-test), n=4.

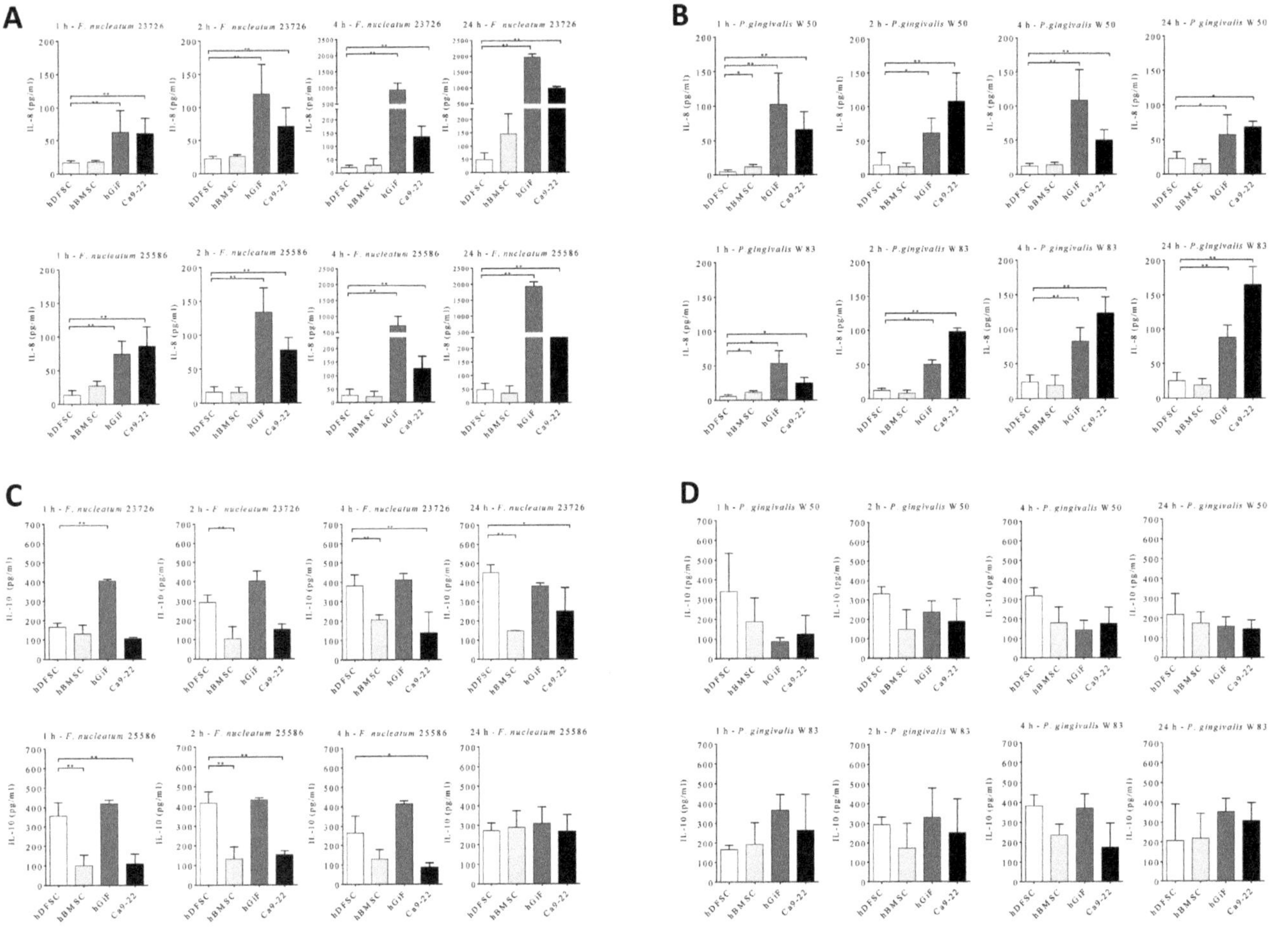

Figure 4. IL-8 and IL-10 secretion. The assays were carried out as described in the text. IL-8 and IL-10 levels were assayed by ELISA. Absorbance was read at 450 nm. Values represent the means ± SD, *$p<0.05$, **$p=0.01$ (T-test), $n=4$. Asterisks indicate statistically significant differences between hDFSC and hBMSC, hGiF, and Ca9-22. Cytokine secretion by hDFSC was compared with secretion by hBMSC, hGiF, and Ca9-22 cells at the same timepoint. **A)** IL-8 measured in the supernatant of the *F.nucleatum* ATCC 23727 and 25586 co-culture after 1, 2, 4, and 24 h. **B)** IL-8 measured in the supernatant of the *P. gingivalis* W50 and W83 co-culture after 1, 2, 4, and 24 h. **C)** IL-10 measured in the supernatant of the *F.nucleatum* ATCC 23727 and 25586 co-culture after 1, 2, 4, and 24 h. **D)** IL-10 measured in the supernatant of the *P. gingivalis* W50 and W83 co-culture after 1, 2, 4, and 24 h.

hGiF co-cultured with *F.nucleatum* ATCC 23726, reaching 1964 pg/ml.

The initial IL-10 response (1 h) was higher than the initial IL-8 response measured for all cells. After 4 h of co-culture, the IL-8 response from hGiF and Ca9-22 increased significantly, whereas the hDFSC and hBMSC remained at almost the same level of IL-10 secretion, but showed only a small increase in IL-8 response.

Discussion

The purpose of the present study was to comparatively characterize interactions between dental stem cell and periodontal pathogens *in vitro*, since in bacterially infected periodontal tissues, a mutual influence of local (progenitor) cells and anaerobic bacteria is highly likely [7]. This, to the best of our knowledge, has not yet been explored in a direct co-culture of hDFSC and periodontal pathogens. A typical challenge described for *in vitro* co-culture experiments with cells and anaerobic bacteria is the limited lifespan of these microorganisms under an aerobic atmosphere [22]. Hence, a co-culture system in which both the cells and anaerobic periodontal pathogenic bacteria were able to survive for a certain time-period was applied and selected cell–bacterial interactions were analyzed.

Host cell invasion is regarded as an important tool of pathogenic bacteria to afford protection from the host immune system, and this invasion contributes to tissue damage [23]. In fact, invasive bacteria have evolved a variety of mechanisms for host cell entry. Initially, the bacteria attach to the host cell membrane, followed by the induction of a series of structural and biochemical changes that facilitate bacterial penetration. Lamont *et al.* [23] demonstrated the invasion of *P. gingivalis* into gingival epithelial cells. It is known that *P. gingivalis* can recognize different host cell types and is capable of targeting different specific and distinct eukaryotic signaling pathways to induce uptake into the cell [24]. Additionally, the study of Han *et al.* [7] showed that *F. nucleatum* is able to adhere to and invade gingival epithelial cells.

In this study, we reveal the differential susceptibility of dental stem cells compared with hBMSC, hGiF, and Ca9-22 towards the very important bacterial pathogenicity traits of host cell adherence and internalization. A significantly lower bacterial adherence and internalization capacity was observed in dental stem cells compared with hGiF and the epithelial cell line Ca9-22.

Variations in the cell-surface structure, as well as differences in the process of invasion, depending on the cell line and its level of differentiation might explain the decreased bacterial attachment and internalization found in hDFSC and hBMSC compared with hGiF or Ca9-22 co-cultures. Furthermore, the expression of varying amounts of specific cell-surface receptors that allow the attachment of pathogens may contribute to the different results for the adherence. Huard-Delcourt *et al.* [25] showed, via FACS-analysis, a different expression of receptors for *P. gingivalis* attachment for the two tested cell lines in that study (KB and gingival epithelial cells). It is not clear what kind of receptors were responsible, but these authors were able to show a saturation plateau for each cell line when all bacteria were bound. Whether PRRs (pathogen recognition receptors) play a role in this context still needs to be determined.

Moreover, a novel stem cell property has been described, which possibly contributes to a reduced bacterial adherence and internalization. BMSCs have recently been shown to impede *in vitro* pathogen growth [1]. Krasnodembskaya *et al.* [26] and Meisel *et al.* [27] described the production of the antibacterial peptide LL-37 by stem cells. If hDFSC also secrete LL-37 remains unclear.

To define differences in the immunological reactions of the cells after co-culture with *F. nucleatum* and *P. gingivalis*, the cytokine release was assessed. IL-8, an important signaling molecule in periodontitis [11], was chosen as being representative for the pro-inflammatory cytokine response. *In vitro*, IL-8 secretion is triggered by invading bacteria, particularly *F. nucleatum*; similar observations for *P. gingivalis* have been made, with different cell lines [28,7]. Sandros *et al.* [16] showed that bacterial binding to the surface of epithelial cells leads to an increased expression of IL-8.

Our results show that after the co-culture of *P. gingivalis* and *F. nucleatum* with hDFSC, the IL-8 secretion was significantly less stimulated compared with hGiF and Ca9-22 cells. All four cell lines under investigation showed the highest IL-8 production after co-culture with *F. nucleatum* ATCC 23726. The decrease in IL-8 levels after 24 h in the *P. gingivalis* W50 co-culture is most likely due to the activity of cytokine degrading enzymes of this species [29]. An anti-inflammatory reaction was analyzed via measurement of the cellular IL-10 response. Initially, all cell types showed a higher IL-10 than IL-8 response. Of note, after 4 h of co-culture, the differentiated cells yielded a significantly higher IL-8 than IL-10 response, whereas the stem cells kept almost the same level of IL-10 secretion, but showed only a small increase in the IL-8 response. This means that hDFSC and hBMSC had higher IL-10 than IL-8 secretion at all measured timepoints.

The differences in the cytokine response between the cells might be rooted in the specificity of the innate immune system that is based on a broad spectrum of PRRs that recognize highly conserved pathogen-associated molecular patterns (PAMPs) [30], including Toll-like receptors (TLRs), the nucleotide-binding oligomerization domain proteins (Nods), and the G-protein-coupled receptors (GPCRs) [28]. By stimulating these receptors, the mitogen-activated protein kinase (MAPK) signaling pathway is activated leading to NF-κB nuclear translocation and resulting in cytokine expression [31]. TLR4 is activated by the LPS of Gram-negative species such as *F. nucleatum*, while *P. gingivalis* LPS signals via TLR2 [28]. Moreover, the fimbriae of *P.gingivalis* and gingipains also activate TLR4 [32–34]. The combined effect of the aforementioned activations and interactions could be responsible, at least in part, for the differences in cell response to various oral bacteria [28].

However, recent reports suggest that human dental mesenchymal stem cells, as well as hBMSC, possess immuno-regulating and anti-inflammatory properties [35,36]. The *in vitro* findings of this work underline the potential immuno-modulating character of local dental progenitor cells in the context of bacterially induced inflammatory processes.

Conclusion

We demonstrate that hDFSC (and hBMSC) show less interaction with periodontal pathogens concerning adherence and internalization. Moreover, in this experimental setup, the cytokine secretion of these cells favored an anti-inflammatory reaction. *In vitro*, we compared hDFSC, hBMSC, differentiated primary gingival fibroblasts, and the permanent tumor cell line Ca9-22 in an anaerobic environment. The results of this study indicate that there are differences between cell lines at various stages of differentiation concerning the tolerance towards bacterial infections. For future potential dental stem cell therapies, the findings presented here might be of clinical relevance.

However, better anaerobic survival of the stem cells could lead to a better chance of tissue regeneration after infection and damage, which seems to be reasonable from an evolutionary perspective. With respect to dental tissues, it can be concluded that hDFSC show a similar behavior to hBMSC in this study.

Supporting Information

Figure S1 Scanning electron microscope images of *F. nucleatum* ATCC 23727 attachment to hDFSC.

Figure S2 Scanning electron microscope images of *F. nucleatum* ATCC 23727 attachment to hDFSC.

Table S1 Adherence and internalization. The attachment and invasion levels are expressed as the percentage of live bacteria retrieved following cell lysis relative to the total number of bacteria at each timepoint. The assays were carried out as described in the text. Attachment and invasion levels of hDFSC and hBMSC were compared with Ca9-22 cells. Values represent the means ± SD, *$p<0.05$, **$p=0.01$ (T-test), $n=4$.

Table S2 IL-8 secretion. IL-8 levels were assayed by ELISA. Absorbance was read at 450 nm. Values represent the means ± SD, *$p<0.05$, **$p=0.01$ (T-test), $n=4$.

Table S3 Il-10 secretion. IL-8 levels were assayed by ELISA. Absorbance was read at 450 nm. Values represent the means ± SD, *$p<0.05$, **$p=0.01$ (T-test), $n=4$.

Acknowledgments

The authors would like to thank M. Frank, Sc.D., Center for Electron Microscopy (EMZ), University of Rostock, for granting access to his core facility.

Author Contributions

Conceived and designed the experiments: HL BK. Performed the experiments: AB KK. Analyzed the data: AB KK. Wrote the paper: AB KK BK HL.

References

1. Auletta J, Deans R, Bartholomew A (2012) Emerging roles for multipotent, bone marrow-derived stromal cells in host defense. Blood 119 (8): 1801–1809.
2. Seo B, Miura M, Gronthos S, Bartold P, Batouli S, et al. (2004) Investigation of multipotent postnatal stem cells from human periodontal ligament. Lancet 364 (9429): 149–155.
3. Morsczeck C, Götz W, Schierholz J, Zeilhofer F, Kühn U, et al. (2005) Isolation of precursor cells (PCs) from human dental follicle of wisdom teeth. Matrix Biol 24 (2): 155–165.
4. Bianco P, Robey P, Simmons P (2008) Mesenchymal Stem Cells: Revisiting History, Concepts, and Assays. Cell Stem Cell 2 (4): 313–319.
5. Honda M, Imaizumi M, Tsuchiya S, Morsczeck C (2010) Dental follicle stem cells and tissue engineering. J Oral Sci 52 (4): 541–552.
6. Morsczeck C, Schmalz G, Reichert T, Völlner F, Galler K, et al. (2008) Somatic stem cells for regenerative dentistry. Clin Oral Investig 12 (2): 113–118.
7. Han Y, Shi W, Huang G, Kinder Haake S, Park N, et al. (2000) Interactions between periodontal bacteria and human oral epithelial cells: Fusobacterium nucleatum adheres to and invades epithelial cells. Infect Immun 68 (6): 3140–3146.
8. Aas J, Paster B, Stokes L, Olsen I, Dewhirst F (2005) Defining the normal bacterial flora of the oral cavity. J Clin Microbiol 43 (11): 5721–5732.
9. Okada H, Murakami S (1998) Cytokine expression in periodontal health and disease. Crit Rev Oral Biol Med 9 (3): 248–266.
10. Takashiba S, Takigawa M, Takahashi K, Myokai F, Nishimura F, et al. (1992) Interleukin-8 is a major neutrophil chemotactic factor derived from cultured human gingival fibroblasts stimulated with interleukin-1 beta or tumor necrosis factor alpha. Infect Immun 60 (12): 5253–5258.
11. Bickel M, others (1993) The role of interleukin-8 in inflammation and mechanisms of regulation. J Periodontol 64 (5 Suppl): 456.
12. Reichert S, Machulla H, Klapproth J, Zimmermann U, Reichert Y, et al. (2008) The interleukin-10 promoter haplotype ATA is a putative risk factor for aggressive periodontitis. J Periodontal Res 43 (1): 40–47.
13. Moore K, de Waal Malefyt R, Coffman R, O'Garra A (2001) Interleukin-10 and the interleukin-10 receptor. Annu Rev Immunol 19 (1): 683–765.
14. Dzink J, Tanner A, Haffajee A, Socransky S (1985) Gram negative species associated with active destructive periodontal lesions. J Clin Periodontol 12 (8): 648–659.
15. Paster B, Boches S, Galvin J, Ericson R, Lau C, et al. (2001) Bacterial diversity in human subgingival plaque. J Bacteriol 183 (12): 3770–3783.
16. Sandros J, Karlsson C, Lappin D, Madianos P, Kinane D, et al. (2000) Cytokine responses of oral epithelial cells to Porphyromonas gingivalis infection. Journal Dent Res 79 (10): 1808–1814.
17. Wada N, Menicanin D, Shi S, Bartold P, Gronthos S (2009) Immunomodulatory properties of human periodontal ligament stem cells. J Cell Physiol 219 (3): 667–676.
18. Gaebel R, Ma N, Liu J, Guan J, Koch L, et al. (2011) Patterning human stem cells and endothelial cells with laser printing for cardiac regeneration. Biomaterials 32 (35): 9218–9230.
19. Haddouti E, Skroch M, Zippel N, Müller C, Birova B, et al. (2009) Human dental follicle precursor cells of wisdom teeth: isolation and differentiation towards osteoblasts for implants with and without scaffolds. Mat.-wiss. u. Werkstofftech 40 (10): 732–737.
20. Pittenger M, Mackay A, Beck S, Jaiswal R, Douglas R, et al. (1999) Multilineage potential of adult human mesenchymal stem cells. Science 284 (5411): 143–147.
21. Kriebel K, Biedermann A, Kreikemeyer B, Lang H (2013) Anaerobic Co-Culture of Mesenchymal Stem Cells and Anaerobic Pathogens-A New In Vitro Model System. PLoS One 8 (11): e78226.
22. Belton C, Izutsu K, Goodwin P, Park Y, Lamont R (2001) Fluorescence image analysis of the association between Porphyromonas gingivalis and gingival epithelial cells. Cellular Microbiology 1 (3): 215–223.
23. Lamont R, Chan A, Belton C, Izutsu K, Vasel D, et al. (1995) Porphyromonas gingivalis invasion of gingival epithelial cells. Infect Immun 63 (10): 3878–3885.
24. Andrian E, Grenier D, Rouabhia M (2006) Porphyromonas gingivalis-epithelial cell interactions in periodontitis. J Dent Res 85 (5): 392.
25. Huard-Delcourt A, Du L, Pellen-Mussi P, Tricot-Doleux S, Bonnaure-Mallet M (1998) Adherence of Porphyromonas gingivalis to epithelial cells, analysis by flow cytometry. Eur J Oral Sci 106 (5): 938–944.
26. Krasnodembskaya A, Song Y, Fang X, Gupta N, Serikov V, et al. (2010) Antibacterial effect of human mesenchymal stem Cells is mediated in part from secretion of the antimicrobial peptide LL-37. Stem Cells 28 (12): 2229–2238.
27. Meisel R, Brockers S, Heseler K, Degistirici O, Bülle H, et al. (2011) Human but not murine multipotent mesenchymal stromal cells exhibit broad-spectrum antimicrobial effector function mediated by indoleamine 2,3-dioxygenase. Leukemia 25 (4): 648–654.
28. Stathopoulou P, Benakanakere M, Galicia J, Kinane D (2010) Epithelial cell pro-inflammatory cytokine response differs across dental plaque bacterial species. Journal of J Clin Periodontol 37 (1): 24–29.
29. Zhang J, Dong H, Kashket S, Duncan M (1999) IL-8 degradation by Porphyromonas gingivalis proteases. Microb Pathog 26 (5): 275–280.
30. Medzhitov R, Janeway J (2000) Innate immune recognition: mechanisms and pathways. Immunol Rev 173 (1): 89–97.
31. Dong C, Davis R, Flavell R (2002) MAP kinases in the immune response. Annu Rev Immunol 20 (1): 55–72.
32. Hajishengallis G, Sojar H, Genco R, DeNardin E (2004) Intracellular signaling and cytokine induction upon interactions of Porphyromonas gingivalis fimbriae with pattern-recognition receptors. Immunol Invest 33 (2): 157–172.
33. Chung W, Dale B (2004) Innate immune response of oral and foreskin keratinocytes: utilization of different signaling pathways by various bacterial species. Infect Immun 72 (1): 352–358.
34. Lourbakos A, Potempa J, Travis J, D'Andrea M, Andrade-Gordon P, et al. (2001) Arginine-specific protease from Porphyromonas gingivalis activates protease-activated receptors on human oral epithelial cells and induces interleukin-6 secretion. Infect Immun 69 (8): 5121.
35. Wada N, Gronthos S, Bartold P (2013) Immunomodulatory effects of stem cells. Periodontology 2000 63 (1): 198–216.
36. Tomic S, Djokic J, Vasilijic S, Vucevic D, Todorovic V, et al. (2011) Immunomodulatory properties of mesenchymal stem cells derived from dental pulp and dental follicle are susceptible to activation by toll-like receptor agonists. Stem Cells Dev 20 (4): 695–708.

19

Identification of Rabbit Annulus Fibrosus-Derived Stem Cells

Chen Liu[1,2], Qianping Guo[1,2], Jun Li[1], Shenghao Wang[2], Yibin Wang[2], Bin Li[1,2]*, Huilin Yang[1,2]*

1 Department of Orthopaedics, The First Affiliated Hospital of Soochow University, Suzhou, Jiangsu, China, **2** Orthopedic Institute, Soochow University, Suzhou, Jiangsu, China

Abstract

Annulus fibrosus (AF) injuries can lead to substantial deterioration of intervertebral disc (IVD) which characterizes degenerative disc disease (DDD). However, treatments for AF repair/regeneration remain challenging due to the intrinsic heterogeneity of AF tissue at cellular, biochemical, and biomechanical levels. In this study, we isolated and characterized a sub-population of cells from rabbit AF tissue which formed colonies *in vitro* and could self-renew. These cells showed gene expression of typical surface antigen molecules characterizing mesenchymal stem cells (MSCs), including CD29, CD44, and CD166. Meanwhile, they did not express negative markers of MSCs such as CD4, CD8, and CD14. They also expressed Oct-4, nucleostemin, and SSEA-4 proteins. Upon induced differentiation they showed typical osteogenesis, chondrogenesis, and adipogenesis potential. Together, these AF-derived colony-forming cells possessed clonogenicity, self-renewal, and multi-potential differentiation capability, the three criteria characterizing MSCs. Such AF-derived stem cells may potentially be an ideal candidate for DDD treatments using cell therapies or tissue engineering approaches.

Editor: Dimitris Kletsas, National Centre for Scientific Research, 'Demokritos', Greece

Funding: This study was supported by National Natural Science Foundation of China (81171479), Natural Science Foundation of Jiangsu Province (BK2011291), Jiangsu Provincial Special Program of Medical Science (BL2012004), and National Undergraduate Training Program for Innovation and Entrepreneurship, Soochow University (201210285044). The funders had no role in study design, data collection and analysis, decision to publish, or preparation of the manuscript.

Competing Interests: The authors have declared that no competing interests exist.

* Email: binli@suda.edu.cn (BL); suzhouspine@163.com (HY)

These authors contributed equally to this work.

Introduction

As the major cause of low back pain which affects about 80% of the population, degenerative disc disease (DDD) has evolved into a serious medical problem and significantly contributes to healthcare costs [1]. Tissue engineering has emerged as a promising approach toward DDD therapy [2]. As a component which plays a critical role in the biomechanical properties of intervertebral disc (IVD), the annulus fibrosus (AF) is essential for confining nucleus pulposus (NP) and maintaining physiological intradiscal pressure [2]. However, despite recent advancements [3–6], major challenge remains toward AF tissue engineering, mainly due to the tremendous complexity of AF tissue at cellular, biochemical, microstructural, and biomechanical levels [7,8].

Cells play a central role in determining the quality of engineered tissues. Currently, tissue engineering of AF mainly involve the use of AF cells [4,9,10], chondrocytes [5], or bone marrow stem cells (BMSCs) [3,6] of various origins. However, due to the ageing of differentiated cells, low cellularity, and the intrinsic phenotype heterogeneity of AF cells, application of AF cells or chondrocytes for AF repair/regeneration is limited [11,12]. Use of BMSCs, which have been overwhelmingly used and shown effectiveness in AF tissue engineering, also confronts with a problem of limited cell availability (only 0.001–0.01% BMSCs in bone marrow aspirates or marrow tissue) [13]. Therefore, seeking new cell sources for AF tissue engineering appears to be necessary.

To date, mesenchymal stem cells (MSCs) have been isolated from a variety of adult tissues and they differ in many ways [14]. As a rule of thumb, MSCs from adult tissues tend to be tissue specific, meaning that MSCs originated from a certain tissue preferentially differentiate into the type of cells residing in this tissue [14–17]. Recently, it has been suggested that stem cell niches are present at the border of the AF and that the stem cells or progenitor cells migrate into the AF [15,16,18]. There have been several lines of evidence implying that stem/progenitor cells exist in AF, such as formation of cartilage, bone, and nerve tissues in AF during IVD degeneration, likely as a result of the differentiation of progenitor cells in AF or NP [8,15,19–21]. Such stem/progenitor cells, if successfully isolated, may be a valuable source for AF cell therapy and tissue engineering due to their AF tissue specificity.

To this end, this study aimed to isolate and characterize stem cells from AF tissue. Such stem cells should possess clonogenicity, self-renewal capability, and multipotency, the common characteristics of MSCs [22]. Since rabbit is a commonly used model for IVD research taking advantage of its moderate size, ease of surgery, and post-surgery analyses [15,16,23], we used rabbit IVDs to isolate a population of AF-derived colony-forming and characterize the properties of these cells. As expected, we found that these cells could self-renew and be readily induced to differentiate into osteocytes, chondrocytes, and adipocytes. Such findings revealed the existence of AF-derived stem cells, which may potentially be a valuable source for repair or regeneration of AF tissue.

Materials and Methods

Isolation of AF-derived Cells

AF samples were isolated from IVDs of female New Zealand white rabbits (6–8 weeks old) (**Fig. S1**) and minced and digested using 150 U/ml Collagenase I (Sigma, Cat.# C0130) in DMEM-LG medium for 4–6 hr. The suspension was then centrifuged at 1000 rpm for 10 min. The cell pellet was re-suspended in DMEM-LG supplemented with 20% FBS, 100 U/ml penicillin, 100 μg/ml streptomycin and plated in 100 mm tissue culture dishes. Cells were maintained in a humidified incubator at 37°C with 5% CO_2. The medium was changed every 2 days until the cells reached sub-confluence, then cells were harvested using 0.25% trypsin-EDTA. Isolation of spleen cells as a control for RT-PCR analysis was described in the file **Methods S1**. The animal surgery protocols were approved by the Institutional Animal Care and Use Committee (IACUC) of Soochow University.

Colony Forming Assay

To test the colony forming capability of AF-derived cells, a colony forming unit-fibroblast (CFU-F) assay was performed. Highly diluted single cell suspensions with cell densities of 200–5,000 cells/ml were plated and cultured for 10 days. Then the cells were washed and fixed with 4% paraformaldehyde for 15 minutes. After further rinsing twice with PBS, cells were stained with 0.07% crystal violet (in ethanol) at 4°C and the colonies were counted.

Polymerase Chain Reaction for Gene Analysis

Total RNA was extracted from cells using Trizol reagent (Invitrogen, Cat.# 15596-026) and then reversely transcribed to cDNA using a reverse transcription kit (Fermentas, Cat.# K1622) after DNase I treatment (Fermentas, Cat.# EN0521). The genes included MSC positive marker genes (CD29, CD44, CD166), MSC negative marker genes (CD4, CD8, CD14), adipocyte specific genes (peroxisome proliferators-activated receptor γ, Lipoprotein lipase), osteocyte specific genes (collagen type I, Runx 2), chondrocyte specific genes (collagen type II, Sox-9) and housekeeping gene GAPDH. All the primer sequences are listed in **Table 1**. The sequences were designed according to published papers or designed by ourselves. All the primers were synthesized by Invitrogen.

Immunofluorescence

Cells were fixed in cold 4% poly-formaldehyde for 15 min, followed with treatment using methanol at −20°C for 5 min. They were then blocked with 4% BSA for 30 min before being incubated with mouse anti-human Oct-4 antibody (1:500, Millipore, Cat.# MAB4401), goat anti-human nucleostemin

Table 1. Sequences of primers for RT-PCR.

Gene	Size (bp)	Primer sequence	Type	Tm (°C)	Gene Bank #
CD29	242	5′-GTCACCAACCGTAGCAA-3′	Forward	58	AY195896.1
		5′-CTCCTCATCTCATTCATCAG-3′	Reverse		
CD44	191	5′-CGATTTGAATATAACCTGCCGC-3′	Forward	63	FJ360436.1
		5′-CGTGCCCTTCTATGAACCCA-3′	Reverse		
CD166	200	5′-GGACAGCCCGAAGGAATACGAA-3′	Forward	63	Y13243.1
		5′-GACACAGGCAGGGAATCACCAA-3′	Reverse		
CD4	273	5′-GATGGAGGTGGAACTGC-3′	Forward	63	NM_001082313.2
		5′-GGAAAGCCCAACACTATG-3′	Reverse		
CD8	126	5′-GGGTGGAAAAGGAGAAGC-3′	Forward	63	L22293.1
		5′-AGGTGAGTGCGGGAGAC-3′	Reverse		
CD14	364	5′-CAGGTGCCTAAGGGACT-3′	Forward	63	NM_001082195.2
		5′-AATAAAGTGGGAAGCGG-3′	Reverse		
GAPDH	107	5′-ACTTTGTGAAGCTCATTTCCTGGTA-3′	Forward	58	L23961
		5′-GTGGTTTGAGGGCTCTTACTCCTT-3′	Reverse		
PPARγ	130	5′-CATTTTCTCAAGCAACAGTC-3′	Forward	54	NM_001082148.1
		5′-CAAAGGAGTGGGAGTGGT-3′	Reverse		
LPL	487	5′-GGCGAGACGCACGAACA-3′	Forward	54	FJ429312.1
		5′-CACCCGCAGTACAAACCCA-3′	Reverse		
Col I	81	5′-CTGACTGGAAGAGCGGAGAGTAC-3′	Forward	58	AY633663
		5′-CCATGTCGCAGAAGACCTTGA-3′	Reverse		
Runx-2	154	5′- CAGGCAGTTCCCAAGCATTTCA-3′	Forward	67	AY598934
		5′- TGGTGGCAGGTAGGTATGGTAGT-3′	Reverse		
Col II	84	5′-TGGGTGTTCTATTTATTTATTGTCTTCCT-3′	Forward	62	S83370
		5′-GCGTTGGACTCACACCAGTTAGT-3′	Reverse		
Sox-9	261	5′-TACGACTGGACGCTGGTGC-3′	Forward	67	AY598935
		5′-CGGGTGGTCTTTCTTGTGCT-3′	Reverse		

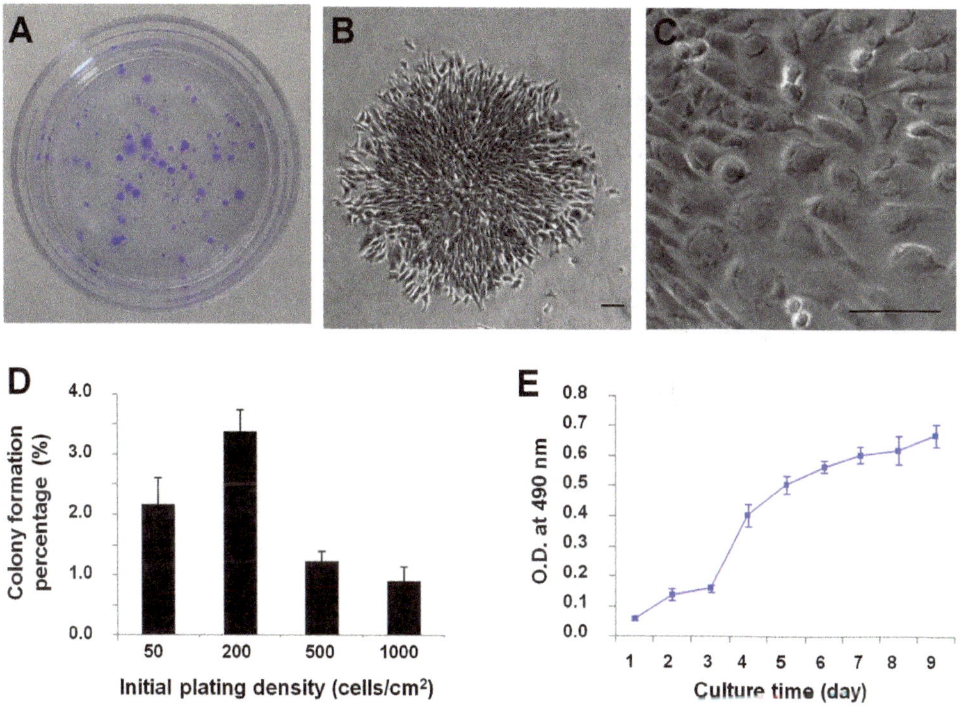

Figure 1. Colony formation and proliferation of rabbit AF-derived cells. (**A**) Total colonies stained with crystal violet at 10 days. (**B**) A representative cell colony. (**C**) The morphology of cells in a colony. (**D**) Colony forming unit assay for AF-derived cells. (**E**) Growth curve of AF-derived cells at passage 3. Scale bars, 100 μm.

antibody (1:250, Neuromics, Cat.# GT15050), and mouse anti-human SSEA-4 antibody (1:200, Invitrogen, Cat.# 41-4000), for staining of Oct-4, nucleostemin, and SSEA-4, respectively. Following that, Cy3-conjugated secondary antibodies were applied. Nucleus staining was accomplished by DAPI staining. Cells were viewed under a fluorescence inverted microscope (EVOS fl, AMG, USA). The detailed protocols are provided in file **Methods S1**.

Induced Differentiation

Passage 2–4 AF-derived cells were seeded at a density of 4×10^4 cells/well in a 24 well plate in basic culture medium (DMEM-LG supplemented with 10% FBS, 100 U/ml penicillin,100 μg/ml streptomycin). They were subjected to induced differentiation by culturing them in osteogenic, chondrogenic, and adipogenic media, respectively. The outcomes were evaluated using Alizarin Red S, and Safranin O, and Oil Red O staining, respectively. The detailed protocols are shown in file **Methods S1**.

Statistical Analysis

All quantitative data are presented as mean ± standard deviation (SD) with no less than three replicates for each experimental condition. Statistical analyses were performed using one-way analysis of variance (ANOVA). Difference with $p<0.05$ was considered statistically significant.

Results

Colony Formation and Proliferation of Rabbit AF-Derived Cells

When the rabbit AF-derived cells were cultured in the growth medium supplemented with 20% FBS, they remained quiescent for 3 to 4 days and then started to form colonies (**Fig. 1**). These colonies differed in sizes with diameter ranging from 1 to 3 mm. The morphology of cells in the colonies also varied, with some of them being cobblestone-like and others being spindle-like (**Fig. S2** and **Fig. S3**). The colony formation capacity was largely dependent on the initial cell seeding density. An initial seeding density of 200 cells/cm^2 was found to result in the highest efficiency of colony formation, in which about 3.4% of plated cells formed colonies (**Fig. 1D**). After 10 to 12 days of culture, cells reached sub-confluence and were harvested and sub-cultured. Cell proliferation capacity was tested using cells at passage three through MTT assay. It was clear that cells began to grow after 1 day and entered log phase at the 3rd day (**Fig. 1E**). The typical population doubling time was 17.8 hrs, indicating the strong self-renewing capacity of the rabbit AF-derived colony-forming cells.

Expression of Stem Cell Markers in AF-Derived Colony-Forming Cells

The gene expression of typical MSC-associated surface antigens in cells was tested using RT-PCR. Clearly, the AF-derived colony-forming cells had strong expression of markers that are usually positive in MSCs, including CD29, CD44 and CD166. Meanwhile, they had little expression of markers which seldom exist in MSCs, including CD4, CD8 and CD14 (**Fig. 2**). As a control, cells from rabbit spleen were found to express all the above genes, which validated the efficacy of the customer-designed primers.

Further, the expression of MSC markers including Oct-4, nucleostemin (NS) and SSEA-4 was examined at protein level using immunofluorescence. As shown in **Fig. 3**, these markers were extensively expressed in the AF-derived cells. Among them, Oct-4 was located both within the nucleus and throughout the

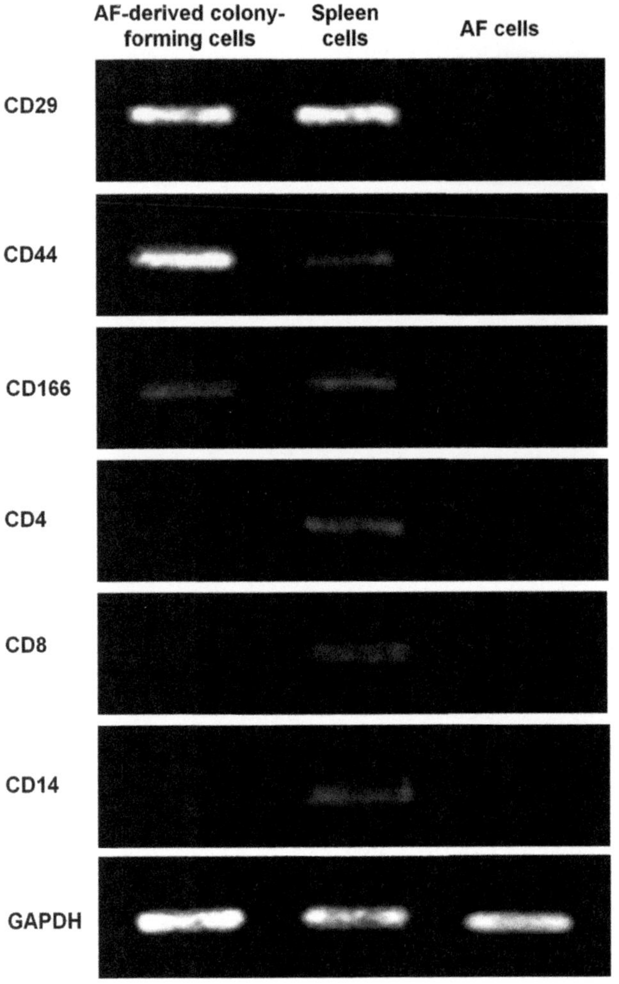

Figure 2. Expression of stem cell markers in AF-derived colony forming cells. AF-derived colony forming cells, but not AF cells, were positive for CD29, CD44, CD166, but negative for CD4, CD8 and CD14. As a positive control, spleen cells expressed all these genes.

cytoplasm; nucleostemin existed exclusively within the nucleus; whereas SSEA-4 was presented within the cytoplasm.

Multi-Potential Differentiation AF-Derived Colony-Forming Cells

Next, the rabbit AF-derived colony-forming cells were subjected to induced differentiation processes including osteogenesis, chondrogenesis, and adipogenesis to examine the multi-differentiation potential of them. When the cells were cultured in osteogenic medium, their morphology started to change at the 5th day. After 3 weeks, calcium deposits were highly visible in the induced cells which were fixed and stained with Alizarin Red S (**Fig. 4A**). In contrast, calcium deposits were rarely seen in control cells which were cultured in basic medium only. In addition, the mRNA expression of osteocytes-specific genes, including *Runx-2* and *Col I*, was examined using RT-PCR. Clearly, expression of both genes was higher in induced cells, indicating that the rabbit AF-derived cells had undergone successful osteogenesis (**Fig. 4B**).

When the AF-derived cells were cultured in chondrogenic differentiation medium, they round up and became chondrocyte-like cells. Strong production of sulfated proteoglycans in the induced cells after 3 weeks was visualized by Safranin O staining (**Fig. 4C**). The control cells cultured in basic medium also showed considerable Safranin O staining, albeit in a much lower level. At mRNA level, expression of chondrocytes-specific genes *Sox-9* and *Col II* was markedly up-regulated in induced cells compared to control cells (**Fig. 4D**).

During adipogenic differentiation, the rabbit AF-derived cells started to secrete small oil droplets, which surrounded the cells and constantly accumulated throughout the induced culture period, after 1 week of induction (**Fig. 4E**). Meanwhile, the expression of adipocyte-specific genes, including *PPAR-γ* and *LPL*, was also markedly enhanced in induced cells compared to control cells (**Fig. 4F**).

Discussion

AF plays an important role in maintaining the physiological structure and function of IVD and its failure contributes to the evolution of DDD [24]. Cell-based therapies, from mere cell injection to tissue engineering, appear to be a promising approach to AF tissue repair or regeneration. Among the several types of cells that have been used for this purpose, MSCs remain to be the most used one because of their easy availability. However, there has been increasing consensus that the differentiation potential of adult MSCs largely varies and is closely dependent on their origin. For instance, synovium-derived MSCs showed better chondrogenic potential compared to MSCs derived from other types of tissues. Adipose-derived MSCs had higher adipogenic potential, while periosteum- and muscle-derived MSCs had better osteogenic potential [14]. Tendon MSCs also preferentially differentiated into tenocyte-like cells [25,26]. Together, such studies indicate that adult MSCs originated from a certain tissue preferentially differentiate into the type of cells residing in this tissue [14,27]. Such tissue-specific stem cells, therefore, are more appropriate cell source for cell therapies and tissue engineering of the specific tissues.

In light of this, we tried to isolate stem cells from rabbit AF, which may potentially be an ideal cell source for cell therapies and tissue engineering of AF, using single-cell colony forming technique. We found that in the best situation, about 3.4% of seeded primary cells formed colonies. Such colony formation efficiency is consistent to those found in other tissues such as tendon [25,26]. The size of colonies and morphology of cells within colonies were heterogeneous, implying that different subpopulations of progenitor cells existed in AF. Such heterogeneity of colony-forming cells echoes the intrinsic cellular heterogeneity within AF. For example, cells at inner AF are chondrocytes-like, while cells at outer AF are fibroblasts-like, contributing to the region-dependent matrix composition of AF [28]. Cell proliferation tests indicated that these AF-derived colony-forming cells were capable of rapid growth, with a typical population doubling time of less than 18 hr. Such cell proliferation capability was kept until at least passage 6, making it possible for *in vitro* cell expansion to obtain sufficient number of AF cells for desired treatments. It should be noted that the regenerative capacity of stem cells markedly deteriorates with the aging of a living organism. Both the self-renewal and differentiation capabilities of adult stem/progenitor cells decrease with the age of animals [29–31]. It is expected that the number of stem/progenitor cells in more mature rabbits is less compared to those in younger species and their functional fitness also weakens. Therefore, young rabbits (6–8 weeks old) were used in this study to assure appropriate characterizations of the AF-derived colony-forming cells.

Further, we checked whether the AF-derived colony-forming cells expressed typical surface antigens of MSCs. Due to the very

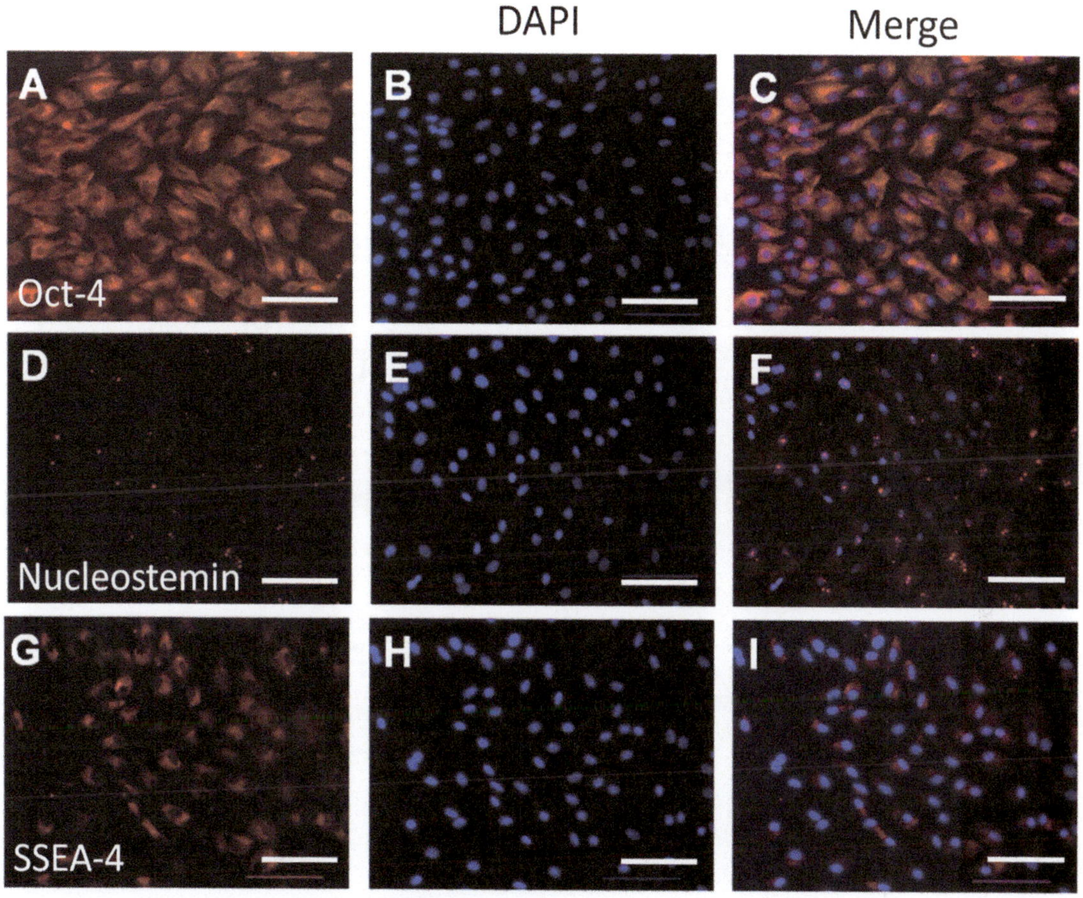

Figure 3. Immunofluorescence assays for stem cell marker protein expression in AF-derived colony forming cells. The AF-derived cells were positive for Oct-4 (**A–C**), nucleostemin (**D–F**) and SSEA-4 (**G–I**). Scale bars, 200 μm.

limited availability of anti-rabbit antibodies, we did not evaluate the protein expression of MSC surface markers using flow cytometry. Instead, we examined the expression of these markers at gene level using RT-PCR. We found that these cells expressed the positive markers of MSCs, including CD29, CD44 and CD166 [32], whereas they seldom expressed the negative markers such as CD4, CD8 and CD14 [33]. As a control, gene expression of cells isolated from rabbit spleen tissue was also examined. These cells were positive for CD4, CD8 and CD14 since spleen tissue contains lymphocytes and monocytes which express such markers [34]. Besides, these cells also expressed positive markers of MSCs [35].

In addition to surface antigens, we also examined the expression of several MSC marker proteins using immunofluorescence to confirm the stemness of AF-derived colony forming cells. Among these markers, Oct-4 is a transcription factor which is essential for maintaining undifferentiated status of pluripotent stem cells by inhibiting tissue-specific genes yet promoting stem cell-specific genes [36]. Nucleostemin is a nucleolar GTP-binding protein which is exclusively expressed in the nucleoli of stem cells, but not in committed and terminally differentiated cells, and is required in maintenance of self-renewal and pluripotency by controlling cell-cycle progression [37]. SSEA-4 is an early embryonic glycolipid antigen which is commonly used as human embryonic stem cell marker and also identifies adult MSCs [38]. We found that the AF-derived cells expressed all these stem cell markers, indicating their undifferentiated status (**Fig. 3**). It is interesting to note that Oct-4 was found to be presented throughout the whole cell instead of being located only in the nucleus as reported in previous studies [39]. Such a seemingly inconsistency may be the result of difference in cell type and origin, as Oct-4 has been found to be expressed in the whole cell of many somatic cells and cancer cells, including human kidney, gastric, mesenchymal, HeLa and MCF-7 cells [40]. In addition, even in the same type of stem cells, the distribution of Oct-4, in nucleus alone or throughout the cytoplasm, may significantly vary depending on the type of commercial antibodies [41].

In this study, we found that the rabbit AF-derived colony forming cells could undergo osteogenesis, chondrogenesis, and adipogenesis and therefore possessed multi-differentiation potential. Such results partially echo the findings from several previous studies based on human AF-derived cells [19,20]. However, our study differs from these studies in that we have, for the first time, specifically isolated the cells that formed colonies, which are presumably stem/progenitor cells of reasonable purity. It is notable that in the chondrogenic test, the control cells, i.e., cells without being subjected to chondrogenic induction, also showed considerable Safranin O staining as well as expression of *Sox-9* and *Col II* genes (**Fig. 4C–D**). This is indeed an indication that such AF-derived colony forming cells have intrinsic tendency to commit to chondrocyte-like cells which constitute the major cell population residing in the fibrocartilage-like AF tissue. Thus, such findings strongly support the significant potential of AF-derived

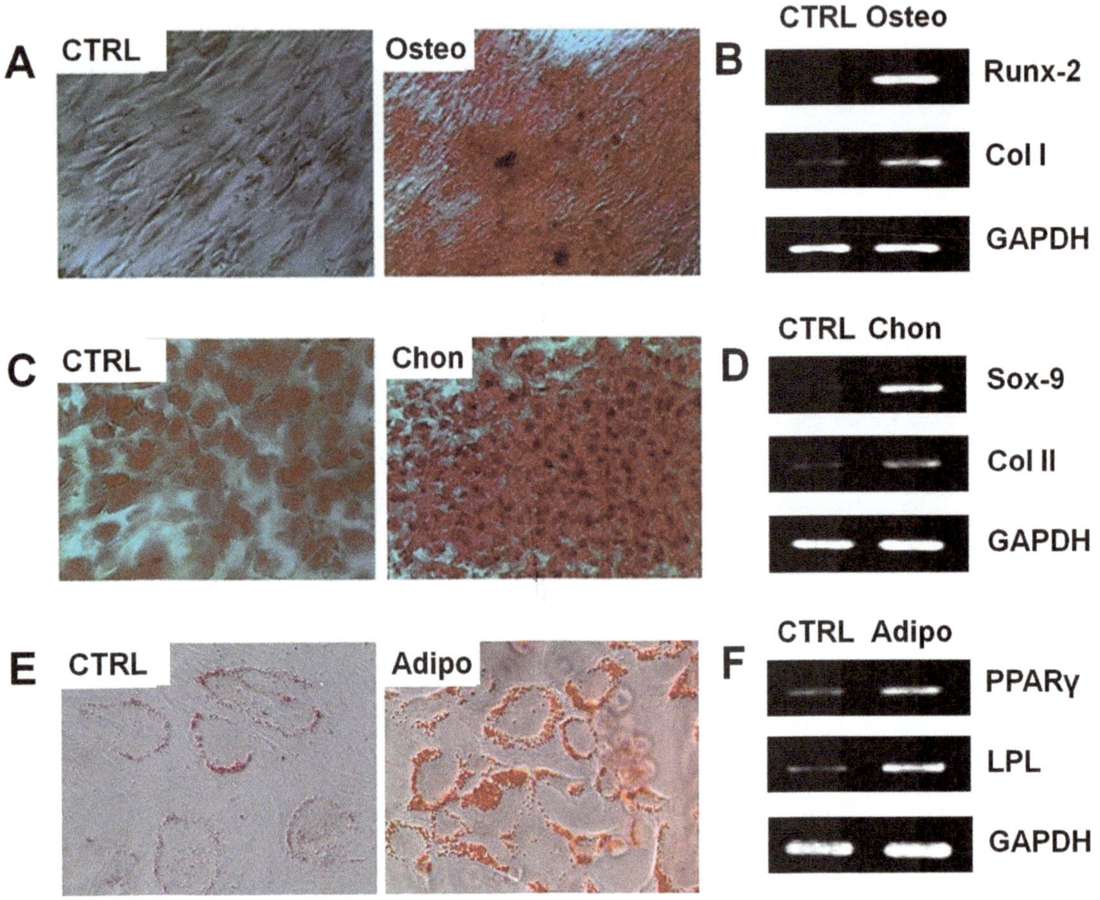

Figure 4. Induced differentiation of AF-derived colony forming cells. (**A–B**) Osteogenic differentiation at 3 weeks. Mineralization was stained with Alizarin red S (**A**). Expression of osteocyte-specific genes, including *Runx-2* and *Col I*, were up-regulated in induced cells as analyzed by RT-PCR (**B**). (**C–D**) Chondrogenic differentiation at 3 weeks. Cells were stained with Safranin O (**C**). Expression of chondrocyte-specific genes *Sox-9* and *Col II* levels was higher in induced cells (**D**). (**E–F**) Adipogenic differentiation at 2 weeks. Secretion of oil droplets (**E**) and expression of adipocyte-specific genes *PPAR-γ* and *LPL* (**F**) were higher in induced cells.

colony forming cells for AF therapies and tissue engineering. It should be noted that without *in vitro* expansion, the AF-derived colony forming cells are generally too few to be directly used for cell therapies. However, after serial passaging a considerable number of cells may be obtained. For example, we usually obtained 20–30 million cells from each rabbit AF segment after three passages, which is enough for a typical cell therapy (0.4–10 million per kg of body weight) [42]. Importantly, the cells showed no apparent difference in morphology, proliferation capacity, stem marker expression and differentiation potential as long as they were within passage six (data not shown). In humans, AF tissues may be obtained through surgeries in which the disc is partially or entirely removed, including discectomy, interbody fusion, or scoliosis treatments [24]. Following that, the AF-derived colony forming cells can be retrieved and expanded for cell therapy or tissue engineering applications, as implied in previous studies [19,20].

In addition, the differentiation potential was independent of the morphology of colony. For example, both chondroblast-like and fibroblast-like colonies could undergo induced differentiation. This implies that while different cell sub-populations may exist in colony forming cells derived from different AF regions, they all have the potential to differentiate into AF cells or various regions. It again signifies the importance of using such colony forming cells for repairing/regenerating the AF tissue, which is intrinsically a heterogeneous tissue and has distinct region-specific characteristics. It should be noted, however, that compared to cells from fibroblast-like colonies, the cells from chondroblast-like colonies showed relatively stronger tendency to differentiate to chondrocyte-like cells, as indicated by the stronger Safranin O staining as well as *Sox-9* and *Col II* gene expression (data not shown). In addition to induced differentiation tests, we found another piece of evidence of the plasticity of AF-derived colony forming cells, i.e., substrate elasticity-dependent differentiation characteristics [43]. When they were cultured on substrates of various stiffness, their gene expression profiles varied in a stiffness-dependent manner. The cells became more chondrocyte-like on relatively compliant substrate with up-regulated *Col II* gene expression, while they became more fibroblast-like on stiffer substrate and markedly expressed more *Col I* gene (**Fig. S4**).

In summary, we have successfully isolated and identified a population of colony forming cells from rabbit AF tissue. These cells possess clonogenicity, self-renewing capability, and multi-differentiation potential, the common characteristics of MSCs. Therefore, we define these cells as AF-derived stem cells (AFSCs). Being capable of rapid *in vitro* expansion and multi-differentiation, the AFSCs may potentially be a valuable cell source for DDD treatments using cell therapy or tissue engineering approaches. It should be noted, however, that findings in this study do not rule out the feasibility of using other cell sources for AF tissue regeneration. In another study, we have found that after certain treatment such as TGF-β3, BMSCs can also be a good cell source for AF regeneration (unpublished data). Future research will explore the lineage commitment of AFSCs toward various types of cells within AF and compare the efficiency as cell source in AF tissue engineering between AFSCs and other commonly used MSCs, including bone marrow stem cells and adipose-derived stem cells.

Supporting Information

Figure S1 Harvest of rabbit AF tissues. (A) A segment of spinal column containing IVDs from T10 through L5 vertebra. (B) IVD harvesting. (C) A whole rabbit IVD. The AF tissue was then separated from the IVD by removing the nucleus pulposus.

Figure S2 Cell colonies formed from rabbit AF cells using single cell culture technique. (A–B) Typical colonies formed from single rabbit AF cells. The cells were seeded at different initial plating densities and were stained at 10 days using crystal violet. (C–D) Gross view of the morphology of cells within cell colonies of various sizes. Scale bars, 100 μm.

Figure S3 Representative close view of the morphology of AF cells within colonies. (A–B) Cells from the inner and outer regions, respectively, of one colony. (C–D) Cells from the inner and outer regions, respectively, of another colony. The cells were stained with crystal violet.

Figure S4 Expression of *Col I* (A) and *Col II* (B) genes in rabbit AF-derived colony forming cells cultured on polyacrylamide hydrogels of different Young's moduli for 1 week.

Methods S1 Supporting methods.

Author Contributions

Conceived and designed the experiments: BL HY. Performed the experiments: QG JL CL SW YW. Analyzed the data: QG JL CL BL. Contributed reagents/materials/analysis tools: QG JL BL. Contributed to the writing of the manuscript: QG JL BL.

References

1. Luo X, Pietrobon R, Sun SX, Liu GG, Hey L (2004) Estimates and patterns of direct health care expenditures among individuals with back pain in the United States. Spine 29: 79–86.
2. Hudson KD, Alimi M, Grunert P, Hartl R, Bonassar LJ (2013) Recent advances in biological therapies for disc degeneration: tissue engineering of the annulus fibrosus, nucleus pulposus and whole intervertebral discs. Curr Opin Biotechnol 24: 872–879.
3. Driscoll TP, Nakasone RH, Szczesny SE, Elliott DM, Mauck RL (2013) Biaxial mechanics and inter-lamellar shearing of stem-cell seeded electrospun angle-ply laminates for annulus fibrosus tissue engineering. J Orthop Res 31: 864–870.
4. Koepsell L, Remund T, Bao J, Neufeld D, Fong H, et al. (2011) Tissue engineering of annulus fibrosus using electrospun fibrous scaffolds with aligned polycaprolactone fibers. J Biomed Mater Res A 99: 564–575.
5. Wan Y, Feng G, Shen FH, Laurencin CT, Li X (2008) Biphasic scaffold for annulus fibrosus tissue regeneration. Biomaterials 29: 643–652.
6. Nerurkar NL, Baker BM, Sen S, Wible EE, Elliott DM, et al. (2009) Nanofibrous biologic laminates replicate the form and function of the annulus fibrosus. Nat Mater 8: 986–992.
7. Li J, Liu C, Guo Q, Yang H, Li B (2014) Regional variations in the cellular, biochemical, and biomechanical characteristics of rabbit annulus fibrosus. PLoS One 9: e91799.
8. Guterl CC, See EY, Blanquer SB, Pandit A, Ferguson SJ, et al. (2013) Challenges and strategies in the repair of ruptured annulus fibrosus. Eur Cell Mater 25: 1–21.
9. Bowles RD, Gebhard HH, Hartl R, Bonassar LJ (2011) Tissue-engineered intervertebral discs produce new matrix, maintain disc height, and restore biomechanical function to the rodent spine. Proc Natl Acad Sci USA 108: 13106–13111.
10. Mizuno H, Roy AK, Vacanti CA, Kojima K, Ueda M, et al. (2004) Tissue-engineered composites of anulus fibrosus and nucleus pulposus for intervertebral disc replacement. Spine 29: 1290–1297.
11. Paesold G, Nerlich AG, Boos N (2007) Biological treatment strategies for disc degeneration: potentials and shortcomings. Eur Spine J 16: 447–468.
12. Horner HA, Roberts S, Bielby RC, Menage J, Evans H, et al. (2002) Cells from different regions of the intervertebral disc: effect of culture system on matrix expression and cell phenotype. Spine 27: 1018–1028.
13. Sabatino M, Ren J, David-Ocampo V, England L, McGann M, et al. (2012) The establishment of a bank of stored clinical bone marrow stromal cell products. J Transl Med 10: 23.
14. Yoshimura H, Muneta T, Nimura A, Yokoyama A, Koga H, et al. (2007) Comparison of rat mesenchymal stem cells derived from bone marrow, synovium, periosteum, adipose tissue, and muscle. Cell Tissue Res 327: 449–462.
15. Henriksson H, Thornemo M, Karlsson C, Hagg O, Junevik K, et al. (2009) Identification of cell proliferation zones, progenitor cells and a potential stem cell niche in the intervertebral disc region: a study in four species. Spine (Phila Pa 1976) 34: 2278–2287.
16. Henriksson HB, Svala E, Skioldebrand E, Lindahl A, Brisby H (2012) Support of concept that migrating progenitor cells from stem cell niches contribute to normal regeneration of the adult mammal intervertebral disc: a descriptive study in the New Zealand white rabbit. Spine (Phila Pa 1976) 37: 722–732.
17. Henriksson HB, Svanvik T, Jonsson M, Hagman M, Horn M, et al. (2009) Transplantation of human mesenchymal stems cells into intervertebral discs in a xenogeneic porcine model. Spine (Phila Pa 1976) 34: 141–148.
18. Barreto Henriksson H, Lindahl A, Skioldebrand E, Junevik K, Tangemo C, et al. (2013) Similar cellular migration patterns from niches in intervertebral disc and in knee-joint regions detected by insitu labeling: an experimental study in the New Zealand white rabbit. Stem Cell Res Ther 4: 104.
19. Risbud MV, Guttapalli A, Tsai TT, Lee JY, Danielson KG, et al. (2007) Evidence for skeletal progenitor cells in the degenerate human intervertebral disc. Spine 32: 2537–2544.
20. Feng G, Yang X, Shang H, Marks IW, Shen FH, et al. (2010) Multipotential differentiation of human anulus fibrosus cells: an invitro study. J Bone Joint Surg Am 92: 675–685.
21. Sakai D, Nakamura Y, Nakai T, Mishima T, Kato S, et al. (2012) Exhaustion of nucleus pulposus progenitor cells with ageing and degeneration of the intervertebral disc. Nat Commun 3: 1264.
22. Bruno S, Bussolati B, Grange C, Collino F, di Cantogno LV, et al. (2009) Isolation and characterization of resident mesenchymal stem cells in human glomeruli. Stem Cells Dev 18: 867–880.
23. Masuda K, Aota Y, Muehleman C, Imai Y, Okuma M, et al. (2005) A novel rabbit model of mild, reproducible disc degeneration by an anulus needle puncture: correlation between the degree of disc injury and radiological and histological appearances of disc degeneration. Spine 30: 5–14.
24. Bron JL, Helder MN, Meisel HJ, Van Royen BJ, Smit TH (2009) Repair, regenerative and supportive therapies of the annulus fibrosus: achievements and challenges. Eur Spine J 18: 301–313.
25. Bi Y, Ehirchiou D, Kilts TM, Inkson CA, Embree MC, et al. (2007) Identification of tendon stem/progenitor cells and the role of the extracellular matrix in their niche. Nat Med 13: 1219–1227.
26. Zhang J, Li B, Wang JH (2011) The role of engineered tendon matrix in the stemness of tendon stem cells invitro and the promotion of tendon-like tissue formation invivo. Biomaterials 32: 6972–6981.
27. Liu TM, Martina M, Hutmacher DW, Hui JH, Lee EH, et al. (2007) Identification of common pathways mediating differentiation of bone marrow- and adipose tissue-derived human mesenchymal stem cells into three mesenchymal lineages. Stem Cells 25: 750–760.
28. Postacchini F, Bellocci M, Massobrio M (1984) Morphologic changes in annulus fibrosus during aging. An ultrastructural study in rats. Spine 9: 596–603.
29. Kohler J, Popov C, Klotz B, Alberton P, Prall WC, et al. (2013) Uncovering the cellular and molecular changes in tendon stem/progenitor cells attributed to tendon aging and degeneration. Aging Cell 12: 988–999.
30. Garcia-Prat L, Sousa-Victor P, Munoz-Canoves P (2013) Functional dysregulation of stem cells during aging: a focus on skeletal muscle stem cells. FEBS J 280: 4051–4062.
31. Yasen M, Fei Q, Hutton WC, Zhang J, Dong J, et al. (2013) Changes of number of cells expressing proliferation and progenitor cell markers with age in rabbit intervertebral discs. Acta Biochim Biophys Sin (Shanghai) 45: 368–376.

32. Boxall SA, Jones E (2012) Markers for characterization of bone marrow multipotential stromal cells. Stem Cells Int 2012: Article ID 975871.
33. Yu Y, Fuhr J, Boye E, Gyorffy S, Soker S, et al. (2006) Mesenchymal stem cells and adipogenesis in hemangioma involution. Stem Cells 24: 1605–1612.
34. Gibbings D, Befus AD (2009) CD4 and CD8: an inside-out coreceptor model for innate immune cells. J Leukoc Biol 86: 251–259.
35. Levesque MC, Heinly CS, Whichard LP, Patel DD (1998) Cytokine-regulated expression of activated leukocyte cell adhesion molecule (CD166) on monocyte-lineage cells and in rheumatoid arthritis synovium. Arthritis Rheum 41: 2221–2229.
36. Greco SJ, Liu K, Rameshwar P (2007) Functional similarities among genes regulated by OCT4 in human mesenchymal and embryonic stem cells. Stem Cells 25: 3143–3154.
37. Qu J, Bishop JM (2012) Nucleostemin maintains self-renewal of embryonic stem cells and promotes reprogramming of somatic cells to pluripotency. J Cell Biol 197: 731–745.
38. Gang EJ, Bosnakovski D, Figueiredo CA, Visser JW, Perlingeiro RC (2007) SSEA-4 identifies mesenchymal stem cells from bone marrow. Blood 109: 1743–1751.
39. Seo KW, Lee SR, Bhandari DR, Roh KH, Park SB, et al. (2009) OCT4A contributes to the stemness and multi-potency of human umbilical cord blood-derived multipotent stem cells (hUCB-MSCs). Biochem Biophys Res Commun 384: 120–125.
40. Tai MH, Chang CC, Kiupel M, Webster JD, Olson LK, et al. (2005) Oct4 expression in adult human stem cells: evidence in support of the stem cell theory of carcinogenesis. Carcinogenesis 26: 495–502.
41. Zuk PA (2009) The intracellular distribution of the ES cell totipotent markers OCT4 and Sox2 in adult stem cells differs dramatically according to commercial antibody used. J Cell Biochem 106: 867–877.
42. English K, French A, Wood KJ (2010) Mesenchymal stromal cells: facilitators of successful transplantation? Cell Stem Cell 7: 431–442.
43. Engler AJ, Sen S, Sweeney HL, Discher DE (2006) Matrix elasticity directs stem cell lineage specification. Cell 126: 677–689.

Paracrine-Mediated Neuroprotection and Neuritogenesis of Axotomised Retinal Ganglion Cells by Human Dental Pulp Stem Cells: Comparison with Human Bone Marrow and Adipose-Derived Mesenchymal Stem Cells

Ben Mead[1,2*], Ann Logan[1], Martin Berry[1], Wendy Leadbeater[1☯¶], Ben A. Scheven[2☯¶]

1 Neurotrauma Research Group, Neurobiology Section, School of Clinical and Experimental Medicine, University of Birmingham, Birmingham, United Kingdom, **2** School of Dentistry, University of Birmingham, Birmingham, United Kingdom

Abstract

We have investigated and compared the neurotrophic activity of human dental pulp stem cells (hDPSC), human bone marrow-derived mesenchymal stem cells (hBMSC) and human adipose-derived stem cells (hAMSC) on axotomised adult rat retinal ganglion cells (RGC) *in vitro* in order to evaluate their therapeutic potential for neurodegenerative conditions of RGC. Using the transwell system, RGC survival and length/number of neurites were quantified in coculture with stem cells in the presence or absence of specific Fc-receptor inhibitors to determine the role of NGF, BDNF, NT-3, VEGF, GDNF, PDGF-AA and PDGF-AB/BB in stem cell-mediated RGC neuroprotection and neuritogenesis. Conditioned media, collected from cultured hDPSC/hBMSC/hAMSC, were assayed for the secreted growth factors detailed above using ELISA. PCR array determined the hDPSC, hBMSC and hAMSC expression of genes encoding 84 growth factors and receptors. The results demonstrated that hDPSC promoted significantly more neuroprotection and neuritogenesis of axotomised RGC than either hBMSC or hAMSC, an effect that was neutralized after the addition of specific Fc-receptor inhibitors. hDPSC secreted greater levels of various growth factors including NGF, BDNF and VEGF compared with hBMSC/hAMSC. The PCR array confirmed these findings and identified VGF as a novel potentially therapeutic hDPSC-derived neurotrophic factor (NTF) with significant RGC neuroprotective properties after coculture with axotomised RGC. In conclusion, hDPSC promoted significant multi-factorial paracrine-mediated RGC survival and neurite outgrowth and may be considered a potent and advantageous cell therapy for retinal nerve repair.

Editor: Eva Mezey, National Institutes of Health, United States of America

Funding: This work was funded by the Rosetrees trust and a Biotechnology and Biological Sciences Research Council (BBSRC) studentship; BM BAS. Grant number BB/F017553/1 (http://www.rosetreestrust.co.uk) (http://www.bbsrc.ac.uk/home/home.aspx). The funders had no role in study design, data collection and analysis, decision to publish, or preparation of the manuscript.

Competing Interests: The authors have declared that no competing interests exist.

* Email: BXM813@bham.ac.uk

☯ These authors contributed equally to this work.

¶ These authors are joint senior authors on this work

Introduction

The axons of retinal ganglion cells (RGC) transmit action potentials along the optic nerve to the superior colliculus (SC) and lateral geniculate nucleus (LGN) that are relayed onwards to the visual cortex. Axotomised RGC die [1,2] so that blindness ensues after traumatic (crush or transection) [3] optic nerve injury. RGC loss is caused by a failure in the supply of neurotrophic factors (NTF; including neurotrophins), retrogradely transported from the SC/LGN neurons, that act as survival signals, ensuring the functional integrity of RGC connections [4–6].

As well as protecting RGC from death, NTF have the potential to promote the regeneration of transected axons and establish re-connection with their targets. The paucity of NTF in the central nervous system (CNS) is one explanation for the lack of axon regeneration compared to the peripheral nervous system (PNS) [2,7] in which successful and functional axon regeneration occurs, largely promoted by Schwann cell-derived NTF [8]. Attempts to promote long distance axon regeneration by the transplantation of peripheral nerve grafts into the CNS have met with some success [9]. For example, grafting a peripheral nerve into the vitreous body after optic nerve crush [8] promotes more RGC axon regeneration in the transected optic nerve than occurs after the removal of Schwann cells before transplantation, suggesting that axotomised RGC regenerate their axons when provided with a constant supply of NTF. However, single NTF supplementation [7], or single dose treatments of NTF such as BDNF [10,11], have proven unsuccessful and sustained delivery of multiple NTF to RGC over extended periods of time is difficult to achieve.

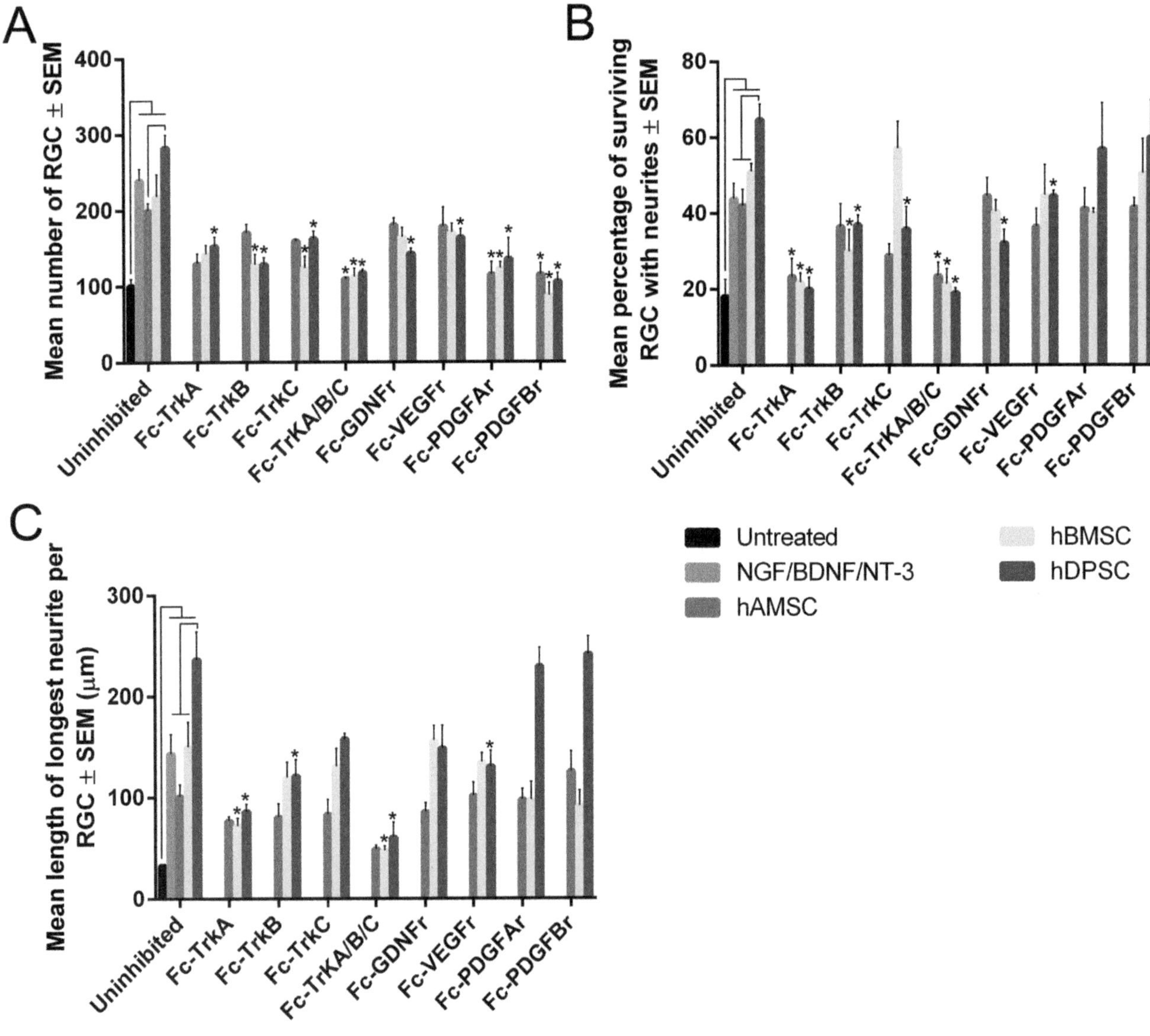

Figure 1. Effects of hDPSC, hBMSC and hAMSC on axotomised RGC in culture. The number of RGC (**A**), percentage of surviving RGC bearing neurites (**B**) and the length of the longest RGC neurite (**C**) in both untreated retinal cultures and after coculture with hDPSC, hBMSC, hAMSC, with or without added exogenous neurotrophins. *Black lines* indicate significant difference at $P<0.05$. The effects of Fc-TrKA, -B, -C, Fc-GDNFr, Fc-VEGFr, Fc-PDGFAr and Fc-PDGFBr inhibitors on RGC survival (**A**) and neuritogenesis (**B, C**) in hDPSC, hBMSC and hAMSC cocultures are also shown (values marked with an *asterisk* indicate significant difference from uninhibited cultures at $P<0.05$).

The vitreous is a relatively accessible immunoprivileged transplantation site [12] that lies directly adjacent to the RGC layer of the retina, allowing diffusion or transport of NTF across the inner limiting membrane to the RGC. Previously, we used intravitreally transplanted genetically modified fibroblasts expressing FGF-2, BDNF and NT-3 to promote RGC survival and axon regeneration after optic nerve crush [7]. Since the translational potential of genetically modified cells is limited, mesenchymal stem cells (MSC), which secrete a large array of NTF, have gained credence as a potential cell therapy for diseased and injured CNS neurons. Human bone marrow-derived mesenchymal stem cells (hBMSC) protect RGC from death in both optic nerve crush [13] and glaucoma experimental models [14–16] through the production of NTF (e.g. platelet-derived growth factor (PDGF) [15]), without differentiation of hBMSC into replacement RGC/RGC-like cells.

We recently demonstrated that rat dental pulp stem cells (DPSC) protected adult rat RGC from death in an optic nerve crush model [17,18]. This effect was significantly greater than that achieved by rat BMSC and mediated through nerve growth factor (NGF), brain-derived neurotrophic factor (BDNF) and neurotrophin 3 (NT-3) via, TrKA, B and C receptor signalling. Our findings were consistent with previous studies showing significant expression [19,20] and secretion of NGF, BDNF and NT-3 by hDPSC [21]. The neuroprotective and axogenic properties of DPSC [17,18] and BMSC [14,15,22] can also be found in other MSC types, in particular adipose-derived mesenchymal stem cells (AMSC) that also secrete multiple NTF [22,23] and promote functional recovery after CNS trauma including spinal cord injury [22,24], stroke [25] and light induced photoreceptor damage [26,27]. However, AMSC have not been tested in a model of RGC death. Comparative analyses of different human MSC is still

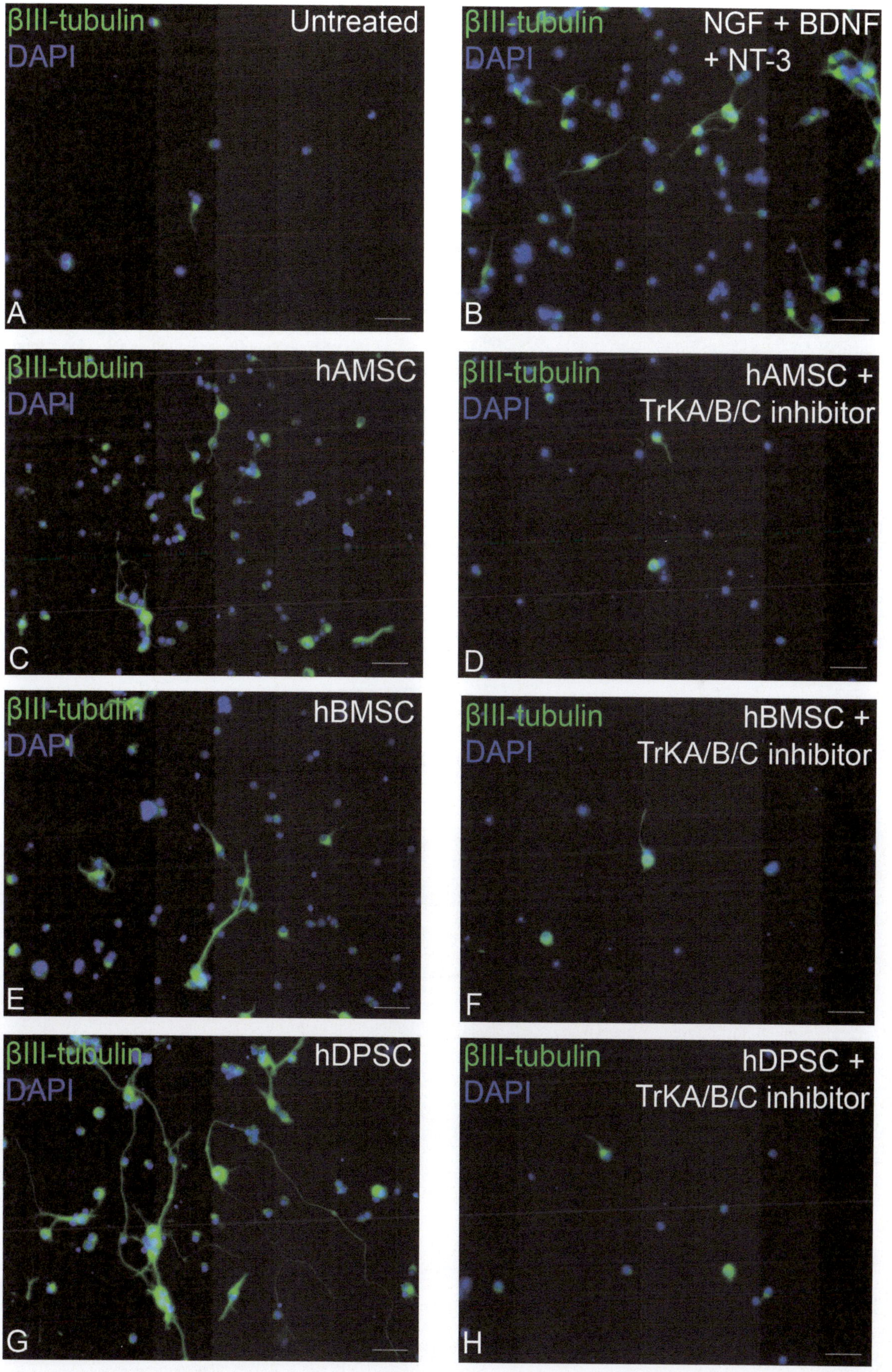
βIII-tubulin
DAPI
Untreated
A
NGF + BDNF
+ NT-3
B
hAMSC
C
hAMSC +
TrKA/B/C inhibitor
D
hBMSC
E
hBMSC +
TrKA/B/C inhibitor
F
hDPSC
G
hDPSC +
TrKA/B/C inhibitor
H

Figure 2. Immunocytochemical staining of RGC after retinal coculture with hDPSC, hBMSC and hAMSC in transwell inserts. *In vitro* RGC cultured either alone (**A**), with exogenous neurotrophins (**B**), with hAMSC (with or without TrK inhibitors (**C**, **D**, respectively)), with hBMSC (with or without TrK inhibitors (**E**, **F**, respectively)) or with hDPSC (with or without TrK inhibitors (**G**, **H**, respectively). All images are representative of the entire culture, nine separate culture wells/treatment, with every three wells using a different animal (*scale bars*: 50 μm), cell nuclei were stained with DAPI.

lacking although important for the determination of the most efficacious paracrine-mediated therapy for the injured retina. Thus, the aim of this study was to evaluate and compare the neuroprotective and neuritogenic effects of hDPSC, hBMSC and hAMSC and to define the stem cell NTF secretome using ELISA and PCR microarray analysis.

Materials and Methods

All reagents were purchased from Sigma (Poole, UK) unless otherwise specified.

DPSC/BMSC/AMSC cultures

hDPSC were purchased from AllCell LLC (Berkeley, CA) and both hBMSC and hAMSC from Lonza (Slough, UK), and each represented pooled samples from 3 donors. The $CD29^+/CD44^+/CD73^+/CD90^+/CD45^-$ (confirmed by supplier) stem cells were seeded into T25/T75 flasks (Corning, Amsterdam, NL) in both a total volume of 5 ml/15 ml DMEM containing 1% penicillin/streptomycin and 10% foetal bovine serum (FBS; Hyclone Laboratories, Logan, UT) and at a density of 1×10^6 cells/2×10^6 cells, respectively. Cultures were maintained at 37°C in 5% CO_2, the supplemented medium was changed every 3 d and the cells were passaged when 80% confluent using 0.05% trypsin/EDTA to lift them from their surface attachment.

Animals

Nine adult male Sprague-Dawley rats weighing 170–200 g (Charles River, Kent, UK) were housed under Home Office guidelines in conditions of 21°C and 55% humidity under a 12 h light and dark cycle, given food/water *ad libitum* and were under constant supervision from trained staff. Animals were killed by "Schedule 1 Methods" i.e. rising concentrations of CO_2 before extraction of retinae. Ethical approval by the University of Birmingham ethical review Sub-Committee for this study was not required due to the *in vitro* nature of the experiment.

Retinal cell coculture

Cell culture 24-well plates (BD Biosciences, Oxford, UK) were pre-coated with 100 μg/ml poly-D-lysine for 60 min and then with 20 μg/ml laminin for 30 min. After culling and ocular dissection, the retinae of nine male Sprague-Dawley were minced in 1.25 ml of papain (20 U/ml; Worthington Biochem, Lakewood, NJ; as per manufacturer's instructions) containing 50 μg/ml of DNase I (62.5 μl; Worthington Biochem) and incubated for 90 min at 37°C. The retinal cell suspension was centrifuged at 300×g for 5 min and the pellet resuspended in 1.575 ml of Earle's balanced salt solution (Worthington Biochem) containing 1.1 mg/ml of reconstituted albumin ovomucoid inhibitor (150 μl; Worthington Biochem) and 56 μg/ml of DNase I (75 μl). After adding to the top of 2.5 ml of albumin ovomucoid inhibitor (10 mg/ml) to form a discontinuous density gradient, the retinal cell suspension was centrifuged at 70×g for 6 min and the cell pellet resuspended in 1 ml of supplemented Neurobasal-A (25 ml Neurobasal-A (Life Technologies, Gibco, Paisley, UK), 1X concentration of B27 supplement (Life Technologies, Invitrogen, Paisley, UK), 0.5 mM of L-glutamine (62.5 μl; Invitrogen) and 50 μg/ml of gentamycin (125 μl; Invitrogen)) and seeded at a density of 125,000 cells/800 μl/well in a 24 well plate. Previous immunocytochemical analysis of these cultures in our lab demonstrates that 60% of these retinal cells are neurons (neurofilament$^+$/βIII-tubulin$^+$), of which 10% are Thy1$^+$ RGC [28].

hDPSC, hBMSC and hAMSC were used at passage 2–4 and plated at a density of 50,000 cells/200 μl onto a 0.4 μm porous cell culture insert (Millipore, Watford, UK) that was placed into each of the 24 wells containing retinal cells to give a total volume of 1 ml of medium/well. Particular wells containing retinal cell cultures were treated either singly or in combination with 5 μg/ml Fc-TrKA, Fc-TrKB and/or Fc-TrKC (R&D Systems, Abingdon, UK) as well as Fc-VEGFr, Fc-GDNFr, Fc-PDGFAr and Fc-PDGFBr fusion protein inhibitors [29] (R&D Systems) which are highly specific inhibitors for the corresponding cognate neurotrophic factor receptors (NTFR). A combination of recombinant human NGF, BDNF and NT-3 was also added to some retinal cell cultures at 60 ng/ml as a positive control. Particular wells containing retinal cells were treated with 0.1 μm, 1 μm and 10 μm of VGF (R&D Systems) instead of hDPSC/hBMSC/hAMSC.

Cocultures were incubated for 3 d at 37°C before immunocytochemical staining of RGC with βIII-tubulin. For this study, large spherical βIII-tubulin$^+$ retinal cells [30] cultured after neuronal isolation from retinae, are referred to as RGC. βIII-tubulin is a reliable and relatively specific marker for RGC [30], although cross reactivity with amacrine cells has been suggested [31]. However, it should be emphasised, that the isolation procedure and growth medium used preferentially selects and yields populations of neuronal cells, we are confident that our findings accurately reflect RGC numbers. All experiments were repeated on 3 separate occasions. Each of the treatment groups in each of the 3 experimental runs comprised 3 replicate wells containing retinal cells harvested from one animal.

Immunocytochemistry

Retinal cells in 24 well plates (BD Biosciences) were fixed in 4% paraformaldehyde (PFA) in phosphate-buffered saline (PBS) for 10 min, washed for 3×10 min of PBS, blocked in blocking solution (0.5% bovine serum albumin (g/ml), 0.3% Tween-20, 15% normal goat serum (Vector Laboratories, Peterborough, UK) in PBS) for 20 min and incubated with primary antibody (anti-rat βIII-tubulin, raised in mouse, #T8660 diluted at 1:500 in antibody diluting buffer (ADB; 0.5% bovine serum albumin, 0.3% Tween-20 in PBS) for 1 h at room temperature. Cells were then washed for 3×10 min in PBS, incubated with the secondary antibody (anti-mouse IgG Fluor 488, raised in goat, 1:400, #A-11001; Life Technologies, Molecular Probes, Paisley, UK) diluted in ADB for 1 h at room temperature, washed for 3×10 min in PBS, mounted in Vectorshield mounting medium containing DAPI (Vector Laboratories) and stored at 4°C.

Microscopy and analysis

For immunocytochemistry, all βIII-tubulin$^+$ RGC, with or without neurites, were counted in each well of the 24 well plates, recording the total number of RGC and the number of RGC with

Table 1. The number of surviving RGC (cells/well), percentage of surviving RGC growing neurites (%) and the length of the longest neurite (μm) after coculture of RGC with hDPSC, hBMSC or hAMSC and inhibition with Fc-TrKA, Fc-TrKB, Fc-TrKC, Fc-TrKA/B/C, Fc-GDNFr, Fc-VEGFr, Fc-PDGFAr and Fc-PDGFBr.

Treatment		Without inhibition	Fc-TrKA	Fc-TrKB	Fc-TrKC	Fc-TrKA/B/C	Fc-GDNFr	Fc-VEGFr	Fc-PDGFAr	Fc-PDGFBr
hDPSC	RGC/well	282.7±17.1	**153.0±12.1**	**129.3±8.5**	**163.3±9.4**	**118.3±4.3**	**143.3±6.6**	**164.7±10.7**	**136.3±27.1**	**106.0±11.3**
	% surviving RGC with neurites	64.8±4.0	**20.0±3.2**	**37.1±2.4**	**35.9±5.8**	**19.2±1.3**	**32.3±3.4**	**44.5±1.4**	57.0±12.1	60.1±9.7
	Length of longest neurite (μm)	236.7±27.6	**86.4±7.0**	**121.6±16.1**	158.2±5.1	**60.6±14.8**	148.9±22.2	**130.8±15.4**	229.5±18.4	241.6±17.6
hBMSC	RGC/well	219±28.4	143.0±11.4	**128.7±13.6**	**124.0±15.4**	**112.7±11.0**	165.3±11.9	171.0±11.6	**123.7±8.0**	**87.7±16.1**
	% surviving RGC with neurites	51.1±2.1	**21.9±2.5**	**30.1±5.7**	57.3±7.0	**21.4±4.1**	40.5±3.0	44.83±8.0	39.9±1.4	50.6±9.0
	Length of longest neurite (μm)	150.7±24.1	**72.0±7.8**	119.9±15.3	131.1±17.6	**47.9±4.2**	156.7±14.2	135.0±8.9	97.4±18.2	91.3±15.0
hAMSC	RGC/well	200.0±10.2	130.0±13.1	171.3±10.4	160.0±2.6	**110.3±1.2**	165.3±11.9	171.0±11.6	**116.3±15.9**	**115.7±14.5**
	% surviving RGC with neurites	42.2±4.3	**23.5±4.8**	36.6±6.0	29.0±3.0	**23.7±3.5**	44.6±4.7	36.6±4.7	41.3±5.3	41.6±2.4
	Length of longest neurite (μm)	101.1±12.1	77.1±4.3	80.8±13.1	83.6±14.2	49.1±3.6	156.7±14.2	135.0±8.9	97.6±10.5	125±19.5

Values significantly ($p<0.05$) different from coculture without inhibition are given in bold (Mean +/− SEM; n = 3).

neurites. Neurite outgrowth was measured in images taken at 20X magnification using an Axiocam HRc camera (Carl Zeiss Ltd, Hertfordshire, UK). Each well was divided into 9 equal sectors and the length of the longest neurite of each RGC in each sector was measured using Axiovision software (Carl Zeiss Ltd). Fluorescently stained cells were analysed by an operator blinded to treatment groups, using a Zeiss Axiovert fluorescent microscope (Carl Zeiss Ltd).

NTF ELISA

To quantify the growth factors and NTF produced by hDPSC, hBMSC and hAMSC, conditioned medium was collected from cultures at passage 2 and 5 after 48 h in culture and assayed using a duoset ELISA kit for human NGF, BDNF and NT-3, VEGF, GDNF, platelet-derived growth factor (PDGF-AA) and PDGF-AB/BB according to the manufacturer's instructions (R&D Systems). Briefly, a standard curve was constructed using NTF standards and test samples of conditioned medium at increasing dilutions, run in duplicate with NTF concentrations extrapolated from the standard curve. Values were normalized to the number of cells in the flask (manually counted *via* haemocytometer) and the volume of medium, and corrected for analyte found in the medium/serum.

Human NTF and NTFR PCR array

The expression of 84 NTF and NTFR genes by hDPSC, hBMSC and hAMSC (passage 2) was assayed using quantitative RT-PCR profiler arrays (PAMM-031) by a commercially run service (Sabiosciences, Qiagen, Hilden, Germany). Housekeeping genes (*b2m* and *hprt1*) were used to normalize the data and the $^{-\Delta Ct}$ compared between cell types and expressed as + or - fold changes. Samples contained one million cells and were run in triplicate.

Statistics

All statistical tests were performed using SPSS 17.0 (IBM SPSS, Inc., Chicago, IL) and data presented as mean ± standard error of the mean (SEM). The Shapiro-Wilkes test was used to ensure all data were normally distributed before parametric testing using a one-way analysis of variance (ANOVA) with a Tukey *post-hoc* test. Statistical differences were considered significant at p values < 0.05. For the qRT-PCR, data were compared by a Student's t test and statistical significance set at $p<0.001$.

Results

hDPSC promoted significantly greater paracrine-mediated neuroprotection and neuritogenesis than hBMSC/hAMSC

hDPSC, hBMSC and hAMSC all promoted a significant increase ($p<0.05$) in the survival of cocultured RGC (282.7±17.1 cells/well; 219±28.4 cells/well; 200.0±10.2 cells/well; respectively) compared with retinal cells cultured alone (100.7±9.5 cells/well); hDPSC, hBMSC and hAMSC neuroprotection was similar to the level obtained after retinal cell culture with recombinant NGF/BDNF/NT-3 (239.7±15.4 cells/well; Figs. 1, 2.). The increase in survival of RGC in hDPSC-treated retinal cultures was significantly greater ($p<0.05$) than that achieved in cocultures with hAMSC ($p<0.05$) but not significantly greater than that seen in cocultures with hBMSC.

Coculture of retinal cells with hDPSC, hBMSC and hAMSC significantly increased ($p<0.05$) the percentage of surviving RGC bearing tubulin^{+} neurites (64.8±4.0%, 51.1±2.1%, 42.2±4.3%,

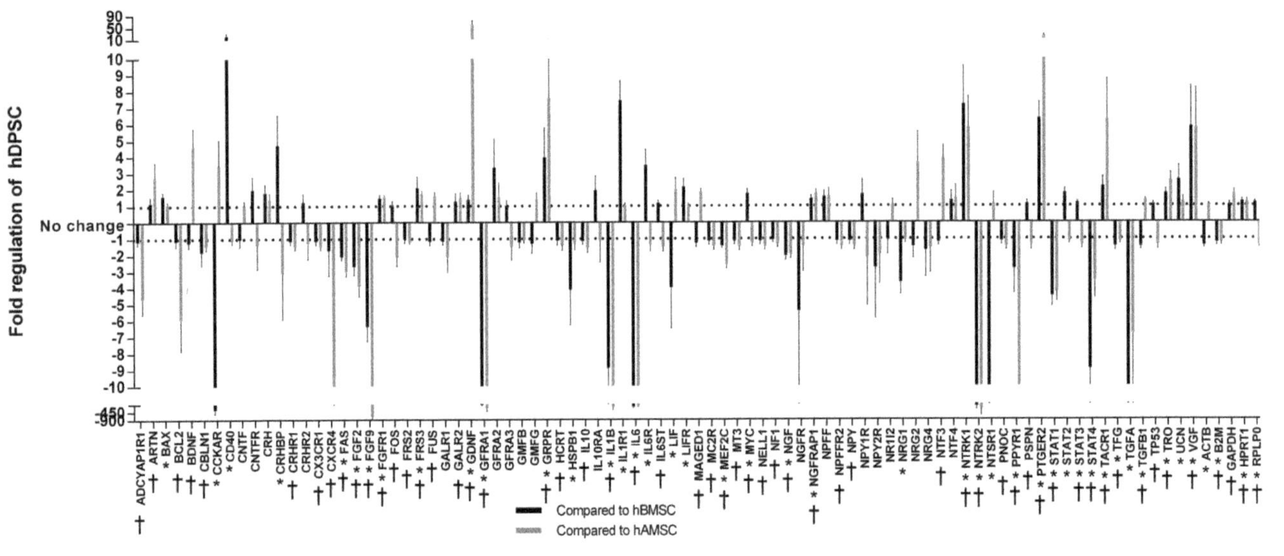

Figure 3. Expression of NTF and NTFR genes by hDPSC, hBMSC and hAMSC. RT-PCR array analysis of 84 genes encoding NTF and NTFR expressed by hDPSC, displayed as fold regulation in comparison to either hBMSC (black bars) or hAMSC (grey bars). The horizontal dotted lines represent fold-changes of ±1 (no difference). Significant differences between hDPSC and hBMSC (*) and hDPSC and hAMSC (†) at $p<0.05$ are labelled on the x axis.

respectively), as well as the length of neurites (236.7 µm±27.6 µm, 150.7 µm±24.1 µm, 101.1 µm±12.1 µm, respectively) compared with retinal cells cultured alone (18.2±4.6%; 32.8 µm±1.4 µm; Fig. 1, 2.). Coculture with hDPSC promoted a significant increase ($p<0.05$) in the number of neurite-bearing RGC as well as neurite length when compared with cocultures with hBMSC and hAMSC, or with retinal cultures exposed to recombinant NGF, BDNF and NT-3. These data confirm that hDPSC, hBMSC and hAMSC have paracrine-mediated neuroprotective and neuritogenic properties, with hDPSC promoting the most significant effects.

NTFR Fc-receptor blockers for multiple NTFR attenuated the neuroprotective and neuritogenic effect of hDPSC/hBMSC/hAMSC

The NTFR blockers Fc-TrKA, Fc-TrKB, Fc-TrKC, Fc-TrKA/B/C, Fc-GDNFr, Fc-VEGFr, Fc-PDGFAr and Fc-PDGFBr significantly attenuated hDPSC mediated neuroprotection and/or neuritogenesis of cocultured RGC (Table. 1; Fig. 1, 2.) compared to uninhibited hDPSC/retinal cell cocultures. Fc-TrKA, Fc-TrKB, Fc-TrKC, Fc-TrKA/B/C, Fc-PDGFAr and Fc-PDGFBr significantly attenuated hBMSC mediated neuroprotection and/or neuritogenesis of cocultured RGC compared to uninhibited hBMSC/retinal cell cocultures. Fc-TrKA/B/C, Fc-PDGFAr and Fc-PDGFBr significantly attenuated hAMSC mediated neuroprotection and/or neuritogenesis of cocultured RGC compared to uninhibited hAMSC/retinal cell cocultures. These data demonstrate that the neuroprotective and neuritogenic effects afforded by each of the stem cell types are mediated through a variety of different NTF.

hDPSC, hBMSC and hAMSC have distinct NTF expression profiles

The PCR array detected 84 NTF and NTFR genes differentially expressed by hDPSC, hBMSC and hAMSC. For example, hDPSC express ≥4 fold higher *cd40, crhbp, grpr, il1r1, ntrk1, ptger2* and *vgf* than hBMSC and ≥4 fold higher *bdnf, gdnf, grpr, nt-3, ptger2, tacr1* and *vgf* than hAMSC (Fig. 3.). hDPSC express ≥4 fold lower *cckar, fgf9, gfra1, hspb1, il1b, il6, ngfr, ntrk2, ntsr1, stat1, stat4 and tgfa* than hBMSC and ≥4 fold lower *adcyap1r1, bcl2, cxcr4, fgf9, gfra1, il1b, il6, ntrk2, ppyr1, stat1, tgfa* than hAMSC. Significant differences ($p<0.001$) are highlighted in the graph (Fig. 3). These data confirm that, despite hDPSC, hBMSC and hAMSC all being designated as mesenchymal stem cells, the NTF secretome is distinct between each stem cell phenotype.

hDPSC secrete multiple NTF at higher levels than hBMSC/hAMSC

The levels of secretion by hDPSC, hBMSC and hAMSC of NGF, BDNF, NT-3, VEGF, GDNF, PDGF-AA, PDGF-AB/BB and CNTF, at passage 2 and 5, are detailed in Table 2, and presented as pg/100,000 cells/48 h. CNTF and PDGF-AB/BB were undetectable in hDPSC/hBMSC/hAMSC conditioned media, while BDNF and NT-3 were undetectable in hAMSC conditioned medium. The hDPSC secreted significantly greater ($p<0.05$) titres of NGF, BDNF and VEGF than hBMSC/hAMSC (Table. 2; Fig. 4.). These data confirm that the hBMSC, hAMSC, and in particular, hDPSC secrete a variety of different NTF.

VGF was neuroprotective for RGC

The differentially and significantly greater transcription of *vgf* in hDPSC compared to hBMSC/hAMSC (Fig. 3.) led to investigation of the neuroprotective and/or proregenerative properties of VGF. VGF promoted a significant increase ($p<0.05$) in the survival of cultured RGC at concentrations of 1 µM (255.5±29.4 cells/well) and 10 µm (263.5±24.4 cells/well), but not at 0.1 µM (148.3±33.1 cells/well), compared to untreated controls (118.7±18.7 cells/well; Fig. 5.).

By contrast, the percentage of surviving RGC bearing neurites in cultures of axotomised retinal neurons did not significantly change after treatment with 0.1 µM (15.8±1.4%), 1 µM (10.0±1.9%) or 10 µM (10.9±1.9%) when compared to untreated controls (17.3±1.3%). The data suggest that VGF is a novel

Table 2. The titre of secreted NGF, BDNF, NT-3, VEGF, GDNF, PDGF-AA, PDGF-AB/BB and CNTF in conditioned media from hDPSC, hBMSC and hAMSC at passage 2 and 5, as determined by ELISA.

		NGF	BDNF	NT-3	VEGF	GDNF	PDGF-AA	PDGF-AB/BB	CNTF
hDPSC	Passage 2	428.0±149.0	141.9±14.4	639.8±134.9	314.7±53.9	113.2±13.1	216.0±19.0	0	0
	Passage 5	331.2±94.8	118.3±9.2	385.3±32.6	287.3±98.4	151.7±51.6	178.9±11.0	0	0
hBMSC	Passage 2	20.4±9.5	44.1±4.8	487.4±65.4	314.7±53.9	161.2±5.2	149.2±7.9	0	0
	Passage 5	37.7±20.7	76.0±12.6	316.2±58.3	251.2±54.9	244.3±94.2	126.8±23.2	0	0
hAMSC	Passage 2	46.1±17.5	0	0	167.6±57.2	35.5±20.2	210.3±87.1	0	0
	Passage 5	33.2±14.0	0	0	66.7±10.5	23.6±10.0	93.9±29.8	0	0

Data are presented as pg/100,000 cells/48 h (Mean +/− SEM; n = 3).

DPSC-derived neuroprotective, but not neuritogenic, factor for RGC at the optimal dose of 1 μm.

Discussion

We have previously shown that rat DPSC promoted neurotrophin-mediated neuroprotection and neuritogenesis/axogenesis of axotomised RGC, both *in vitro* and *in vivo* [17]. Interestingly, rat DPSC promoted significantly greater neuroprotection and axon regeneration than rat BMSC, which reportedly protect RGC from death in models of optic nerve crush and glaucoma [14,17]. We aimed to determine if the more potent neurotrophic properties were replicated by hDPSC, when compared to hBMSC and hAMSC and confirm that, like rat DPSC, hDPSC secreted multiple neuroprotective and axogenic NTF that protected RGC from death and promoted the growth of their neurites to a significantly greater extent than either hAMSC or hBMSC, which both have distinct NTF profiles as determined by RT-PCR and ELISA. Thus, we conclude that hDPSC may constitute an effective paracrine-mediated cellular therapy for retinal, and potentially other CNS injuries [17].

To determine the neuroprotective and neuritogenic effects of hDPSC-, hBMSC- and hAMSC- derived NTF, we cultured these human-derived stem cells with RGC. Similar to our previous results using rat stem cells [17], hDPSC and hBMSC stimulated RGC survival and neuritogenesis to levels greater than those seen in untreated control retinal cultures. Moreover, hDPSC were more neuroprotective and neuritogenic than hBMSC (although the former measure did not reach statistical significance), which corroborates our previous findings [17] as well as those of others showing greater functional restoration when hDPSC, as opposed to hBMSC, were transplanted into spinal cord lesion sites [32]. The neuroprotective and neuritogenic properties of hAMSC were significantly less than hDPSC and this was correlated with lower secreted levels of NGF, BDNF, NT-3, VEGF and GDNF observed by ELISA.

We used Fc-NTFR fusion protein blockers to examine the mechanism of the hDPSC-, hBMSC- and hAMSC-mediated neuroprotection and neuritogenesis. The neuroprotective effect of hDPSC was significantly reduced after the addition of each individual Fc-NTFR, confirming the contribution of stem cell-derived NGF, BDNF, NT-3, GDNF, VEGF and PDGF-AA/AB/BB. Similar observations were seen with hBMSC, suggesting that the mechanisms for neuroprotection/neuritogenesis are similar and that the reduced neuroprotective effect of hBMSC compared to hDPSC is explained by the reduced neurotrophic profile. Moreover, hAMSC had a similar RGC protective/regenerative potency to hBMSC yet, owing to a lack of BDNF secretion, Fc-TrKB had no effect on hAMSC-mediated RGC survival or neuritogenesis. Interestingly, Fc-PDGFBr reduced hDPSC−/hBMSC−/hAMSC-mediated neuroprotection despite no PDGF-AB/BB being detected in conditioned media by ELISA. This might be explained by the previous observation that PDGF-AB/BB was secreted at very low levels by hBMSC (40-fold < PDGF-AA) [15] and therefore expected to fall below the detectable range for the ELISA used. Thus, the efficacy of Fc-PDGFBr could be attributable either to the potency of low levels of PDGF-AB/BB or to PDGF-AB/BB secreted by glia present in the retinal cultures in response to hDPSC−/hBMSC−/hAMSC-derived growth factor stimulation.

The secretion of multiple NTF by hDPSC confirms previous findings showing NTF gene expression by hDPSC [19,20,32] as well as NTF secretion by rat-derived DPSC [17,33] and neurotrophins by hDPSC [21]. In concert with our findings using

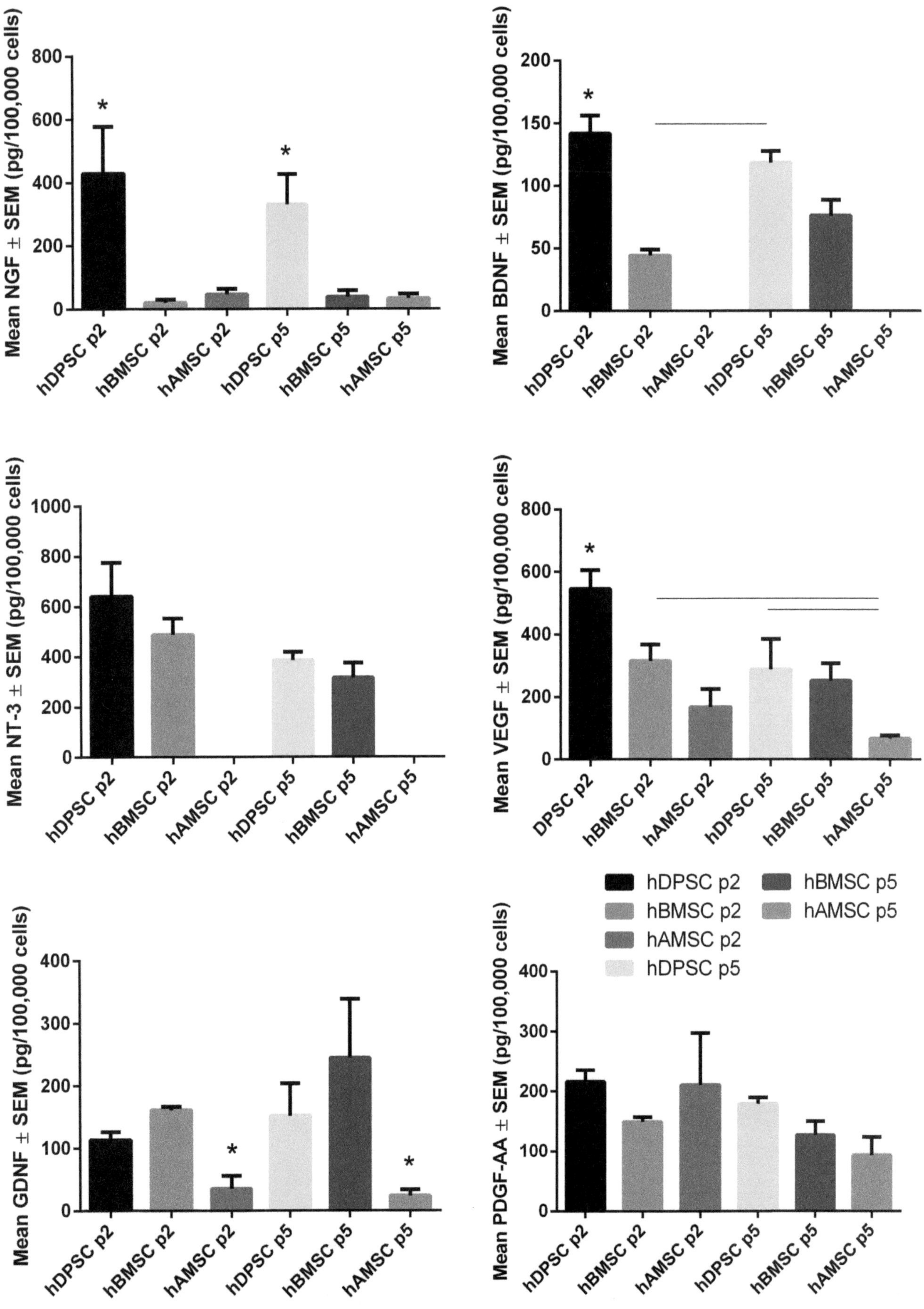
Mean NGF ± SEM (pg/100,000 cells)
0
200
400
600
800
*
*
hDPSC p2
hBMSC p2
hAMSC p2
hDPSC p5
hBMSC p5
hAMSC p5
Mean BDNF ± SEM (pg/100,000 cells)
0
50
100
150
200
*
hDPSC p2
hBMSC p2
hAMSC p2
hDPSC p5
hBMSC p5
hAMSC p5
Mean NT-3 ± SEM (pg/100,000 cells)
0
200
400
600
800
1000
hDPSC p2
hBMSC p2
hAMSC p2
hDPSC p5
hBMSC p5
hAMSC p5
Mean VEGF ± SEM (pg/100,000 cells)
0
200
400
600
800
*
DPSC p2
hBMSC p2
hAMSC p2
hDPSC p5
hBMSC p5
hAMSC p5
Mean GDNF ± SEM (pg/100,000 cells)
0
100
200
300
400
*
*
hDPSC p2
hBMSC p2
hAMSC p2
hDPSC p5
hBMSC p5
hAMSC p5
hDPSC p2
hBMSC p2
hAMSC p2
hDPSC p5
hBMSC p5
hAMSC p5
Mean PDGF-AA ± SEM (pg/100,000 cells)
0
100
200
300
400
hDPSC p2
hBMSC p2
hAMSC p2
hDPSC p5
hBMSC p5
hAMSC p5

Figure 4. NGF, BDNF, NT-3, CNTF, VEGF, GDNF, PDGF-AA and PDGF-AB/BB titres by ELISA in hDPSC, hBMSC and hAMSC. hDPSC-, hBMSC- and hAMSC-conditioned media (passage 2 and 5) collected after 48 h of cell culture. (n = 3; *asterisks* indicate significant differences from all other samples/*black lines* indicate significant difference between specific samples at $p<0.05$). CNTF and PDGF-AB/BB were undetectable in all samples.

rat DPSC [17], hDPSC secreted significantly more NGF and BDNF than hBMSC. Interestingly, compared to previous data on rat DPSC, hDPSC secreted higher titres of NGF but lower titres of BDNF [17]. Our results confirmed those in the literature on the extensive NTF secretory profile of hBMSC including NGF, BDNF, NT-3 and GDNF [17,34–36], adding VEGF to the list of known secreted factors but failed to detect CNTF [32]. hAMSC secrete NGF, BDNF, NT-3, GDNF and VEGF [22,23]. However in the present study, we have not detected the secretion of BDNF and NT-3, possibly because the titre was below the detectable range of the ELISA or that none was secreted in this study. Our results also support the findings of recent studies by others showing that hBMSC secrete PDGF-AA [15] and extend this observation to show that hDPSC and hAMSC share this property.

To further elucidate the relative neurotrophic activities of hDPSC, hBMSC and hAMSC, we conducted a RT-PCR array of 84 NTF/NTFR genes. The data showed that all three stem cell types have distinct NTF gene expression profiles. In particular, we found that hDPSC expressed prostaglandin E2 receptor (*ptger2*) at 6 and 10 fold higher than both hBMSC and hAMSC, respectively. Ptger2 stimulates the synthesis and release of neurotrophins from multiple cell types [37–39]. hDPSC also expressed over 100-fold lower interleukin-6 (*il6*) than hBMSC and hAMSC. The cytokine IL6 is neuroprotective after binding to the gp130 receptor [40]

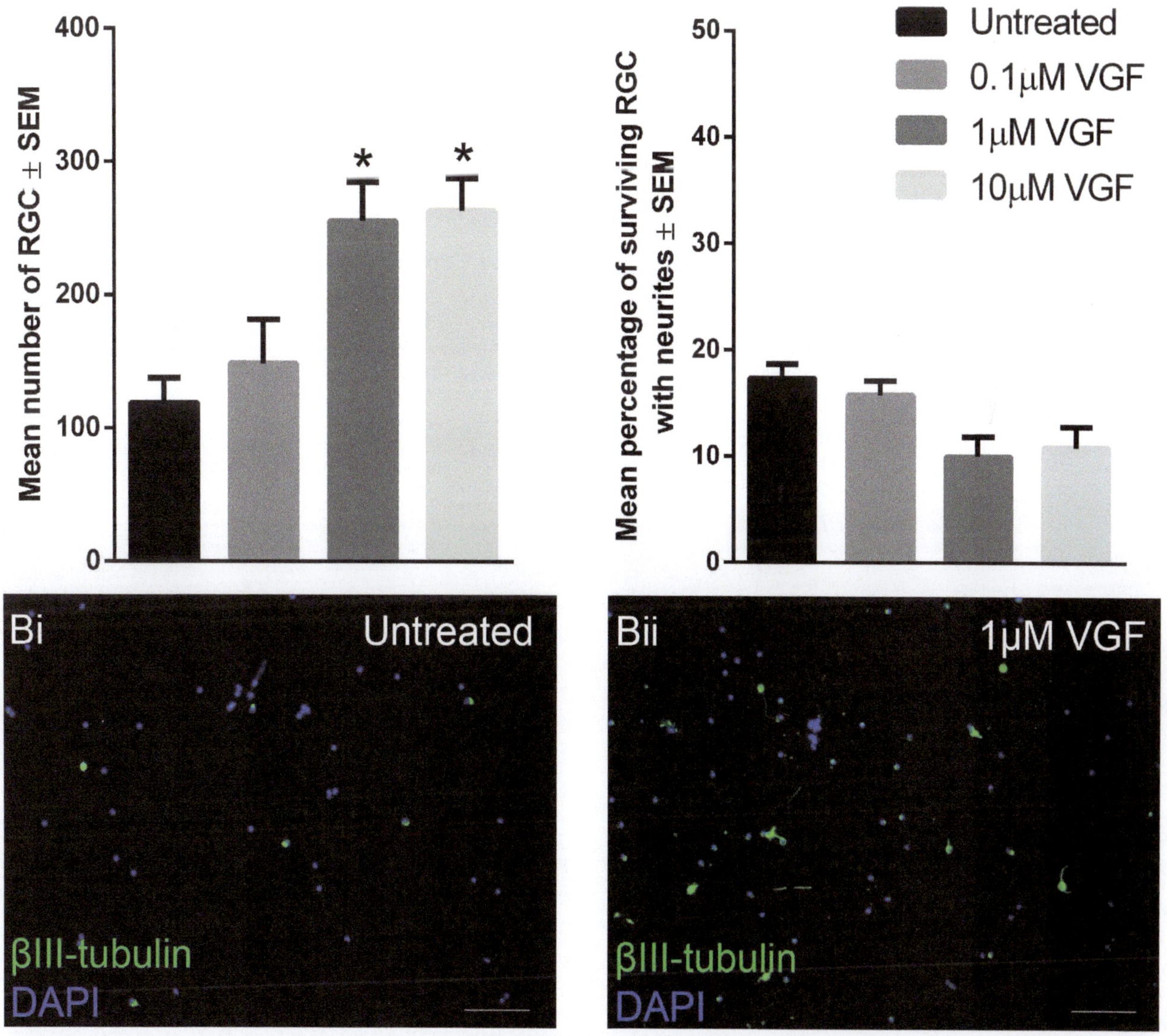

Figure 5. The effects of VGF on RGC in retinal cultures. The total number of surviving RGC as well as the percentage of RGC bearing neurites either cultured alone, or with 0.1 μM, 1 μM and 10 μM VGF (**A**). *Asterisks* indicate significant difference from other treatment groups at $p<0.05$. RGC cultured either alone (**Bi**) or with 1 μm VGF (**Bii**). All images are representative of the entire culture, nine separate culture wells/treatment with every three wells using different animals (*scale bars*: 100 μm), cell nuclei were stained with DAPI.

and promotes axon regeneration after activating the JAK/STAT3 pathway [41]. As a pro-inflammatory cytokine it is likely there are other IL6-mediated effects not fully realised in our *in vitro* paradigm. Finally, hBMSC express 6-fold higher fibroblast growth factor-9 (*fgf9*) than hDPSC, whereas hAMSC express over 700-fold higher *fgf9* than hDPSC. FGF9, unlike other FGF isoforms, stimulates the survival of RGC by binding to FGF receptor-3 [42], possibly highlighting a distinct mechanism for hAMSC-induced RGC neuroprotection. These data reinforce the notion that the stem cell origin is critical in determining their selection and application as cellular therapies for the treatment of particular neurological conditions.

NTF analyses by ELISA corresponded well with the RT-PCR microarray data, a point particularly well illustrated by the correlative BDNF and GDNF protein and mRNA data. Interestingly NGF values for protein and mRNA were paradoxical; demonstrating lower levels of NGF gene expression in hDPSC compared to hBMSC/hAMSC, while the titres of NGF protein in the conditioned medium from hDPSC cultures was significantly higher than that in the medium from hBMSC/hAMSC. These findings underline some discrepancies between gene expression and NTF protein secretion, which may be explained by differences in either the timing of sampling (PCR reflecting a snap-shot event while protein levels are cumulative) or an abundance of pre-existing stores of NGF in hDPSC.

The results presented here support the assertion that the RGC survival effects of the MSC are mediated in part by PDGF-AA [15]. Interestingly, inhibitors to the PDGFr did not significantly reduce RGC neuritogenesis, suggesting that MSC-derived PDGF is important for neuroprotection but not neuritogenesis in RGC.

RGC neurite outgrowth appeared to be particularly dependent on hDPSC-, hBMSC- and hAMSC-derived NGF. Thus, considering the enhanced secretion of NGF by hDPSC, may explain why hDPSC are more neuritogenic than hBMSC and hAMSC. BDNF and NT-3 are also neuritogenic for RGC, since neurite outgrowth was suppressed by the cognate Fc-TrK inhibitors. Noteworthy, hAMSC promoted very little neuritogenesis, although the response was significantly reduced when all three TrK receptors were simultaneously blocked.

Finally, this study is the first to identify VGF as a novel factor expressed at higher levels (>4 fold) in hDPSC than in hBMSC or hAMSC. VGF is a peptide present in the CNS and PNS [43] that protects motor neurons in animal models of amyotrophic lateral sclerosis [44] and increases the survival of cerebellar granule cells after serum deprivation [45]. We also showed that VGF was active in the retinal culture model at a concentration similar to that previously reported as active in other models of CNS injury [45]. In particular, VGF was significantly RGC neuroprotective, but not neuritogenic, suggesting that VGF may be involved in hDPSC-mediated RGC neuroprotection. Thus we propose that VGF may be a novel therapeutic NTF for RGC neuroprotection.

In conclusion, our results show that hDPSC is a more potent stem cell type for paracrine-mediated neuroprotection and regeneration of RGC than either hBMSC or hAMSC. The hDPSC-mediated neuroprotection and neuritogenesis is achieved through the paracrine effect of multiple secreted NTF, including PDGF (neuroprotection) and NGF (axon regeneration). Moreover, VGF is identified as a novel RGC neuroprotective factor expressed by hDPSC. hDPSC may represent an effective and advantageous cellular therapy for retinal nerve protection and repair.

Author Contributions

Conceived and designed the experiments: BM AL WL BAS. Performed the experiments: BM. Analyzed the data: BM. Contributed reagents/materials/analysis tools: BM AL MB WL BAS. Contributed to the writing of the manuscript: BM AL MB WL BAS.

References

1. Berkelaar M, Clarke DB, Wang YC, Bray GM, Aguayo AJ (1994) Axotomy results in delayed death and apoptosis of retinal ganglion cells in adult rats. J Neurosci 14: 4368–4374.
2. Berry M, Ahmed Z, Lorber B, Douglas M, Logan A (2008) Regeneration of axons in the visual system. Restorative Neurology and Neuroscience 26: 147–174.
3. Sarkies N (2004) Traumatic optic neuropathy. Eye 18: 1122–1125.
4. Quigley HA (1999) Neuronal death in glaucoma. Progress in Retinal and Eye Research 18: 39–57.
5. Quigley HA, McKinnon SJ, Zack DJ, Pease ME, Kerrigan-Baumrind LA, et al. (2000) Retrograde axonal transport of BDNF in retinal ganglion cells is blocked by acute IOP elevation in rats. Invest Ophthalmol Vis Sci 41: 3460–3466.
6. Dawbarn D, Allen SJ (2003) Neurotrophins and neurodegeneration. Neuropathol Appl Neurobiol 29: 211–230.
7. Logan A, Ahmed Z, Baird A, Gonzalez AM, Berry M (2006) Neurotrophic factor synergy is required for neuronal survival and disinhibited axon regeneration after CNS injury. Brain 129: 490–502.
8. Berry M, Carlile J, Hunter A (1996) Peripheral nerve explants grafted into the vitreous body of the eye promote the regeneration of retinal ganglion cell axons severed in the optic nerve. Journal of Neurocytology 25: 147–170.
9. Richardson PM, McGuinness UM, Aguayo AJ (1980) Axons from CNS neurons regenerate into PNS grafts. Nature 284: 264–265.
10. Ko ML, Hu DN, Ritch R, Sharma SC, Chen CF (2001) Patterns of retinal ganglion cell survival after brain-derived neurotrophic factor administration in hypertensive eyes of rats. Neurosci Lett 305: 139–142.
11. Ko ML, Hu DN, Ritch R, Sharma SC (2000) The combined effect of brain-derived neurotrophic factor and a free radical scavenger in experimental glaucoma. Invest Ophthalmol Vis Sci 41: 2967–2971.
12. Jiang LQ, Streilein JW (1991) Immune privilege extended to allogeneic tumor cells in the vitreous cavity. Invest Ophthalmol Vis Sci 32: 224–228.
13. Levkovitch-Verbin H, Sadan O, Vander S, Rosner M, Barhum Y, et al. (2010) Intravitreal injections of neurotrophic factors secreting mesenchymal stem cells are neuroprotective in rat eyes following optic nerve transection. Invest Ophthalmol Vis Sci 51: 6394–6400.
14. Johnson TV, Bull ND, Hunt DP, Marina N, Tomarev SI, et al. (2010) Neuroprotective Effects of Intravitreal Mesenchymal Stem Cell Transplantation in Experimental Glaucoma. Investigative Ophthalmology & Visual Science 51: 2051–2059.
15. Johnson TV, Dekorver NW, Levasseur VA, Osborne A, Tassoni A, et al. (2013) Identification of retinal ganglion cell neuroprotection conferred by platelet-derived growth factor through analysis of the mesenchymal stem cell secretome. Brain.
16. Johnson TV, Martin KR (2012) Cell transplantation approaches to retinal ganglion cell neuroprotection in glaucoma. Curr Opin Pharmacol.
17. Mead B, Logan A, Berry M, Leadbeater W, Scheven BA (2013) Intravitreally transplanted dental pulp stem cells promote neuroprotection and axon regeneration of retinal ganglion cells after optic nerve injury. Invest Ophthalmol Vis Sci 54: 7544–7556.
18. Mead B, Logan A, Berry M, Leadbeater W, Scheven BA (2014) Dental pulp stem cells, a paracrine-mediated therapy for the retina. Neural Regeneration Research 9: 577–578.
19. Nosrat CA, Fried K, Lindskog S, Olson L (1997) Cellular expression of neurotrophin mRNAs during tooth development. Cell and Tissue Research 290: 569–580.
20. Nosrat IV, Widenfalk J, Olson L, Nosrat CA (2001) Dental Pulp Cells Produce Neurotrophic Factors, Interact with Trigeminal Neurons in Vitro, and Rescue Motoneurons after Spinal Cord Injury. Developmental Biology 238: 120–132.
21. Martens W, Sanen K, Georgiou M, Struys T, Bronckaers A, et al. (2013) Human dental pulp stem cells can differentiate into Schwann cells and promote and guide neurite outgrowth in an aligned tissue-engineered collagen construct in vitro. The FASEB Journal.
22. Zhou Z, Chen Y, Zhang H, Min S, Yu B, et al. (2013) Comparison of mesenchymal stem cells from human bone marrow and adipose tissue for the treatment of spinal cord injury. Cytotherapy.
23. Kalbermatten DF, Schaakxs D, Kingham PJ, Wiberg M (2011) Neurotrophic activity of human adipose stem cells isolated from deep and superficial layers of abdominal fat. Cell and Tissue Research 344: 251–260.
24. Arboleda D, Forostyak S, Jendelova P, Marekova D, Amemori T, et al. (2011) Transplantation of Predifferentiated Adipose-Derived Stromal Cells for the Treatment of Spinal Cord Injury. Cellular and Molecular Neurobiology 31: 1113–1122.

25. Kang SK, Lee DH, Bae YC, Kim HK, Baik SY, et al. (2003) Improvement of neurological deficits by intracerebral transplantation of human adipose tissue-derived stromal cells after cerebral ischemia in rats. Experimental Neurology 183: 355–366.
26. Sugitani S, Tsuruma K, Ohno Y, Kuse Y, Yamauchi M, et al. (2013) The potential neuroprotective effect of human adipose stem cells conditioned medium against light-induced retinal damage. Experimental Eye Research 116: 254–264.
27. Tsuruma K, Yamauchi M, Sugitani S, Otsuka T, Ohno Y, et al. (2014) Progranulin, a Major Secreted Protein of Mouse Adipose-Derived Stem Cells, Inhibits Light-Induced Retinal Degeneration. Stem Cells Translational Medicine 3: 42–53.
28. Suggate EL, Ahmed Z, Read ML, Eaton-Charnock K, Douglas MR, et al. (2009) Optimisation of siRNA-mediated RhoA silencing in neuronal cultures. Molecular and Cellular Neuroscience 40: 451–462.
29. Douglas MR, Morrison KC, Jacques SJ, Leadbeater WE, Gonzalez AM, et al. (2009) Off-target effects of epidermal growth factor receptor antagonists mediate retinal ganglion cell disinhibited axon growth. Brain 132: 3102–3121.
30. Sullivan KF (1988) Structure and Utilization of Tubulin Isotypes. Annual Review of Cell Biology 4: 687–716.
31. Lorber B, Tassoni A, Bull ND, Moschos MM, Martin KR (2012) Retinal ganglion cell survival and axon regeneration in Wld(S) transgenic rats after optic nerve crush and lens injury. Bmc Neuroscience 13.
32. Sakai K, Yamamoto A, Matsubara K, Nakamura S, Naruse M, et al. (2012) Human dental pulp-derived stem cells promote locomotor recovery after complete transection of the rat spinal cord by multiple neuro-regenerative mechanisms. Journal of Clinical Investigation 122: 80–90.
33. Gale Z, Cooper PR, Scheven BAA (2011) Effects of Glial Cell Line-derived Neurotrophic Factor on Dental Pulp Cells. Journal of Dental Research 90: 1240–1245.
34. Dormady SP, Bashayan O, Dougherty R, Zhang XM, Basch RS (2001) Immortalized multipotential mesenchymal cells and the hematopoietic microenvironment. J Hematother Stem Cell Res 10: 125–140.
35. Wilkins A, Kemp K, Ginty M, Hares K, Mallam E, et al. (2009) Human bone marrow derived mesenchymal stem cells secrete brain-derived neurotrophic factor which promotes neuronal survival in vitro. Stem Cell Res 3: 63–70.
36. Chen Q, Long Y, Yuan X, Zou L, Sun J, et al. (2005) Protective effects of bone marrow stromal cell transplantation in injured rodent brain: synthesis of neurotrophic factors. J Neurosci Res 80: 611–619.
37. Cruz Duarte P, St-Jacques B, Ma W (2012) Prostaglandin E2 contributes to the synthesis of brain-derived neurotrophic factor in primary sensory neuron in ganglion explant cultures and in a neuropathic pain model. Exp Neurol 234: 466–481.
38. Kanda N, Koike S, Watanabe S (2005) Prostaglandin E2 enhances neurotrophin-4 production via EP3 receptor in human keratinocytes. J Pharmacol Exp Ther 315: 796–804.
39. Hutchinson AJ, Chou CL, Israel DD, Xu W, Regan JW (2009) Activation of EP2 prostanoid receptors in human glial cell lines stimulates the secretion of BDNF. Neurochem Int 54: 439–446.
40. Jung JE, Kim GS, Chan PH (2011) Neuroprotection by interleukin-6 is mediated by signal transducer and activator of transcription 3 and antioxidative signaling in ischemic stroke. Stroke 42: 3574–3579.
41. Leibinger M, Muller A, Gobrecht P, Diekmann H, Andreadaki A, et al. (2013) Interleukin-6 contributes to CNS axon regeneration upon inflammatory stimulation. Cell Death Dis 4: e609.
42. Kinkl N, Ruiz J, Vecino E, Frasson M, Sahel J, et al. (2003) Possible involvement of a fibroblast growth factor 9 (FGF9)-FGF receptor-3-mediated pathway in adult pig retinal ganglion cell survival in vitro. Mol Cell Neurosci 23: 39–53.
43. Salton SRJ, Ferri GL, Hahm S, Snyder SE, Wilson AJ, et al. (2000) VGF: A novel role for this neuronal and neuroendocrine polypeptide in the regulation of energy balance. Frontiers in Neuroendocrinology 21: 199–219.
44. Shimazawa M, Tanaka H, Ito Y, Morimoto N, Tsuruma K, et al. (2010) An Inducer of VGF Protects Cells against ER Stress-Induced Cell Death and Prolongs Survival in the Mutant SOD1 Animal Models of Familial ALS. Plos One 5.
45. Severini C, Ciotti MT, Biondini L, Quaresima S, Rinaldi AM, et al. (2008) TLQP-21, a neuroendocrine VGF-derived peptide, prevents cerebellar granule cells death induced by serum and potassium deprivation. Journal of Neurochemistry 104: 534–544.

21

A Novel *In Vitro* Method for Detecting Undifferentiated Human Pluripotent Stem Cells as Impurities in Cell Therapy Products Using a Highly Efficient Culture System

Keiko Tano[1,2], Satoshi Yasuda[2], Takuya Kuroda[2], Hirohisa Saito[3], Akihiro Umezawa[1], Yoji Sato[2,4,5,6]*

1 Department of Reproductive Biology, National Research Institute for Child Health and Development, Tokyo, Japan, **2** Division of Cellular & Gene Therapy Products, National Institute of Health Sciences, Tokyo, Japan, **3** National Research Institute for Child Health and Development, Tokyo, Japan, **4** Department of Quality Assurance Science for Pharmaceuticals, Graduate School of Pharmaceutical Sciences, Nagoya City University, Aichi, Japan, **5** Department of Cellular & Gene Therapy Products, Graduate School of Pharmaceutical Sciences, Osaka University, Osaka, Japan, **6** Department of Translational Pharmaceutical Sciences, Graduated School of Pharmaceutical Sciences, Kyushu University, Fukuoka, Japan

Abstract

Innovative applications of cell therapy products (CTPs) derived from human pluripotent stem cells (hPSCs) in regenerative medicine are currently being developed. The presence of residual undifferentiated hPSCs in CTPs is a quality concern associated with tumorigencity. However, no simple *in vitro* method for direct detection of undifferentiated hPSCs that contaminate CTPs has been developed. Here, we show a novel approach for direct and sensitive detection of a trace amount of undifferentiated human induced pluripotent stem cells (hiPSCs) using a highly efficient amplification method in combination with laminin-521 and Essential 8 medium. Essential 8 medium better facilitated the growth of hiPSCs dissociated into single cells on laminin-521 than in mTeSR1 medium. hiPSCs cultured on laminin-521 in Essential 8 medium were maintained in an undifferentiated state and they maintained the ability to differentiate into various cell types. Essential 8 medium allowed robust hiPSC proliferation plated on laminin-521 at low cell density, whereas mTeSR1 did not enhance the cell growth. The highly efficient culture system using laminin-521 and Essential 8 medium detected hiPSCs spiked into primary human mesenchymal stem cells (hMSCs) or human neurons at the ratio of 0.001%–0.01% as formed colonies. Moreover, this assay method was demonstrated to detect residual undifferentiated hiPSCs in cell preparations during the process of hMSC differentiation from hiPSCs. These results indicate that our highly efficient amplification system using a combination of laminin-521 and Essential 8 medium is able to detect a trace amount of undifferentiated hPSCs contained as impurities in CTPs and would contribute to quality assessment of hPSC-derived CTPs during the manufacturing process.

Editor: Graca Almeida-Porada, Wake Forest Institute for Regenerative Medicine, United States of America

Funding: This work was supported by Research Grants from the Japanese Ministry of Health, Labour and Welfare (H23-SAISEI-IPPAN-004, H23-SAISEI-IPPAN-005, H24-IYAKU-SHITEI-027, H25-JITSUYOKA(SAISEI)-IPPAN-008, and Marketing Authorization Facilitation Program for Innovative Therapeutic Products), http://www.mhlw.go.jp/english/policy/other/research-projects/index.html. The funders had no role in study design, data collection and analysis, decision to publish, or preparation of the manuscript.

Competing Interests: The authors have declared that no competing interests exist.

* Email: yoji@nihs.go.jp

Introduction

Cell therapy products (CTPs) are expected to offer promising treatments for serious and life-threating diseases for which no adequate therapy is currently available. An increasing number of CTPs derived from human pluripotent stem cells (hPSCs), i.e. induced pluripotent stem cells (hiPSCs) and embryonic stem cells (hESCs), are being developed for regenerative medicine/cell therapy because of their infinite self-renewal capacity and their ability to differentiate into various types of cells. Quality assessment of CTPs is critical to ensure their safety and efficacy for clinical application [1]. CTPs derived from hPSCs possibly include the cells of interest and also other cells such as undifferentiated cells, precursor cells and other differentiated cells. The presence of residual undifferentiated cells in CTPs derived from hPSCs is one of the most serious concerns for tumorigenicity because the undifferentiated hPSCs have a capacity to form teratoma in animals [1–4]. Hentze et al. previously reported that hundreds of undifferentiated hESCs were enough to produce a teratoma in immunodeficient SCID mice [5]. We cannot exclude the possibility that a trace amount of residual undifferentiated hPSCs in CTPs cause ectopic tissue formation, tumor development and/or malignant transformation after transplantation. Therefore, establishment of a detection method for residual undifferentiated cells is necessary for the safety and quality assessment of CTPs derived from hPSCs.

An *in vivo* teratoma formation assay is the only method to directly assess tumorigenicity of undifferentiated cells, but this assay is costly and time-consuming [2,3]. Several *in vitro* methods, such as flow cytometry and quantitative real-time PCR (qRT-PCR) analysis, can also detect residual undifferentiated hPSCs in CTPs [2,3]. Our previous report has shown that flow cytometry using anti-TRA-1-60 antibody and qRT-PCR using a specific probe and primers for *LIN28* mRNA can detect as low as 0.1% and 0.002% undifferentiated hiPSCs spiked into retinal pigment epithelial (RPE) cells, respectively [3]. However, both of these methods have the disadvantage of detecting undifferentiated cell marker expression but not functionally undifferentiated cells *per se*. The soft agar colony formation assay is commonly used to detect tumorigenic cells with a property of anchorage-independent growth. However, this assay is not appropriate for the detection of hPSCs because they undergo apoptosis associated with dissociation into single cells [3,6]. At present, there is no simple method to directly detect a trace amount of hPSCs *in vitro*.

Recently, some cell culture matrices have been reported to sustain self-renewal of dissociated hPSCs without apoptosis [7,8]. We focused on a culture system enabling hPSC cell growth without apoptosis and developed a direct *in vitro* method for detecting a trace amount of undifferentiated hPSCs in CTPs. Laminin-521, a laminin isoform that is normally expressed in hESCs, is known to stimulate robust hPSC proliferation in an undifferentiated state in combination with mTeSR1 medium [7]. In the present study, we present a novel approach to detect undifferentiated hiPSCs contaminating CTPs through efficient amplification using a laminin-521-based cell culture system with Essential 8 medium [9] instead of mTeSR1 medium.

Materials and Methods

Cell culture

The hiPSC lines, 201B7, 253G1 and 409B2, were provided by the RIKEN BRC through the Project for Realization of Regenerative Medicine and the National Bio-Resource Project of the MEXT, Japan [17–19]. hiPSCs were first cultured on mitomycin C-treated SNL cells (a mouse fibroblast STO cell line expressing a neomycin-resistance gene cassette and LIF) in primate ES cell medium (ReproCell, Kanagawa, Japan) supplemented with 4 ng/ml human basic fibroblast growth factor (bFGF; R&D Systems, Inc., Minneapolis, USA). hiPSC colonies were passaged as small clumps once every 5–6 days using CTK solution (ReproCell) and STEMPRO EZPassage (Invitrogen, Carlsbad, CA, USA). hiPSCs were then passaged onto Matrigel-coated dishes with mTeSR1 (Stem Cell Technologies, Vancouver, CAN) for at least 2 passages before plating on laminin-521 or directly subcultured onto laminin-521-coated dishes. Subculture on laminin-521-coated dishes was performed as follows: near-confluent cells were treated with 0.5 mM EDTA/D-PBS for 6–7 minutes at 37°C. Cells were pipetted to achieve single-cell suspension and centrifuged at 30× *g* for 4 minutes. After centrifugation, the cell pellet was suspended in Essential 8 medium (Life Technologies, USA) and seeded at $2-3\times10^4$ cells/cm^2 on laminin-521-coated dishes. Cells were grown in Essential 8 medium at 37°C in a 5% CO_2 atmosphere and passaged once in 3–4 days. Primary human mesenchymal stem cells (hMSCs) were purchased from Lonza and cultured in MSCGM medium (Lonza, Walkersville, MO, USA). hMSCs at passage 7 were used in this study. Primary human neurons were purchased from ScienCell Research Laboratories (Carlsbad, CA, USA).

Cell proliferation assay

hiPSCs were dissociated into single cells and seeded on matrix-coated plates at a density of 3×10^4 cells/cm^2 or at the indicated density as described below. Tissue culture plates (BD Falcon, NJ, USA) were coated with laminin-521 (BioLamina, Sundbyberg, Sweden) dissolved in D-PBS at 4 μg/cm^2 at 37°C for 2 h. Control plates were coated with Matrigel (BD Biosciences, MA, USA) at 16 μg/cm^2. Viable cells were quantified every 24 h using CyQUANT Cell Proliferation Assay Kit (Life Technologies) according to the manufacturer's instructions.

Quantitative RT-PCR

Total RNA was treated with DNase I and isolated using RNeasy Mini Kit (Qiagen Hilden, Germany) according to the manufacturer's instructions. Quantitative RT-PCR was performed using the QuantiTect Probe one-step RT-PCR Kit (Qiagen) on StepOnePlus Real Time PCR system (Life Technologies). Gene expression levels were normalized to *GAPDH* expression levels, which were quantified using TaqMan human GAPDH control reagents (Life Technologies). Primers and probes were obtained from Sigma-Aldrich. The sequences of primers and probes are listed in Table S1.

Teratoma assay

Teratoma formation experiments were performed by injecting 253G1 cells (1×10^6 cells/testis) that were cultured with laminin-521 and Essential 8 medium (passage 36), into the testes of severe combined immunodeficiency (SCID) mice at the age of 8 weeks under pentobarbital anesthesia. The mice were sacrificed with an overdose of pentobarbital 10 weeks after the transplant, and the isolated teratoma was fixed in 10% formalin. The paraffin-embedded section was stained with hematoxylin and eosin (HE). Animal experiments were performed at UNITECH Co., Ltd. (Chiba, Japan) in accordance with the animal ethical committee's approval (Permit Number: KIS-130712i-20 at UNITECH Co., Ltd. and 444 at NIHS).

Differentiation assay

Differentiation of hiPSCs into three germ layers was performed as follows: 253G1 cells were plated on laminin-521 at a density of 3×10^4 cells/cm^2 in Essential 8 medium and expanded until they were nearly confluent. A) Ectoderm lineage differentiation: Neural hiPSCs differentiation was performed according to the previously reported protocol with some modifications [10]. Briefly, culture medium was changed from Essential 8 to DMEM/F12 medium containing 20% knockout serum replacement (KSR, Life Technologies), 10 μM SB431542 (Sigma-Aldrich) and 500 ng/ml Noggin (R&D systems). After 4 days of differentiation, SB431542 was withdrawn and increasing amounts of N2 medium (25%, 50%, or 75%) was added to the KSR medium every 2 days. From day 10 of differentiation, the medium was changed to N2B27 medium without bFGF containing 500 ng/ml Noggin, and cells were cultured for 15 days. B) Mesoderm lineage differentiation: hiPSCs were cultured for 15 days in DMEM/F12 containing 10% FBS, 2 mM L-glutamine 1% nonessential amino acids (Life Technologies) and 0.1 mM β-mercaptoethanol [8]. C) Endoderm lineage differentiation: hepatic differentiation of hiPSCs was performed according to the previously reported protocol with some modifications [11]. On the first day of differentiation, the medium was replaced with RPMI1640 (Sigma-Aldrich) containing B27 supplement (Life Technologies), 100 ng/ml activin A (R&D systems), 50 ng/ml Wnt3a (R&D systems) and 1 mM sodium butyrate (NaB) (Sigma-Aldrich). On the following 2 days, NaB was omitted from the medium. After 3 days of differentiation, the medium was replaced with knockout-DMEM containing 20% KSR, 1 mM L-glutamine, 1% nonessential amino acids, and 1% DMSO for 5 days.

Differentiation of 253G1 cells into MSCs was performed according to the previously reported protocol with some modifications [14]. On the first day of differentiation, 253G1 cells subcultured on laminin-521 in Essential 8 medium were dissociated into single cells and suspended in EB formation medium (AggreWell Medium, Stem Cell Technologies) with 10 μM Y-27632 (Wako, Japan), a ROCK inhibitor for generation of embryoid bodies (EBs). Cells (1×10^6) were then added to a well of the AggreWell Plate and incubated for 24 h at 37°C in a 5% CO_2 atmosphere. After 24 h, EBs were plated on 35-mm dishes (BD) in Stemline II (Sigma-Aldrich) supplemented with 50 ng/ml BMP4 (R&D systems) and 50 ng/ml VEGF (R&D systems) for 2 days. Medium was changed to Stemline II containing BMP4 (50 ng/ml), VEGF (50 ng/ml) and bFGF (22.5 ng/ml), and the EBs were cultured for 2 days. EBs were dissociated into single cells and replated in Methocult H4536 (Stem Cell Technologies) containing Growth Enhancement Media Supplement (EX-CYTE, Millipore), 50 ng/ml VEGF, 50 ng/ml Flt3-ligand (R&D systems), 50 ng/ml thrombopoietin (R&D systems) and 30 ng/ml bFGF for hemangioblast formation. After 8 days, cultures were harvested and plated as defined passage 0 in MSC growth medium (αMEM+ 20% FBS) on Matrigel.

Immunofluorescence staining

Immunofluorescence staining was performed as follows: cells were fixed with 4% paraformaldehyde in PBS for 15 minutes. After washing three times with PBS, cells were permeabilized with 0.1% Triton X-100 in PBS for 10 minutes, and then blocked with Blocking One (nacalai tesque, Kyoto, Japan) at 4°C over night. Cells were incubated with primary antibody against α-fetoprotein (AFP, 1:400; Dako) for 30 minutes at room temperature, with antibody against smooth muscle actin (SMA, 1:400; Sigma-Aldrich), βIII tubulin (0.5 μg/ml; abcam) or TRA-1-60 (1:200; Millipore) for 1 hour at room temperature, or with antibody against CD105 (1:200; abcam) at 4°C over night. After washing with PBS three times, cells were incubated with secondary antibody conjugated with Alexa Fluor 488 (Invitrogen) for 30 minutes at RT. VECTASHIELD mounting medium with DAPI (VECTOR) was used for nuclear staining. The samples were examined using an Olympus IX71 microscope equipped with cellSens Standard software (Olympus).

Embryoid body formation

Embryoid bodies were generated from hiPSCs using AggreWell 800 plates (Stem Cell Technologies) according to the manufacturer's instructions, with some modifications. hiPSCs dissociated with 0.5 mM EDTA were collected and suspended in EB formation medium (AggreWell Medium, Stem Cell Technologies) supplemented with 10 μM Y-27632 (Wako). The cells were added to each well (5×10^5 cells/well) in the AggreWell plate and incubated for 24 h at 37°C in a 5% CO_2 atmosphere. EBs were harvested from AggreWell plate and cultured in 35-mm dishes (BD) with primate ES cell culture medium (ReproCell) without bFGF. The medium was changed every 3 days. After 10 days of incubation, total RNA was isolated from EBs. The expression levels of each differentiation marker were determined using quantitative RT-PCR, as described above.

Statistics

Statistical analysis was performed using SigmaPlot 12.5 Software (Systat Software Inc., CA). The data were analyzed using two-way ANOVA or two-way repeated-measures ANOVA followed by a Bonferroni t-test as a post hoc test. A probability below 0.05 was considered significant.

Results

Essential 8 medium promotes hiPSCs cell growth on laminin-521

hPSCs are known to easily undergo apoptosis induced by dissociation [6]. To achieve an efficient hPSC cell growth, we determined the optimal culture conditions that allow robust proliferation of hiPSCs dissociated into single cells. Here, we focused on laminin-521 as a cell culture matrix, which permits survival of dissociated hPSCs without a Rho-associated protein kinase (ROCK) inhibitor [7]. mTeSR1 medium is conventionally used to culture dissociated hiPSCs on dishes coated with laminin-521 [7], and other hPSC media besides mTeSR1 have not been fully characterized with laminin-521. To examine effects of medium on hPSC cell growth on laminin-521, we compared the hiPSC growth rate using conventional mTeSR1 medium with Essential 8 medium, with optimized components [9]. After subculture on Matrigel, 253G1 hiPSCs were dissociated into single cells and seeded on laminin-521-coated plates at a density of 3×10^4 cells/cm^2. No difference was observed between mTeSR1 and Essential 8 medium in the cell number of hiPSCs cultured on laminin-521 at 24 h after plating (Figure 1A and 1B). However, the cells cultured on laminin-521 in Essential 8 medium showed rapid expansion compared to those in the mTeSR1 medium, and they reached nearly confluent at 72 h after plating (Figure 1A). Cell growth quantification revealed that the number of cells in Essential 8 medium was 3-fold higher than those in mTeSR1 medium at 72 h after plating (Figure 1B). Similar results were also obtained using another hiPSC line, 201B7 (Figure S1A). These results suggest that Essential 8 medium promotes hiPSC proliferation more rapidly than mTeSR1 medium when grown on laminin-521. When dissociated hiPSCs were cultured on Matrigel, Essential 8 medium did not significantly promote cell proliferation compared with mTeSR1 medium (Figure 1B and Figure S1A). We found that Essential 8 medium enhanced the hiPSC growth rate even on LM511-E8, which is a fragment of laminin 511 [8], but this effect was weaker than that on laminin-521 (Figure S1B). Taken together, these results suggest that a combination of laminin-521 and Essential 8 medium is a potent cell culture system for efficient amplification of dissociated hiPSCs *in vitro* compared to other culture systems. Therefore, we decided to develop a novel method for detecting undifferentiated hPSCs using a combination of laminin-521 and Essential 8 medium.

Culture system using laminin-521 and Essential 8 medium maintains the undifferentiated state and pluripotency of hiPSCs

We next examined whether a culture system using laminin-521 and Essential 8 medium maintains undifferentiated states through serial passages. 253G1 cells were dissociated into single cells and sequentially subcultured on laminin-521 in Essential 8 medium. 253G1 cells exhibited vigorous proliferation for more than 30 passages under these conditions (data not shown). Quantitative RT-PCR analysis revealed that the expression levels of undifferentiated hPSC markers (*OCT3/4*, *NANOG*, *SOX2* and *LIN28*) in 253G1 cells cultured with laminin-521 and Essential 8 medium were similar to those cultured with Matrigel and mTeSR1 medium (Figure 2A). Moreover, the serial passages with laminin-521 and Essential 8 medium did not have any effect on expression of the undifferentiated markers. We also examined the effects of subculture on laminin-521 in Essential 8 medium using other hiPSC lines, 201B7 and 409B2, and showed that these cells expressed undifferentiated markers through serial passages (Figure S2A-B). We next tested the pluripotency of hiPSCs cultured on laminin-521 in Essential 8 medium using lineage-specific

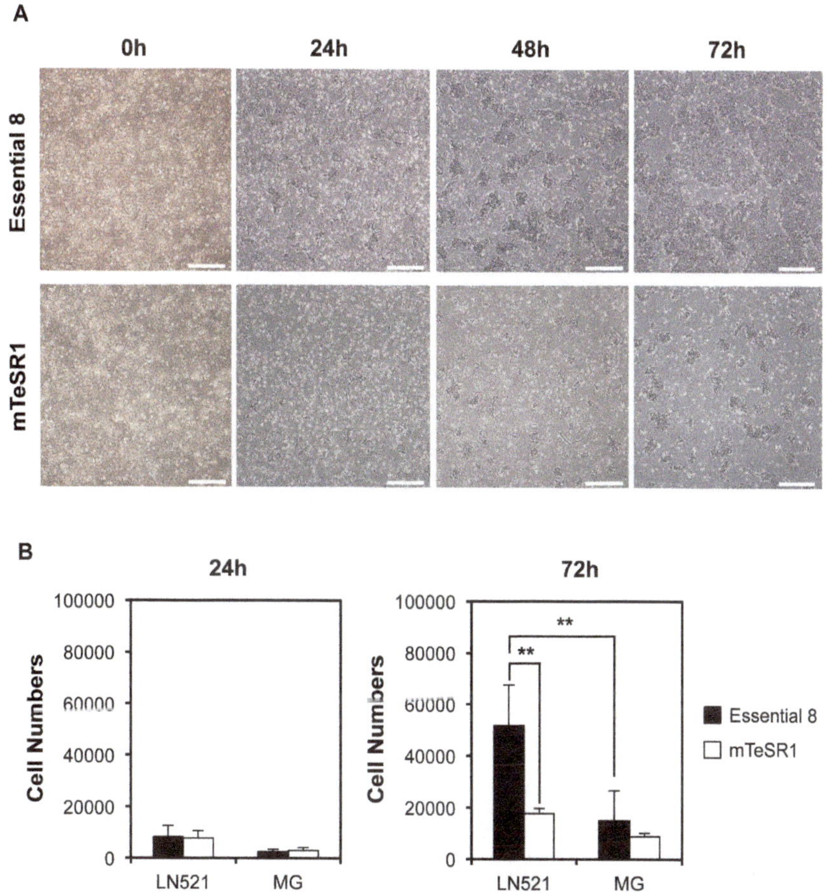

Figure 1. Robust proliferation of 253G1 cells cultured on laminin-521 in Essential 8 medium. (A) Morphology of the 253G1 cells expanded on laminin-521 in Essential 8 or mTeSR1 medium after dissociation into single cells. Scale bars, 500 μm. (B) Quantification of the number of dissociated 253G1 cells expanded on laminin-521 or Matrigel in Essential 8 or mTeSR1 medium. Data are presented as the mean ± standard deviation (SD) of three independent experiments (**$P<0.01$, two-way ANOVA followed by Bonferroni t-test as post-hoc test). LN521, laminin-521; MG, Matrigel.

differentiation protocols. Immunofluorescence analysis using antibodies specific for markers of three germ layers clearly demonstrated that 253G1 cells subcultured with laminin-521 and Essential 8 medium could selectively differentiate into endoderm, mesoderm and ectoderm expressing AFP, α-SMA and βIII tubulin, respectively (Figure 2B). To further examine pluripotency of hiPSCs subcultured with laminin-521 and Essential 8 medium, we facilitated spontaneous differentiation of cells grown as aggregates (embryoid bodies). Embryoid bodies derived from 253G1 cells increased gene expression of differentiated markers for all three germ layer lineages (Figure 2C), consistent with the observation obtained using 201B7 and 409B2 cells (Figure S2C-D). We also confirmed that 253G1 cells cultured with laminin-521 and Essential 8 medium were engrafted in testes of SCID mice and formed teratomas that involved all three germ layers (Figure 2D). These results strongly suggest that a culture system using laminin-521 and Essential 8 medium supports the undifferentiated state and pluripotency of hiPSCs through serial passages.

Essential 8 medium enables hiPSCs to proliferate rapidly from low cell density on laminin-521

Higher sensitivity is expected to ensure the accuracy and reliability of the detection of trace amounts of undifferentiated cells. To achieve higher sensitivity, culture system using laminin-521 and Essential 8 should have a capacity to rapidly expand hPSCs even at a low cell density. Therefore, we next tested whether Essential 8 medium promotes expansion of the hiPSCs on laminin-521 plated at a low cell density. 253G1 cells were seeded into laminin-521-coated plates at a density of 3.2×10^4 cells/cm^2, 1.6×10^4 cells/cm^2 and 8.0×10^3 cells/cm^2, and grown until nearly confluent. Cells grown in mTeSR1 reduced proliferative capacity as seeding density became lower. Conversely, cells cultured in Essential 8 medium showed robust propagation over a prolonged period of time even when they were seeded at low cell density (Figure 3A-C). Similarly, 201B7 and 409B2 cells seeded at lower density in Essential 8 medium also showed robust proliferation compared to the cells in mTeSR1 (Figure 3D-I). Essential 8 medium also promoted cell growth when hiPSCs were plated at lower density of 800 cells/cm^2 (Figure S3). These results indicate that Essential 8 medium promotes expansion of hiPSCs plated on laminin-521 at a low cell density. Thus, a culture system using laminin-521 and Essential 8 medium is considered to be well suited for direct detection of trace amounts of undifferentiated cells.

Culture system using laminin-521 and Essential 8 medium is useful for direct detection of undifferentiated hiPSCs contained in somatic cells

Residual undifferentiated cells contaminating hPSC-based CTPs are a quality concern associated with tumorigencity [1–4].

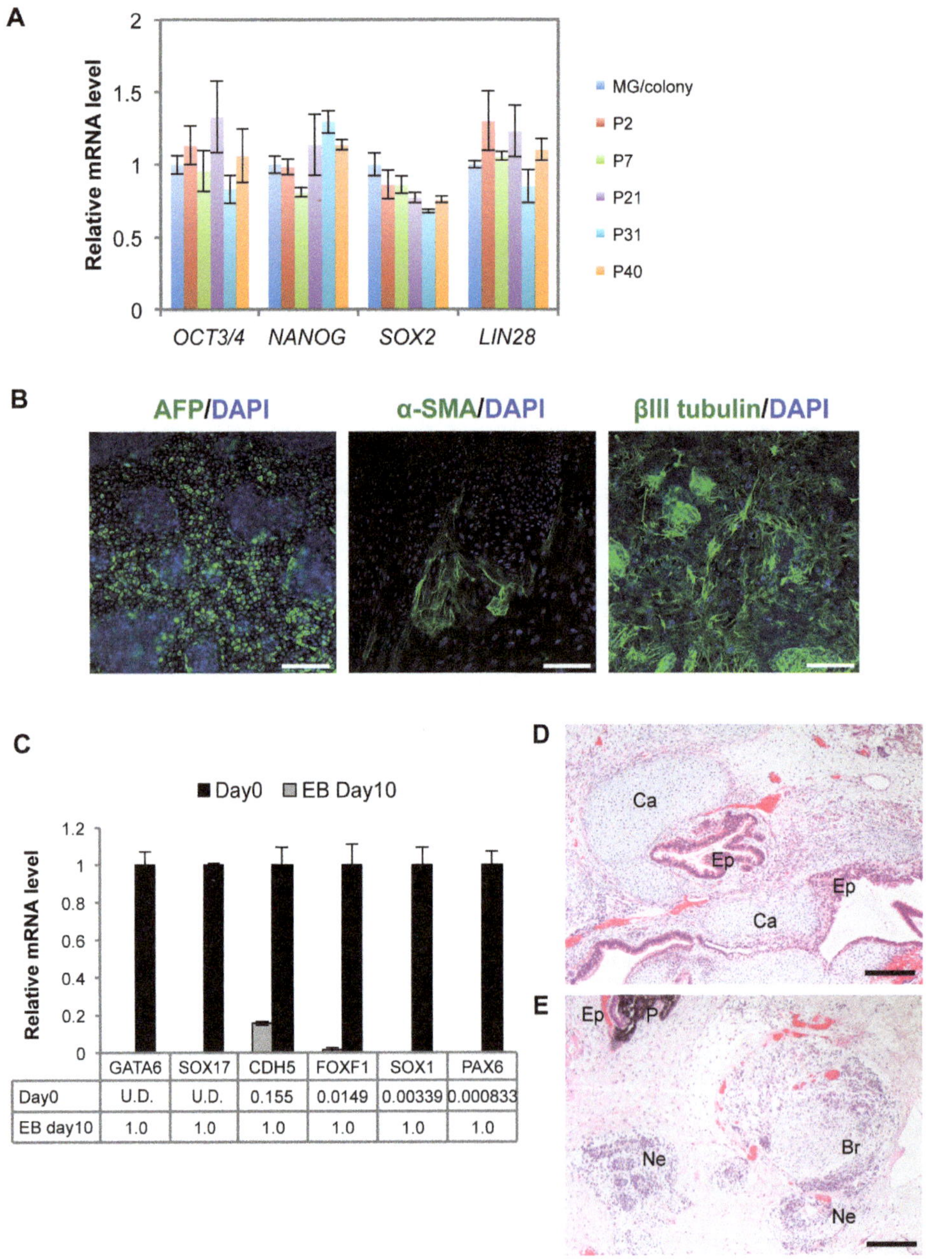

Figure 2. Characterization of 253G1 cells subcultured on laminin-521 in Essential 8 medium. (A) Expression levels of undifferentiated cell markers (*OCT3/4, NANOG, SOX2 and LIN28*) in 253G1 cells subcultured on laminin-521 in Essential 8 were determined using qRT-PCR. Relative mRNA expression levels are presented as ratios to the level of that in control cells on Matrigel. Results are the mean ± SD (n = 3). (B) *In vitro* differentiation analysis of 253G1 cells subcultured on laminin-521 in Essential 8 medium. Immunostaining of the markers for three germ layers are shown: endoderm (alpha-fetoprotein (AFP)), mesoderm (α-smooth muscle actin (SMA)) and ectoderm (βIII tubulin). Scale bars, 200 μm. (C) Expression levels of differentiated cell markers in embryoid bodies (EBs) derived from 253G1 cells: endoderm (*GATA6, SOX17*), mesoderm (*CDH5, FOXF1*), ectoderm (*SOX1, PAX6*). Relative mRNA expression levels are presented as ratios to the level of that in control cells (EBs at Day 10). Results are the mean ± SD (n = 3). (D-E) Teratomas derived from 253G1 cells cultured on laminin-521 in Essential 8 medium are shown. Hematoxylin and eosin staining showed the features of three germ layers: Ep, epithelium-like tissue (endoderm); Ca, cartilage (mesoderm); Ne, neural rosette-like tissue (ectoderm); P, pigmented neuroectodermal resembling meranocyte (ectoderm); Br, brain-like tissue (ectoderm). Scale bars, 200 μm.

To determine whether a culture system using laminin-521 and Essential 8 medium can detect a trace amount of undifferentiated hiPSCs in CTPs, we spiked dissociated hiPSCs into primary human somatic cells and cultured these cells on laminin-521 in Essential 8 medium. As a model of the somatic cells, we employed human mesenchymal stem cells (hMSCs), because "off-the-shelf" hMSCs derived from hPSCs are a promising CTP [12–14]. We spiked 409B2 cells (1%, 1000 cells; 0.1%, 100 cells; 0.01%, 10 cells) into 1×10^5 hMSCs and plated these cells onto laminin-521-coated wells. hiPSCs were co-cultured with hMSCs on laminin-521-coated dishes in Essential 8 medium and formed distinctive colonies (Figure 4A). At day 7 after plating, we detected 362, 42.5 and 6 colonies (the mean of duplicate measurements) in 1%, 0.1% and 0.01% spiked samples, respectively (Figure 4B). We did not

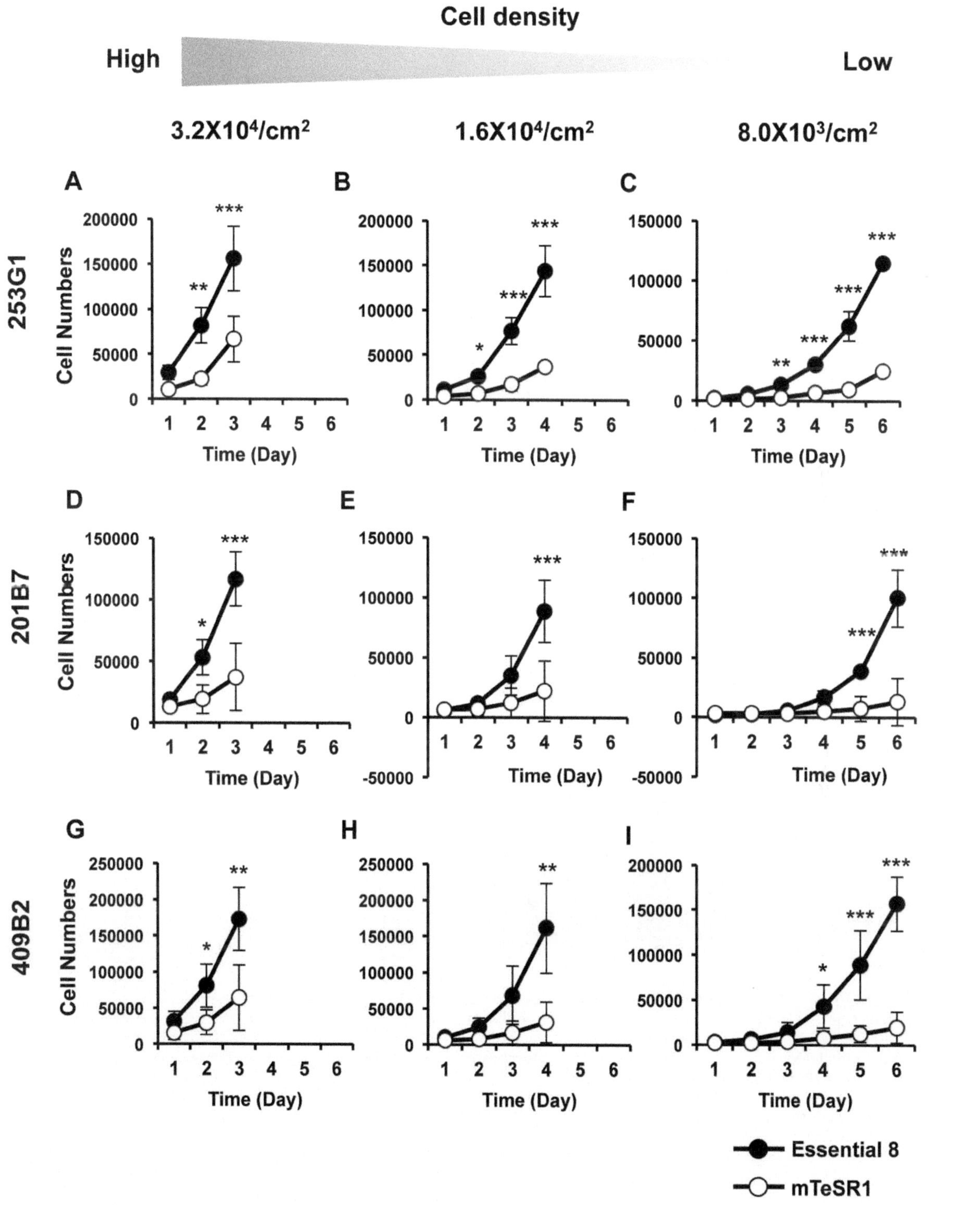

Figure 3. Rapid cell proliferation of hiPSCs plated at low cell density on laminin-521 in Essential 8 medium. (A-I) Quantification of the number of 253G1, 201B7 and 409B2 cells expanded on laminin-521 in Essential 8 or mTeSR1 medium. Cell numbers were counted every 24 h after plating at 3.2×10^4 cells/cm^2 (A, D, G), 1.6×10^4 cells/cm^2 (B, E, H) and 8.0×10^3 cells/cm^2 (C, F, I), respectively. Data are presented as the mean ± standard deviation (SD) of three independent experiments (*$P<0.05$, **$P<0.01$, *** $P<0.001$, two-way repeated-measures ANOVA followed by a Bonferroni post-hoc test).

find any colonies when only hMSCs were cultured on laminin-521 in Essential 8 medium. In addition, immunofluorescence staining with anti-TRA-1-60 antibody showed that these colonies formed in an undifferentiated state (Figure 4A), suggesting that colonies derived from hiPSCs were formed in an hMSC monolayer under conditions with laminin-521 and Essential 8 medium. We also

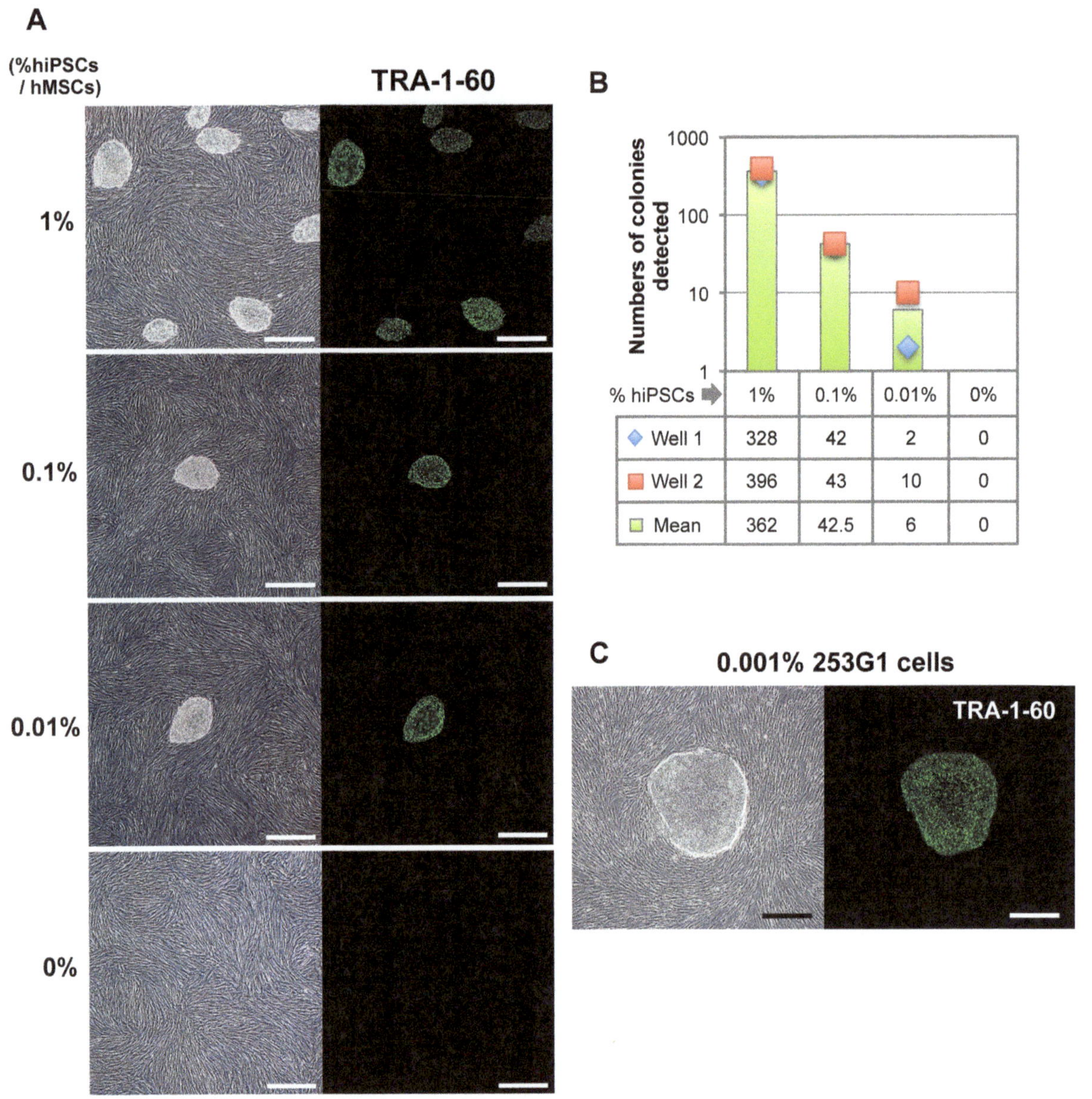

% hiPSCs	1%	0.1%	0.01%	0%
Well 1	328	42	2	0
Well 2	396	43	10	0
Mean	362	42.5	6	0

Figure 4. Detection of hiPSCs spiked into hMSCs on the culture system using laminin-521 and Essential 8 medium. (A) Morphologies of forming colonies derived from 409B2 cells spiked into hMSCs are shown (images in the left). 409B2 cells (1%, 1000 cells; 0.1%, 100 cells; 0.01%, 10 cells; 0%, 0 cells) were spiked into hMSCs (100,000 cells) and co-cultured on laminin-521-coated wells in 6-well plates in Essential 8 medium for 7 days. Expression of the undifferentiated marker, TRA-1-60, in these colonies was assessed using immunofluorescence staining (images in the right). Each experiment was carried out in duplicate. Scale bars, 500 μm. (B) Numbers of the colonies detected in each spiked sample in (A) are shown. Data are present as raw data in each well (shown by plots) or the mean of well 1 and well 2 (shown by bar graphs). (C) Morphology of a forming colony derived from 253G1 cells spiked into hMSCs at the ratio of 0.001% (6 hiPSCs to 600,000 hMSCs) is shown (images in the left). Mixture of those cells was co-cultured on a 100-mm cell culture dish coated with laminin-521 in Essential 8 medium for 9 days. Forming colony was stained with anti TRA-1-60 antibody (images in the right). Experiment was carried out in duplicate. Scale bars, 500 μm.

tested another hiPSC line, 253G1, for undifferentiated cells spiked into hMSCs. We found that 253G1 cells spiked into hMSCs at the ratio of 1% and 0.1% formed approximately 100 and 20 colonies, respectively, on laminin-521 in Essential 8 medium (Figure S4). We detected one colony when 253G1 cells were spiked into hMSCs at a ratio of 0.01% or 0.001% and co-cultured on a laminin-521-coated dish in Essential 8 medium (Figure S4 and Figure 4C). Taken together, our culture system using laminin-521 and Essential 8 medium allows the direct detection of 0.001%–0.01% hiPSCs in hMSCs as a result of efficient cell amplification.

We also confirmed that no colonies were detected when a mixture of hiPSCs and hMSCs were cultured on laminin-521 in MSCGM medium instead of Essential 8 medium. In the absence of laminin-521, several colonies were detected in Essential 8 when hMSCs contained 1% hiPSCs but not when hMSCs contained 0.1% and 0.01% hiPSCs (data not shown). These results suggest that laminin-521 is required to detect trace amounts of hiPSCs in hMSCs (less than 0.1%).

To know whether this culture system also works in detecting trace amounts of hiPSCs contaminating other types of cells besides

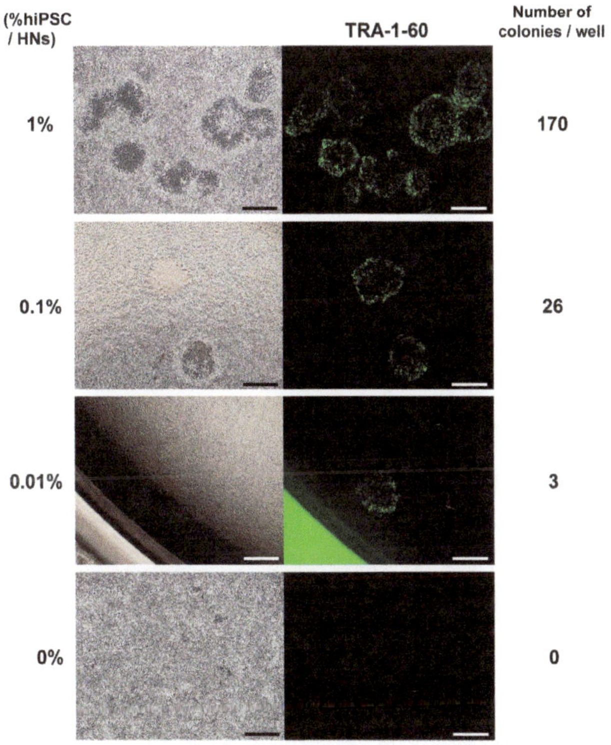

Figure 5. Detection of hiPSCs spiked into human neurons on the culture system using laminin-521 and Essential 8 medium. Morphologies of forming colonies derived from 253G1 cells spiked into human neurons are shown (images in the left). 253G1 cells (1%, 1000 cells; 0.1%, 100 cells; 0.01%, 10 cells; 0%, 0 cells) were spiked into human neurons (100,000 cells) and co-cultured on laminin-521-coated wells in 12-well plates in Essential 8 medium for 6 days. Forming colonies were stained with anti TRA-1-60 antibody (images in the right). HNs, human neurons. Scale bars, 500 μm.

hMSCs, we next tested colony formation of hiPSCs spiked into primary human neurons. Spiked 253G1 cells were co-cultured with human neurons on laminin-521 in Essential 8 medium and clearly formed colonies (Figure 5), which is consistent with the observation using hiPSCs spiked into hMSCs. We detected 170, 26 and 3 colonies that were positive for TRA-1-60 when 253G1 cells were spiked into 1×10^5 human neurons at the ratio of 1, 0.1 and 0.01%, respectively. There was no colony when only human neurons were cultured on our system. These results suggest that this culture system is also useful for detection of trace amounts of hiPSCs not only in hMSCs but also in other types of cells such as human neurons. We also confirmed that no colonies were formed on the well that was not coated with laminin-521 even when human neurons containing 10% hiPSCs were plated (data not shown), indicating that formation of the colonies derived from hiPSCs in human neurons is dependent on laminin-521.

Culture system using laminin-521 and Essential 8 medium has a capacity for direct detection of residual undifferentiated cells contained in differentiating hiPSC cultures

Finally, we examined whether this culture system using laminin-521 and Essential 8 medium is applicable in direct detection of residual hiPSCs contained in differentiated cells derived from hiPSCs. We attempted to differentiate 253G1 cells into MSCs as described in Materials and Methods (Figure 6A). Using this protocol, we observed attached cells with fibroblast-like morphology at the stage of passage 0 MSCs. We confirmed that approximately 20% of these attached cells were positive for staining with anti-CD105 antibody, a MSC marker antibody (Figure S5). During the differentiation process of 253G1 cells into MSCs, we examined the expression levels of residual pluripotency markers in the cell cultures. qRT-PCR analysis revealed that expression of *OCT3/4*, *NANOG* and *LIN28* mRNA were clearly decreased in a time-dependent manner, however, expression levels at the same time point varied markedly among those genes (Figure 6B). In the cells at day 6 of differentiation, mRNA levels of *OCT3/4*, *NANOG* and *LIN28* were 7.3%, 4.8% and 86.4% of the control at day 1, respectively (Figure 6B). At day 14 of differentiation, although *OCT3/4* and *LIN28* were still at detectable levels of 2.6% and 17.2% of control cells, respectively, *NANOG* expression was not detected. These results indicate that the population of residual hiPSCs in differentiating cells, when estimated by the qRT-PCR data, greatly varies and depends on the pluripotency marker gene employed for the estimation. In addition, it is also possible that all the qRT-PCR signals were derived from partially differentiated cells, not from fully undifferentiated cells. To examine colony formation of residual undifferentiated cells in differentiating cell culture, cells at day 6 were dissociated into single cells and replated on laminin-521 in Essential 8 medium. Small cell clusters began to emerge 4 days after plating, rapidly expanded and formed colonies on laminin-521 in Essential 8, while other types of cells gradually decreased their numbers (Figure 6D). After 8 days of culture, 9.5 colonies (the mean of duplicate measurements) were formed from differentiating cells (5×10^4) (Figure 6C) and they were all positive for TRA-1-60 (Figure 6D), indicating that the colonies were derived from residual undifferentiated cells in the differentiating cell cultures. These results suggest that the culture method using a combination of laminin-521 and Essential 8 directly detects residual undifferentiated cells by highly efficient cell amplification. Based on our finding that approximately 0.3 and 6.7 colonies were formed from 1×10^4 MSCs containing 0.01% and 0.1% of 253G1 cells, respectively, in this culture system (Figure S4), and assuming that the sensitivity of the system for hPSCs in EBs are comparable to that in MSCs, the population of the undifferentiated cells in the differentiating cell cultures on day 6 (1.9 colonies/10^4 cells) was estimated to be in between 0.01% and 0.1%. When we tested colony formation using cell cultures on day 14 of differentiation, no colonies were detected on laminin-521 in Essential 8 medium (Figure 6C and data not shown), suggesting that the population of the residual hiPSCs was less than 0.01%.

Discussion

A method to detect residual undifferentiated hPSCs contained in CTPs is required to evaluate product quality during manufacturing processes. In the present study, we propose a novel method to detect a trace amount of undifferentiated hPSCs by highly efficient amplification of those cells *in vitro*. We showed that Essential 8 medium significantly promotes cell growth of hiPSCs dissociated into single cells on laminin-521 compared with the conventional medium, mTeSR1. In addition, Essential 8 medium allowed robust proliferation of hiPSCs even at low cell density on laminin-521. We also demonstrated that 0.001%–0.01% hiPSCs spiked into primary hMSCs were clearly detected and formed colonies on laminin-521 in Essential 8. Similarly, we confirmed that 0.01% hiPSCs spiked into primary human neurons were also detectable on this system. Moreover, we showed that residual undifferentiated hiPSCs contained in differentiating cells were

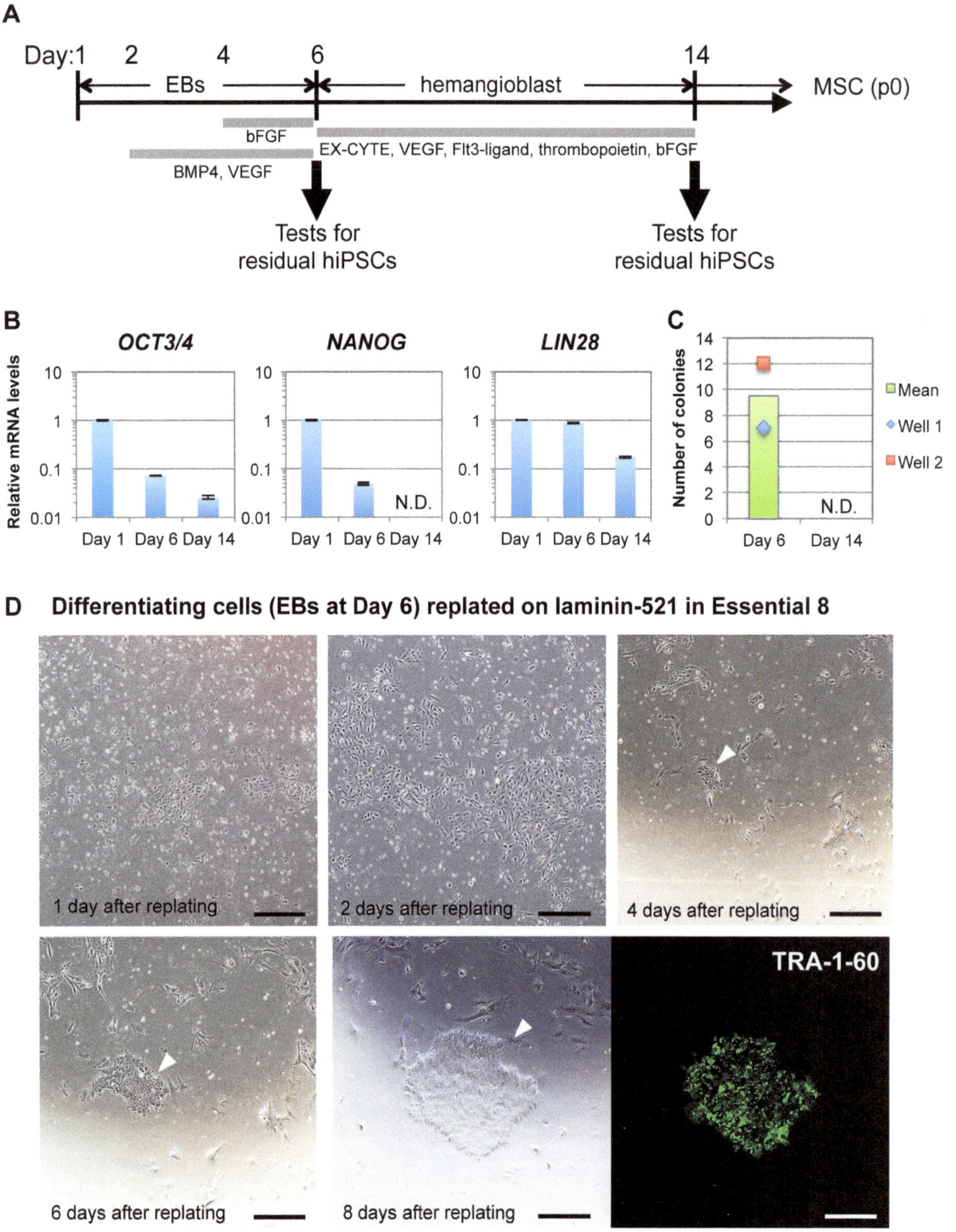

Figure 6. Detection of residual undifferentiated cells contained in differentiating cell cultures. (A) Differentiation scheme of 253G1 cells into MSCs is shown. (B) Expression levels of undifferentiated cell markers (*OCT3/4, NANOG* and *LIN28*) in each cell culture were determined using qRT-PCR. Relative mRNA expression levels are presented as ratios to the level of that in 253G1 cells at Day 1. Results are the mean ± SD (n = 3). (C) Numbers of the forming colonies derived from residual undifferentiated cells in differentiating cell culture at Day 6 or Day 14 are shown. Experiments were carried out in duplicate. Data are present as raw data in each well (shown by plots) or the mean of well 1 and well 2 (shown by bar graphs). (D) Phase contrast images of forming colonies derived from residual undifferentiated cells are shown. Cells at Day 6 of differentiation (EBs) were dissociated into single cells by Accutase and cultured on laminin-521-coated wells in Essential 8 medium (5×10^4/well). After 4 days of culture, small clusters emerged and then started to grow rapidly. Finally, they formed colonies that were positive for TRA-1-60 (shown by immunofluorescence staining, green). Arrowheads indicate a colony derived from same origin. Scale bars, 500 μm.

detectable by forming colonies on laminin-521 in Essential 8 in the process of hMSC differentiation. These results indicate that a culture system utilizing a combination of laminin-521 and Essential 8 medium provides a direct and highly sensitive method for detecting undifferentiated hPSCs. To our knowledge, this is the first report to show a direct and highly-sensitive *in vitro* method for detecting undifferentiated hPSCs as impurities in CTPs.

In this study, highly efficient amplification of undifferentiated hPSCs has been uniquely applied to quality control of CTPs. Amplified hPSC colonies were visible using phase-contrast microscopy and also immunofluorescence staining using pluripotency antibodies, which enabled direct detection of hPSCs contaminating CTPs. Our method distinguished between undifferentiated cells and other cells *in vitro*, and overcame the disadvantage of other *in vitro* methods such as flow cytometry and qRT-PCR. The flow cytometry analysis detects known marker molecules expressed in undifferentiated hPSCs using antibodies and proteins. Signals originating from non-specific detection commonly affect sensitivity of the assay as background. Our *in vitro* method can lower the background arising from non-specific detection and is expected to specifically detect residual undifferentiated hPSCs in CTPs. The qRT-PCR method is highly sensitive and can rapidly quantify undifferentiated cell contamination in CTPs. However, in the present study, gene expression levels of pluripotency markers during the differentiation process of hiPSCs into MSCs varied markedly among those marker genes (Figure 6B). Moreover, there remains a possibility that expression signals of marker genes were not derived from totally undifferentiated hPSCs, but from partially differentiated cells. Indeed, the expression level of *LIN28* did not decrease so much during the differentiation as those of the other genes, which was not obviously associated with the differentiation status of the cells in EBs on Day 6 (Figure 6B), although we have previously reported that *LIN28* was a useful marker for monitoring the level of residual hiPSCs in RPE cells derived from hiPSCs [3]. Thus, it is difficult to determine the presence of residual hiPSCs simply by qRT-PCRs. In contrast, direct detection method using the highly efficient amplification system can clearly detect the presence of intact undifferentiated cells. Based on the result from direct detection of residual hiPSCs when tested the cells on Day 6 of differentiation (approximately 0.01%–0.1%) (Figure 6C-D), it is conceivable that the qRT-PCR signals for the pluripotency marker genes (Figure 6B) are partly derived from residual hiPSCs but mainly derived from partially differentiated cells. Similarly, in the case of the cells at Day 14 of differentiation, the majority of the qRT-PCR signals of *OCT3/4* and *LIN28* (Figure 6B) are considered to be attributable to partially differentiating cells but not to intact hiPSCs. Combination of the *in vitro* methods including our cell culture method would mutually support useful quality assessment of CTPs to detect undifferentiated hPSCs.

In addition to the detection of undifferentiated cells, this culture system using laminin-521 and Essential 8 medium allows further characterization of the undifferentiated cells if they are maintained *in vitro* or inoculated into immunodeficient animals. Analyses for the properties of the residual undifferentiated cells would be necessary not only for the quality assessment of CTPs, but also for improvement of quality specifications of hPSCs as a raw/intermediate material for production of CTPs.

Here, we showed that our culture system is able to detect 0.01% of 409B2 hiPSCs and 0.001% of 253G1 hiPSCs, both of which were spiked into hMSCs (Figure 4). The detection sensitivity for hiPSCs spiked into hMSCs was different between the two hiPSC lines, although such a difference in cell growth on laminin-521 was not found between these two cell lines (Figure 3). This difference may be attributable to the difference in the growth potential of hPSCs in the specific environment provided by CTPs. Kanemura *et al.* have recently demonstrated that hiPSCs co-cultured with iPSC-derived RPE undergo apoptosis by pigment epithelium-derived factor (PEDF) secreted from hiPSC-derived RPE [15], showing that CTPs themselves have the potential to affect cell growth of hPSCs. In the present study, the influence of the co-culture system with hMSCs to the proliferation of hiPSCs might have been different between the two cell lines.

The mechanism by which laminin-521 and Essential 8 medium enhance hiPSCs cell proliferation remains unclear. Rodin *et al.* have recently shown that addition of E-cadherin to laminin-521 permitted the efficient clonal expansion of hESCs [7]. E-cadherin is known to be the primary cell-cell adhesion molecule and essential for hESC survival [16]. We observed that anti-E-cadherin antibody decreased growth potential of hiPSCs under our experimental conditions (data not shown). Therefore, E-cadherin signaling may play some important roles in the rapid cell growth on laminin-521 in Essential 8 medium.

Tumorigenicity is one of the major safety concerns for CTPs derived from hPSCs that are transplanted into patients. However, testing strategies for the tumorigenicity of hPSC-derived CTPs have not yet been established. Here, we introduced a novel testing method for directly detecting a trace amount of undifferentiated hPSCs *in vitro*. The ability of each tumorigenicity-associated test should be taken into consideration to evaluate tumorigenicity of residual undifferentiated hPSCs as impurities in products. *In vivo* tumorigenicity tests using immunodeficient animals can detect tumorigenic cells including undifferentiated hPSCs, but this method is costly and time-consuming. The flow cytometry analysis and qRT-PCR are rapid, but these methods indirectly detect tumorigenic cells depending on marker molecules. Risk of tumorigenicity in hPSCs-derived CTPs should be assessed, based on the results from an appropriate combination of these tumorigenicity-associated tests. Our novel method will contribute to establishment of the testing strategies for tumorigenicity in products, following evaluation of the quality of CTPs derived from hPSCs for the future regenerative medicine/cell therapy.

Supporting Information

Figure S1 (A) Quantification of the number of dissociated 201B7 cells expanded on laminin-521 or Matrigel in Essential 8 or mTeSR1 medium. Data are presented as the mean ± standard deviation (SD) of three independent experiments (**$P<0.01$, two-way ANOVA followed by a Bonferroni post-hoc test). LN521, laminin-521. MG, Matrigel. (B) Quantification of the number of dissociated 201B7 cells expanded on laminin-521 or LM511-E8 in Essential 8 or mTeSR1 medium. Results are presented as the mean ±SD (n = 3) (***$P<0.001$, two-way ANOVA followed by a Bonferroni post-hoc test).

Figure S2 (A-B) Expression levels of undifferentiated markers (*OCT3/4, NANOG, SOX2* and *LIN28*) in 201B7 cells (A) and 409B2 cells (B) subcultured on laminin-521 in Essential 8 were determined using qRT-PCR. Relative mRNA expression levels are presented as ratios to the level of that in control cells subcultured on Matrigel in mTeSR1 medium by colony passage. Results are presented as the mean ± SD (n = 3). (C-D) Expression levels of markers for the differentiation of embryoid bodies (EBs) derived from 201B7 cells (C) and 409B2 cells (D): endoderm (*GATA6, SOX17*), mesoderm (*CDH5, FOXF1*), and ectoderm (*SOX1, PAX6*). Relative mRNA expression levels are presented as ratios to the level of that in control cells (EBs at day 10). Results are presented as the mean ± SD (n = 3).

Figure S3 Quantification of the number of 253G1 cells expanded on laminin-521 in Essential 8 or mTeSR1 medium. Cell numbers were counted at day 6, 9, and 12 after plating at 8.0×10^3 cells/cm^2 or 8.0×10^2 cells/cm^2.

Figure S4 Morphologies of forming colonies derived from 253G1 cells spiked into hMSCs are shown (images in the left). 253G1 cells (1%, 300 cells; 0.1%, 30 cells; 0.01%, 3 cells; 0%, 0 cells) were spiked into hMSCs (30,000 cells) and co-cultured on 12-well plates coated with laminin-521 in Essential 8 medium for 9 days. Expression of the undifferentiated cell marker, TRA-1-60, in these colonies was assessed using immunofluorescence staining (images to the right). Each experiment was carried out in duplicate.

Figure S5 Phase contrast images of the cells at day 18 of differentiation (at the stage of passage 0 MSCs) are shown. Expression of MSC marker, CD105, in these cells was examined using immunofluorescence staining (images to the right). Arrowheads indicate the cells that were positive for CD105.

Table S1 Sequences of the primers and probes for qRT-PCR.

Acknowledgments

This work was supported by Research Grants from the Japanese Ministry of Health, Labour and Welfare (H23-SAISEI-IPPAN-004, H23-SAISEI-IPPAN-005, H24-IYAKU-SHITEI-027, H25-JITSUYOKA(SAISEI)-IP-PAN-008, and Marketing Authorization Facilitation Program for Innovative Therapeutic Products).

Author Contributions

Conceived and designed the experiments: KT SY YS. Performed the experiments: KT. Analyzed the data: KT SY TK HS AU YS. Contributed reagents/materials/analysis tools: KT SY TK HS AU YS. Wrote the paper: KT SY YS. Acquired the funding: HS AU YS.

References

1. Bailey AM (2012) Balancing tissue and tumor formation in regenerative medicine. Sci Transl Med 4: 147fs28. Rev.
2. Kuroda T, Yasuda S, Sato Y (2013) Tumorigenicity studies for human pluripotent stem cell-derived products. Biol. Pharm. Bull 36: 189–192.
3. Kuroda T, Yasuda S, Kusakawa S, Hirata N, Kanda Y, et al. (2012) Highly sensitive in vitro methods for detection of residual undifferentiated cells in retinal pigment epithelial cells derived from human iPS cells. PLoS One 7: e37342.
4. Food and Drug Administration (2008) Cellular Therapies derived from Human Embryonic Stem Cells-Considerations for Pre-Clinical Safety Testing and Patient Monitoring-April 2008. Cellular, Tissue, and Gene Therapies Advisory Committee (CTGTAC) Meeting #45. Available: <http://www.fda.gov/ohrms/dockets/ac/08/briefing/2008-0471B1_1.pdf>.
5. Hentze H, Soong PL, Wang ST, Phillips BW, Putti TC, et al. (2009) Teratoma formation by human embryonic stem cells: evaluation of essential parameters for future safety studies. Stem Cell Res 2: 198–210.
6. Watanabe K, Ueno M, Kamiya D, Nishiyama A, Matsumura M, et al. (2007) A ROCK inhibitor permits survival of dissociated human embryonic stem cells. Nat. Biotechnol 25: 681–686.
7. Rodin S, Antonsson L, Niaudet C, Simonson OE, Salmela E, et al. (2014) Clonal culturing of human embryonic stem cells on laminin-521/E-cadherin matrix in defined and xeno-free environment. Nat. Commun 5: 3195.
8. Miyazaki T, Futaki S, Suemori H, Taniguchi Y, Yamada M, et al. (2012) Laminin E8 fragments support efficient adhesion and expansion of dissociated human pluripotent stem cells. Nat. Commun 3: 1236.
9. Chen G, Gulbranson DR, Hou Z, Bolin JM, Ruotti V, et al. (2011) Chemically defined conditions for human iPSC derivation and culture. Nat. Methods 8: 424–429.
10. Chambers SM, Fasano CA, Papapetrou EP, Tomishima M, Sadelain M, et al. (2009) Highly efficient neural conversion of human ES and iPS cells by dual inhibition of SMAD signaling. Nat. Biotechnol 27: 275–280.
11. Kajiwara M, Aoi T, Okita K, Takahashi R, Inoue H, et al. (2012) Donor-dependent variations in hepatic differentiation from human-induced pluripotent stem cells. Proc Natl Acad Sci U S A 109: 12538–12543.
12. Jung Y, Bauer G, Nolta JA (2012) Concise review: Induced pluripotent stem cell-derived mesenchymal stem cells: progress toward safe clinical products. Stem Cells 30: 42–47.
13. Chen YS, Pelekanos RA, Ellis RL, Horne R, Wolvetang EJ, et al. (2012) Small molecule mesengenic induction of human induced pluripotent stem cells to generate mesenchymal stem/stromal cells. Stem Cells Transl Med 1: 83–95.
14. Kimbrel EA, Kouris NA, Yavanian G, Chu J, Qin Y, et al. (2014) Mesenchymal stem cell population derived from human pluripotent stem cells displays potent immunomodulatory and therapeutic properties. Stem Cells Dev 23: 1611–1624.
15. Kanemura H, Go MJ, Nishishita N, Sakai N, Kamao H, et al. (2013) Pigment epithelium-derived factor secreted from retinal pigment epithelium facilitates apoptotic cell death of iPSC. Sci Rep 3: 2334.
16. Xu Y, Zhu X, Hahm HS, Wei W, Hao E, et al. (2010) Revealing a core signaling regulatory mechanism for pluripotent stem cell survival and self-renewal by small molecules. Proc Natl Acad Sci U S A 107: 8129–8134.
17. Takahashi K, Tanabe K, Ohnuki M, Narita M, Ichisaka T, et al. (2007) Induction of pluripotent stem cells from adult human fibroblasts by defined factors. Cell 131: 861–872.
18. Nakagawa M, Koyanagi M, Tanabe K, Takahashi K, Ichisaka T, et al. (2008) Generation of induced pluripotent stem cells without Myc from mouse and human fibroblasts. Nat. Biotechnol 26: 101–106.
19. Okita K, Matsumura Y, Sato Y, Okada A, Morizane A, et al. (2011) A more efficient method to generate integration-free human iPS cells. Nat. Methods 8: 409–412.

Permissions

All chapters in this book were first published in PLOS ONE, by The Public Library of Science; hereby published with permission under the Creative Commons Attribution License or equivalent. Every chapter published in this book has been scrutinized by our experts. Their significance has been extensively debated. The topics covered herein carry significant findings which will fuel the growth of the discipline. They may even be implemented as practical applications or may be referred to as a beginning point for another development.

The contributors of this book come from diverse backgrounds, making this book a truly international effort. This book will bring forth new frontiers with its revolutionizing research information and detailed analysis of the nascent developments around the world.

We would like to thank all the contributing authors for lending their expertise to make the book truly unique. They have played a crucial role in the development of this book. Without their invaluable contributions this book wouldn't have been possible. They have made vital efforts to compile up to date information on the varied aspects of this subject to make this book a valuable addition to the collection of many professionals and students.

This book was conceptualized with the vision of imparting up-to-date information and advanced data in this field. To ensure the same, a matchless editorial board was set up. Every individual on the board went through rigorous rounds of assessment to prove their worth. After which they invested a large part of their time researching and compiling the most relevant data for our readers.

The editorial board has been involved in producing this book since its inception. They have spent rigorous hours researching and exploring the diverse topics which have resulted in the successful publishing of this book. They have passed on their knowledge of decades through this book. To expedite this challenging task, the publisher supported the team at every step. A small team of assistant editors was also appointed to further simplify the editing procedure and attain best results for the readers.

Apart from the editorial board, the designing team has also invested a significant amount of their time in understanding the subject and creating the most relevant covers. They scrutinized every image to scout for the most suitable representation of the subject and create an appropriate cover for the book.

The publishing team has been an ardent support to the editorial, designing and production team. Their endless efforts to recruit the best for this project, has resulted in the accomplishment of this book. They are a veteran in the field of academics and their pool of knowledge is as vast as their experience in printing. Their expertise and guidance has proved useful at every step. Their uncompromising quality standards have made this book an exceptional effort. Their encouragement from time to time has been an inspiration for everyone.

The publisher and the editorial board hope that this book will prove to be a valuable piece of knowledge for researchers, students, practitioners and scholars across the globe.

List of Contributors

Tomasz Kropiwnicki
Department of Neurosurgery, The Children's Memorial Health Institute, Warsaw, Poland

Elzbieta Jurkiewicz and Marcin Roszkowski
Department of Radiology, Magnetic Resonance Unit, The Children's Memorial Health Institute, Warsaw, Poland

Krystyna Domanska-Janik and Barbara Lukomska
NeuroRepair Department, Mossakowski Medical Research Centre, Polish Academy of Sciences, Warsaw, Poland

Miroslaw Janowski
NeuroRepair Department, Mossakowski Medical Research Centre, Polish Academy of Sciences, Warsaw, Poland
Department of Neurosurgery, Mossakowski Medical Research Centre, Polish Academy of Sciences, Warsaw, Poland
Russell H. Morgan Department of Radiology and Radiological Science, Division of Magnetic Resonance Research, The Johns Hopkins University School of Medicine, Baltimore, Maryland, United States of America
Cellular Imaging Section and Vascular Biology Program, Institute for Cell Engineering, The Johns Hopkins University School of Medicine, Baltimore, Maryland, United States of America

Jeff W. M. Bulte
Russell H. Morgan Department of Radiology and Radiological Science, Division of Magnetic Resonance Research, The Johns Hopkins University School of Medicine, Baltimore, Maryland, United States of America
Cellular Imaging Section and Vascular Biology Program, Institute for Cell Engineering, The Johns Hopkins University School of Medicine, Baltimore, Maryland, United States of America
Department of Chemical and Biomolecular Engineering, The Johns Hopkins University School of Medicine, Baltimore, Maryland, United States of America
Department of Biomedical Engineering, The Johns Hopkins University School of Medicine, Baltimore, Maryland, United States of America

Hiroto Sasaki and Hiroyuki Honda
Department of Biotechnology, Graduate School of Engineering, Nagoya University, Nagoya, Aichi, Japan

Ichiro Takeuchi
Department of Computer Science/Scientific and Engineering Simulation, Graduate School of Engineering, Nagoya Institute of Technology, Nagoya, Aichi, Japan

Mai Okada
Department of Basic Medicinal Sciences, Graduate School of Pharmaceutical Sciences, Nagoya University, Nagoya, Aichi, Japan

Rumi Sawada
Division of Medical Devices, National Institute of Health Sciences, Setagaya-ku, Tokyo, Japan

Yasujiro Kiyota
Nikon Corporation, Chiyoda-ku, Tokyo, Japan

Kei Kanie and Ryuji Kato
Department of Biotechnology, Graduate School of Engineering, Nagoya University, Nagoya, Aichi, Japan
Department of Basic Medicinal Sciences, Graduate School of Pharmaceutical Sciences, Nagoya University, Nagoya, Aichi, Japan

Gregory V. Goldmacher and Daniel Y. Lee
Farber Institute for Neurosciences, Thomas Jefferson University, Philadelphia, Pennsylvania, United States of America

Robert Rosenwasser
Department of Neurological Surgery, Thomas Jefferson University, Philadelphia, Pennsylvania, United States of America

Rena Nasser and Sargon Yigit
Farber Institute for Neurosciences, Thomas Jefferson University, Philadelphia, Pennsylvania, United States of America
Department of Neuroscience, Thomas Jefferson University, Philadelphia, Pennsylvania, United States of America

Lorraine Iacovitti
Farber Institute for Neurosciences, Thomas Jefferson University, Philadelphia, Pennsylvania, United States of America
Department of Neuroscience, Thomas Jefferson University, Philadelphia, Pennsylvania, United States of America
Department of Neurological Surgery, Thomas Jefferson University, Philadelphia, Pennsylvania, United States of America

Shinsuke Yuasa, Dai Kusumoto, Akira Kunitomi, Yuki Saito, Kojiro Yae, Yoshikazu Kishino, Marina Okada, Hisayuki Hashimoto, Makoto Takei, Toru Egashira, Masaki Kodaira, Yusuke Kuroda, Atsushi Tanaka, Shinichiro Okata, Tomoyuki Suzuki, Mitsushige Murata, Jun Fujita and Keiichi Fukuda
Department of Cardiology, Keio University School of Medicine, Shinjuku-ku, Tokyo, Japan

Tomohisa Seki and Shugo Tohyama
Department of Cardiology, Keio University School of Medicine, Shinjuku-ku, Tokyo, Japan
Japan Society for the Promotion of Science, Chiyoda-ku, Tokyo, Japan

Naoki Tajiri, Yuji Kaneko and Cesar V. Borlongan
Center of Excellence for Aging & Brain Repair, Department of Neurosurgery and Brain Repair, University of South Florida College of Medicine, Tampa, Florida, United States of America

David M. Quach, Yuji Kaneko, Stephanie Wu, David Lee, Tina Lam, Ken L. Hayama, Thomas G. Hazel and Karl Johe
Neuralstem, Inc., Rockville, Maryland, United States of America

Michael C. Wu
Neurodigitech, LLC., San Diego, California, United States of America

Gemma N. Jones, Dafni Moschidou, Tamara-Isabel Puga-Iglesias, Katarzyna Kuleszewicz and Pascale V. Guillot
Institute of Reproductive and Developmental Biology, Imperial College London, London, United Kingdom

Maximilien Vanleene and Sandra J. Shefelbine
Department of Bioengineering, Imperial College London, London, United Kingdom

George Bou-Gharios
Kennedy Institute of Rheumatology, University of Oxford, London, United Kingdom

Nicholas M. Fisk
UQ Centre for Clinical Research, University of Queensland, Brisbane, Queensland, Australia

Anna L. David
Prenatal Cell and Gene Therapy Group, Institute for Women's Health, University College London, London, United Kingdom

Paolo De Coppi
Surgery Unit, UCL Institute of Child Health, London, United Kingdom

Fan Yang
Translational Biomarker Discovery Core, Beckman Research Institute, City of Hope Comprehensive Cancer Center, Duarte, California, United States of America

Sangkil Nam, Robin Zhao and David A. Horne
Molecular Medicine, Beckman Research Institute, City of Hope Comprehensive Cancer Center, Duarte, California, United States of America

Christine E. Brown and Renate Starr
Cancer Immunotherapeutics & Tumor Immunology, Beckman Research Institute, City of Hope Comprehensive Cancer Center, Duarte, California, United States of America

Linda H. Malkas
Molecular and Cellular Biology, Beckman Research Institute, City of Hope Comprehensive Cancer Center, Duarte, California, United States of America

Richard Jove
Vaccine & Gene Therapy Institute of Florida, Port St. Lucie, Florida, United States of America

Robert J. Hickey
Translational Biomarker Discovery Core, Beckman Research Institute, City of Hope Comprehensive Cancer Center, Duarte, California, United States of America
Molecular Pharmacology, Beckman Research Institute, City of Hope Comprehensive Cancer Center, Duarte, California, United States of America

Hala O. El-Mesallamy and Nadia M. Hamdy
Biochemistry Department, Faculty of Pharmacy, Ain-Shams University, Cairo, Egypt

Mohamed R. Diab and Sarah M. Dardir
Research Department, Holding Company for Biological Products and Vaccines (VACSERA), Giza, Egypt

Roberta Barbizan, Mateus V. Castro and Alexandre L. R. Oliveira
Laboratory of Nerve Regeneration, Department of Structural and Functional Biology, University of Campinas - UNICAMP, Campinas, São Paulo, Brazil
Center for the Study of Venoms and Venomous Animals (CEVAP), São Paulo State University (UNESP - Univ Estadual Paulista), Botucatu, São Paulo, Brazil

Yu Pan, Shuai Liu, Haifei Wu, Jing Lv and Yifan Zhang
Department of Nuclear Medicine, Ruijin Hospital, School of Medicine, Shanghai Jiao Tong University, Shanghai, China

Xiaoqian Xu
Department of Hematology, Changhai Hospital, Secondary Military Medical University, Shanghai, China

Jincheng Wu, Mahboubeh Rahmati Rostami and Emmanuel S. Tzanakakis
Department of Chemical and Biological Engineering, Tufts University, Massachusetts, United States of America

Diana P. Cadavid Olaya
Department of Chemical and Biological Engineering, State University of New York at Buffalo, Buffalo, New York, United States of America

Heng Zeng, Lanfang Li and Jian-Xiong Chen
Department of Pharmacology and Toxicology, University of Mississippi Medical Center, Jackson, Mississippi, United States of America

Qiang Li., Xinmin Zhou, Yongquan Shi, Jinge Li, Linhua Zheng, Lina Cui, Jun Zhang, Lu Wang, Zheyi Han, Ying Han and Daiming Fan
State Key Laboratory of Cancer Biology, Xijing Hospital of Digestive Diseases, The Fourth Military Medical University, Xi'an, Shaanxi Province, China

Horng-Jyh Harn
Department of Pathology, China Medical University Hospital, Taichung, Taiwan

Ying-Jiun Chien, Cheng-Hsuan Chang and Chien-Yu Hsu
Graduate Institute of Basic Medical Science, China Medical University, Taichung, Taiwan

Woei-Cherng Shyu and Shinn-Zong Lin
Center for Neuropsychiatry, China Medical University Hospital, Taichung, Taiwan

Shih-Ping Liu
Department of Pathology, China Medical University Hospital, Taichung, Taiwan
Center for Neuropsychiatry, China Medical University Hospital, Taichung, Taiwan

Ru-Huei Fu
Center for Neuropsychiatry, China Medical University Hospital, Taichung, Taiwan
Graduate Institute of Immunology, China Medical University, Taichung, Taiwan

Yu-Chuen Huang and Shih-Yin Chen
Genetics Center, Department of Medical Research, China Medical University Hospital, Taichung, Taiwan
Graduate Institute of Chinese Medical Science, China Medical University, Taichung, Taiwan

Ning Li
Department of Physiology and Cell Biology, The Ohio State University, Columbus, Ohio, United States of America

Zeeshan Pasha and Muhammad Ashraf
Department of Pharmacology, University of Illinois at Chicago, Chicago, Illinois, United States of America

Thomas Frauenfelder, Hatem Alkadhi and Andreas Boss
Institute of Diagnostic Radiology, University Hospital Zurich, Zurich, Switzerland

Steffen M. Zeisberger
Swiss Centre for Regenerative Medicine, University of Zurich, Zurich, Switzerland

Luc Behr and Sebastien Sammut
IMM RECHERCHE, Institute Mutualiste Montsouris, Paris, France

Jacques Scherman
Clinic for Cardiovascular Surgery, University Hospital of Zurich, Zurich, Switzerland

Ruth Schwartländer and Viola Vogel
Department of Health Science and Technology, Laboratory for Biologically Oriented Materials, ETH Zurich, Zurich, Switzerland

Peter Vogt
Department of Pathology, University Hospital of Zurich, Zurich, Switzerland

Benedikt Weber, Petra Wolint and Chad E. Brokopp
Swiss Centre for Regenerative Medicine, University of Zurich, Zurich, Switzerland
Department of Surgical Research, University Hospital of Zurich, Zurich, Switzerland

Jürg Grünenfelder and Volkmar Falk
Department of Surgical Research, University Hospital of Zurich, Zurich, Switzerland
Clinic for Cardiovascular Surgery, University Hospital of Zurich, Zurich, Switzerland

Maximilian Y. Emmert and Simon P. Hoerstrup
Swiss Centre for Regenerative Medicine, University of Zurich, Zurich, Switzerland
Department of Surgical Research, University Hospital of Zurich, Zurich, Switzerland
Clinic for Cardiovascular Surgery, University Hospital of Zurich, Zurich, Switzerland

Bo Qu, Li-Xia Zhao, Hui Xing, Li-Ying Lian, Jia-Zhao Tong, Bei-Bei Wang and Shi-Zhu Jin
Department of Gastroenterology and Hepatology, The Second Affiliated Hospital, Harbin Medical University, Harbin, China

Guo-Rong Xin
Anorectal Department, The First Affiliated Hospital, JiaMuSi Medical University, JiaMuSi, China

Hai-Yan Jiang
Infections Department, JiaMuSi Central Hospital, JiaMuSi, China

Anne Biedermann, Katja Kriebel and Hermann Lang
Department of Operative Dentistry and Periodontology, University of Rostock, Rostock, Germany

Bernd Kreikemeyer
Institute of Med. Microbiology, Virology and Hygiene, University of Rostock, Rostock, Germany

Jun Li
Department of Orthopaedics, The First Affiliated Hospital of Soochow University, Suzhou, Jiangsu, China

Shenghao Wang and Yibin Wang
Orthopedic Institute, Soochow University, Suzhou, Jiangsu, China

Chen Liu, Qianping Guo, Bin Li and Huilin Yang
Department of Orthopaedics, The First Affiliated Hospital of Soochow University, Suzhou, Jiangsu, China
Orthopedic Institute, Soochow University, Suzhou, Jiangsu, China

Ann Logan, Martin Berry and Wendy Leadbeater
Neurotrauma Research Group, Neurobiology Section, School of Clinical and Experimental Medicine, University of Birmingham, Birmingham, United Kingdom

Ben A. Scheven
School of Dentistry, University of Birmingham, Birmingham, United Kingdom

Ben Mead
Neurotrauma Research Group, Neurobiology Section, School of Clinical and Experimental Medicine, University of Birmingham, Birmingham, United Kingdom
School of Dentistry, University of Birmingham, Birmingham, United Kingdom

Satoshi Yasuda and Takuya Kuroda
Division of Cellular & Gene Therapy Products, National Institute of Health Sciences, Tokyo, Japan

Hirohisa Saito
National Research Institute for Child Health and Development, Tokyo, Japan

Akihiro Umezawa
Department of Reproductive Biology, National Research Institute for Child Health and Development, Tokyo, Japan

Keiko Tano
Department of Reproductive Biology, National Research Institute for Child Health and Development, Tokyo, Japan

Division of Cellular & Gene Therapy Products, National Institute of Health Sciences, Tokyo, Japan

Yoji Sato
Division of Cellular & Gene Therapy Products, National Institute of Health Sciences, Tokyo, Japan
Department of Quality Assurance Science for Pharmaceuticals, Graduate School of Pharmaceutical Sciences, Nagoya City University, Aichi, Japan
Department of Cellular & Gene Therapy Products, Graduate School of Pharmaceutical Sciences, Osaka University, Osaka, Japan
Department of Translational Pharmaceutical Sciences, Graduated School of Pharmaceutical Sciences, Kyushu University, Fukuoka, Japan

Index

A

Adipocyte, 9, 48, 51, 192, 194

Adipogenesis, 191, 194-195, 198

Adipogenic, 8-12, 14, 16-20, 47-48, 51-52, 58, 95, 163-164, 193-194, 196

Adrenergic, 35-36

Agitation, 104-109, 111, 113-114

Ameliorate, 27, 39, 46, 102, 135-136, 183

Amurensis, 62-63, 65

Amyotrophic, 6, 39, 46, 208

Anaerobic, 184-190

Anagnostopoulos, 102, 135

Anatomical, 1, 24, 81, 178

Angiogenesis, 30, 67, 69, 78, 80,

115-120, 122, 124-125, 150, 153, 155-159, 174

Angiopoietin, 30, 125, 150, 152

Antibodies, 24, 32, 38, 44, 46, 53, 57-59, 63-64, 68, 117, 127, 132, 141, 146-147, 150, 158, 163, 169, 175, 185, 193, 195, 219

Anticancer, 62-63, 65, 69

Antioxidant, 76-77, 79, 124

Apoptosis, 20, 30, 62-64, 66-70, 113, 116-120, 122-125, 138, 142, 145, 149, 154-156, 158, 162, 182, 208, 211-212, 219

Arrhythmia, 31, 37, 162, 164

Arteriole, 117, 119, 122

Astrocytes, 26, 28-30, 92

Atmosphere, 63, 127, 184-186, 188, 211-212

Autologous, 2, 4, 6, 47-48, 55, 57, 60-61, 71, 78, 94, 102, 135, 158, 172-173, 183

Autophagy, 20, 118-119, 123-125

B

Bacterial, 80, 184-190

Baculovirus, 94-95, 97-103

Berbamine, 62-63, 65, 67, 69-70

Bioreactor, 104, 106-109, 112-115

Biotechnol, 7, 20, 37, 60, 79, 102-103, 115, 147, 172, 197, 220

C

Capillary, 28, 82, 117, 119, 122

Carbohydrate, 72, 79

Carcinoma, 45-46, 53, 63, 70, 147, 184-185

Cardiomyocyte, 31-37, 124, 155, 157-158, 166

Cardiomyopathy, 36, 125, 149, 160, 172-173

Cardiovascular, 36, 61, 94, 116, 160, 168-169, 171

Chemokines, 28-29, 134, 136

Chemotherapy, 62, 65, 69

Chondrocyte, 9, 192, 194-196

Chondrogenic, 8-12, 14, 16-20, 114, 163-164, 193-196

Chromosome, 48-49, 129, 178

Cytokines, 28-29, 79, 113, 126, 132, 134-137, 145, 147, 149, 156, 158, 184

Cytoplasm, 34, 48, 53, 194-195

Cytotherapy, 6, 60-61, 102, 135, 208

Cytotoxicity, 67, 69, 94, 102

D

Deacetylase, 120, 124-125

Degeneration, 81, 84, 90, 92-93, 125, 174, 178, 191, 197, 209

Degradation, 124-125, 156, 190

Dehydrated, 32, 83, 132, 185

Dehydrogenase, 80, 150

Diagnostic, 48, 55, 117, 160

Diaminobenzidine, 42, 132

Dimethylthiazol, 64, 150

E

Elasticity, 20, 196, 198

Electrophoresis, 57, 64, 132, 175

Electrophysiological, 31, 35-36

Embryogenesis, 50, 60

Embryonic, 6, 20, 31-32, 34, 36-37, 49-51, 53-55, 60-61, 68, 70-71, 80, 94-95, 98, 102-105, 114-116, 135, 137, 145-147, 195, 198, 210, 220

Endocrinol, 60, 79, 103, 135

Endothelial, 30, 45, 49, 52, 78, 80-81, 104, 112, 115-116, 124-125, 132-135, 149, 151, 155-159, 172, 176, 190

Epithelial, 21, 52, 56, 78, 136, 155, 174, 176-177, 183-185, 188-190, 211, 220

Erythropoietin, 76, 80

Extracellular, 5, 21, 32, 63, 79, 92, 156, 197

Extraction, 10, 18, 20, 38, 57-59, 82, 129, 175-176, 202

F

Fetoprotein, 48, 132, 212, 214

Fibroblast, 20, 32-34, 36-37, 61, 63, 95, 146-147, 150, 152, 156, 159, 192, 196, 208-209, 211, 217

Fibronectin, 32, 49, 52, 61, 79

Fluorochrome, 160, 169-170

Formaldehyde, 42, 83, 162, 192

G

Gastroenterology, 135-136, 174, 183

Gingivalis, 184-186, 188-190

Glutamine, 23, 32, 58-59, 63, 78-80, 95, 146, 202, 211

Granulocyte, 127, 177

H

Haematopoietic, 50, 172

Haemocytometer, 57, 203

Haemodialysate, 71-72, 79-80

Hematopoiesis, 79, 115, 135-136, 172

Hematoxylin, 38, 42, 57, 73, 78, 132, 211, 214

Hepatitis, 99, 135-136, 148

Hepatocellular, 63, 70, 135, 147

Hepatocytes, 54, 126, 133-134, 136, 159, 172

Hepatogenic, 48, 52, 54, 58-59, 61

Hydroxyproline, 71, 73-74, 80

Hypertrophy, 79, 92, 116-117, 119, 124-125, 155, 157

I

Immunogenicity, 38, 45, 94, 102

Immunological, 6, 30, 48, 63, 102, 189

Intestinal, 101, 136, 174-183

Intracellular, 34-35, 57, 63, 80, 96, 98, 100-101, 117, 119, 124, 167, 190, 198

Intracerebral, 6, 30, 38-40, 42, 44, 209

Intramyocardial, 117-118, 122, 149-150, 158, 160, 168-172

Intraspinal, 46, 86, 90, 92

Ipsilateral, 3, 24, 28, 42, 82-84, 86, 88-89, 92

Ischaemic, 29, 125, 172-173

L

Leukocytes, 79, 176-177

Lymphocytes, 26, 28, 37, 63, 163, 167, 171, 195

M

Macrophages, 78, 80, 163, 167, 169, 171

Mammalian, 70-71, 93, 99-100, 102-103, 113, 115-116, 137, 145, 147

Mechanism, 22, 31, 38, 45, 63, 79, 124, 134, 157, 174, 184, 205, 208, 219-220

Membranes, 60, 64, 115, 161, 175

Mercaptoethanol, 32, 58, 95, 146, 211

Mesenchymal, 6, 8-10, 18, 20-21, 30, 47, 53, 59-61, 71, 79-82, 89-90, 92-94, 97, 102-103, 116, 125-126, 135-137, 158-162, 167-176, 178, 182-185, 190-191, 195, 197-200, 208-211, 214, 220

Mesodermal, 48, 50, 54, 61, 137

Metalloproteinases, 136, 156, 158

Methylation, 21, 51, 55, 58

Microglial, 26, 29-30, 86, 89, 92

Microglobulin, 160, 167, 169

Mitochondria, 116, 120, 124-125, 155

Molecular, 2, 8, 18-21, 32, 34-35, 37, 61, 69-70, 72, 78, 95, 98, 102, 117, 122, 124, 150, 169, 174, 177, 179, 182-183, 185, 189, 197, 202, 208-209

Monoclonal, 24, 36-37, 132, 146-147, 175, 185

Motoneurons, 39, 46, 81, 83-84, 86, 88, 90, 92-93, 208

Myocardial, 80, 101, 116-117, 119, 121-122, 124-125, 136, 150, 153-173

N

Neovascularization, 78, 124, 135, 149, 156-157

Neuritogenesis, 199-200, 203-205, 208

Neurobiol, 30, 45, 61, 93, 208

Neurogenic, 2, 45, 47-48, 51-52, 54, 59

Neuroprotection, 29, 46, 81, 84, 86, 90, 92-93, 199, 203-205, 208-209

Neurospheres, 62-69

Neurotrophic, 45, 81-82, 86, 89-90, 92-93, 199-200, 205, 207-209

Neurotrophins, 93, 199-200, 202, 205, 207-208

Neutrophils, 26, 28-29, 78-79, 184

Nitrocellulose, 59, 72, 117, 175

Nucleatum, 184-190

Nucleostemin, 191-195, 198

O

Oligonucleotides, 63, 68

Osteoblasts, 20, 47, 185, 190

Osteosarcoma, 20, 63, 68-70

P

Pancreatic, 6, 54, 70, 80, 102, 115

Paraformaldehyde, 19, 24, 32, 41, 132, 146, 150, 175, 192, 202, 212

Paramagnetic, 2, 161, 169

Pathology, 36, 51, 79, 126, 137, 160

Penicillin, 32, 58-59, 146, 175, 193, 202

Pentobarbital, 162, 175, 211
Pericardium, 161, 165-166
Periodontitis, 184-185, 189-190
Peroxidase, 42, 63-64, 80, 117, 132
Pharmaceuticals, 63, 72, 82, 210
Pharmacology, 62, 116, 149
Phosphorylation, 62, 68-69, 113, 119, 123-125, 147
Platelets, 71, 73, 79, 113
Pluripotency, 48-50, 52-54, 59-60, 99, 103-104, 107, 111-112, 137-139, 141, 145, 147, 195, 198, 212-213, 217, 219
Polymerase, 32, 57-58, 62, 66, 70, 84, 129, 175-176, 192
Polypeptides, 157, 175
Porphyromonas, 184, 190

R
Radioiodide, 94, 96-98, 100-101
Radionuclide, 94-95, 98-102
Reperfusion, 23, 124-125, 158
Restoration, 3, 27, 61, 91-92, 133, 149, 156, 158, 205

S
Symporter, 94-95, 99, 102-103

T
Therapeutics, 29, 55, 104-105, 114
Thoracotomy, 150, 161, 165-166
Tomography, 95-96, 101, 163, 168, 172
Transduction, 2, 21, 60, 70, 102-103, 142, 147, 157
Tumorigenicity, 94, 102, 210, 219-220